Semiparallel Submanifolds in Space Forms

Ülo Lumiste

Semiparallel Submanifolds in Space Forms

Springer

Ülo Lumiste
Institute of Pure Mathematics
University of Tartu
Tartu 50409
Estonia
ulo.lumiste@ut.ee

ISBN: 978-1-4419-2389-9 e-ISBN: 978-0-387-49913-0
DOI: 10.1007/978-0-387-49913-0

Mathematics Subject Classification (2000): 53-02, 53B25, 53-C35, 53C40

Printed on acid-free paper

springer.com

Contents

0

Introduction

Among Riemannian manifolds, the most interesting and most important for applications are the symmetric ones. From the local point of view, they were introduced independently by P. A. Shirokov [Shi 25] and H. Levy [Le 25] as Riemannian manifolds with covariantly constant (also called parallel) curvature tensor field R, i.e., with

$$\nabla R = 0, \tag{0.1}$$

where ∇ is the Levi-Civita connection [L-C 17]. An extensive theory of symmetric Riemannian manifolds was worked out by É. Cartan in [Ca 26]. He showed that a Riemannian manifold M has parallel R if and only if every point x has a normal neirhbourhood such that all geodesic symmetries with respect to x are isometries.

If for each point $x \in M$ there exists an involutive isometry s_x of M for which x is an isolated fixed point, then M is called a (globally) symmetric space. The closure of the group of isometries generated by $\{s_x : x \in M\}$ in the compact-open topology is a Lie group G that acts transitively on the symmetric space; hence the typical isotropy subgroup H at a point of M is compact, and $M = G/H$.

The classical examples are connected complete Riemannian manifolds with constant sectional curvature c, called space forms (see [Wo 72], Section 2.4).

Later, a similar development took place in the geometry of submanifolds in space forms, where a fundamental role is played by the first (or metric) form g (as the induced Riemannian metric) and the second fundamental form h. Besides the Levi-Civita connection ∇, with $\nabla g = 0$, a normal connection $\nabla^\perp$ is also defined. The submanifolds with parallel fundamental form, i.e., with

$$\bar{\nabla} h = 0, \tag{0.2}$$

where $\bar{\nabla}$ is the pair of ∇ and $\nabla^\perp$, deserve special attention. Due to the Gauss identity, each of them is intrinsically a locally symmetric Riemannian manifold.

The first result here was given by V. F. Kagan [Ka 48], who showed that in Euclidean space E^3, the surfaces with parallel h are open subsets of planes, round spheres, and circular cylinders $S^1 \times E^1$. All of these have nonnegative Gaussian

Ü. Lumiste, *Semiparallel Submanifolds in Space Forms*,
DOI 10.1007/978-0-387-49913-0_1,

curvature. The surfaces of negative constant Gaussian curvature in E^3 are therefore examples of submanifolds which are intrinsically locally symmetric, but have nonparallel h.

The hypersurfaces with parallel h in E^n were determined by U. Simon and A. Weinstein [SW 69]. Some new examples of surfaces with parallel h in E^4 were given by C.-S. Houh [Ho 72]: the Clifford tori $S^1 \times S^1$ and the Veronese surfaces. The general theory of submanifolds M^m with parallel h in E^n was initiated by J. Vilms [Vi 72], who showed, in particular, that each of them has totally geodesic Gauss image. Normally flat submanifolds with parallel h in Euclidean spaces and spheres were classified by R. Walden [Wa 73].

A properly developed theory was worked out by D. Ferus [Fe 74, 80]. He proved that a submanifold M^m with parallel h in E^n has the property of local extrinsic symmetry, in the sense that every point has a neighborhood invariant under reflection of E^n with respect to the normal subspace at this point; also conversely, an M^m with this property has parallel h. This was proved in general, for M^m in a Riemannian manifold N^n, by W. Strübing [St 79]. Therefore, the submanifolds with parallel h, especially the complete ones, were called symmetric submanifolds by Ferus (and then by others); here extrinsically was meant, but often not explicitly stated. The other important result of Ferus was that a general symmetric submanifold in E^n reduces to a product of irreducible symmetric submanifolds, each of which (except possibly a Euclidean subspace) lies in a sphere, is minimal in it, and can be obtained as the standard immersion of a Riemannian symmetric R-space. Conversely, each such standard immersion gives a symmetric submanifold; and the products of these immersions (possibly including a Euclidean subspace) exhaust all symmetric submanifolds in E^n. These results gave a classification of such submanifolds in terms of special chapters of the theory of Lie groups and symmetric spaces. All of these submanifolds can be considered as symmetric orbits.

This classification was then extended to submanifolds with parallel h in space forms by M. Takeuchi [Ta 81], who found it more suitable here to use the term parallel submanifolds. This term has become more popular, especially when the local point of view has been considered.

The theory of parallel submanifolds is concisely treated in recent monographic works by B.-Y. Chen [Ch 2000] (Chapter 8), Ü. Lumiste [Lu 2000] (Sections 5–7), and by J. Berndt, S. Console, and C. Olmos [BCO 2003] (Section 3.7: "Symmetric submanifolds").

Already in the first investigations of symmetric Riemannian manifolds [Shi 25] and [Ca 26], it was noted that these manifolds must also satisfy the integrability condition

$$R(X, Y) \cdot R = 0 \tag{0.3}$$

of the differential system $\nabla R = 0$. (Here X and Y are tangent vector fields, and $R(X, Y)$ is considered as a field of linear operators, acting on R.) Riemannian manifolds with this point-wise condition were considered separately by É. Cartan in [Ca 46]. His investigations were continued by A. Lichnerowicz [Li 52, 58] and R. Couty [Co 57]. The term semisymmetric for Riemannian manifolds M satisfying this condition was introduced by N. S. Sinyukov [Si 56, 62], who showed the importance of this condition in the theory of geodesic mappings of Riemannian manifolds (see [Si 79], Chapter 2, Section 3).

A fruitful impulse for investigations of manifolds of this class was given by K. Nomizu in [No 68], who conjectured that all complete irreducible n-dimensional Riemannian manifolds ($n \geq 3$) satisfying $R(X, Y) \cdot R = 0$ are locally symmetric, i.e., that they must also satisfy $\nabla R = 0$. This conjecture was supported by the result that for a Riemannian manifold, $\nabla^k R = 0$ with $k > 1$ implies $\nabla R = 0$, proved for the compact case in [Li 58], and for the complete case in [NO 62]; and this is also valid in general (cf. [KN 63], Vol. 1, Remark 7). However, Nomizu's conjecture was eventually refuted. Namely, in [Ta 72] a hypersurface in E^4 was constructed satisfying $R(X, Y) \cdot R = 0$ but not $\nabla R = 0$. A counterexample of arbitrary dimension was given in [Sek 72].

Semisymmetric Riemannian manifolds were classified by Z. I. Szabó, locally, in [Sza 82]. He showed that for every semisymmetric Riemannian manifold M, there exists an everywhere dense open subset U of M, such that around every point of U, the manifold is locally isometric to a space that is the direct product of an open subset of a Euclidean space and of infinitesimally irreducible simple semisymmetric leaves, each of which is either (i) locally symmetric, or (ii) locally isometric to an elliptic, a hyperbolic, a Euclidean, or a Kählerian cone, or (iii) locally isometric to a space foliated by Euclidean leaves of codimension 2 (or to a two-dimensional manifold, in the case $\dim M = 2$).

These classification results of Szabó were presented briefly in the book [BKV 96], whose main purpose was to summarize recent results on semisymmetric Riemannian manifolds of subclass (iii); these are now called Riemannian manifolds of conullity two, and may be considered the most interesting among semisymmetric Riemannian manifolds.

Parallel submanifolds were likewise later placed in a more general class of submanifolds, generalizing the parallel ones in the same sense as locally symmetric Riemannian manifolds (i.e., with $\nabla R = 0$) were generalized by semisymmetric Riemannian manifolds (i.e., with $R(X.Y) \cdot R = 0$). Namely, the integrability condition of the differential system $\bar{\nabla} h = 0$ is

$$\bar{R}(X, Y) \cdot h = 0, \tag{0.4}$$

where $\bar{R}$ is the curvature operator of the connection $\bar{\nabla} = \nabla \oplus \nabla^{\perp}$, and X, Y are tangent vector fields, as above. This condition in fact already came up in [Fe 74a] and then in [BR 83]. The general concept of submanifolds in E^n satisfying (0.4) was introduced by J. Deprez [De 85], who called them *semiparallel*. He proved that all of them are, intrinsically, semisymmetric Riemannian manifolds and gave a classification of semiparallel surfaces in E^n. In [De 86], he also classified semiparallel hypersurfaces, and in [De 89], summarized these first results.

The investigation of semiparallel submanifolds was continued by the author in [Lu 87a, 88a, b, 89a–c, 90a–e], etc., then by F. Dillen in [Di 90b, 91b], [DN 93], and A. C. Asperti in [As 93], [AM 94]. The first summaries were published in [Lu 91f] and then in the monographic article [Lu 2000a] (whose review in *Mathematical*

Reviews (see [MR 2000j: 53071]) is concluded by A. Bucki as follows: "The author's contribution to the theory of submanifolds with parallel fundamental form with his more than forty papers on the subject is colossal"). Currently the monograph [Lu 2000a] is no longer completely up to date; several new results have been added to the theory since then.

The present book will give a more complete survey of the theory of semiparallel submanifolds and of some generalizations in space forms. Semiparallel submanifolds are treated here mainly as second-order envelopes of symmetric orbits.

The book consists of twelve chapters. The first three chapters are preparatory in character. In Chapter 1, the necessary background for subsequent chapters is given using frame bundles (i.e., the Cartan moving frame method) and exterior differential calculus, together with vector and tensor bundles. Basic facts from the theories of space forms and of symmetric and semisymmetric Riemannian manifolds are covered.

In Chapter 2, the general theory of smooth submanifolds in space forms is developed. The second fundamental form h is introduced, together with its higher-order generalizations, their fundamental identities, and the corresponding normal and osculating subspaces are covered. This is done by using orthonormal frames suitably adapted to the submanifold.

In Chapter 3, the theory of parallel submanifolds is developed. Here the specifics of their Gauss maps, their local extrinsic symmetry, Ferus's decomposition theorem and its connection with symmetric R-spaces are presented. The most important examples of complete parallel submanifolds are also given: Segre, Plücker, and Veronese submanifolds.

All of this is in preparation for the main subject, which is the investigation of semiparallel submanifolds. These are introduced in Chapter 4, where some characterizations for their class and several subclasses are given. It is emphasized that (0.4) is a pointwise condition and therefore can be treated purely algebraically. The decomposition theorem for semiparallel submanifolds is also dealt with in the same manner. The analytic fact, that these submanifolds are characterized by the integrability condition of the differential system (0.2) for parallel submanifolds, is interpreted geometrically in the theorem from [Lu 90a], stating that every semiparallel submanifold is a second-order envelope of parallel submanifolds; such envelopes are found for Segre submanifolds, as examples (extending the result of [Lu 91a]).

Chapter 5 is devoted to normally flat semiparallel submanifolds. This class includes all semiparallel submanifolds of principal codimension 1, in particular hypersurfaces, and also semiparallel submanifolds of principal codimension 2 in space forms of nonpositive curvature. A general geometric description is given for normally flat semiparallel submanifolds as immersed warped products of spheres.

Semiparallel submanifolds of low dimensions are considered in Chapters 6 (surfaces) and 7 (three-dimensional submanifolds). They are all classified; the submanifolds of the most general class are described as second-order envelopes of Veronese submanifolds. It is shown that each two-dimensional holomorphic Riemannian manifold can be immersed isometrically into (pseudo-)Euclidean space of dimension ≥ 7, as a surface of this most general class of semiparallel surfaces; but this does not generalize to three dimensions. Some general classes of semiparallel three-dimensional submanifolds are investigated, consisting of second-order envelopes of three-dimensional Segre submanifolds (logarithmic spiral tubes) and of products of Veronese surfaces and plane curves of constant curvature.

In Chapter 8, the decomposition theorems are given: for general parallel and semiparallel submanifolds, for normally flat 2-parallel submanifolds, and for submanifolds with flat van der Waerden–Bortolotti connection. Here the concept of main symmetric orbit is introduced; this is a standardly imbedded symmetric R-space and is minimal in some sphere. The most general semisymmetric submanifold in Euclidean space is locally the second-order envelope of products of main symmetric orbits, some circles and a plane. This is a consequence of the result of [Lu 90a] and of Ferus's famous results [Fe 80].

In Chapter 9, the concept of umbilic-like main symmetric orbit is introduced and studied. A main symmetric orbit is said to be umbilic-like if every second-order envelope of submanifolds congruent or similar to this orbit is a single such orbit; a sphere is an elementary example. Here all known results about umbilic-like main symmetric orbits are presented; the Segre orbits were already investigated from this point of view in Section 4.6 (see Theorem 4.6.1). For the second-order envelope of the family of these main orbits, a differential system is formulated, and then investigated by Cartan's method of differential prolongation. This investigation for Plücker orbits, showing their umbilic-likeness, is carried out in detail. For the other symmetric orbits of the Plücker action, the unitary orbits, this investigation is technically very complicated; only the general scheme is given here and illustrated completely for a model case. For the m-dimensional Veronese orbit, it is shown that in Euclidean space $E^{\frac{1}{2}m(m+3)+1}$, this orbit is not umbilic-like. For the other symmetric orbit of the Veronese action, the Veronese–Grassmann orbit, its umbilic-likeness is asserted, but space and technical complications preclude giving all the details. The general scheme of proof is given, some essential intermediate results are obtained, and the complete proof is illustrated for a model case.

In Chapter 10, it is proved first that a product of umbilic-like main symmetric orbits in Euclidean space is also umbilic-like. This result gives the possibility of extending the description of normally flat semiparallel submanifolds as warped products of spheres to general semiparallel submanifolds, i.e., considering them also as warped products.

Chapter 11 is devoted to semiparallel immersions of semisymmetric Riemannian manifolds, and seeks answers to the problem: can such a manifold be immersed isometrically as a semiparallel submanifold? The answer is positive for dimension $m = 2$, as already shown in Chapter 6. The problem is investigated for dimensions $m > 2$. First, it is proved that if an m-dimensional semiparallel submanifold in E^n is generated by $(m-2)$-dimensional planes, then it is intrinsically a Riemannian manifold of conullity two of the planar type; the other types (i.e., hyperbolic, parabolic, or elliptic type) are not possible. Also, for normally flat semiparallel submanifolds M^m in E^n, it is shown that if such a submanifold is intrinsically of conullity two, then it is of planar type. The same holds for all semiparallel three-dimensional submanifolds. Therefore, it can be conjectured that perhaps this assertion is true in general. The chapter concludes with a theorem that makes this conjecture very plausible. At least, it is certain that there exist semisymmetric Riemannian manifolds, namely of conullity two, that cannot be immersed into Euclidean space as semiparallel submanifolds.

In Chapter 12, some generalizations are considered. First, the k-semiparallel submanifolds for $k > 1$ are introduced and studied. Their relation to envelopes of order k of some family of k-parallel submanifolds is investigated. It is proved that there exist 2-semiparallel submanifolds that are nontrivial, i.e., not parallel and not locally Euclidean; namely, every normally flat semiparallel submanifold (see Chapter 5) turns out to be 2-semiparallel. However, the study of k-semiparallel (in particular 2-semiparallel) submanifolds is still in its initial phase; a complete classification is given only for 2-semiparallel surfaces in space forms (see Section 12.3).

Two other generalizations, namely, the recurrent and the (recently introduced) pseudoparallel submanifolds are discussed briefly in Section 12.4.

Some generalizations have been made by extending the semiparallel condition from the second fundamental form h to some tensor fields (including mixed fields) that are derived from h and the metric form g by some tensor calculus operations. Results of V. Mirzoyan are presented, where the idea of enveloping by corresponding parallel submanifolds is used. These results are illustrated with examples involving surfaces with parallel or semiparallel mean curvature vector field, or normal curvature tensor field, or Ricci tensor field, etc. Hypersurfaces with semiparallel Ricci tensor field are studied in particular, mainly in connection with the famous Ryan's problem: do there exist any hypersurfaces M^m in E^{m+1} with semiparallel Ricci tensor field, that are not intrinsically semisymmetric Riemannian manifolds? It is shown that Mirzoyan's classification result in [Mi 99] covers all known results about this problem, including an affirmative answer for dimension $m \geq 5$ (see [Lu 2002b]). Some special results about the extended Ryan's problem for normally flat submanifolds are also given. The book concludes with a proof that, among the submanifolds of codimension 2 in E^n, there exist normally flat submanifolds that are intrinsically semisymmetric but not semiparallel. This gives additional support to the conjecture stated above (cf. Chapter 11).

The author is grateful to the Estonian Science Foundation for support during the research work; results are summarized in this book. He also expresses his sincere gratitude to Jaak Vilms for valuable help with editing the text of the book and to grandson Imre for technical assistance.

July 2007

Ülo Lumiste
University of Tartu, Estonia

1

Preliminaries

1.1 Real Spaces with Bilinear Metric

Let A be a point set and G a group with identity element e. A map $A \times G \to A$, $(x, g) \mapsto x \circ g$ with $x \circ e = x$, $(x \circ g_1) \circ g_2 = x \circ (g_1 g_2)$ defines a (right) *action* of the group G on A, also called a (right) *G-action* on A. One also says that G acts on A as a *transformation group*. The action is *effective* if $e \in G$ is the only element of G with the property: $x \circ g = x$ for arbitrary $x \in A$; *transitive*, if for every two $x_1, x_2 \in A$ there exists an element $g \in G$, so that $x_2 = x_1 \circ g$; *simply transitive*, if this g is unique for every $(x_1, x_2) \in A \times A$.

The set $H_x = \{h \in G \mid x \circ h = x\}$ is a subgroup of G, called the *isotropy subgroup* of $x \in A$. Obviously $H_{x \circ g} = g^{-1} H_x g$.

The set $G_x = \{y \in A \mid \exists g, y = x \circ g\}$ is called the *orbit* of x under the G-action on A. The orbits are equivalence classes: $x_1 \sim x_2 \Leftrightarrow \exists g \in G, x_2 = x_1 \circ g$. They form the factor set of this equivalence, called the *orbit set*. A G-action on A induces a transitive G-action on every orbit. Obviously, transitivity of the G-action means that there is only one orbit.

Let G be the additive group of vectors of an n-dimensional real vector space V^n and let there be given an effective transitive and simply transitive action of this G on A. This means that

1. if $x \in A$, $v \in V^n$, there exists $y = x \circ v \in A^n$ (one also denotes this by $v = \vec{xy}$),
2. $(x \circ v) \circ w = x \circ (v + w) = x \circ (w + v) = (x \circ w) \circ v$ (i.e., if $\vec{xy} = \vec{wz}$, then $\vec{xw} = \vec{yz}$),
3. for $x, y \in A$ there exists a unique $v \in V^n$ so that $v = \vec{xy}$.

Then A is called a real n-dimensional *affine space*, denoted by A^n, and V^n is said to be *the vector space of A^n*.

Let T be an m-dimensional vector subspace of V^n. The above action of V^n on A^n induces an action of T on A^n. Every orbit of the latter action is called an m-dimensional *affine subspace* of A^n, or briefly, an *m-plane* in A^n. Intrinsically it is an m-dimensional affine space, and T is called the *direction vector subspace* of this m-plane.

Ü. Lumiste, *Semiparallel Submanifolds in Space Forms*,
DOI 10.1007/978-0-387-49913-0_2, © Springer Science+Business Media, LLC 2009

Let $V^n \times V^n \to \mathbb{R}$, $(v_1, v_2) \mapsto \langle v_1, v_2 \rangle$ be a nondegenerate bilinear form, called a *bilinear metric* or, equivalently, a *scalar product*. Two vectors v_1 and v_2 are said to be *orthogonal*, if $\langle v_1, v_2 \rangle = 0$. This is denoted by $v_1 \perp v_2$.

If T is an m-dimensional vector subspace of V^n and the scalar product induces a nondegenerate bilinear form on it, then T is called a *regular* subspace, otherwise a *singular* subspace. For a regular subspace T the set $T^\perp = \{v \mid v \perp w$ for every $w \in T\}$ is also a regular subspace, called the *orthogonal complement* of T; here $V^n = T \oplus T^\perp$, thus $T^\perp$ is $(n-m)$-dimensional.

A real n-dimensional affine space A^n whose vector space V^n is equipped with a scalar product as above is called a *space with bilinear metric*. If the scalar product is symmetric, i.e., $\langle v_1, v_2 \rangle = \langle v_2, v_1 \rangle$ for arbitrary $(v_1, v_2) \in V^n \times V^n$, the space is called *(pseudo-)Euclidean space* ${}_sE^n$; here s is the number of negative coefficients in the canonical representation of the quadratic form $\langle v, v \rangle$. In particular, if this form is positive definite, then the space is *Euclidean space* $E^n (= {}_0E^n)$, otherwise *pseudo-Euclidean space* ${}_sE^n$, $s > 0$ (as is seen, in the latter case without the round brackets around "pseudo-").[1] In particular, ${}_1E^n$ is *Lorentz space*, for $n = 4$ also called *Minkowski space*, the spacetime of the special relativity theory.

An m-plane in ${}_sE^n$ is said to be *regular* if its direction vector subspace is regular, otherwise it is said to be *singular*, in particular *isotropic*, if the scalar product vanishes identically.

In pseudo-Euclidean space ${}_sE^n$ a regular m-plane can be Euclidean or pseudo-Euclidean. In relativity theory, especially the case $s = 1$, such m-planes are also called, correspondingly, *spacelike* or *timelike*, and a singular m-plane is called *light-like*.

In general, if the bilinear form is not nondegenerate but is symmetric, then A^n is called a *semi-Euclidean space*; for instance, every singular m-plane in pseudo-Euclidean space ${}_sE^n$ is an example of such a semi-Euclidean space.

1.2 Moving Frames

Let A^n be a real affine space with vector space V^n. Let $\varepsilon^0 = (e_1^0, \dots, e_n^0)$ be a basis of V^n and o a point of A^n. The pair (o, ε^0) is called a *frame* of A^n with *origin* o and *basis vectors* e_I^0, $I \in \{1, \dots, n\}$. Every basis ε^0 determines an isomorphism $V^n \to \mathbb{R}^n$, $v \mapsto (v^1, \dots, v^n)$ with $v = v^I e_I^0$. (Henceforth the Einstein summation convention is used, that is, the right-hand side actually means that $\sum_{I=1}^n v^I e_I^0 = v^1 e_1^0 + \cdots + v^n e_n^0$.) Every frame determines a homeomorphism $A^n \to \mathbb{R}^n$, $x \mapsto (x^1, \dots, x^n)$ with $\vec{ox} = x^I e_I^0$.

Considering the set of all frames (x, ε) of A^n one defines a (right) action of the general linear group $GL(n, \mathbb{R})$ (i.e., the multiplicative group of all real nonsingular $n \times n$-matrices $A = (A_I^J)$) on this set by $(x, \varepsilon) \circ A = (x, \varepsilon A)$; here εA is the product

[1] Note that some authors use slightly different terminology, e.g., in [Ra 53] (pseudo-)Euclidean spaces are called Euclidean, Euclidean spaces are called properly Euclidean; in [KN 63, 69] pseudo-Euclidean spaces are called indefinite Euclidean.

of $1 \times n$- and $n \times n$- matrices ε and A, i.e., $(\varepsilon A)_I = e_J A^J_I$. This introduces on this set a principal bundle structure with base A^n and structural group $GL(n, \mathbb{R})$ (see [KN 63], Chapter I, Section 5). Every fibre (i.e., orbit of the action) is the set of all frames having the same origin x. This principal bundle is called the *frame bundle* of A^n and realizes the idea of a *moving frame* of É. Cartan (an arbitrary element of this bundle is considered here as a moving frame in A^n; see [IL 2003]).

With respect to a fixed frame ε^0, every moving frame ε in A^n is determined by the coordinates x^I of its origin x, according to $\vec{ox} = x^I e^0_I$, and by the elements X^J_I of the matrix in $e_I = e^0_J X^J_I$, where $I, J, \dots \in \{1, \dots, n\}$.

Note that the differential $d(\vec{ox})$ does not depend on the choice of the origin o, because $\vec{\acute{o}x}$ and $\vec{ox}$ differ only by the constant vector $\vec{o\acute{o}}$. Therefore, $d(\vec{ox})$ can be denoted simply by dx. Also, the point $x \in A^n$ can be identified with its radius vector $\vec{ox}$ from the fixed origin o.

One can calculate

$$dx = e_I \omega^I, \quad de_I = e_J \omega^J_I, \tag{1.2.1}$$

where

$$\omega^I = (X^{-1})^I_J dx^J, \quad \omega^J_I = (X^{-1})^K_I dX^J_K, \quad (X^{-1})^K_J X^J_I = \delta^K_I. \tag{1.2.2}$$

The formulas (1.2.1) are called the *infinitesimal displacement equations* of the moving frame. The differential 1-forms (1.2.2), called the *infinitesimal displacement* 1*-forms*, satisfy the equations

$$d\omega^I = \omega^J \wedge \omega^I_J, \quad d\omega^J_I = \omega^K_I \wedge \omega^J_K, \tag{1.2.3}$$

which are obtained by exterior differentation from (1.2.1) (see [Ste 64], Chapter III, Section 1; [IL 2003], Section B.2) and thus are necessary and sufficient conditions for the complete integrability of (1.2.1). Here (1.2.3) are called the *structure equations* of A^n.

In a real space with bilinear metric one can introduce for every frame the matrix $g = (g_{IJ})$, where $g_{IJ} = \langle e_I, e_J \rangle$. By differentiation, one obtains the relation

$$dg_{IJ} = g_{KJ} \omega^K_I + g_{IK} \omega^K_J. \tag{1.2.4}$$

If the space is ${}_sE^n$, the frame bundle can be reduced to the principal bundle of *orthonormal frames*, characterized by $g_{IJ} = \epsilon_I \delta_{IJ}$, where ϵ_I is -1 for s values of I and 1 for the remaining $n - s$ values of I, and δ_{IJ} is the Kronecker delta. The structural group of the above bundle is the pseudo-orthogonal group ${}_sO(n, \mathbb{R})$; in the cases $s = 0$, $s = 1$, and $s = n - 1$, respectively, this is the orthogonal group $O(n, \mathbb{R})$, and the Lorentz groups ${}_1O(n, \mathbb{R})$ and ${}_{n-1}O(n, \mathbb{R})$ (the last two are isomorphic; see [Wo 72], Section 2.4).

For the bundle of orthonormal frames, the relation (1.2.4) reduces to

$$\varepsilon_J \omega^J_I + \varepsilon_I \omega^I_J = 0 \qquad \text{(no sum!)}. \tag{1.2.5}$$

In the case of E^n, i.e., when $s = 0$, the matrix ω^J_I is skew-symmetric and gives an arbitrary element of the Lie algebra of the orthogonal group $O(n, \mathbb{R})$. For ${}_1E^n$, when $s = 1$, one obtains the same for the Lorentz group ${}_1O(n, \mathbb{R})$.

1.3 (Pseudo-)Riemannian Manifolds

The (pseudo-)Euclidean space ${}_sE^n$ is a special case of the more general concept of a (pseudo-)Riemannian manifold ${}_sN^n$. This is a real n-dimensional differentiable manifold with a smooth field g of symmetric scalar products in the tangent vector spaces. Here the constant natural number s has the same meaning as in ${}_sE^n$. For a local section

$$(x, \varepsilon) = (x; e_1, \dots, e_n)$$

of the frame bundle on ${}_sN^n$ and two tangent vector fields, $X = e_I X^I$ and $Y = e_J Y^J$, one has $\langle X, Y\rangle = g_{IJ} X^I Y^J$, where $g_{IJ} = \langle e_I, e_J\rangle$ are the components of the metric tensor field on ${}_sN^n$, denoted also by g. In the particular case when $\langle X, X\rangle$ is positive definite, the (pseudo-)Riemannian manifold $N^n (= {}_0N^n)$ is called a *Riemannian manifold*, otherwise, a *pseudo-Riemannian manifold* (cf. footnote 1).

A linear connection ∇ on a (pseudo-)Riemannian manifold (see, e.g., [KN 63], Chapter III), which has the property that g is covariantly constant with respect to ∇, i.e., $\nabla g = 0$, is called a *(pseudo-)Riemannian* (in particular *Riemannian*, or *pseudo-Riemannian*) *connection*. Componentwise, the last condition is

$$\nabla g_{IJ} \equiv dg_{IJ} - g_{KJ}\omega_I^K - g_{IK}\omega_J^K = 0, \tag{1.3.1}$$

where $\omega = (\omega_I^J)$ is the matrix field of connection 1-forms of ∇. It is well known that every (pseudo-)Riemannian manifold has a unique (pseudo-)Riemannian connection ∇ without torsion, called the *Levi-Civita connection.*[2]

The elements $\{\omega^I\}$ of the coframe bundle on ${}_sN^n$ and the connection 1-forms ω_I^J of the Levi-Civita connection satisfy the structure equations

$$d\omega^I = \omega^J \wedge \omega_J^I, \quad d\omega_I^J = \omega_I^K \wedge \omega_K^J + \Omega_I^J, \tag{1.3.2}$$

where

$$\Omega_I^J = -R^J_{I,KL}\omega^K \wedge \omega^L \tag{1.3.3}$$

are the *curvature* 2-*forms* of the Levi-Civita connection ∇. Here the coefficients $R^J_{I,KL}$ are the components of the *curvature tensor field* R of ∇. By exterior differentiation of (1.3.1), one obtains via (1.3.2) the equality

$$\Omega_{IJ} + \Omega_{JI} = 0, \tag{1.3.4}$$

where $\Omega_{IJ} = g_{IP}\Omega_J^P = -R_{IJ,KL}\omega^K \wedge \omega^L$ and so $R_{IJ,KL} = g_{IP}R^P_{J,KL}$. Then exterior differentiation of (1.3.2) yields the relations

$$\omega^J \wedge \Omega_J^I = 0, \quad d\Omega_I^J = \Omega_K^J \wedge \omega_I^K - \omega_K^J \wedge \Omega_I^K, \tag{1.3.5}$$

[2] In some recent books, e.g., [Pe 98], historical terminology is disregarded and the Levi-Civita connection is called simply the (pseudo-)Riemannian connection. Sometimes these two terms are considered as equivalent, and then the (pseudo-)Riemannian connection as defined above is called the *metric connection* (see, e.g., [KN 63], Chapter IV; also [Li 55], Section 52; [He 62], Chapter I, Section 9).

which are equivalent to the identities

$$R^I_{J,KL} + R^I_{K,LJ} + R^I_{L,JK} = 0, \quad \nabla_P R^I_{J,KL} + \nabla_K R^I_{J,LP} + \nabla_L R^I_{J,PK} = 0 \quad (1.3.6)$$

(the *Bianchi identities*), where $(\nabla_P R^I_{J,KL})\omega^P = \nabla R^I_{J,KL}$ and

$$\nabla R^I_{J,KL} = dR^I_{J,KL} - R^I_{P,KL}\omega^P_J - R^I_{J,PL}\omega^P_K - R^I_{J,KP}\omega^P_L + R^P_{J,KL}\omega^I_P. \quad (1.3.7)$$

The first identities (1.3.6) and the consequences $R_{IJ,KL} + R_{JI,KL} = 0$ from (1.3.4) imply

$$R_{IJ,KL} = R_{KL,IJ}. \quad (1.3.8)$$

A (pseudo-)Riemannian manifold ${}_sN^n$ of dimension $n > 2$ is said to be a manifold of *constant curvature* if its curvature forms can be represented as $\Omega^J_I = cg_{IK}\omega^J \wedge \omega^K$. Then from (1.3.5) it follows that $dc \wedge \omega^J \wedge \omega^K = 0$, and since $dc = c_I\omega^I$ this gives $c_I\omega^I \wedge \omega^J \wedge \omega^K = 0$. Due to the supposition $n > 2$, for every value of I there exist values of J and K such that $\omega^I \wedge \omega^J \wedge \omega^K \neq 0$. Therefore, $c_I = 0$, and thus $c = \text{const}$. This constant c is called the *curvature* of such a ${}_sN^n$ (cf. [Wo 72], 2.2.7).

The structure equations for a Riemannian manifold of constant curvature c are, due to (1.3.2),

$$d\omega^I = \omega^J \wedge \omega^I_J, \quad d\omega^J_I = \omega^K_I \wedge \omega^J_K + cg_{IK}\omega^J \wedge \omega^K. \quad (1.3.9)$$

1.4 Standard Models of Space and Spacetime Forms

The space ${}_sE^n$ is the simplest n-dimensional (pseudo-)Riemannian manifold of zero curvature.

A connected complete Riemannian manifold of constant curvature c is called a *space form* (see [Wo 72], Section 2.4). Their standard models, denoted by $N^n(c)$, are as follows:

- for $c = 0$, the Euclidean space E^n,
- for $c > 0$,

$$S^n(c) = \{x \in E^{n+1} \mid \langle \vec{ox}, \vec{ox} \rangle = r^2\},$$

  which is the sphere with a real radius $r = 1/\sqrt{c}$ and with center at the origin o,
- for $c < 0$, a connected component of

$$H^n(c) = \{x \in {}_1E^{n+1} \mid \langle \vec{ox}, \vec{ox} \rangle = -r^2\},$$

  which is the sphere in Lorentz space ${}_1E^{n+1}$ with imaginary radius $r = i/\sqrt{|c|}$ and with center at the origin o.

Note that $H^n(c)$ consists of two connected components, each of which is a *hyperbolic* (or *Lobachevsky–Bolyai*) space.

The Minkowski space ${}_1E^4$ (the special case of Lorentz space for $n+1=4$), which is the spacetime of the special relativity theory, is a simple case of pseudo-Riemannian space ${}_sN^n$ of constant curvature c, namely, the case of $s=1$, $n=4$, $c=0$.

In general a connected complete pseudo-Riemannian space ${}_sN^n$ of constant curvature c is called a *spacetime form* and is denoted by ${}_sN^n(c)$. The standard models are ${}_sE^n$ and the connected components of ${}_sS^n(c)$ and ${}_sH^n(c)$, where the latter two are defined similarly to $S^n(c)$ and $H^n(c)$, with E^{n+1} and ${}_1E^{n+1}$ replaced, respectively, with ${}_sE^{n+1}$ and ${}_{s+1}E^{n+1}$ (see [Wo 72], Section 2.4).

Here the special cases are *de Sitter spacetime* ${}_sS^n(c)$ and *anti-de Sitter spacetime* ${}_sH^n(c)$, which for $n=4$ and $s=1$ (resp. $s=2$) are the simplest nonflat spacetime models for general relativity theory (see [HE 73], [PR 86]).

The moving frame bundle of ${}_\sigma E^{n+1}$, where σ is s or $s+1$, can be adapted to a standard (pseudo-)Riemannian model ${}_sN^n(c)$ as follows.

For every frame it is supposed that

(1) $x\in {}_sN^n(c)$, i.e., $\langle\vec{ox},\vec{ox}\rangle=c^{-1}=\text{const}$,
(2) $e_{n+1}\parallel\vec{ox}$, i.e., $e_{n+1}=-\sqrt{|c|}\vec{ox}$ and therefore $g_{n+1,n+1}=\langle e_{n+1},e_{n+1}\rangle=|c|c^{-1}=\operatorname{sign}c$,
(3) $e_1,\dots,e_n$ are orthogonal to e_{n+1}, therefore tangent to ${}_sN(c)$, so that $g_{I,n+1}=0$ $(I=1,\dots,n)$.

Differentiation of the equality in (1) gives $\langle dx,e_{n+1}\rangle=0$; hence $\omega^{n+1}=0$. Similarly from the equalities in (2) and (3) one obtains

$$\omega_{n+1}^{n+1}=0,\quad \omega_{n+1}^{I}=-\sqrt{|c|}\omega^I,\quad \omega_I^{n+1}=\operatorname{sign}c\sqrt{|c|}g_{IK}\omega^K, \tag{1.4.1}$$

where $I,J,\dots$ are in $\{1,\dots,n\}$ and the last relation holds due to (1.2.4) and the equality in (3).

For such a frame bundle adapted to ${}_sN^n(c)$, the relations (1.2.1) and (1.2.3) imply (writing them for dimension $n+1$ and using (1.4.1)) that

$$dx=e_I\omega^I,\quad de_I=e_J\omega_I^J-xcg_{IK}\omega^K, \tag{1.4.2}$$

$$d\omega^I=\omega^J\wedge\omega_J^I,\quad d\omega_I^J=\omega_I^K\wedge\omega_K^J+cg_{IK}\omega^J\wedge\omega^K, \tag{1.4.3}$$

where now $I,J,\dots\in\{1,\dots,n\}$ and (1.2.4) hold. Recall that the radius vector $\vec{ox}$ from the center o of the sphere ${}_sS^n(c)$ (resp. ${}_sH^n(c)$) is being denoted here simply by x, and so $\langle dx,dx\rangle=g_{IJ}\omega^I\omega^J$.

For the (pseudo-)Euclidean space ${}_sE^n\subset{}_sE^{n+1}$ one must take $e_{n+1}=\text{const}$. This leads to the particular case of the formulas (1.4.2) and (1.4.3), obtained by $c=0$ (and thus to (1.2.1) and (1.2.3)). So the formulas above are universal for all standard models of space and spacetime forms.

Remark 1.4.1. The standard models of spacetime forms ${}_sN^n(c)$ can also be treated by means of projective geometry as follows.

Every such model lies in ${}_\sigma E^{n+1}$ with fixed origin at the center of the model ${}_sN^n(c)$; here $\sigma=s$ or $s+1$. There is a one-to-one correspondence between $\mathbb{R}^{n+1}$

and ${}_{\sigma}E^{n+1}$. The projectivization of $\mathbb{R}^{n+1}$ gives the real projective space $P^n(\mathbb{R})$ and then the asymptotic cone of ${}_sN^n(c)$ gives the absolute quadric ${}_sQ^{n-1} \subset P^n(\mathbb{R})$, which determines the projective metric of curvature c. Two vectors of ${}_{\sigma}E^{n+1}$ are orthogonal iff the corresponding points of $P^n(\mathbb{R})$ are polar with respect to ${}_sQ^{n-1}$. The q-dimensional totally geodesic submanifolds of ${}_sN^n(c)$ (the q-dimensional great spheres) can then be interpreted as projective q-planes of $P^n(\mathbb{R})$. This simplifies the understanding of the geometry of ${}_sN^n(c)$ and will be used often below. Note that a projective q-plane is the intersection of the model sphere ${}_sN^n(c)$ with a $(q+1)$-plane through the origin in ${}_{\sigma}E^{n+1}$.

1.5 Symmetric (Pseudo-)Riemannian Manifolds

A vector field $X = e_I X^I$ on a (pseudo-)Riemannian manifold ${}_sN^n$ is said to be *parallel* along a curve in ${}_sN^n$, if $\nabla X = 0$ on this curve, where $\nabla X = e_I(\nabla X^I)$ and $\nabla X^I = dX^I + X^J\omega^I_J$. A curve in ${}_sN^n$ is a *geodesic* if its tangent vector field is parallel along the curve. It is well known that a geodesic with nonzero arclength s, defined by $ds^2 = g_{IJ}\omega^I\omega^J$, is locally a curve of stationary length between any two of its points.

A (pseudo-)Riemannian manifold ${}_sN^n$ is said to have *parallel curvature tensor field* R if $\nabla R = 0$ on ${}_sN^n$, or more explicitly, if $\nabla R^I_{J,KL} = 0$ (i.e., if (0.1) is satisfied), where the left side is defined by (1.3.7).

Let U_{x_0} be a normal neighborhood of a point $x_0 \in_s N^n$, i.e., every point $x \in U_{x_0}$ is connected to x_0 by only one geodesic of ${}_sN^n$ which lies in U_{x_0}. Suppose this curve to be nonisotropic (i.e., with nonzero arclength) and take on it the point x', which is at the same real or imaginary distance from x_0 as x, but on the other side, one gets the *geodesic symmetry map* with respect to x_0. A pseudo-Riemannian manifold ${}_sN^n$ is *locally symmetric* if each of its points x_0 has a normal neighborhood whose geodesic symmetry map with respect to x_0 is an isometry.

É. Cartan proved the following relationship between these properties (also in the more general case of affinely connected manifolds).

Theorem 1.5.1 ([Ca 26] and [He 62], Chapter IV, Section 1). *A (pseudo-)Riemannian manifold ${}_sN^n$ is locally symmetric if and only if its curvature tensor field R is parallel on ${}_sN^n$.*

A Riemannian manifold N^n is said to be globally symmetric if every point x_0 is an isolated fixed point of an involutive isometry s_{x_0} of N^n; here involutive means that $s_{x_0}^2 = \mathrm{Id}$. It follows that x_0 has a normal neighborhood on which s_{x_0} is a geodesic symmetry map (see [He 62], Chapter IV, Section 3). Thus a globally symmetric N^n is also locally symmetric, and vice versa, every complete simply connected locally symmetric Riemannian manifold is globally symmetric (see [He 62], Chapter IV, Section 5). More generally, for every point x_0 of a locally symmetric Riemannian manifold N^n there exist a globally symmetric Riemannian manifold $\tilde{N}^n$, an open neighborhood U_{x_0} of x_0 in N^n, and an isometry φ mapping U_{x_0} onto an open neighborhood of the point $\varphi(x_0)$ in $\tilde{N}^n$.

The manifold $\tilde{N}^n$ is diffeomorphic to the homogeneous space G/K, where G is the identity component of the Lie group of isometries of $\tilde{N}^n$ and K is the compact subgroup of isometries with fixed point x_0; the diffeomorphism $G/K \to \tilde{N}^n$ is given by $gK \mapsto g \circ x_0$, $g \in G$ (see [He 62], Chapter IV, Section 3).

In turn, let G be a connected Lie group, K a closed subgroup with compact $\mathrm{Ad}_G(K)$, and γ an analytic automorphism of G such that $(K_\gamma)_0 \subset K \subset K_\gamma$, where K_γ is the set of fixed points of γ and $(K_\gamma)_0$ is its identity component. Then for every G-invariant Riemannian structure on G/K this G/K is a globally symmetric Riemannian manifold (see [He 62], Chapter IV, Section 3). In this case (G, K) is called a *Riemannian symmetric pair*.

These results reduce the study of globally symmetric Riemannian manifolds $\tilde{N}^n$ to the study of Riemannian symmetric pairs (G, K) by means of Lie group theory.

Remark 1.5.2. In general, symmetric pseudo-Riemannian manifolds have not been studied so thoroughly as the Riemannian ones. É. Cartan [Ca 26] had noted that these types of manifolds with solvable isometry group exist. The case of dimension 4 was then studied in [Wal 46] and [Wal 50] (see also [Ab 71]). In [Ro 49b], [Fed 56], [Fed 59] symmetric pseudo-Riemannian manifolds with simple groups of isometries were classified; in [Be 57] the case of semisimple groups was also included. The classification problem for four-dimensional symmetric Einsteinian spaces with Lorentzian signature and of the first type was solved by A.Z. Petrov [Pe 66]. In [CML 68], all symmetric four-dimensional spaces of signature ± 2 were listed. A complete classification of the spaces of signature 2 with solvable transvection group was given in [CP 70]; see also [Ast 73].

Example 1.5.3. Comparing the structure equations (1.3.2) and (1.4.3), one sees that for the standard models ${}_sN^n(c)$ of spacetime forms the curvature 2-forms are

$$\Omega_I^J = c g_{IK} \omega^J \wedge \omega^K, \quad c = \text{const}.$$

Thus $R^J_{I,KL} = -c g_{IK} \delta^J_L$. From (1.2.4) $\nabla g_{IK} = 0$; also $\nabla \delta^J_L = d\delta^J_L - \delta^J_P \omega^P_L + \delta^P_L \omega^J_P = 0$, and this leads to $\nabla R^J_{I,KL} = 0$. Hence, every ${}_sN^n(c)$ is a locally symmetric (pseudo-)Riemannian manifold; actually it is also globally symmetric (see [Wo 72], Chapter 11).

Example 1.5.4. The manifold of all q-dimensional vector subspaces of a p-dimensional real vector space $\mathbb{R}^p$ is called the *Grassmann manifold* and denoted by $G(q, \mathbb{R}^p)$ (see, e.g., [Sha 88], Chapter 1, Section 4). Let a pseudo-Euclidean metric of index k be given in $\mathbb{R}^p$ by the metric tensor $g_{\lambda\mu}$ and consider the manifold of all regular q-dimensional vector subspaces of index l. This manifold is called the *Grassmann manifold of regular subspaces* and is denoted by ${}_{l,k}G^{q,p}$.

For an element of ${}_{l,k}G^{q,p}$ considered as a subspace, the orthonormal basis $\{e_\lambda\}$ in $\mathbb{R}^p$ $(1 \le \lambda \le p)$ can be chosen so that e_a and e_u $(1 \le a \le q; q+1 \le u \le p)$ are vectors belonging to this subspace and to its orthogonal complement, respectively. Thus the subspace is determined by the simple q-vector $e_1 \wedge e_2 \wedge \cdots \wedge e_q$.

Let $\wedge^q(\mathbb{R}^p)$ be the space of antisymmetric $(q, 0)$-tensors (see [Ste 64], Chapter I, Section 4). This $\wedge^q(\mathbb{R}^p)$ is a vector space, for which the simple q-vectors $e_{\lambda_1} \wedge \cdots \wedge e_{\lambda_q}$

with $\lambda_1 < \cdots < \lambda_q$ form a basis. From the infinitesimal displacement equations (1.2.1) it follows that

$$d(e_1 \wedge e_2 \wedge \cdots \wedge e_q) = \sum_{a,u} (e_1 \wedge \cdots \wedge e_{a-1} \wedge e_u \wedge e_{a+1} \wedge \cdots \wedge e_m)\omega_a^u, \quad (1.5.1)$$

because for an orthonormal basis $\omega_a^a = 0$ (no sum; see (1.2.5)), so that the 1-forms ω_a^u play the role of ω^I in the first formula of (1.2.1). Now the argument used in [Lu 92a], [Maa 74] can be applied. There it is shown that the pseudo-Riemannian structure on ${}_{l,k}G^{q,p}$ is given by

$$ds^2 = g^{ab} g_{uv} \omega_a^u \omega_b^v,$$

where $1 \le a, b \le q$; $q+1 \le u, v \le p$ (see also [Ha 65]) and that it is Einstein of constant scalar curvature; for $k = l = 0$ this is established in [Le 61]. Thus ω_a^u generates a moving coframe on ${}_{l,k}G^{q,p}$ and for this the first structure equations (1.3.2) must hold. On the other hand,

$$d\omega_a^u = \omega_a^b \wedge \omega_b^u + \omega_a^v \wedge \omega_v^u = \omega_b^v \wedge (-\omega_a^b \delta_v^u + \delta_a^b \omega_v^u),$$

so that the role of ω_J^I in (1.3.2) is played by the 1-forms in the last parentheses above. Since $d\omega_J^I$ is now

$$d(-\omega_a^b \delta_v^u + \delta_a^b \omega_v^u) = -(\omega_a^c \wedge \omega_c^b + \omega_a^w \wedge \omega_w^b)\delta_v^u + \delta_a^b(\omega_v^c \wedge \omega_c^u + \omega_v^w \wedge \omega_w^u)$$

and $\omega_J^K \wedge \omega_K^I$ is

$$(-\omega_c^b \delta_v^w + \delta_c^b \omega_v^w) \wedge (-\omega_a^c \delta_w^u + \delta_a^c \omega_w^u) = -\omega_a^c \wedge \omega_c^b \delta_v^u + \omega_v^w \wedge \omega_w^u \delta_a^b,$$

the curvature 2-forms Ω_J^I in (1.3.2) for ${}_{l,k}G^{q,p}$ are

$$-\omega_a^w \wedge \omega_w^b \delta_v^u + \delta_a^b \omega_v^c \wedge \omega_c^u.$$

Now (1.2.5) implies that $\omega_w^b = -\varepsilon_b \varepsilon_w \omega_b^w$, so that these curvature 2-forms are

$$(\delta_a^d g^{bc} g_{wx} \delta_v^u - \delta_a^b g^{cd} g_{vw} \delta_x^u)\omega_d^w \wedge \omega_c^x,$$

where $g^{bc} = \varepsilon_b \delta^{bc}$, $g_{vw} = \varepsilon_v \delta_{vw}$, etc., and the indices w, x and c, d run through the same values as u, v and a, b, respectively. The reduced coefficients are the components of the curvature tensor $R^I_{J,KL}$ of ${}_{l,k}G^{q,p}$. This and the above expressions for ω_J^I imply that for the pseudo-Riemannian connection ∇^G of ${}_{l,k}G^{q,p}$ the equations $\nabla^G R^I_{J,KL} = 0$ hold (cf. (1.3.7)).

Consequently, the following statement holds.

Theorem 1.5.5. *The Grassmann manifold ${}_{l,k}G^{q,p}$ of regular subspaces is a locally symmetric pseudo-Riemannian manifold, which is Einstein of constant scalar curvature.*

In the particular case $q = 2$, this ${}_{l,k}G^{2,p}$ is called the *Plücker manifold*.

The same conclusion also holds for $k = 0$ (thus also $l = 0$); then "pseudo-" is to be omitted and the corresponding Grassmann manifold is denoted simply by $G^{q,p}$. A projective space treatment of most of these results for Grassmann manifolds with polar normalization can be found in [AG 96], Chapter 6, Section 6.6; see also [Ro 49a].

Grassmann manifolds are also globally symmetric, as shown in [Wo 72] (for the Riemannian case, see Section 9.2, where a corresponding Riemannian symmetric pair is used; for the pseudo-Riemannian case, cf. Section 12.2).

Remark 1.5.6. Some generalizations of symmetric Riemannian spaces have been made by Fedenko [Fed 77] and Kowalski [Kow 80]. In [KoK 87] Kowalski's approach is transferred to the geometry of submanifolds M^m in E^n; in [CMR 94] the same is for M^m in $N^n(c)$.

Another generalization is made by Deszcz in [Des 92] and for submanifolds in [ALM 99, 2002], [LT 2006] (see Section 12.4).

1.6 Semisymmetric (Pseudo-)Riemannian Manifolds

According to Theorem 1.5.1 the class of locally symmetric pseudo-Riemannian manifolds is analytically characterized by the system of differential equations $\nabla R = 0$ for the components of the curvature tensor field R (cf. with (0.1)). More explicitly, due to (1.3.7) this system is

$$dR^I_{J,KL} - R^I_{P,KL}\omega^P_J - R^I_{J,PL}\omega^P_K - R^I_{J,KP}\omega^P_L + R^P_{J,KL}\omega^I_P = 0. \tag{1.6.1}$$

The integrability condition of this system can be obtained by exterior differentiation, using the structure equations (1.3.2), which leads to the equations

$$R^I_{P,KL}\Omega^P_J + R^I_{J,PL}\Omega^P_K + R^I_{J,KP}\Omega^P_L - R^P_{J,KL}\Omega^I_P = 0. \tag{1.6.2}$$

Replacing Ω^P_J with the expressions $R^P_{J,QS}\omega^Q \wedge \omega^S$ given in (1.3.3), and collecting the terms before $\omega^Q \wedge \omega^S$, one obtains a system of purely algebraic (quadratic) equations for the components of R. Contracting the left sides of these equations with coordinates of two linearly independent tangent vectors $X = e_Q X^Q$, $Y = e_S Y^S$ and considering $R^P_{J,QS}X^Q Y^S = R^P_J(X, Y)$ as the entries of the matrix of a linear operator $R(X, Y)$ acting on R, this algebraic system can be written concisely as (cf. with (0.3))

$$R(X, Y) \cdot R = 0. \tag{1.6.3}$$

The system (1.6.2), or the equivalent system (1.6.3), was already found as the integrability condition of (1.6.1) in the first investigations by P. A. Shirokov and É. Cartan about symmetric spaces (see [Shi 25], [Ca 26]). A natural generalization of these spaces was considered by É. Cartan, who in [Ca 46] introduced the Riemannian manifolds satisfying (1.6.3). His investigations were continued by A. Lichnerowicz [Li 52], [Li 58] and R. Couty [Co 57].

What follows is a short survey of the results about the Riemannian manifolds satisfying (1.6.3). More detailed information can be found in the monograph [BKV 96].

The term *semisymmetric* for manifolds satisfying the condition (1.6.3) was introduced by N. S. Sinyukov [Si 56, 62], who showed the importance of this condition in the theory of geodesic mappings of Riemannian manifolds (see [Si 79], Chapter 2, Section 3).

A fruitful impulse for investigations of manifolds of this class was given by K. Nomizu, who in [No 68] conjectured that all complete, irreducible n-dimensional Riemannian manifolds ($n \geq 3$) satisfying $R(X, Y) \cdot R = 0$ are locally symmetric, i.e., they also satisfy $\nabla R = 0$. This conjecture was supported by the result that for a Riemannian manifold $\nabla^k R = 0$ yields $\nabla R = 0$, which was proved for the compact case in [Li 58] and for the complete case in [NO 62] (and it is also valid in general; cf. [KN 63], Vol. 1, Remark 7). However, Nomizu's conjecture was eventually refuted. Namely, in [Ta 72] a hypersurface in E^4 was constructed satisfying $R(X, Y) \cdot R = 0$ but not $\nabla R = 0$; and a counterexample of arbitrary dimension was given in [Sek 72]. Nevertheless, by adding some further conditions to $R(X, Y) \cdot R = 0$, the conjecture becomes true; such additional conditions were given in [ST 70], [Tan 71], [Fu 72]. For instance, it is shown in these papers that it suffices to add $\nabla C = 0$, $S =$ const, where C is the tensor of conformal curvature and S is the scalar curvature (cf. also [Sek 75], [Sek 77]).

For pseudo-Riemannian manifolds the term *semisymmetric* was used (for the case of Lorentzian signature) by V. R. Kaigorodov [Kai 78] in the course of investigations on the curvature structure of spacetime (cf. also [Kai 83]).

Let a (pseudo-)Riemannian space be a direct product of the same kind of spaces. Then the frame bundle can be adapted so that the basis vectors are successively tangent to the mutually orthogonal components of the product. Then $R^I_{J,KL}$ are zero if two of the indices I, J, K, L are indices of basis vectors tangent to different components. A straightforward calculation shows that if (1.6.2) is satisfied for every component, then it is also satisfied for the direct product. The same holds if (1.6.2) replaced by $\nabla R = 0$, i.e., by (1.6.1). Thus the direct product of semisymmetric (resp. symmetric) (pseudo-)Riemannian manifolds is a semisymmetric (resp. symmetric) (pseudo-)Riemannian manifold.

The local classification of semisymmetric Riemannian manifolds was given by Z. I. Szabó, locally in [Sza 82] and then globally in [Sza 85]. First he proved by means of the infinitesimal or the local holonomy group that for every semisymmetric Riemannian manifold M^m there exists a dense open subset U such that around the points of U the manifold M^m is locally isometric to a direct product of semisymmetric manifolds $M_0 \times M_1 \times \cdots \times M_r$, where M_0 is an open part of a Euclidean space and the manifolds M_i, $i > 0$, are infinitesimally irreducible simple semisymmetric leaves. Here a semisymmetric M is called a *simple leaf* if at each of its points x the primitive holonomy group determines a simple decomposition $T_x M = V_x^{(o)} + V_x^{(1)}$, where this group acts trivially on $V_x^{(0)}$ and there is only one subspace $V_x^{(1)}$ that is invariant for this group. A simple leaf is said to be infinitesimally irreducible if at least at one point the infinitesimal holonomy group acts irreducibly on $V_x^{(1)}$.

The dimension $\nu(x) = \dim V_x^{(0)}$ is called the *index of nullity* at x and $u(x) = \dim M - \nu(x)$ the *index of conullity* at x.

The classification theorem of Szabó asserts the following (according to the formulation given in [BKV 96]).

Theorem 1.6.1. *For every semisymmetric Riemannian manifold there exists an everywhere dense open subset U such that around every point of U the manifold is locally isometric to a space that is the direct product of an open part of a Euclidean space and of infinitesimally irreducible simple semisymmetric leaves, each of which is one of the following:*

(a) *if $\nu(x) = 0$ and $u(x) > 2$, then locally symmetric (hence locally isometric to a symmetric space);*
(b) *if $\nu(x) = 1$ and $u(x) > 2$, then locally isometric to an elliptic, a hyperbolic or a Euclidean cone;*
(c) *if $\nu(x) = 2$ and $u(x) > 2$, then locally isometric to a Kählerian cone;*
(d) *if $\nu(x) = \dim M - 2$ and $u(x) = 2$, then locally isometric to a space foliated by Euclidean leaves of codimension* 2 (*or to a two-dimensional manifold, this in the case when* $\dim M = 2$).

The following examples give more detailed descriptions (according to [Sza 82] and [BKV 96]) of these product components, some of them in the general (pseudo)-Riemannian situation.

Example 1.6.2 (*for case* (a)). Every symmetric (pseudo)-Riemannian space is also semisymmetric.

Indeed, the condition (1.6.1) yields (1.6.2), and thus (1.6.3) too, because they are the integrability conditions of (1.6.1).

Example 1.6.3 (*for case* (b)). Consider $\mathbb{R}_+ \times \mathbb{R}^{n-1}$ with the standard coordinate system $(x^0, x^1, \ldots, x^{n-1})$ and the Riemannian metric given by

$$ds^2 = (dx^0)^2 + (x^0 + C)^2[(dx^1)^2 + \cdots + (dx^{n-1})^2].$$

This Riemannian manifold is called the *Euclidean cone* and it is semisymmetric.

Example 1.6.4 (*for case* (b)). Let $S^{n-1}(c)$ be a sphere with center o in E^n and v a point in E^{n+1} such that the straight line ov is orthogonal to $E^n \subset E^{n+1}$. The *elliptic cone* is a hypersurface in E^{n+1} described by the straight half-lines emanating from v (the vertex) and intersecting the points of $S^{n-1}(c)$. With its induced metric, this hypersurface is intrinsically an n-dimensional Riemannian manifold that turns out to be semisymmetric. This induced metric has an expression similar to the metric in the previous example: in the square brackets one takes here the standard metric of $S^{n-1}(c)$.

Example 1.6.5 (*for case* (b)). Let the expression in square brackets be replaced by the standard metric of the $(n-1)$-dimensional hyperbolic space. Then the Riemannian manifold with this metric is called the *hyperbolic cone*.

This space can also be represented in the following way. Let $H^{n-1}(c)$ be a sphere with center o and imaginary radius $i/\sqrt{|c|}$ in ${}_1E^n \subset {}_1E^{n+1}$. Its connected component is intrinsically a hyperbolic space (see Section 1.3). Choose a point v in ${}_1E^{n+1}$ so that the straight line ov is orthogonal to ${}_1E^n$, and consider the hypersurface in ${}_1E^{n+1}$ defined by the straight half-lines emanating from v and intersecting the points of a connected component of $H^{n-1}(c)$. This hypersurface is in fact a hyperbolic cone.

Example 1.6.6 (*for case* (c)). The Kählerian cones are the complex analogs of Examples 1.6.4, 1.6.5 and 1.6.6 (see [BKV 96]).

Example 1.6.7 (*for case* (d)). Every two-dimensional (pseudo-)Riemannian manifold is semisymmetric.

Indeed, if $n = 2$, then (1.3.4) and (1.3.8) imply that $R_{IJ,KL}$ is nonzero only if $(IJ, KL) = (12, 12)$ or a permutation of $(12, 12)$. This easily yields (1.6.3).

Example 1.6.8 (*case* (d) *in general*). In the Szabó classification a special role is played by the n-dimensional Riemannian manifolds foliated by $(n-2)$-dimensional Euclidean spaces. They are characterized as those manifolds, whose tangent vector spaces are orthogonal products $V_x^{(0)} + V_x^{(1)}$, where $V_x^{(1)}$ are of dimension 2 at every point, and $V_x^{(0)}$ define a foliation of dimension $(n-2)$ with Euclidean spaces as leaves.

These foliated manifolds were treated implicitly in [Sza 82], without considering any explicit expressions for their metrics. They were considered as solutions of a certain integrable system of nonlinear partial differential equations. A more detailed analysis was given by Kowalski in [Kow 96] for the three-dimensional case ($n = 3$); see also [BKV 96].

In [BKV 96] (cf. the remark concluding Chapter 8) the situation was described as follows. The general solution of the basic system of partial differential equations given by Szabó depends formally on $\frac{1}{2}(n-2)(n+3)+4$ arbitrary functions of two variables and $\frac{1}{2}(n-2)(n+3)$ arbitrary functions of one variable. In dimension 3, this means seven functions of two variables and three functions of one variable. But a more detailed analysis has shown that in fact three arbitrary functions of two variables suffice to parametrise the corresponding spaces. The exact number of arbitrary functions of two variables that parametrise local isometry classes of foliated semisymmetric manifolds in dimension n remained an unsolved problem in [BKV 96]. It was noted only that the first explicit examples depending on one arbitrary function of two variables were constructed in [KoTV 90] and [KoTV 92]. The new approach, given by O. Kowalski in dimension 3, was then generalized by E. Boeckx [Bo 95] to arbitrary dimension n. These results are summarized in [BKV 96], where these manifolds are called Riemannian manifolds *of conullity two*, motivated by case (d).

Remark 1.6.9. For four-dimensional semisymmetric Riemannian manifolds an elementary classification can be given independently from Szabó's (which is indirect and relies on some essential results from other sources, for instance, a theorem of Kostant). This elementary classification is given in [Lu 96e], [Lu 96f]. The result is as follows.

Theorem 1.6.10. *Locally, every four-dimensional semisymmetric Riemannian manifold is one of the following:*

(a) *a locally Euclidean manifold,*
(b) *a space of nonzero constant curvature,*
(c) *a locally symmetric space other than* (a) *and* (b),
(d) *the direct product of two two-dimensional spaces,*
(e) *locally isometric to an elliptic or hyperbolic cylinder* (*i.e., direct product* $S^3 \times \mathbb{R}$ *or* $H^3 \times \mathbb{R}$) *or to a Euclidean, elliptic or hyperbolic cone,*
(f) *space foliated by two-dimensional totally geodesic and locally Euclidean leaves that are transversally flat along themselves and normally flat in sections normal to them.*

The proof is given in [Lu 96f] and is based on an elementary algebraic classification of semisymmetric curvature operators [Lu 96e]. Chern bases are used to minimize the number of nonzero components of the curvature tensor. This considerably simplifies the semisymmetric condition as a system of quadratic equations on these components.

Note that in the pseudo-Riemannian case the problem of detailed classification of semisymmetric manifolds is currently still open, to the author's knowledge.

Remark 1.6.11. In [Kow 96] (and then in [BKV 96]) the following terminology is used for semisymmetric Riemannian manifolds of types (a)–(d): the manifolds of type (a) are said to be of "trivial" class, types (b) and (c) of "exceptional" class, and of type (d) "typical" class.

For three-dimensional manifolds of this last class, O. Kowalski introduced (in a preprint of 1991 and published afterwards in [Kow 96]) the geometric concept of *asymptotic foliation*, which was generalized by E. Boeckx [Bo 95] to arbitrary dimensions.

An $(m-1)$-dimensional submanifold M^{m-1} of a manifold M^m of conullity two is called an *asymptotic leaf* if it is generated by $(m-2)$-dimensional Euclidean leaves of M^m and if its tangent spaces are parallel along each Euclidean leaf with respect to the Levi-Civita connection ∇ of M^m.

An *asymptotic distribution* on M^m is an $(m-1)$-dimensional distribution that is integrable and whose integral submanifolds are asymptotic leaves. The integral manifolds of an asymptotic distribution determine a foliation of M^m, called an *asymptotic foliation*.

For an M^m of conullity two, the adapted frame bundle and corresponding coframes can be chosen so that the Euclidean leaves are determined by $\omega^a = 0$, $a, b, \dots \in \{1, 2\}$. Since this last differential system is totally integrable, $d\omega^1$ and $d\omega^2$ must vanish as an algebraic consequence of $\omega^1 = \omega^2 = 0$ (due to the Frobenius theorem, second version; see [Ste 64]). This together with the fact that Euclidean leaves are totally geodesic, because M is a simple leaf, yields, due to (1.3.2),

$$\omega_u^1 = A_u\omega^1 + B_u\omega^2, \quad \omega_u^2 = C_u\omega^1 + F_u\omega^2; \tag{1.6.4}$$

where $u, v, \dots \in \{3, \dots, m\}$.

Let the unit vector $X = e_1 \cos\varphi + e_2 \sin\varphi$ be taken so that $\mathrm{span}\{X, e_3, \dots, e_m\}$ is the tangent plane of an asymptotic leaf. Then, $\nabla_{e_u} X = \nabla_X e_u + [e_u, X]$ must belong to the tangent plane of this asymptotic leaf for every value of u. Since the tangent distribution of these leaves is a foliation, this tangent plane contains $[e_u, X]$. Thus this plane must also contain

$$\nabla_X e_u = \nabla_{e_1} e_u \cos\varphi + \nabla_{e_2} e_u \sin\varphi = (\omega_u^k(e_1)e_k)\cos\varphi + (\omega_u^k(e_2)e_k)\sin\varphi.$$

Hence

$$(A_u e_1 + C_u e_2)\cos\varphi + (B_u e_1 + F_u e_2)\sin\varphi$$

must belong to $\mathrm{span}\{X, e_3, \dots, e_m\}$ and therefore must be a multiple of $X = e_1 \cos\varphi + e_2 \sin\varphi$. This last condition is equivalent to

$$B_u \sin^2\varphi + (A_u - F_u)\cos\varphi \sin\varphi - C_u \cos^2\varphi = 0.$$

But along the asymptotic leaf, $\omega^1 \sin\varphi = \omega^2 \cos\varphi$, so that the above condition reduces to

$$C_u(\omega^1)^2 + (F_u - A_u)\omega^1\omega^2 - B_u(\omega^2)^2 = 0.$$

According to [Kow 96], [BKV 96] a foliated M is said to be *planar* if it admits infinitely many asymptotic foliations. If it admits just two (or one, or none, respectively) asymptotic foliations, it is said to be *hyperbolic* (or *parabolic*, or *elliptic*, respectively).

2

Submanifolds in Space Forms

2.1 A Submanifold and Its Adapted Frame Bundle

Submanifolds will be considered in the context of differentiable manifolds of class C^∞ (see [Ste 64], Chapter II; [KN 63], Chapter I), or more precisely, in the context of (pseudo-)Riemannian manifolds (see [KN 69], Chapter VII; [Ch 73b], [Ch 2000], [BCO 2003]). It is worth mentioning that the introduction of [Ch 2000] contains a brief survey of the long history of the differential geometry of submanifolds.[1] Recent developments in submanifold theory are described in the introduction of [BCO 2003].

Let $f : M^m \to {}_sN^n(c)$ be an isometric immersion of class C^∞ of an m-dimensional (pseudo-)Riemannian manifold into an n-dimensional space form (or spacetime form, if $s > 0$), $n > m$, taken as the standard model ${}_sN^n(c)$ (see Section 1.4). Then $f(M^m)$ is a *submanifold* in ${}_sN^n(c)$ (see [KN 63], Chapter VII, also [Ch 73b] and [Ch 2000], for the case of Riemannian manifolds). Such a submanifold will be denoted simply by M^m, i.e., f is considered as the inclusion map.

For such a submanifold M^m its *tangent vector space* T_xM^m at an arbitrary point $x \in M^m$ is a regular vector subspace of $T_x({}_sN^n(c))$ and therefore has an orthogonal complement $T_x^\perp M^m$ in the latter, which is an $(n-m)$-dimensional regular vector space, called the *normal vector space* of the submanifold M^m at x.

If, for the case $s > 0$ and at an arbitrary point $x \in M^m$, the tangent vector space T_xM^m is spacelike (resp. timelike), then the submanifold M^m in ${}_sN^n(c)$ is also said to be spacelike (resp. timelike).

If $c \neq 0$ then ${}_sN^n(c) \subset {}_\sigma E^{n+1}$ (recall that $\sigma = s$ for $c > 0$ and $\sigma = s+1$ for $c < 0$). Thus an orthogonal complement $T_x^{*\perp}M^m$ of T_xM^m in $T_x({}_\sigma E^{n+1})$ is defined, called the *outer normal vector space* of M^m at x; obviously $T_x^{*\perp}M^m$ is the span of $T_x^\perp M^m$ and of $x = -(\sqrt{|c|})^{-1}e_{n+1}$, which are mutually orthogonal. If $c = 0$, then $N^n(0) = E^n$ and the designation *outer* is superfluous.

[1] One should note, however, that omitted in this survey are some newer historical investigations shedding light, in particular, on the emerging role of M. Bartels, F. Minding, and K. Peterson of the 19th century differential geometric school at the University of Tartu (Dorpat) (see [Stru 33], [GLOP 70], [Rei 73], [Ph 79], [Lu 96g], [Lu 97a], [Lu 99b]).

Ü. Lumiste, *Semiparallel Submanifolds in Space Forms*,
DOI 10.1007/978-0-387-49913-0_3,

The union of all tangent (normal or outer normal) vector spaces constitutes the *tangent* (resp. *normal* or *outer normal*) *vector bundle* of M^m, denoted by TM^m (resp. $T^{\perp}M^m$ or $T^{*\perp}M^m$). Its sections are the *tangent* (resp. *normal* or *outer normal*) *vector fields* on M^m.

In this book the method of frame bundles and exterior differential calculus is used. For a submanifold M^m in ${}_sN^n(c)$ the frame bundle adapted to ${}_sN^n(c)$ can be reduced to the subbundle of frames adapted to M^m as follows (see [KN 69], Chapter VII, Section 1). Let $x \in M^m$, let the first m basis vectors $e_1, \dots, e_m$ (in general, e_i, where $i, j, \dots \in \{1, \dots, m\}$) belong to T_xM^m and the next $n-m$ basis vectors $e_{m+1}, \dots, e_n$ (in general, e_α, where $\alpha, \beta, \dots \in \{m+1, \dots, n\}$) to $T_x^{\perp}M^m$. Then $g_{i\alpha} = 0$, and due to (1.2.4)

$$g_{\beta\alpha}\omega_i^\beta + g_{ik}\omega_\alpha^k = 0, \tag{2.1.1}$$

$$dg_{ij} = g_{kj}\omega_i^k + g_{ik}\omega_j^k, \quad dg_{\alpha\beta} = g_{\gamma\beta}\omega_\alpha^\gamma + g_{\alpha\gamma}\omega_\beta^\gamma. \tag{2.1.2}$$

Since the differential dx of the radius vector of the point $x \in M^m$ (recall that it is also denoted by x) belongs to T_xM^m, the first equation of (1.4.2) reduces to $dx = e_i\omega^i$, which means

$$\omega^\alpha = 0. \tag{2.1.3}$$

The submanifold M^m can be considered as an integral submanifold in ${}_sN^n(c)$ of this differential system (2.1.3).

From (1.4.3) and (2.1.3) it follows that $\omega^i \wedge \omega_i^\alpha = 0$, and now Cartan's lemma (see [Ste 64], Chapter I, Section 4; [BCGGG 91], p. 320; [IL 2003], p. 314) gives

$$\omega_i^\alpha = h_{ij}^\alpha\omega^j, \quad h_{ij}^\alpha = h_{ji}^\alpha. \tag{2.1.4}$$

Therefore, from (1.4.2)

$$de_i = e_j\omega_i^j + (e_\alpha h_{ij}^\alpha - xcg_{ij})\omega^j, \tag{2.1.5}$$

and so for an arbitrary vector field $X = e_iX^i$ in the tangent vector bundle TM^m one has

$$dX = e_i(dX^i + X^j\omega_j^i) + (e_\alpha h_{ij}^\alpha - xcg_{ij})X^i\omega^j,$$

where the right-hand side is a sum of a tangent component and an outer normal component. Now if the point x is considered fixed, so that $dx = 0$ and thus all $\omega^i = 0$, one must also have $dX = 0$. Hence in the tangent component, the expressions in the first set of parentheses must be linear combinations of these ω^j. In other words, $dX^i + X^j\omega_j^i = \nabla_jX^i\omega^j$. The expression ∇_jX^i is the covariant derivative of the (1,0)-tensor field X^i on M^m with respect to the Levi-Civita connection ∇ of M^m, and thus ω_j^i are the connection 1-forms of ∇.

In the normal component, the coefficients h_{ij}^α, taken from (2.1.4), constitute a mixed tensor field, called the *second fundamental tensor* of M^m in ${}_sN^n(c)$. This mixed tensor field determines the *second fundamental form* (denoted by h) of M^m in ${}_sN^n(c)$ with values in $T_x^{\perp}M^m$.

To describe the relationship between this tensor and form, let another tangent vector field $Y = e_j Y^j$ be given on M^m and let t be the parameter of its integral curve such that $dx/dt = Y$. Then $\omega^j = Y^j dt$, and in the normal component of dX/dt with respect to ${}_sN^n(c)$ one has $h(X, Y) = e_\alpha h^\alpha_{ij} X^i Y^j$ (cf. [KN 69], Chapter VII, Section 3 and [Ch 73b], Chapter 2, Section 1).

With respect to ${}_\sigma E^{n+1}$ the normal component of de_i has the vector-valued coefficients

$$h_{ij} - xcg_{ij} = h^*_{ij}, \tag{2.1.6}$$

where $h_{ij} = e_\alpha h^\alpha_{ij}$, so that (2.1.5) is

$$de_i = e_j \omega^j_i + h^*_{ij}\omega^j. \tag{2.1.7}$$

The coefficients h^*_{ij} define a bilinear symmetric form with values in $T^{*\perp}_x M^m$, called the *outer second fundamental form* of M^m and denoted by h^*, i.e.,

$$h^*(X, Y) = h(X, Y) - xc\langle X, Y\rangle,$$

where $\langle X, Y\rangle = g_{ij} X^i Y^j$ is the scalar product of X and Y.

The usual lowering and raising of indices can be used by means of g_{ij}, $g_{\alpha\beta}$ and g^{ij}, $g^{\alpha\beta}$, where $g^{ik} g_{kj} = \delta^i_j$, $g^{\alpha\gamma} g_{\gamma\beta} = \delta^\alpha_\beta$. For instance, if one defines $h^{\ i}_{\alpha\ k} = g^{ij} h^\beta_{jk} g_{\beta\alpha}$, then $h^{\ i}_{\alpha\ k}\xi^\alpha$ gives the *shape (or Weingarten) operator* A_ξ of M^m in $N^n(c)$, which can also be defined by $\langle A_\xi(X), Y\rangle = \langle \xi, h(X, Y)\rangle$ (see, e.g., [KN 69], Chapter VII, Section 3, [BCO 2003], 2.1).

For the normal basis vectors e_α of the frame adapted to M^m one has, due to (1.4.2), (2.1.1), and (2.1.4),

$$de_\alpha = e_i(-h^{\ i}_{\alpha\ k}\omega^k) + e_\beta \omega^\beta_\alpha;$$

hence, for a normal vector field $\xi = e_\alpha \xi^\alpha$,

$$d\xi = e_\alpha(d\xi^\alpha + \xi^\beta \omega^\alpha_\beta) - e_i h^{\ i}_{\alpha\ k}\omega^k \xi^\alpha.$$

Here in the normal component of $d\xi$ the coefficients $d\xi^\alpha + \xi^\beta\omega^\alpha_\beta = \nabla^\perp \xi^\alpha$ give the covariant derivative of the normal (1,0)-tensor field ξ^α on M^m with respect to the *normal connection* $\nabla^\perp$ of M^m in $N^n(c)$, with the *connection* 1*-forms* ω^α_β. From (1.4.3) one obtains

$$d\omega^j_i = \omega^k_i \wedge \omega^j_k + \Omega^j_i, \quad d\omega^\beta_\alpha = \omega^\gamma_\alpha \wedge \omega^\beta_\gamma + \Omega^\beta_\alpha, \tag{2.1.8}$$

where

$$\Omega^j_i = \omega^\alpha_i \wedge \omega^j_\alpha + cg_{ik}\omega^j \wedge \omega^k, \quad \Omega^\beta_\alpha = \omega^i_\alpha \wedge \omega^\beta_i \tag{2.1.9}$$

are called the *curvature 2-forms* of ∇ and $\nabla^\perp$, respectively. Making substitutions from (2.1.1) and (2.1.4) and denoting

$$R^{\ j}_{i,pq} = (\langle h_{i[p}, h^j_{q]}\rangle + cg_{i[p}\delta^j_{q]}) = \langle h^*_{i[p}, h^{*j}_{q]}\rangle, \quad R^{\beta}_{\alpha,pq} = h^{\ i}_{\alpha\ [p} h^\beta_{q]i} \tag{2.1.10}$$

(see (2.1.6); the square brackets denote alternation of the indices p and q) one obtains the following expressions of these curvature 2-forms:

$$\Omega_i^j = -R^j_{i,pq}\omega^p \wedge \omega^q, \quad \Omega_\alpha^\beta = -R^\beta_{\alpha,pq}\omega^p \wedge \omega^q \tag{2.1.11}$$

(cf. [Li 55], Sections 41 and 53).

In particular, if M^m coincides with ${}_sN^n(c) \subset {}_\sigma E^{n+1}$, the set of values of index α is empty and hence $h_{ij} = 0$; hence from (2.1.10) $R^j_{i,pq} = cg_{i[p}\delta^j_{q]}$, where now $i, j, \dots \in \{1, \dots, n\}$; recall that c is related to the radius r of the sphere ${}_sN^n(c)$ by $r^2 = c^{-1}$.

Applying exterior differentiation to (2.1.2) and using (2.1.8), one gets

$$g_{kj}\Omega_i^k + g_{ik}\Omega_j^k = 0, \quad g_{\gamma\beta}\Omega_\alpha^\gamma + g_{\alpha\gamma}\Omega_\beta^\gamma = 0. \tag{2.1.12}$$

Thus, for (pseudo-)Riemannian submanifolds in space or spacetime forms, the matrices $\Omega = (\Omega_{ij})$ and $\Omega^\perp = (\Omega_{\alpha\beta})$, where $\Omega_{ij} = g_{ik}\Omega_j^k$ and $\Omega_{\alpha\beta} = g_{\alpha\gamma}\Omega_\beta^\gamma$ are obtained from the curvature 2-forms by lowering of the upper indices, are skew-symmetric matrices.

Remark 2.1.1. The apparatus of the Riemannian connection ∇ together with the corresponding absolute differential calculus was worked out by E. B. Christoffel, R. Lipschitz, G. Ricci, T. Levi-Civita (see [Sch 24], [L-C 25]).The normal connection $\nabla^\perp$ of a submanifold was introduced by É. Cartan in his Sorbonne lectures of the 1920s (see, e.g., [Ca 60]; its curvature 2-forms are considered as components of the torsion of a submanifold) and investigated afterwards by D. I. Perepelkin [Per 35], F. Fabricius-Bierre [Fa 36], among others.

For submanifolds L. van der Waerden [vdWa 27] and E. Bortolotti [Bo 27] worked out a special notation scheme, called the D-symbolics in [SchStr 35]. Subsequently, the pair of ∇ and $\nabla^\perp$ was denoted by $\bar\nabla$ and called the *van der Waerden–Bortolotti connection* of the submanifold M^m in $N^n(c)$ (see [Ch 73b], [Lu 2000a]).

A modern treatment of these topics by means of adapted frames and coframes bundles can be found, for $s = 0$, e.g., in [Lu 2000a]; here (and in what follows) a more general treatment is given for the case $s > 0$.

2.2 Higher-Order Fundamental Forms

Equations (2.1.4) defining the second fundamental form h of a submanifold M^m in ${}_sN^n(c)$, via the corresponding mixed tensor field h^α_{ij}, are the starting point for introducing the higher-order fundamental forms of M^m. Here what we call the differential prolongation method can be used as follows.

Applying exterior differentiation to (2.1.4) and using (1.4.3) and (2.1.3), one gets

$$\bar\nabla h^\alpha_{ij} \wedge \omega^j = 0, \tag{2.2.1}$$

where

$$\bar{\nabla} h^{\alpha}_{ij} = dh^{\alpha}_{ij} - h^{\alpha}_{kj}\omega^k_i - h^{\alpha}_{ik}\omega^k_j + h^{\beta}_{ij}\omega^{\alpha}_{\beta} \tag{2.2.2}$$

are the components of the covariant differential of h^{α}_{ij} with respect to the van der Waerden–Bortolotti connection $\bar{\nabla}$.

Applying Cartan's lemma to (2.2.1), one now gets

$$\bar{\nabla} h^{\alpha}_{ij} = h^{\alpha}_{ijk}\omega^k, \quad h^{\alpha}_{ijk} = h^{\alpha}_{ikj}. \tag{2.2.3}$$

The coefficients of ω^k on the right-hand side are components of the covariant derivative of the second fundamental tensor with respect to $\bar{\nabla}$. This derivative is called the *third-order* (or simply *third*) *fundamental tensor* of M^m in ${}_sN^n(c)$. From the second equations of (2.1.4) and (2.2.3) it follows that this tensor is symmetric with respect to all three lower indices and therefore is uniquely given by the *third-order* (or simply *third*) *fundamental form*[2] $\bar{\nabla} h$ of M^m, defined similarly to the second fundamental form by

$$(\bar{\nabla} h)(X, Y, Z) = e_{\alpha} h^{\alpha}_{ijk} X^i Y^j Z^k.$$

Proceeding to the second step of the differential prolongation process, one obtains from (2.2.3) by exterior differentiation

$$\bar{\nabla} h^{\alpha}_{ijk} \wedge \omega^k = \bar{\Omega} \circ h^{\alpha}_{ij}, \tag{2.2.4}$$

where

$$-\bar{\Omega} \circ h^{\alpha}_{ij} = h^{\alpha}_{kj}\Omega^k_i + h^{\alpha}_{ik}\Omega^k_j - h^{\beta}_{ij}\Omega^{\alpha}_{\beta}, \tag{2.2.5}$$

and $\bar{\nabla} h^{\alpha}_{ijk}$ is an expression similar to that in (2.2.2) but with an additional term for the index k, and the sum taken over l.

The operator $\bar{\Omega}$ in (2.2.5) is called the *curvature 2-form operator* of the van der Waerden–Bortolotti connection $\bar{\nabla}$.

Making substitutions from (2.1.11) into (2.2.5) one can write equation (2.2.4) in the form $(\dots)^{\alpha}_{ijk} \wedge \omega^k = 0$ and then use Cartan's lemma to obtain

$$\bar{\nabla} h^{\alpha}_{ijk} = h^{\alpha}_{ijkl}\omega^l, \tag{2.2.6}$$

where h^{α}_{ijkl} are now no longer symmetric with respect to the indices k and l, in general, because here $(\dots)^{\alpha}_{ijk}$ is $\bar{\nabla} h^{\alpha}_{ijk}$ with some complementary terms.

This process can be continued, giving the following sequence:

$$\begin{aligned}
(2.2.6) &\Rightarrow \bar{\nabla} h^{\alpha}_{ijkl} \wedge \omega^l = \bar{\Omega} \circ h^{\alpha}_{ijk} \Rightarrow \bar{\nabla} h^{\alpha}_{ijkl} = h^{\alpha}_{ijklp}\omega^p \Rightarrow \cdots \\
&\Rightarrow \bar{\nabla} h^{\alpha}_{ijp_1\dots p_s} \wedge \omega^{p_s} = \bar{\Omega} \circ h^{\alpha}_{ijp_1\dots p_{s-1}} \\
&\Rightarrow \bar{\nabla} h^{\alpha}_{ijp_1\dots p_s} = h^{\alpha}_{ijp_1\dots p_s p_{s+1}}\omega^{p_{s+1}}
\end{aligned}$$

[2] One should not confuse this form with the third fundamental form of the classical theory of surfaces M^2 in E^3, which is $\langle de_3, de_3\rangle$ and is used mainly in connection with the spherical map of M^2 (see, e.g., [Ka 48], Section 63; [Fav 57], Part 2, Section IV.11).

$$\Rightarrow \bar{\nabla} h^{\alpha}_{ijp_1\dots p_{s+1}} \wedge \omega^{p_{s+1}} = \bar{\Omega} \circ h^{\alpha}_{ijp_1\dots p_s} \Rightarrow \cdots . \tag{2.2.7}$$

The h^{α}_{ijk} are the coefficients of the third-order fundamental form $\bar{\nabla} h$ of M^m, and in general $h^{\alpha}_{ijp_1\dots p_s}$ are the coefficients of the $(s+2)$-*order fundamental tensor* $\bar{\nabla}^s h$ of the submanifold M^m in ${}_sN^n(c)$. The operator $\bar{\Omega}$ works as in (2.2.5)—for every lower index there is a term on the right-hand side.

Here $\bar{\nabla}^s h$ can be treated, like h and $\bar{\nabla} h$ above, as a vector-valued $(s+1)$-form $(TM^m)^{s+2} \to T^{\perp}M^m$ with values in $T^{\perp}M^m$, called the $(s+2)$-*order fundamental form* (see, e.g., [DN 93]). For this, the components of $\bar{\nabla}^s h$ have to be contracted with the coordinates $X_1^i, X_2^j, X_3^{p_1}, \dots, X_{s+2}^{p_s}$ of some $s+2$ tangent vector fields $X_1, \dots, X_{s+2}$.

The above deduction can also be done in its outer version, i.e., with respect to ${}_{\sigma}E^{n+1}$. The normal component of dX in ${}_{\sigma}E^{n+1}$ has vector-valued coefficients, shown in (2.1.5), which determine the *outer second fundamental form* h^*. Since $x = -(\sqrt{|c|})^{-1} e_{n+1}$, one has here $h^*_{ij} = e_{\alpha^*} h^{*\alpha^*}_{ij}$, where $\alpha^* = m+1, \dots, n+1$. Therefore,

$$h^{*\alpha}_{ij} = h^{\alpha}_{ij}, \quad h^{*(n+1)}_{ij} = (\operatorname{sign} c)\sqrt{|c|} g_{ij}, \tag{2.2.8}$$

where sign c is 1 for $c > 0$ and -1 for $c < 0$.

The *outer shape operator* is determined for $\xi^* = \xi + \xi^{n+1} e_{n+1}$ by

$$\langle A^*_{\xi^*}(X), Y\rangle = \langle \xi^*, h^*(X,Y)\rangle = \langle A_{\xi}(X), Y\rangle + c(\sqrt{|c|})^{-1}\xi^{n+1}\langle X, Y\rangle.$$

Due to (1.4.1) and (2.1.2) one has

$$\omega_i^{n+1} = \operatorname{sign} c\sqrt{|c|} g_{ik}\omega^k, \quad \omega^i_{n+1} = -\sqrt{|c|}\omega^i, \quad \omega^{\alpha}_{n+1} = \omega^{n+1}_{\alpha} = 0;$$

therefore,

$$\Omega^{n+1}_{\alpha} = \omega^i_{\alpha} \wedge \omega^{n+1}_i = -g_{\alpha\beta} h^{\beta}_{jk}\omega^k \wedge (\operatorname{sign} c\sqrt{|c|}\omega^j) = 0, \tag{2.2.9}$$

because $h^{\alpha}_{jk} = h^{\alpha}_{kj}$, and

$$\bar{\nabla}^* h^{*\alpha}_{ij} = \bar{\nabla} h^{\alpha}_{ij}, \quad \bar{\nabla}^* h^{*(n+1)}_{ij} = 0,$$

where $\bar{\nabla}^*$ is the van der Waerden–Bortolotti connection of M^m in ${}_{\sigma}E^{n+1}$, i.e., the outer version of $\bar{\nabla}$.

Now (2.2.3) in its outer version shows that $h^{*(n+1)}_{ijk} = 0$, thus the only essential coefficients of the *third-order outer fundamental form* $\bar{\nabla}^* h^*$ are the coefficients of $\bar{\nabla} h$, its other coefficients (with upper index $n+1$) being zero.

The procedure above in its outer version leads to the $(s+2)$-*order outer fundamental form* $\bar{\nabla}^{*s} h^*$, whose only essential coefficients are the same as those of $\bar{\nabla}^s h$, the others (with upper index $n+1$) being zero.

Remark 2.2.1. It appears that the higher-order fundamental forms were first introduced by P. Del-Pezzo in [DelP 1886], and then used by several authors (see [SchStr 35], Band II, Section 13). There has been renewed interest in them more recently, e.g., by V. Mirzoyan in [Mi 78b], [Mi 91a, b, d], by F. Dillen in [Di 90a, b], [Di 91a, c], [Di 92], and by Hyun and Takagi in [HT 97] (see also [Lu 2000a], I.3).

2.3 Fundamental Identities

The coefficients $R^j_{i,pq}$ on the right side of the first formula (2.1.11) are components of the *curvature tensor* R of the Riemannian connection ∇ and therefore belong to the inner geometry of M^m. The first equation (2.1.10) is the *Gauss identity*—an equation connecting the curvature tensor R of ∇ (an inner geometric object) with the second fundamental form h.

The coefficients $R^\beta_{\alpha,pq}$ are components of the *normal curvature tensor* $R^\perp$, which describes curvature of the normal connection $\nabla^\perp$. The second equation (2.1.10) is the *Ricci identity* for $\nabla^\perp$.

If $R = 0$ (resp. $R^\perp = 0$) then M^m in ${}_sN^n(c)$ is said to be *intrinsically* (resp. *normally*) *flat*. Note that with respect to ${}_\sigma E^{n+1}$ the condition $R^{*\perp} = 0$ is equivalent to $R^\perp = 0$, because according to (2.1.11) and (2.2.9) $R^{*(n+1)}_{\alpha,pq} = 0$. Thus normal flatness of M^m in ${}_sN^n(c)$ implies the same in ${}_\sigma E^{n+1}$ and vice versa.

In (2.2.3) the coefficients h^α_{ijk} are components of $\bar{\nabla}h$; they are sometimes also denoted by $\bar{\nabla}_k h^\alpha_{ij}$. The second equation in (2.2.3) yields $\bar{\nabla}_k h^\alpha_{ij} = \bar{\nabla}_j h^\alpha_{ik}$, the classical *Peterson–Mainardi–Codazzi identity*.[3] Therefore, the mixed tensor with these components is sometimes also called the *Codazzi tensor*.[4]

Similarly h^α_{ijkl} in (2.2.6) are the components of $\bar{\nabla}^2 h = \bar{\nabla}(\bar{\nabla}h)$, also denoted by $\bar{\nabla}_l \bar{\nabla}_k h^\alpha_{ij}$. Substituting this into (2.2.4) and using (2.1.9), one obtains

$$\bar{\nabla}_{[k}\bar{\nabla}_{l]} h^\alpha_{ij} = R^p_{i,kl} h^\alpha_{pj} + R^p_{j,kl} h^\alpha_{ip} - R^\alpha_{\beta,kl} h^\beta_{ij}. \tag{2.3.1}$$

It is easy to see that h^α_{ij} can be replaced here by the components of the outer version h^*.

The preceding equation is the *general Ricci identity* for the alternated second covariant derivative of h with respect to the connection $\bar{\nabla}$. Note that its left side is obtained from the coefficients h^α_{ijkl} of the *fourth-order fundamental form* $\bar{\nabla}^2 h$ after alternation of the last two lower indices. So the two equal sides of (2.3.1) characterize the nonsymmetricity of $\bar{\nabla}^2 h$; recall that these components h^α_{ijkl} are symmetric with respect to the first three lower indices, due to the Peterson–Mainardi–Codazzi identity.

From (2.2.7) one can see that the components of the higher-order fundamental forms $\bar{\nabla}^3 h, \ldots, \bar{\nabla}^s h, \ldots$ are, in general, also nonsymmetric with respect to the lower indices, excluding the first three indices.

2.4 Osculating and Normal Subspaces of Higher Order

Let $u^1, \ldots, u^m$ be local coordinates around a point x on a (pseudo-)Riemannian submanifold M^m in ${}_sN^n(c)$. Let this point be identified, as above, with its radius

[3] For the classical case, when $c = 0$, $m = 2$, and $n = 3$, this identity was first given in a preliminary form in 1853 by K. Peterson in his Dorpat (now Tartu, Estonia) dissertation, then in 1856 by G. Mainardi, and in a more modern form in 1860 by D. Codazzi (see [GLOP 70], [Rei 73], [Ph 79], [Lu 97a]).

[4] See, e.g., [Ka 48], Sections 52 and 55, where this tensor in the classical case is denoted by π_{ijk}.

vector from a fixed origin o in ${}_\sigma E^{n+1}$ (if $c \neq 0$ this o is the center of the standard model of the space or spacetime form). The radius vector x is a function of u^i and so its variation can be expressed as

$$\Delta x = dx + \frac{1}{2}d^2x + \cdots + \frac{1}{s!}d^sx + \cdots$$

due to Taylor's formula. Here $dx = x_i du^i = e_j \omega^j$ belongs to the tangent vector space $T_x M^m$. If $e_j = \xi^i_j x_i$, then $du^i = \xi^i_j \omega^j$, $de_j = d\xi^i_j x_i + \xi^i_j x_{ij} du^j$ and comparison with

$$de_j = e_k \omega^k_j + h^*_{jk}\omega^k, \tag{2.4.1}$$

which follows from (1.4.2), (2.1.5), and (2.1.6), shows that the normal component of x_{ij} belongs to the vector subspace span$\{h^*_{jk}\}$ at x. This subspace is called the *first-order outer normal subspace* of M^m at x and is denoted by ${}^{(1)}T^{*\perp}_x M^m$. Recall that $h^*_{jk} = h_{jk} - xcg_{jk}$, $\quad h_{jk} = e_\alpha h^\alpha_{jk}$. Therefore, this subspace lies in the linear span of x and span$\{h_{jk}\}$, which are mutually orthogonal. Here span$\{h_{jk}\}$ is the *first-order* or *(principal) normal subspace* ${}^{(1)}T^\perp_x M^m$ of M^m in ${}_sN^n(c)$ at x, and it is the orthogonal complement of x in span$\{x, h^*_{jk}\}$.

In general the subspace ${}^{(1)}T^{*\perp}_x M^m$ is (pseudo-)Euclidean. Thus ${}^{(1)}T^\perp_x M^m$ is also (pseudo-)Euclidean.

The last subspace contains the special vector $H = \frac{1}{m}g^{ij}h_{ij}$, which is invariantly connected with the point x. This vector is called the *mean curvature vector* at x.

Now some special classes of submanifolds can be introduced. If $H = 0$ at every point $x \in M^m$, then M^m is called a *minimal submanifold.* (Sometimes, when $s > 0$, the term *maximal* is also used in this context). If $h^\alpha_{ij} = \chi^\alpha g_{ij}$ at a point x, then x is said to be an *umbilic point* and if $\chi^\alpha = 0$, a *flat point.* By contracting with g^{ij} one sees here that $\chi^\alpha = H^\alpha$; so at an umbilic point $h(X, Y) = H\langle X, Y\rangle$. If all points of the submanifold are umbilic, the submanifold is called *totally umbilic.* A submanifold all of whose points are flat (i.e., on which h_{ij} is identically zero) is called *totally geodesic* and is obviously minimal.

The linear span of the tangent subspace $T_x M^m$ and the first-order outer normal subspace ${}^{(1)}T^{*\perp}_x M^m$ is called the *first-order outer osculating subspace* of M^m at the point x and denoted by ${}^{(1)}O^*_x M^m$. In general it is (pseudo-)Euclidean. It is clear that $dx + \frac{1}{2}d^2x$ belongs to this subspace.

The linear span of $T_x M^m$ and ${}^{(1)}T^\perp_x M^m$ is called the *first-order osculating subspace* ${}^{(1)}O_x M^m$ of M^m in ${}_sN^n(c)$ at x. Obviously it is spanned by all e_i and h_{jk}, and therefore it is the orthogonal complement of the radius vector x in the span of ${}^{(1)}O^*_x M^m$ and x.

Similar to the above, one can establish that the partial derivative x_{ijk} in $d^3x = x_{ijk}du^i du^j du^k$ belongs to the vector subspace spanned by the first-order outer osculating subspace ${}^{(1)}O^*_x M^m$ and span$\{h_{ijk}\}$, where $h_{ijk} = e_\alpha h^\alpha_{ijk}$. This vector subspace is called the *second-order outer osculating subspace* ${}^{(2)}O^*_x M^m$ of M^m at the point x; it contains $dx + \frac{1}{2}d^2x + \frac{1}{3!}d^3x$. In general it is (pseudo-)Euclidean. The orthogonal complement of ${}^{(1)}O^*_x M^m$ in it is called the *second-order outer normal subspace* of M^m at x and denoted by ${}^{(2)}T^{*\perp}_x M^m$.

The orthogonal complement of the radius vector x in the span$\{{}^{(2)}O^*_x M^m,\ x\}$ is called the *second-order osculating subspace* ${}^{(2)}O_x M^m$ of M^m in ${}_sN^n(c)$ at the point x. The orthogonal complement of ${}^{(1)}O_x M^m$ in ${}^{(2)}O_x M^m$ is called the *second-order normal subspace* ${}^{(2)}T^{\perp}_x M^m$ of M^m at x.

Repeating this argument, one obtains the *sth-order outer osculating subspace* ${}^{(s)}O^*_x M^m$, which is spanned by the $(s-1)$th-order outer osculating subspace ${}^{(s-1)}O^*_x M^m$ and span$\{h_{i_1 i_2 \ldots i_s}\}$ and contains $dx + \frac{1}{2}d^2x + \cdots \frac{1}{s!}d^s x$. In general both of these subspaces are (pseudo-)Euclidean. The orthogonal complement of ${}^{(s-1)}O^*_x M^m$ in ${}^{(s)}O^*_x M^m$ is called the *sth-order outer normal subspace* ${}^{(s)}T^{*\perp}_x M^m$ of M^m at x.

The corresponding subspaces without the designation *outer* are defined similarly (see, e.g., [CGR 90] for the Euclidean case $E^n = {}_0N^n(0)$).

A submanifold M^m in ${}_sN^n(c)$ is said to be *regular* if its normal subspaces of all orders have constant dimensions, denoted below by $m_1, \ldots, m_s, \ldots$, and are (pseudo-)Euclidean. Then the frame bundle adapted to M^m can be specialized by means of these subspaces as follows. In choosing the vectors e_α, one takes the first m_1 normal basis vectors e_{α^1}, $\alpha^1 \in \{m+1, \ldots m+m_1\}$ to be in the first-order normal subspace, the next m_2 vectors e_{α^2}, $\alpha^2 \in \{m+m_1+1, \ldots, m+m_1+m_2\}$ to be in the second-order normal subspace, etc. Consequently, among the components h^α_{ij}, those with upper indices α^s, $s > 1$ are zero; among h^α_{ijk}, those with upper indices α^s, $s > 2$ are zero, etc. Thus equations (2.2.3) yield $h^{\beta^1}_{ij}\,\omega^{\alpha^2}_{\beta^1} = h^{\alpha^2}_{ijk}\omega^k$; here the matrix of coefficients on the left side has maximal rank m_1. It follows that $\omega^{\alpha^s}_{\beta^1} = 0$, $s > 2$, and

$$\omega^{\alpha^2}_{\beta^1} = \chi^{\alpha^2}_{\beta^1 k}\omega^k. \tag{2.4.2}$$

The same process can be repeated for higher orders and gives, in general,

$$\omega^{\alpha^t}_{\beta^s} = 0, t > s+2, \tag{2.4.3}$$

$$\omega^{\alpha^{s+1}}_{\beta^s} = \chi^{\alpha^{s+1}}_{\beta^s k}\,\omega^k. \tag{2.4.4}$$

The orbits of the higher-order outer osculating (resp. normal) vector subspaces at $x \in M^m$ in ${}_sE^{n+1}$ are called the *higher-order outer osculating* (resp. *normal*) *planes* of M^m at x. Analogous planes without the designation *outer* can be similarly defined for the projective structure of ${}_sN^n(c)$ (see Remark 1.6.1).

Remark 2.4.1. In Section 2.1 the space form ${}_sN^n(c)$ can be replaced by an arbitrary (pseudo-)Riemannian manifold ${}_sN^n$. Then one has a (pseudo-)Riemannian submanifold M^m in ${}_sN^n$. The same arguments as in Section 2.1 and 2.2 lead to the definition of the second fundamental form h; note that the difference between (1.4.3) and (1.3.2) concerning the curvature 2-forms does not affect the arguments involved. One can likewise introduce the classes of minimal, totally umbilic, and totally geodesic submanifolds, as above (cf., e.g., [Ch 2000], for the case $s = 0$).

In Section 3.2 below, the following assertion will be important. Recall that a submanifold M^m of N^n is totally geodesic if and only if its second fundamental form is identically zero. An equivalent formulation of this condition is that the covariant differential in N^n of every vector field Y tangent to M^m has zero normal component, i.e., is tangent to M^m (see [KN 69], Chapter VII, Prop. 8.9).

3

Parallel Submanifolds

3.1 Parallel and k-Parallel Submanifolds

A tensor field F on a (pseudo-)Riemannian manifold is said to be *parallel* if its covariant derivative vanishes (i.e., if $\nabla F = 0$; see, e.g., [KN 63], Chapter III, Section 2). The same terminology is used for mixed tensor fields on submanifolds M^m in ${}_sN^n(c)$ (then $\bar{\nabla}$ is used in place of ∇), and for differential forms on M^m as well.

A submanifold M^m in ${}_sN^n(c)$ whose second fundamental form h is parallel, i.e., (0.2) is satisfied: $\bar{\nabla}h = 0$ (equivalently, whose third fundamental form vanishes), is called a *parallel submanifold.*

Parallel surfaces M^2 in E^3 were defined by V. Kagan in [Ka 48], Section 55 as surfaces with vanishing Codazzi tensor (see Chapter 2, footnote[4]). It was shown that they are open parts of planes (totally geodesic), spheres (totally umbilic) or round cylinders. Note that the latter can be considered as products of straight lines E^1 and circles S^1 (which are the only parallel curves in E^3). These surfaces have the property that every chord makes equal angles with the normal vectors at its endpoints, where these are directed to the same side of the surface. J. Dubnov and l. Beskin showed in [DB 59] that this geometric property holds only for these surfaces with vanishing Codazzi tensor.

Interest in parallel submanifolds in E^n arose again in the 1970s. At first they were called *submanifolds with parallel second fundamental form.* Hypersurfaces M^{n-1} in E^n with this property were investigated in [SW 69]. For general submanifolds, this condition was explicitly stated in the paper [ChdCK 70] on minimal submanifolds of spheres with second fundamental form of constant length. New examples of such surfaces M^2—the Clifford tori $S^1 \times S^1$ in $S^3(c) \subset E^4$ and the Veronese surfaces in $S^4(c) \subset E^5$—were given in [Hou 72] (cf. Section 3.3 below). Then followed [Vi 72], [Wa 73], and fundamental investigations by D. Ferus [Fe 74a, b, c].

The concept can be generalized to the higher-order fundamental forms as follows.

Let M^m be a regular submanifold in ${}_sN^n(c)$. It is seen from (2.2.7) that if its $(s+2)$-order fundamental form vanishes, then the fundamental forms of orders $(s+3)$, $(s+4)$, … also vanish, i.e., $\bar{\nabla}^s h = 0 \Rightarrow \bar{\nabla}^t h = 0$ for $t \in \{s+1, s+2, \dots\}$.

Ü. Lumiste, *Semiparallel Submanifolds in Space Forms*,
DOI 10.1007/978-0-387-49913-0_4,

If the first $\bar{\nabla}^s h$ that vanishes is of order k, then M^m is said to be *k-parallel* (cf. [Di 91c]).

Here the value $k = 0$ can be included by defining $\bar{\nabla}^0 h = h$. Thus, a 0-parallel M^m is simply a totally geodesic M^m in $N^n(c)$. The next simplest cases are the 1-parallel submanifolds, i.e., parallel submanifolds that are not totally geodesic.

The condition $\bar{\nabla} h = 0$, which characterizes the parallel submanifolds, is equivalent to $h^\alpha_{ijk} = 0$ (see (2.2.3)), or, due to (2.2.2), to

$$dh^\alpha_{ij} = h^\alpha_{kj}\omega^k_i + h^\alpha_{ik}\omega^k_j - h^\beta_{ij}\omega^\alpha_\beta. \tag{3.1.1}$$

For the vector-valued second fundamental form $h_{ij} = e_\alpha h^\alpha_{ij}$, and also for $h^*_{ij} = h_{ij} - xcg_{ij}$, a straightforward computation shows that

$$\nabla h_{ij} + e_k(g^{kl}\langle h_{ij}, h_{lp}\rangle)\omega^p = \nabla h^*_{ij} + e_k(g^{kl}\langle h^*_{ij}, h^*_{lp}\rangle)\omega^p = e_\alpha h^\alpha_{ijk}\omega^i, \tag{3.1.2}$$

where

$$\nabla h_{ij} = dh_{ij} - h_{kj}\omega^k_i - h_{ik}\omega^k_j, \tag{3.1.3}$$

and the same for h^*_{ij}. This yields the following (see [Lu 96a], for $s = 0$ [Lu 96b]; cf. also [Tak 81]).

Proposition 3.1.1. *The following conditions are equivalent:*

(1) *M^m is parallel as a submanifold in ${}_sN^n(c)$;*
(2) *in the case $c \neq 0$, M^m is parallel as a submanifold in ${}_\sigma E^{n+1}$ (recall $\sigma = s$ for $c > 0$ and $\sigma = s + 1$ for $c < 0$);*
(3) *$(\nabla h_{ij})_x \in T_x M^m$ for every point $x \in M^m$;*
(4) *$(\nabla h^*_{ij})_x \in T_x M^m$ for every point $x \in M^m$.*

Parallel submanifolds have been studied in many papers (see, e.g., [Ak 76], [Mi 78a], [Fe 80], [Na 80], [Re 81], [Ma 81], [Ma 83a, b], [Re 83]), and there is now a well-developed theory of parallel submanifolds in space forms; a summary is given in [Lu 2000a] and also in [BCO 2003], Section 3.7.

Kagan's result concerning spheres in E^3, mentioned at the beginning of this section, can be generalized as follows.

Proposition 3.1.2. *Every totally umbilic submanifold M^m with $m > 1$ in ${}_sN^n(c)$ is a parallel submanifold with flat $\nabla^\perp$.*

Proof. A totally umbilic submanifold is characterized by $h^\alpha_{ij} = H^\alpha g_{ij}$ (see Section 2.4), where $H^\alpha = \frac{1}{m} g^{ij} h^\alpha_{ij}$, and so $\bar{\nabla} h^\alpha_{ij} = (\nabla^\perp H^\alpha) g_{ij}$. Here $\nabla^\perp H^\alpha = H^\alpha_k \omega^k$; hence now in (2.2.3) $h^\alpha_{ijk} = H^\alpha_k g_{ij}$, and thus $H^\alpha_k g_{ij} = H^\alpha_i g_{kj}$; by contracting with g^{kj} one obtains $H^\alpha_i = H^\alpha_i m$. Since $m > 1$, the last equation implies $H^\alpha_i = 0$ and also $\bar{\nabla} h^\alpha_{ij} = 0$, i.e., the submanifold M^m is parallel.

Moreover, now $\omega^\alpha_i = H^\alpha g_{ij}\omega^j$ and therefore due to (2.1.9) it follows that $\Omega^\beta_\alpha = 0$, i.e., $\nabla^\perp$ is indeed flat.

For the inner geometry of parallel submanifolds, the following result holds.

Proposition 3.1.3. *Every parallel submanifold M^m in ${}_sN^n(c)$ is intrinsically a locally symmetric Riemannian manifold.*

Proof. The Gauss identity (2.1.10) gives

$$\nabla R^j_{i,kl} = \langle \nabla h^*_{i[p}, h^*_{q]k} \rangle g^{kj} + \langle h^*_{i[p}, \nabla h^*_{q]k} \rangle g^{kj},$$

because $\nabla g^{kj} = 0$. Due to Proposition 3.1.1, part (4), the multiplicands in both scalar products in the above expression for $\nabla R^j_{i,kl}$ are orthogonal; thus $\nabla R^j_{i,kl} = 0$, or briefly, $\nabla R = 0$.

For the extrinsic geometry of parallel submanifolds, one has the following proposition.

Proposition 3.1.4. *Suppose a parallel submanifold M^m in ${}_sN^n(c)$ has all first-order outer osculating subspaces ${}^{(1)}O^*_x M^m$ of the same constant dimension $m + m_1$. Then this subspace is independent of $x \in M^m$, and M^m is contained in an $(m + m_1)$-plane of ${}_\sigma E^{n+1}$ with this vector subspace. If $c \neq 0$ and $s = 0$, this plane intersects $N^n(c)$ in an $N^{m+m_1-1}(c_1)$ containing M^m, and $c_1 > c$.*

Proof. Recall that a parallel M^m is characterized by $h^\alpha_{ijk} = 0$. Now (1.4.2), (2.1.3), (2.4.1) and (3.1.2) give

$$dx = e_i \omega^i, \quad de_i = e_j \omega^j_i + h^*_{ij} \omega^j, \tag{3.1.4}$$

$$dh^*_{jk} = -e_i (g^{il} \langle h^*_{jk}, h^*_{lp} \rangle) \omega^p + h^*_{ik} \omega^i_j + h^*_{ji} \omega^i_k. \tag{3.1.5}$$

Hence ${}^{(1)}O^*_x M^m$, spanned by e_i and h^*_{jk}, is invariant under arbitrary motion of the adapted frame on M^m, i.e., it is independent of the point $x \in M^m$. One can choose the frame so that among e_α some $n - m - m_1$ vectors e_{α_1}, orthogonal to ${}^{(1)}O_x M^m$, are constant. Obviously $\langle e_i, e_{\alpha_1} \rangle = 0$ holds, and it implies $d\langle x, e_{\alpha_1} \rangle = 0$. Thus $\langle x, e_{\alpha_1} \rangle = c_{\alpha_1} (= \text{const})$ and the radius vector of the point $x \in M^m$ satisfies $n - m - m_1$ linearly independent linear equations. These equations define an $(m + m_1)$-dimensional plane in ${}_\sigma E^{n+1}$, that intersects ${}_sN^n(c)$ in an ${}_tN^{m+m_1-1}(c_1)$, and that contains M^m.

If $c \neq 0$ and $s = 0$, this plane does not pass through the center of the model sphere $N^n(c)$, since then the vector x would be orthogonal to all e_i in σE^{n+1} and could not then be their linear combination. This implies that $c_1 > c$.

Propositions 3.1.2 and 3.1.4 together imply the following result.

Proposition 3.1.5. *Every totally umbilic submanifold M^m $(m > 1)$ in $N^n(c)$ is a standard model $N^m(c_1)$ or its open part.*

Proof. Indeed, $h^\alpha_{ij} = H^\alpha g_{ij}$, $H \neq 0$ imply here that $m_1 = 1$; hence M^m is contained in an $N^m(c_1)$.

Remark 3.1.6. Suppose $c \geq 0$. Then $N^n(c)$ is a Euclidean space E^n (if $c = 0$) or a sphere $S^n(c) \subset E^{n+1}$ (if $c > 0$), and therefore a totally umbilic M^m in it is simply a sphere $S^m(c_1)$ with $c_1 > c$, or an open part of the latter.

The situation is more interesting if $c < 0$, where a connected component of $N^n(c) \subset {}_1E^{n+1}$ is the hyperbolic space $H^n(c)$. An easy calculation shows that the outer mean curvature vector $H^* = H - cx$ of a totally umbilic M^m satisfies $dH^* = -\|H^*\|^2 dx$, due to $\nabla^\perp H^\alpha = 0$. If $\|H^*\| \neq 0$, the point with radius vector $z = x + \|H^*\|^{-2} H^*$ is a fixed point of ${}_1E^{n+1}$, and $\|z - x\| = \|H^*\|^{-1} = \text{const}$. If $\|z\| < 0$, then M^m is a sphere in $H^n(c)$, but if $\|z\| > 0$, then M^m is an equidistant submanifold (i.e., consists of points equidistant from a hyperplane of an $H^{m+1}(c_1)$). If $\|H^*\|$ tends to 0, then M^m tends to a horosphere (i.e., to an orthogonal submanifold of straight lines, mutually parallel in the Lobachevsky sense).

The following extension of Proposition 3.1.4 for k-parallel submanifolds holds.

Proposition 3.1.7. *Suppose a k-parallel submanifold M^m in ${}_sN^n(c)$ has all kth-order outer osculating subspaces ${}^{(k)}O^*_x M^m$ of the same constant dimension $m + m_1 + \cdots + m_k$. Then this subspace is independent of $x \in M^m$, and M^m is contained in an $(m + m_1 + \cdots + m_k)$-plane of ${}_\sigma E^{n+1}$ with this vector subspace. If $c \neq 0$, this plane intersects ${}_sN^n(c)$ in an ${}_tN^{m+m_1+\cdots+m_k-1}(c_1)$ containing M^m.*

Proof. k-parallel submanifolds are characterized by $h^\alpha_{ijp_1\ldots p_k} = 0$, which due to (2.2.7) implies $h^\alpha_{ijp_1\ldots p_l} = 0$ for all $l \geq k$, but not for $l < k$. Now (3.1.4) holds as before, but on the right side of (3.1.5) one must add $h_{ijp_1}\omega^{p_1}$, where $h_{ijp_1} = h^\alpha_{ijp_1} e_\alpha$ (see (3.1.2) and (3.1.3)). In a similar manner one gets

$$dh_{ijp_1} = -e_k(g^{kl}\langle h_{ijp_1}, h_{lp}\rangle)\omega^p + h_{pjp_1}\omega^p_j + h_{ipp_1}\omega^p_j + h_{ijp}\omega^p_{p_1} + h_{ijp_1p_2}\omega^{p_2},$$

and so on, until finally

$$dh_{ijp_1\ldots p_{k-1}} = h_{qjp_1\ldots p_{k-1}}\omega^q_i + \cdots + h_{ijp_1\ldots q}\omega^q_{p_{k-1}}.$$

Hence the vector space ${}^{(k)}O^*_x M^m$, spanned by all $e_i, h^*_{ij}, h_{ijp_1}, \ldots, h_{ijp_1\ldots p_{k-1}}$, does not depend on the point $x \in M^m$.

The rest of the proof is essentially the same as for Proposition 3.1.4.

3.2 Examples: Segre and Plücker Submanifolds

A new example of parallel submanifolds can be obtained by means of the Segre map, known from algebraic geometry. The Segre map is introduced in [Mum 76], Section 2B, as the imbedding of the product $P^p \times P^q$ of projective spaces into P^{pq+p+q} defined in homogeneous coordinates by $(u^{i_1}, v^{i_2}) \longmapsto (w^{i_1 i_2})$, where $w^{i_1 i_2} = u^{i_1} v^{i_2}$.

Now let the real projective spaces P^p and P^q be converted into elliptic spaces, considered as spheres of the same radius r, with antipodal points identified. Then the Segre map can be seen as

$$S^p(k) \times S^q(k) \longrightarrow \mathbb{R}^{(p+1)(q+1)}, \quad (u^{i_1}, v^{i_2}) \longmapsto e_{(i_1,i_2)} u^{i_1} v^{i_2},$$

where $k = r^{-2} =$ const, and $e_{(i_1,i_2)}$ are vectors of an orthonormal basis of Euclidean space $\mathbb{R}^{(p+1)(q+1)}$; here i_1 and i_2 run over $\{1, \dots, p\}$ and $\{p+1, \dots, p+q\}$, respectively. Therefore,

$$\|e_{(i_1,i_2)} u^{i_1} v^{i_2}\| = \sqrt{\sum (u^{i_1} v^{i_2})^2} = \sqrt{\sum (u^{i_1})^2 \cdot \sum (v^{i_2})^2} = k^{-1},$$

showing that the image actually does lie in $S^{pq+p+q}(c)$, with $c = k^2$.

The image of this Segre map is called the *Segre submanifold* and denoted by $S_{(p,q)}(k)$. It is also called the *Segre orbit*, because the image is obviously the orbit of a Lie subgroup of the isometry group $O((p+1)(q+1), \mathbb{R})$ of $S^{pq+p+q}(k^2)$, this subgroup being isomorphic to $O(p+1, \mathbb{R}) \times O(q+1, \mathbb{R})$.

Theorem 3.2.1. *The Segre submanifold $M^m \equiv S_{(p,q)}(k)$, $m = p+q$, is a complete parallel submanifold in $S^{pq+m}(k^2) \subset E^{(p+1)(q+1)}$.*

Proof. The completeness of $S_{(p,q)}(k)$ follows directly from its definition. Geometrically it is characterized as an m-dimensional submanifold in $S^{pq+m}(c)$, $m = p+q$, having two families of generating great p-dimensional (resp. q-dimensional) spheres of $S^{pq+m}(c)$, totally orthogonal at every point $x \in M^m$; they are defined by $v^{i_2} =$ const (resp. $u^{i_1} =$ const).

Let the bundle of orthonormal frames adapted to this M^m be adapted further so that at x the vectors e_{i_1} (resp. the vectors e_{i_2}) are tangent to the generating sphere, determined by v^{i_2}=const (resp. $u^{i_1} =$ const). Then due to (2.1.5) $de_i = e_j \omega_i^j + h^*_{ij} \omega^j$, where in (2.1.6) now $g_{ij} = \delta_{ij}$. The generating great spheres are determined by $\omega^{j_2} = 0$ (resp. $\omega^{j_1} = 0$). Therefore, mod ω^{j_2} must be $de_{i_1} = e_{j_1} \omega_{i_1}^{j_1} + xc\omega^{i_1}$, so that $h_{i_1 j_1} = 0$ and $\omega_{i_1}^{j_2} = 0 \pmod{\omega^{k_2}}$. Hence $\omega_{i_1}^{j_2} = \Gamma_{i_1 k_1}^{j_2} \omega^{k_2}$. Here the roles of the subindices 1 and 2 can be exchanged. This, together with $\omega_{i_1}^{j_2} + \omega_{j_2}^{i_1} = 0$, leads to $\omega_{i_1}^{j_2} = 0$. Substituting this into (2.1.7) via $i = i_1$ and $j = j_2$, one obtains $\Omega_{i_1}^{j_2} = 0$, which, due to (2.1.9) and (2.1.10), leads to

$$\langle h_{i_1 j_2}, h_{i_1 j_2} \rangle = c, \quad \langle h_{i_1 j_2}, h_{k_1 l_2} \rangle = 0 \quad (i_1 \neq k_1 \quad \text{or} \quad j_2 \neq l_2).$$

It follows that the unit vectors $e_{(i_1 j_2)} = k^{-1} h_{i_1 j_2}$ are mutually orthogonal and they can be chosen as frame vectors e_α, so that now the pair-index $(i_1 j_2)$ plays the role of α. Recall that along with $h_{i_1 j_1} = 0$ one also has $h_{i_2 j_2} = 0$. Therefore, (2.1.4) now appear as

$$\omega_{k_1}^{(i_1 j_2)} = k\delta_{k_1}^{i_1} \omega^{j_2}, \quad \omega_{k_2}^{(i_1 j_2)} = k\delta_{k_2}^{j_2} \omega^{i_1}. \tag{3.2.1}$$

Then by exterior differentiation and Cartan's lemma, one obtains

$$\omega_{(k_1 l_2)}^{(i_1 j_2)} = \delta_{k_1}^{i_1} \omega_{l_2}^{j_2} + \omega_{k_1}^{i_1} \delta_{l_2}^{j_2}. \tag{3.2.2}$$

This leads, via (2.1.1), to

$$de_{(i_1j_2)} - e_{(k_1j_2)}\omega_{i_1}^{k_1} - e_{(i_1l_2)}\omega_{j_2}^{l_2} = -k(e_{i_1}\omega^{j_2} + e_{j_2}\omega^{i_1}).$$

It remains here merely to replace $e_{(i_1j_2)}$ by $k^{-1}h_{i_1j_2}$, in order to see that the left side gives the only nonzero components of $k^{-1}\nabla h_{ij}$, and then to become convinced that condition (3) of Proposition 3.1.1 is satisfied. This completes the proof.

Note that the Segre submanifold $S_{(p,q))}(k)$ in $S^{pq+p+q}(k^2)$ can be locally given, with respect to the adapted orthonormal frame bundle introduced above, by the Pfaffian system consisting of the equations $\omega^{(i_1j_2)} = 0$, $\omega_{i_1}^{j_2} = 0$, (3.2.1), and (3.2.2). It is easy to check that this system is totally integrable.

This Segre submanifold $S_{(p,q)}(k)$ is, of course, a Riemannian parallel submanifold. But it is possible to introduce its pseudo-Riemannian version.

For this one uses the following identity:

$$\left[-\sum_{a=1}^{k}(u^a)^2 + \sum_{s=k+1}^{p}(u^s)^2\right]\cdot\left[-\sum_{b=1}^{l}(v^b)^2 + \sum_{t=l+1}^{q}(v^t)^2\right]$$
$$= -\sum_{a,t}(u^a v^t)^2 - \sum_{b,s}(u^s v^b)^2 + \sum_{a,b}(u^a v^b)^2 + \sum_{s,t}(u^s v^t)^2,$$

which shows that the Segre map can also be considered as

$${}_kN^p(\kappa) \times {}_lN^q(\kappa) \longrightarrow \mathbb{R}^{(p+1)(q+1)}, \quad (u^{i_1}, v^{i_2}) \longmapsto e_{(i_1,i_2)}u^{i_1}v^{i_2},$$

where, e.g., ${}_kN^p(\kappa)$ is the standard model of a spacetime form, as introduced in Section 1.4 (i.e., either ${}_kS^p(\kappa)$ or ${}_{k-1}H^p(\kappa)$), and $e_{(i_1,i_2)}$ are vectors of an orthonormal basis in pseudo-Euclidean space $\mathbb{R}^{(p+1)(q+1)}$ of index $s = k(q-l) + l(p-k)$. The image lies in ${}_sN^{pq+p+q}(c)$, with $c = \kappa^2$, and is denoted by ${}^{kl}S_{pq}(\kappa)$. Moreover, ${}_kN^p(\kappa)$ and ${}_lN^q(\kappa)$ map into its generating great spheres.

The proof of Theorem 3.2.1 also works to prove this generalization, the only difference being that the indices $i_1, j_1, \ldots$ now run over the union of the ranges of a and s, and $i_2, j_2, \ldots$ run over the union of the ranges of b and t.

Therefore, the following generalization of Theorem 3.2.1 can be formulated.

Theorem 3.2.2. *The pseudo-Riemannian Segre submanifold ${}^{kl}S_{pq}(\kappa)$ is a complete parallel submanifold in ${}_sN^{pq+p+q}(c)$.*

One more example of parallel submanifolds can be given by considering the Grassmann manifold ${}_{l,k}G^{q,p}$ of regular subspaces, introduced in Example 1.5.4, as imbedded in $\wedge^q(\mathbb{R}^p)$. This vector space has the pseudo-Euclidean metric defined by

$$\langle e_{\lambda_1} \wedge \cdots \wedge e_{\lambda_q}, e_{\mu_1} \wedge \cdots \wedge e_{\mu_q}\rangle = r^2 g_{\lambda_1\mu_{[1}} \cdots g_{\lambda_{|q|}\mu_{q]}} = r^2 \det \|g_{\lambda_\rho\mu_\sigma}\|,$$

where $r = \text{const}$ and $\rho, \sigma \in \{1, \ldots, q\}$ (cf. [Ste 64], Chapter 1, Section 4). The Grassmann manifold, imbedded in $\wedge^q(\mathbb{R}^p)$ with this metric, will be denoted below by ${}_{l,k}G^{p,q}(r)$. If $\{e_\lambda\}$, with $\lambda \in \{1, \ldots, p\}$, is an orthonormal basis of $\mathbb{R}^p$, then $e_{\lambda_1} \wedge \cdots \wedge e_{\lambda_q}$ and $e_{\mu_1} \wedge \cdots \wedge e_{\mu_q}$ are orthogonal when $\{\lambda_1, \ldots, \lambda_q\} \neq \{\mu_1, \ldots \mu_q\}$; and each of them has length squared equal to $\pm r^2$. Recall that the dimension of $\wedge^q(\mathbb{R}^p)$ is $n = C(p,q) = \binom{p}{q}$.

Theorem 3.2.3. *The Grassmann manifold* ${}_{l,k}G^{q,p}(r)$ *of regular subspaces imbedded in* $\wedge^q(\mathbb{R}^p)$ *is a minimal submanifold of a space form, and if* $q = 2$ *this submanifold (called the Plücker submanifold) is a parallel submanifold.*

Proof. The above imbedding can be considered as an inclusion. Then $e_1 \wedge \cdots \wedge e_q$ in Example 1.5.4 can be identified with $x \in {}_{l,k}G^{q,p}(r)$. The length squared of x in $\wedge^q(\mathbb{R}^p)$ is $\pm r^2$; thus the submanifold ${}_{l,k}G^{q,p}(r)$ is contained in the standard model of a space form. Moreover, equation (1.5.1) can be written in the form

$$dx = e_{1\ldots \overset{a}{u} \ldots q}\theta^u_a. \tag{3.2.3}$$

Here the simple q-vector before ω^u_a in (1.5.1) is denoted by $re_{1\ldots \overset{a}{u} \ldots q}$ and $\theta^u_a = r\omega^u_a$.

A straightforward computation leads to

$$\begin{aligned} de_{1\ldots \overset{a}{u} \ldots q} = & -xg_{uv}\omega^v_a + e_{1\ldots \overset{b}{v} \ldots q}(-\omega^a_b\delta^v_u + \delta^a_b\omega^v_u) \\ & + \sum_{b\neq a}^{v\neq u} e_{1\ldots \overset{a}{u} \ldots \overset{b}{v} \ldots q}(\omega^v_b). \end{aligned} \tag{3.2.4}$$

Here

$$e_{1\ldots \overset{a}{u} \ldots \overset{b}{v} \ldots q} = \frac{1}{r}(e_1 \wedge \cdots \wedge e_{a-1} \wedge e_u \wedge e_{a+1} \wedge \cdots \wedge e_{b-1} \wedge e_v \wedge e_{b+1} \wedge \cdots \wedge e_q)$$

are normal to the submanifold and their lengths squared are ± 1 in $\wedge^q(\mathbb{R}^p)$. It follows that after multiplying them by r, they take the role of h_{ij} in the geometry of the standard model (see (2.1.4)). The mean curvature vector H of the submanifold under consideration is zero because the summation in the last term of (3.2.4) is taken as $b \neq a$ and $v \neq u$. Hence the submanifold is minimal in this model.

Let $q = 2$. Then the point x of the corresponding Plücker submanifold ${}_{l,k}G^{2,p}(r)$ is represented by $e_1 \wedge e_2$ and the tangent vectors are represented by $e_u \wedge e_2$ and $e_1 \wedge e_u$, where u runs through $\{3, \ldots, p\}$. The latter will be denoted below by $rE_u = e_u \wedge e_2$ and $rE_{\bar{u}} = e_1 \wedge e_u$, where r is a positive constant. Then (3.2.3) and (3.2.4) are

$$dx = E_u\theta^u + E_{\bar{u}}\theta^{\bar{u}}, \tag{3.2.5}$$

where $\theta^u = r\omega^u_1$, $\theta^{\bar{u}} = r\omega^u_2$, and

$$dE_u = -\frac{1}{r^2}x\theta^u + E_v\omega^v_u + E_{\bar{u}}\omega^2_1 + \frac{1}{r}E_{[uv]}\theta^{\bar{v}}, \tag{3.2.6}$$

$$dE_{\bar{u}} = -\frac{1}{r^2}x\theta^{\bar{u}} + E_{\bar{v}}\omega^v_u - E_u\omega^2_1 - \frac{1}{r}E_{[uv]}\theta^v, \tag{3.2.7}$$

where $E_{[uv]} = \frac{1}{r}e_u \wedge e_v = -\frac{1}{r}e_v \wedge e_u = -E_{[vu]}$ are normal to the submanifold, together with $x = e_1 \wedge e_2$. It is seen that the components of the vector-valued second fundamental form are

$$h_{uv} = -\frac{1}{r^2}\delta_{uv}x, \quad h_{\bar{u}\bar{v}} = -\frac{1}{r^2}\delta_{\bar{u}\bar{v}}x, \quad h_{u\bar{v}} = h_{\bar{v}u} = E_{[uv]}. \tag{3.2.8}$$

A straightforward computation shows that

$$dE_{[uv]} = \frac{1}{r}[-E_u\theta^{\bar{v}} + E_v\theta^{\bar{u}} + E_{\bar{u}}\theta^v - E_{\bar{v}}\theta^u] + E_{[wv]}\omega_u^w + E_{[uw]}\omega_v^w. \tag{3.2.9}$$

This, together with (3.2.8), shows that condition (3) in Proposition 3.1.1 is satisfied for ${}_{l,k}G^{2,p}(r)$. Thus this Plücker submanifold is parallel.

This completes the proof.

Remark 3.2.4. The assertion in Theorem 3.2.3 concerning the Plücker submanifold is proved (for the particular case of $k = l = 0$) in [Lu 92a], [Lu 96b].

If $q \geq 3$ and $p - q \geq 3$, then the Grassmann submanifold considered above is not parallel (see [Lu 92a]). Indeed, if, for instance, $q = p - q = 3$, then $d(e_u \wedge e_v \wedge e_3)$ has a nonzero component $\omega_3^w(e_u \wedge e_v \wedge e_w)$ outside of the first-order outer normal subspace. But due to Proposition 3.1.1 that is impossible for a parallel submanifold.

3.3 Example: Veronese Submanifold

Let us consider in the spacetime form ${}_sN^n(c)$ a complete (pseudo-)Riemannian submanifold M^m that is characterized by the following properties (see Section 2.4):

(1) M^m is regular (in the sense that its first-order normal subspaces have constant dimension m_1 and regular metric),
(2) M^m lies in its first-order outer osculating subspace of dimension $m + m_1$,
(3) M^m is intrinsically of constant curvature and all its isometries are induced by the isometries of the ambient space, taken as this first-order osculating subspace.

The set of all isometries of such an M^m forms a $\frac{1}{2}m(m+1)$-parameter (pseudo-)orthogonal group, whose Lie algebra consists of the matrices of the 1-forms ω^i and ω_i^j which determine the infinitesimal displacement of the tangent part $\{x, e_i\}$ of an orthonormal moving frame adapted to M^m. Indeed, here the conditions analogous to (1.2.5) are satisfied (with i, j in place of I, J); therefore there are $m + \frac{1}{2}m(m-1) = \frac{1}{2}m(m+1)$ such 1-forms, and they are linearly independent.

For the submanifold M^m in ${}_sN^n(c)$, the equations of (2.4.1) hold, where e_k and h^*_{jk} are orthogonal. Due to property (3), all h^*_{jk} must form a rigid configuration, and therefore all $\langle h^*_{jk}, h^*_{lp}\rangle$ must be constant for all the displacements above. Due to (3.1.2) and (3.1.3) (with upper *),

$$d\langle h^*_{jk}, h^*_{lp}\rangle = \langle h^*_{ik}, h^*_{lp}\rangle\omega^i_j + \langle h^*_{ji}, h^*_{lp}\rangle\omega^i_k + \langle h^*_{jk}, h^*_{ip}\rangle\omega^i_l + \langle h^*_{jk}, h^*_{li}\rangle\omega^i_p$$
$$+ \langle h_{ijk}, h^*_{lp}\rangle\omega^i + \langle h^*_{jk}, h_{ilp}\rangle\omega^i,$$

where $h_{ijk} = e_\alpha h^\alpha_{ijk}$, and the left side must now vanish. Hence the expression on the right side must also vanish since ω^i and ω^j_i $(i < j)$ are linearly independent, due to $\varepsilon_j \omega_i^j + \varepsilon_i \omega_j^i = 0$; recall that here $\varepsilon_1, \ldots, \varepsilon_m$ are independently +1 or -1, where the number of -1's depends on the index of the metric of M^m. For ω_i^j $(i < j)$, this implies that for every four different values of a, b, c, d,

$$\begin{aligned}
\langle h^*_{aa}, h^*_{aa}\rangle &= \varepsilon_a \varepsilon_b [2\langle h^*_{ab}, h^*_{ab}\rangle + \langle h^*_{aa}, h^*_{bb}\rangle],\\
\langle h^*_{aa}, h^*_{ab}\rangle &= \langle h^*_{aa}, h^*_{bc}\rangle = \langle h^*_{ab}, h^*_{ac}\rangle = \langle h^*_{ab}, h^*_{cd}\rangle = 0,\\
\varepsilon_b \langle h^*_{ab}, h^*_{ab}\rangle &= \varepsilon_c \langle h^*_{ac}, h^*_{ac}\rangle,\\
\varepsilon_b \langle h^*_{aa}, h^*_{bb}\rangle &= \varepsilon_c \langle h^*_{aa}, h^*_{cc}\rangle.
\end{aligned}$$

From the first equation it follows that $\langle h^*_{aa}, h^*_{aa}\rangle = \langle h^*_{bb}, h^*_{bb}\rangle$, so that all $\langle h^*_{ii}, h^*_{ii}\rangle$ have the same value. Similarly from the last two equations it follows that $\varepsilon_i \varepsilon_j \langle h^*_{ij}, h^*_{ij}\rangle$ $(i \neq j)$ have the same value; let it be denoted by α, and likewise $\varepsilon_i \varepsilon_j \langle h^*_{ii}, h^*_{jj}\rangle$ $(i \neq j)$ will be denoted below by β. Here α and β are constants.

Now the first equation becomes $\langle h^*_{ii}, h^*_{ii}\rangle = 2\alpha + \beta$, and all the equations can be summarized as

$$\varepsilon_i \varepsilon_k \langle h^*_{ij}, h^*_{kl}\rangle = (\delta_{ik}\delta_{jl} + \delta_{il}\delta_{jk})\alpha + \delta_{ij}\delta_{kl}\beta. \tag{3.3.1}$$

Moreover, for linearly independent ω^i, the vanishing of $d\langle h^*_{jk}, h^*_{lp}\rangle$ gives

$$\langle h_{ijk}, h^*_{lp}\rangle = -\langle h^*_{jk}, h_{ilp}\rangle.$$

Due to the symmetry in (2.1.4) and (2.2.3), this can be continued to

$$-\langle h^*_{jk}, h_{lip}\rangle = \langle h^*_{ip}, h_{ljk}\rangle = \langle h_{ljk}, h^*_{ip}\rangle,$$

so that every index of the first triplet can be exchanged by every index of the second pair. Thus $\langle h_{ijk}, h^*_{lp}\rangle = \langle h_{lpk}, h^*_{ij}\rangle = -\langle h_{ijk}, h^*_{lp}\rangle$, and so $\langle h_{ijk}, h^*_{lp}\rangle = 0$. This implies that h_{ijk} is orthogonal to every vector of the first-order outer normal subspace.

Due to property (2) above, the h_{ijk} belong to the same first-order outer normal subspace, which has a regular metric due to property (1). This implies that $h_{ijk} = 0$, and therefore the submanifold under consideration is indeed parallel.

Hence (3.1.5) holds, where now $g^{il} = \varepsilon_i \delta^{il}$ (no sum!), and (3.3.1) is to be considered. Therefore,

$$e_i(g^{il}\langle h^*_{jk}, h^*_{lp}\rangle) = \sum_i e_i(\varepsilon_i \langle h^*_{ip}, h^*_{jk}\rangle) = \sum_i e_i(\varepsilon_j[(\delta_{ij}\delta_{pk} + \delta_{ik}\delta_{pj})\alpha + \delta_{ip}\delta_{jk}\beta]),$$

so that (3.1.5) reduces to

$$dh^*_{jk} = -\varepsilon_j[(e_j\omega^k + e_k\omega^j)\alpha + \delta_{jk}e_p\omega^p\beta] + h^*_{lk}\omega^l_j + h^*_{jl}\omega^l_k, \tag{3.3.2}$$

where $e_p\omega^p = dx$. After interchanging j and k it is seen from this that $\varepsilon_j = \varepsilon_k = \varepsilon$ for every pair j, k. Hence the inner metric of the submanifold M^m is either positive or negative definite.

Moreover, in (3.3.1) now $\varepsilon_i \varepsilon_k = \varepsilon^2 = 1$, and so due to (2.1.10) and (2.1.11),

$$\Omega_i^j = \varepsilon(\alpha - \beta)\omega^i \wedge \omega^j. \tag{3.3.3}$$

Using exterior differentiation in (3.3.2), and making use of (2.4.1), (1.2.3), (2.1.3), and (3.3.3), one eventually obtains

$$(2\alpha - \beta)[h^*_{jl}\omega^l \wedge \omega^k + h^*_{kl}\omega^l \wedge \omega^j] = 0.$$

Here one has two possibilities: Either (1) $2\alpha - \beta \neq 0$ and $[h^*_{jl}\delta_i^k + h^*_{kl}\delta_i^j]\omega^l \wedge \omega^i = 0$ or (2) $\beta = 2\alpha$.

In the first case $h^*_{jl}\delta_i^k + h^*_{kl}\delta_i^j - h^*_{ji}\delta_l^k - h^*_{ki}\delta_l^j = 0$; after summing by $k = l$, this leads to $h^*_{ij} = H^*\delta_{ij}$, i.e., to a totally umbilic submanifold M^m (see Section 2.4).

In the second case now $\varepsilon_i = \varepsilon_k = \varepsilon = \pm 1$, so due to (3.3.1) one has

$$\langle h^*_{ij}, h^*_{kl} \rangle = (\delta_{ik}\delta_{jl} + \delta_{il}\delta_{jk} + 2\delta_{ij}\delta_{kl})\alpha. \tag{3.3.4}$$

If $\alpha \neq 0$, then the Gramian matrix of the vectors $h^*_{11}, h^*_{12}, \ldots, h^*_{m-1,m}, h^*_{mm}$ is nonzero, and hence all these vectors are linearly independent. Thus the first-order outer normal subspace has maximal possible dimension $\frac{1}{2}m(m+1)$.

Due to $\beta = 2\alpha$, now (3.3.3) reduces to $\Omega_i^j = -\varepsilon\alpha\omega^i \wedge \omega^j$, so that the submanifold M^n has constant curvature $\varepsilon\alpha$.

As a result, the following statement can be formulated.

Theorem 3.3.1. *Suppose M^m in ${}_sN^n(c)$ is a complete regular (pseudo-)Riemannian submanifold of constant curvature, lying in its first-order outer osculating subspace, and having the property that all its inner isometries are induced by isometries of the ambient space. Then M^m is a parallel submanifold with positively or negatively definite inner metric and either* (1) *it is totally umbilic, or* (2) *its first-order outer osculating subspace, containing M^m, has maximal possible dimension $\frac{1}{2}m(m+3)$, and* (3.3.4) *holds, where $\varepsilon\alpha$ is the constant curvature of M^m.*

The parallel submanifold of the second case (2) is called the *Veronese submanifold* because there is a direct connection with the Veronese map known in algebraic geometry (see, e.g., [Sha 88], Chapter I.4).

Remark 3.3.2. For $s = c = 0$, i.e., for a Riemannian submanifold M^m in E^n, Theorem 3.3.1 was established by R. Mullari [Mu 62b]. In his terminology, a submanifold of constant curvature having the property that all of its inner isometries are induced by the isometries of the ambient space is called a *submanifold of maximal symmetry*.

Remark 3.3.3. A more general treatment of the Veronese submanifold, without the assumption of definiteness of the inner metric, is given in [Blo 86] (see Section 4.7 below).

3.4 Parallel Submanifolds and the Gauss Map

An active study of parallel submanifolds of arbitrary dimension was started in [SW 69], where hypersurfaces M^m with parallel second fundamental form in Euclidean space E^{m+1} were described. There followed [Er 71], [YI 71], [YI 72], which also considered submanifolds M^m with $\bar{\nabla} h = 0$ in E^n.

The first general geometric result for arbitrary submanifolds M^m with $\bar{\nabla} h = 0$ in E^n was given by J. Vilms in [Vi'72] and is connected with the Gauss map of a submanifold.

For a submanifold M^m in ${}_kE^n$, the Gauss map is defined as the mapping $M^m \to {}_{l,k}G^{m,n}$ into the Grassmann manifold ${}_{l,k}G^{m,n}$ of m-dimensional vector subspaces in an n-dimensional (pseudo-)Euclidean vector space, which maps a point $x \in M^m$ into $T_x M^m$ considered as a point of ${}_{l,k}G^{m,n}$.

The following theorem is given in [Vi 72] for the particular case $k = l = 0$. Here a different proof is given, using the framework of an adapted orthogonal frame bundle.

Theorem 3.4.1. *The image of a parallel submanifold $M^m \subset {}_kE^n$ under the Gauss map is a totally geodesic submanifold of ${}_{l,k}G^{m,n}$.*

Proof. The tangent plane of M^m at x is spanned by the basis vectors $e_1, \dots, e_m$ of an adapted frame. If a point x and its image in ${}_{l,k}G^{m,n}$, determined by the simple m-vector $e_1 \wedge \cdots \wedge e_m$, are identified, then one sees via (3.2.3) and (1.5.1) that the tangent vector space of the image of M^m is spanned by the simple m-vectors $(e_1 \wedge \cdots \wedge e_{i-1} \wedge e_\alpha \wedge e_{i+1} \wedge \cdots \wedge e_m)h^\alpha_{ij}$; note that here i and α play the same role as a and u, respectively, in (3.2.3) and (1.5.1). Let the m-vector in parentheses be denoted by $e_{1\ldots\overset{i}{\alpha}\ldots m}$, as in (3.2.3), and put $x = e_1 \wedge \cdots \wedge e_m$. A straightforward computation leads to

$$de_{1\ldots\overset{i}{\alpha}\ldots m} = -x g_{\alpha\beta}\omega^\beta_i + e_{1\ldots\overset{k}{\beta}\ldots m}(-\omega^i_k\delta^\beta_\alpha + \delta^i_k\omega^\beta_\alpha) + \Sigma^{\beta\neq\alpha}_{k\neq i} e_{1\ldots\overset{i}{\alpha}\ldots\overset{k}{\beta}\ldots q}(\omega^\beta_k) \quad (3.4.1)$$

(cf. (3.2.4)). Therefore, the differential of the m-vector $e_{1\ldots\overset{i}{\alpha}\ldots m}h^\alpha_{ij}$ tangent to the image has, in the inner geometry of ${}_{l,k}G^{m,n}$, only components tangent to the same image. Indeed, in the right-hand side of this differential only the terms

$$e_{1\ldots\overset{k}{\beta}\ldots m}(-\omega^i_k\delta^\beta_\alpha + \delta^i_k\omega^\beta_\alpha)h^\alpha_{ij} + e_{1\ldots\overset{i}{\alpha}\ldots m}dh^\alpha_{ij}$$

are to be considered in this inner geometry, since all other terms are normal to ${}_{l,k}G^{m,n}$ as a submanifold of $\wedge^m(\mathbb{R}^n)$; recall, the latter is the vector space of all m-vectors of the vector space $\mathbb{R}^n$ of ${}_kE^n$, equipped with the usual metric (cf. Example 1.5.4, or [Ste 64], Chapter 1, Section 4). If M^m is parallel, then (3.1.1) holds, and the terms to be considered reduce to

$$e_{1\ldots\overset{i}{\alpha}\ldots m}h^\alpha_{ik}\omega^k_j = (e_1 \wedge \cdots \wedge e_{i-1} \wedge h_{ik} \wedge e_{i+1} \wedge \ldots e_m)\omega^k_j,$$

whence they are tangent to the image of M^m. Hence this image is indeed totally geodesic.

Remark 3.4.2. Parallel submanifolds are not the only ones with totally geodesic Gauss map. In [CY 83] all surfaces ($m = 2$) in E^n with totally geodesic Gauss map are classified. The same problem for arbitrary m is studied in [CY 84] (see also [PK 86]). A generalization to submanifolds with totally umbilical Gauss map is given in [KP 87]. A generalized Gauss map for parallel submanifolds is introduced in [Na 90]. Recently, the theorem by J. Vilms [Vi 72] (see Theorem 3.4.1 for $k = l = 0$) has been generalized in [JR 2006] to the case of parallel submanifolds M^m in an arbitrary n-dimensional Riemannian manifold N considering the Grassmann bundle over N.

3.5 Parallel Submanifolds and Local Extrinsic Symmetry

An important contribution to the theory of parallel submanifolds was given by D. Ferus [Fe 74a, b, c], [Fe 80]. In several aspects he followed É. Cartan's theory of locally symmetric Riemannian spaces and developed its extrinsic analogue.

The property of a space that each of its geodesic symmetry maps is an isometry of a neighborhood (see Section 1.5) has the following extrinsic analogue.

Let M^m be a submanifold in a space form ${}_sN^n(c)$, taken as a standard model in ${}_\sigma E^{n+1}$ (see Section 1.4). At every point $x \in M^m$ the tangent and normal vector subspaces T_xM^m and $T_x^\perp M^m$ of M^m are defined as vector subspaces of $T_x[{}_sN^n(c)]$. If $c \neq 0$ there is also the outer normal vector subspace $T_x^{*\perp}M^m$ of M^m in ${}_\sigma E^{n+1}$, spanned by $T_x^\perp M^m$ and by the radius vector of x with origin at the center o of the model sphere. The m- and $(n-m+1)$-planes in ${}_\sigma E^{n+1}$ through x with vector subspaces T_xM^m and $T_x^{*\perp}M^m$, respectively, are the *tangent* and *outer normal planes* of M^m at x. Note that here $T_x^\perp M^m$ is the tangent vector subspace at x of the intersection of ${}_sN^n(c)$ with the outer normal plane; this intersection is the *normal (n–m)-plane* of M^m in ${}_sN^n(c)$ at x.

A submanifold M^m of ${}_sN^n(c)$ is said to have *local extrinsic symmetry* if each of its points x_0 has a normal neighborhood whose geodesic symmetry map with respect to x_0 is induced by a reflection in ${}_sN^n(c)$ with respect to the normal $(n–m)$-plane at x_0.

By a reflection one means here that vectors $\vec{x_0x}$ and $\vec{x_0x'}$ from x_0 to x and to x' in ${}_\sigma E^{n+1}$ have the same component in the outer normal $(n-m+1)$-plane at x_0, and their components in the tangent m-plane at x_0 differ only by sign.

D. Ferus [Fe 80] showed (for submanifolds in E^n) the following relationship between the properties of being parallel and of having local extrinsic symmetry. (In the more general setting of submanifolds in a Riemannian manifold, the same was done by W. Strübing [Str 79].)

Theorem 3.5.1. *If a submanifold M^m in ${}_sN^n(c)$ has local extrinsic symmetry then it is parallel, i.e., $\bar{\nabla}h = 0$ holds.*

Proof. Let a geodesic curve between x and x' be given by $x(s)$ so that $x(0) = x_0$, $x(s) = x$, $x(-s) = x'$, where s is the arclength parameter. Then

$$x(s) = x(0) + \frac{d}{ds}x(0)s + \frac{1}{2}\frac{d^2}{ds^2}x(0)s^2 + \frac{1}{6}\frac{d^3}{ds^3}x(0)s^3 + (\ldots)s^4.$$

Since $\frac{d}{ds}x$ is a vector tangent to M^m with constant length ± 1, it is a linear combination of the tangent basis vectors e_i of an adapted frame. Thus there exist coefficients X^i so that $\frac{d}{ds}x = e_i X^i$. It follows that

$$\frac{d^2}{ds^2}x = \frac{1}{ds}d(e_i X^i) = \frac{1}{ds}[(e_j\omega_i^j + e_\alpha h_{ij}^\alpha \omega^j)X^i + e_i dX^i]$$

$$= \frac{1}{ds}e_i \nabla X^i + e_\alpha h_{ij}^\alpha X^i X^j,$$

using the facts that $dx = (e_j X^j)ds$ yields $\omega^j = X^j ds$ and that $\nabla X^i = dX^i + X^j\omega_j^i$ is the covariant derivative of X^i with respect to the Riemannian connection ∇.

For a geodesic curve one has $\nabla X^i = 0$; hence $dX^i = -X^j\omega_j^i$. This yields

$$\frac{d^3}{ds^3}x = \frac{1}{ds}d(e_\alpha h_{ij}^\alpha X^i X^j),$$

and the normal component of this vector is

$$\frac{1}{ds}(e_\beta \omega_\alpha^\beta h_{ij}^\alpha X^i X^j + e_\alpha dh_{ij}^\alpha X^i X^j - e_\alpha h_{ij}^\alpha X^k \omega_k^i X^j - e_\alpha h_{ij}^\alpha X^i X^k \omega_k^j)$$

$$= \frac{1}{ds}(e_\alpha \bar{\nabla} h_{ij}^\alpha X^i X^j) = e_\alpha h_{ijk}^\alpha X^i X^j X^k.$$

It follows that the normal component of the vector $x(s) - x(0)$ is

$$\frac{1}{2}e_\alpha h_{ij}^\alpha X^i X^j s^2 + \frac{1}{6}e_\alpha h_{ijk}^\alpha X^i X^j X^k s^3 + (\ldots)s^4.$$

The difference between this normal component and that of $x(-s) - x(0)$ is

$$\frac{1}{3}e_\alpha h_{ijk}^\alpha X^i X^j X^k s^3 + (\ldots)s^4.$$

In the case of extrinsic local symmetry, this difference must be zero for every $\frac{d}{ds}x = e_i X^i$. Dividing by s^3 and taking $s \to 0$ one obtains $h_{ijk}^\alpha X^i X^j X^k = 0$. Due to symmetry of coefficients with respect to i, j, k, this is equivalent to $\bar{\nabla} h_{ij}^\alpha = h_{ijk}^\alpha \omega^k = 0$, i.e., to $\bar{\nabla} h = 0$. This completes the proof.

Remark 3.5.2. The tangential component of $x(s) - x(0)$ is

$$(e_i X^i)s - \frac{1}{6}(e_i h_{\alpha j}^i h_{kl}^\alpha X^j X^k X^l)s^3,$$

if one disregards the term $(\ldots)s^4$. It is seen that the tangential component of $x(-s) - x(0)$ differs only by sign, and it does not matter here if $\bar{\nabla} h_{ij}^\alpha$ is zero or not. Thus the converse of the preceding theorem holds, up to this order of approximation.

Actually this converse is true without any recourse to an approximation.

Theorem 3.5.3. *An M^m in ${}_sN^n(c)$ is extrinsically locally symmetric if and only if it is a parallel submanifold.*[1]

Proof. The proof is given by W. Strübing [Str 79] (in the more general case of a submanifold M^m of an arbitrary Riemannian manifold) and it relies on a more detailed study of the geodesic symmetry map, supposing that h is parallel.

Remark 3.5.4. The result of Theorem 3.5.3 was stated differently in [Str 79], [Fe 80], and [BR 83], namely that the submanifolds with $\bar{\nabla}h = 0$, especially the connected complete ones, are (*extrinsically*) *symmetric*. The name *parallel* was introduced by M. Takeuchi [Tak 81], and it is now more popular, especially when the local point of view is taken.

3.6 Complete Parallel Irreducible Submanifolds as Standard Imbedded Symmetric *R*-Spaces

Theorem 3.5.1 states that if a submanifold M^m in ${}_sN^n(c)$ has local extrinsic symmetry, then it is parallel. Also the converse holds (see Remark 3.5.2 and Theorem 3.5.3).

Now for such an M^m, one can show by considering reflections of ${}_sN^n(c)$ with respect to normal subspaces of M^m, and using the invariance of the latter, at least locally, by these reflections, that if M^m is complete and simply connected, then it is an orbit of a Lie group of isometries of ${}_sN^n(c)$. This allows parallel submanifolds to be described in the framework of the theory of Lie groups and symmetric spaces (see, e.g., [He 78]).

The main problem here concerns conditions on a symmetric space which allow it to be imbedded in ${}_sN^n(c)$ as a parallel submanifold and the consequent nature of the imbedding.

This problem was completely solved by Ferus [Fe 74a–c, 80] for $s = c = 0$ (i.e., for $N^n(c) = E^n$); the result was then extended to $N^n(c)$ with $c \neq 0$ by Takeuchi [Tak 81] and by Backes and Reckziegel [BR 83].

Following [Fe 80], one introduces first a special class of symmetric spaces and their imbeddings.

Let G be a real connected semisimple Lie group of noncompact type with finite center. Let $\mathbf{g} = \mathbf{k} + \mathbf{p}$ be a Cartan decomposition of its Lie algebra, and K the corresponding maximal compact subgroup. Let $0 \neq \eta \in \mathbf{p}$, and

$$K_0 := \{k \in K \,|\, \mathrm{Ad}(k)\eta = \eta\}.$$

Then

$$f: \; M := K/K_0 \to \mathbf{p}, \quad [k] \mapsto \mathrm{Ad}(k)\eta$$

[1] In the classical special case $m = 2, n = 3, s = 0$ an equivalent assertion was established already in 1958 by J. Dubnov and I. Beskin, as mentioned in Section 3.1 above; also see footnote 5 in [Lu 2000a].

is an imbedding in the Euclidean space **p** with metric given by the Killing form of **g**. If here $(\operatorname{ad}\eta)^3 = \operatorname{ad}\eta$ (or, equivalently, $(\operatorname{ad}\eta)$ is a semisimple endomorphism of **g** with eigenvalues $-1, 0, 1$), then the induced Riemannian metric turns M into a Riemannian symmetric space. Such an M is called a *symmetric R-space*, and f is said to be its *standard imbedding*. If f is followed by an (affine) conformal map into some Euclidean space, this composition will also be called a standard imbedding.

A submanifold M^m in E^n is said to be *irreducible* if M^m is not a Riemannian product $M^{m_1} \times \cdots \times M^{m_r}$ of more than one component, each of which is imbedded in its own subspace E^{n_ρ}, $\rho \in \{1, \dots, r\}$, and the latter are totally orthogonal in E^n. Otherwise the submanifold M^m in E^n is said to be *reducible* or the *product* of submanifolds M^{m_ρ} in E^{n_ρ}.

The main result of Ferus is as follows.

Theorem 3.6.1. *A submanifold M^m in E^n is irreducible and parallel if and only if it is an open part of a standard imbedded symmetric R-space.*

The proof in [Fe 74a–c, 80] is not easy and will not be reproduced here. This theorem reduces the classification of parallel submanifolds M^m in E^n to the classification of symmetric R-spaces and their standard imbeddings.

New presentations of the proof have been given in [EH 95] and very recently in the monograph [BCO 2003], Section 3.7, where the term *locally symmetric submanifold* is preferred, and this theorem is also extended to submanifolds of general space forms, using the approach of [Tak 81]. Standard imbedded symmetric R-spaces are considered in [BCO 2003] as orbits of s-representations.

There exist some classifications of symmetric R-spaces (see, e.g., [KNa 64, 65], [TK 68], [Tai 68], [Ko 68]), but the information about the second fundamental forms of their standard imbeddings is not sufficient for further study.

Some new types of parallel submanifolds in E^n were studied as orbits of actions in E^n of the isometry subgroups, which generate the Veronese or Plücker orbits, in particular.

Let the special orthogonal group $SO(m+1, \mathbb{R})$ act in $E^{\frac{1}{2}m(m+3)}$ so that among its orbits there are the m-dimensional Veronese orbits in concentric hyperspheres; note that the common center of the latter is also called the center of these Veronese orbits. It was shown in [Lu 95a] (see also the discussion in Section 9.6 of this book) that the only parallel orbits in $E^{\frac{1}{2}m(m+3)}$ are the m-dimensional Veronese orbits, which form two cones with a common vertex in the center, and the $(l+1)(m-l)$-dimensional *Veronese–Grassmann orbits*, each of which consists of the centers of the l-dimensional totally geodesic submanifolds (they are the Veronese submanifolds) of an m-dimensional Veronese orbit.

Similarly, the Plücker action of $SO(p, \mathbb{R})$ in $E^{\frac{1}{2}p(p-1)}$ can be introduced as an action whose orbit set contains a Plücker orbit (i.e., a complete submanifold of normed simple bivectors in the space of all bivectors of $\mathbb{R}^p$ with natural Euclidean metric; see Section 3.2). It is shown in [Lu 96b] (see also the reproduction in Section 9.3 of this book) that the only parallel orbits of this action are the Plücker orbits and, for even $p = 2q$, the orbits (called the *unitary orbits*; see Section 9.3), which are the standard imbedded symmetric spaces $SO(2q, \mathbb{R})/SU(q, \mathbb{C})$.

Remark 3.6.2. Submanifolds with parallel second fundamental form h have also been studied in other spaces with structural groups, different from the real space forms (see the surveys in [Lu 2000a], Section 23, and [Ch 2000], Section 8). The results can be summarized as follows.

For Kähler submanifolds of a complex space form of constant holomorphic sectional curvature c, it was proved in [Kon 74] that if the submanifold has parallel h and $c \leq 0$, then the submanifold is totally geodesic (see also [Kon 75]). For $c > 0$, Nakagawa and Takagi in [NT 76] gave a full classification of the Kähler submanifolds with parallel h in complex projective space; note that this classification is formulated independently of Ferus's papers [Fe 74a–c], [Fe 80]. The special properties of complex geometry made it possible to use more direct methods for this classification. Four types of parallel submanifolds are obtained in arbitrary dimensions, in addition to the totally geodesic ones: the complex quadrics, and the Veronese, Plücker, and Segre submanifolds. There are two types in special dimensions: standard imbedded $SO(10)/U(5)$ (complex dimension 10) and $E_6/\operatorname{Spin}(10) \times T$ (complex dimension 16). Some new characterizations of these six types of Kähler submanifolds were then given in [Ros 84], [Ros 85], [Ros 86] and by Udagawa [Ud 86]. Totally real parallel submanifolds were investigated in [Kon 75], [Na 81].

One should note that for real space forms and Riemannian submanifolds the situation is more complicated. To the real versions of the four above-mentioned types there can be added, for example, the Veronese–Grassmann submanifolds and the unitary submanifolds (as standard imbedded $SO(2q, \mathbb{R})/SU(q, \mathbb{C})$ for arbitrary q). There exist more exceptional types, in addition to $E_6/\operatorname{Spin}(10) \times T$, for example, $F_4/\operatorname{Spin}(9)$ (the Cayley projective plane), $E_7/E_6 \times T$, E_6/F_4 (see, e.g., [Ko 68]).

Parallel submanifolds of symmetric spaces other than real and complex space forms were studied in [Na 80], [Ts 85a–c]; the last two papers treat submanifolds of quaternion projective space and of the Cayley projective plane, respectively. Characterizations of these submanifolds in the above-mentioned spaces by bounding of the length squared of h were given in [CGa 89] and [CGl 90]. Parallel submanifolds in Sasakian space forms have been investigated in [Pi 89], in Heisenberg space in [Be 99], in symplectic affine space in [Pa 2000a, b], [PS 86], and in the real special linear group $SL(1, R)$ in [BeD 2002]. More information about symmetric orbits in Riemannian and Hermitian symmetric spaces, as well as about parallel submanifolds in complex and quaternionic space forms, concerning especially the recent works by Naitoh, and Tsukada is given in [BCO 2003], Section 9.3 and 9.4, as well as in the papers [Os 2002] and [Bern 2003]. Recently, in [BENT 2005] the classification problem of symmetric submanifolds in Riemannian symmetric spaces has been finished.

Parallel pseudo-Riemannian submanifolds in spacetime forms with indefinite metric have also been studied in the past few years. Their study in the context of Lie groups and symmetric spaces was developed in depth by Naitoh [Na 84] (see also [Na 86]). Some classification problems in low codimensions were solved by Magid [Mag 84]. The decomposition and description results of Ferus were extended to this case by Blomstrom [Blo 85]. A detailed classification of parallel and semiparallel timelike surfaces in a Lorentzian spacetime form was obtained in [Lu 97b].

Remark 3.6.3. In [San 85] C. U. Sanchez introduced the notion of an extrinsic k-symmetric submanifold in E^n and gave a classification for odd k. Furthermore, in [San 92] he proved that the extrinsic k-symmetric submanifolds are essentially characterized by the property of having parallel second fundamental form with respect to the canonical connection of k-symmetric space. This implies that every complete extrinsic k-symmetric submanifold is an orbit of an s-representation (see [BCO 2003], Remark 7.2.8).

4

Semiparallel Submanifolds

4.1 The Semiparallel Condition and Its Special Cases

Since a parallel submanifold M^m of ${}_sN^n(c)$ satisfies $\bar{\nabla}h = 0$, both sides of equation (2.3.1): $\bar{\nabla}_{[k}\bar{\nabla}_{l]}h^{\alpha}_{ij} = R^p_{i,kl}h^{\alpha}_{pj} + R^p_{j,kl}h^{\alpha}_{ip} - R^{\alpha}_{\beta,kl}h^{\beta}_{ij}$ must be zero. The vanishing of the left side says that the coefficients $\bar{\nabla}_k\bar{\nabla}_l h^{\alpha}_{ij} = h^{\alpha}_{ijkl}$ of the fourth fundamental form $\bar{\nabla}^2 h$ are symmetric with respect to the last two upper indices, and therefore via the Peterson–Mainardi–Codazzi identity (cf. Section 2.3) also with respect to all four upper indices. This leads to a special class of submanifolds.

Proposition 4.1.1. *For a submanifold M^m in ${}_sN^n(c)$ the following conditions are equivalent:*

(1) *the fourth fundamental form $\bar{\nabla}^2 h$ is symmetric with respect to all its arguments;*
(2) *the outer fourth fundamental form $\bar{\nabla}^{*2}h$ is symmetric with respect to all its arguments;*
(3) *both the equation*

$$R^p_{i,kl}h^{\alpha}_{pj} + R^p_{j,kl}h^{\alpha}_{ip} - R^{\alpha}_{\beta,kl}h^{\beta}_{ij} = 0 \tag{4.1.1}$$

and its outer version hold;

(4) *both the equation*

$$h^{\alpha}_{pj}\Omega^p_i + h^{\alpha}_{ip}\Omega^p_j - h^{\beta}_{ij}\Omega^{\alpha}_{\beta} = 0, \tag{4.1.2}$$

and its outer version hold.

Proof. Conditions (1) and (2) are equivalent due to Section 2.2, conditions (1) and (3) due to (2.3.1), and conditions (3) and (4) due to (2.1.11).

One sees from (2.3.1) that condition (3) can be considered as the integrability condition of the Pfaffian system $\bar{\nabla}h^{\alpha}_{ij} = 0$, which characterizes parallel submanifolds. The same can be said, of course, for the other conditions stated in Proposition 4.1.1.

Ü. Lumiste, *Semiparallel Submanifolds in Space Forms*,
DOI 10.1007/978-0-387-49913-0_5, © Springer Science+Business Media, LLC 2009

A submanifold M^m of ${}_sN^n(c)$ satisfying one of the conditions of Proposition 4.1.1, each of which is equivalent to (0.4), is said to be *semiparallel*.[1]

For the intrinsic geometry of semiparallel submanifolds, one has the following (see [De 85]).

Proposition 4.1.2. *Every semiparallel submanifold M^m of ${}_sN^n(c)$ is intrinsically a semisymmetric Riemannian manifold.*

Proof. The semiparallel condition (4.1.2) and the Gauss identity (2.1.7) yield, after some calculations, the semisymmetric condition (1.6.2), written now for the inner geometry of M^m, i.e., instead of $I, J, \ldots$ one has $i, j, \ldots$.

The converse does not hold. Every two-dimensional (pseudo-)Riemannian manifold is semisymmetric (see Example 1.6.7), but every (pseudo-)Riemannian surface M^2 of ${}_sN^n(c)$ is not semiparallel, as can be seen from the classification of two-dimensional (pseudo-)Riemannian submanifolds, given for $s = c = 0$ in [De 85], for $s = 0$, $c \neq 0$ in [Me 91], [AM 94], and for $s = 1$, $c = 0$ in [Lu 97b]; see also Chapter 6.

There are some special classes of semiparallel submanifolds.

Proposition 4.1.3. *Every parallel submanifold M^m of ${}_sN^n(c)$ is semiparallel.*

Proof. Indeed, $\bar{\nabla}h = 0$ is equivalent to $\bar{\nabla}_k h_{ij} = 0$, and by (2.3.1) this implies the semiparallel condition.

It is clear that the converse of this proposition is not true.

The semiparallel condition (4.1.2) is also satisfied in case the van der Waerden–Bortolotti connection $\bar{\nabla}$ is flat, i.e., its curvature 2-forms vanish:

$$\Omega_i^j = 0, \quad \Omega_\alpha^\beta = 0. \tag{4.1.3}$$

This gives the following.

Proposition 4.1.4. *Every submanifold M^m of ${}_sN^n(c)$ with flat $\bar{\nabla}$ is semiparallel.*

For the classical case of M^2 in E^3 this proposition implies that every developable surface is semiparallel but not parallel in general, due to a result of V. F. Kagan [Ka 48] (see Section 3.1; in fact, among nontrivial developable surfaces in E^3, only the round cylinders are parallel).

One more special class consists of the submanifolds with parallel third fundamental form $\bar{\nabla}h$, i.e., with $\bar{\nabla}_k(\bar{\nabla}_l h_{ij}^\alpha) = 0$. From (2.3.1) it then follows immediately that (4.1.1) is satisfied. This yields the following.

[1] This term was introduced by J. Deprez [De 85, 86]]. In some other early papers on this class of submanifolds, the designation *semisymmetric* was used instead (see [Lu 87a], [Lu 88a, b], [Lu 89b, c], [Lu 90a, b, d, e], [Lu 91d, f], [Lu 92a], [LR 90], [LR 92], [Mi 91c]); this name was motivated by the term *symmetric* used by Ferus in [Fe 80] for submanifolds with parallel second fundamental form (see also Remark 3.5.4); in these papers, the qualification *extrinsically* was understood but not explicitly stated.

Proposition 4.1.5. *Every submanifold M^m of ${}_sN^n(c)$ with parallel third fundamental form $\bar{\nabla}h$ (i.e., a 2-parallel submanifold) is semiparallel.*

These submanifolds are not in general parallel. For example, the cylinder on a Cornu spiral (clothoid) in E^3 has parallel third fundamental form, but is not parallel due to Kagan's result cited above (this also follows from a result given below in Section 6.8).

A submanifold M^m of ${}_sN^n(c)$ can also be considered in the outer context, i.e., with respect to the ambient ${}_\sigma E^{n+1}$ containing ${}_sN^n(c)$ (see Section 1.4). The outer version of fundamental forms was considered in Section 2.2 above.

Proposition 4.1.6. *A submanifold M^m in ${}_sN^n(c)$ is semiparallel if and only if it is semiparallel as a submanifold of the ambient ${}_\sigma E^{n+1}$.*

Proof. Suppose (4.1.2) is satisfied. In the outer version, the superscript α is replaced by α^* (see Section 2.2), and the index value $m+1$ must be added. Then (2.2.8) and (2.2.9) show that the equation to be added to (4.1.2) is satisfied. The converse also follows easily.

Theorem 4.1.7. *In a space form ${}_sN^n(c)$ with $c \leq 0$, suppose M^m is a semiparallel submanifold that is minimal (i.e., has $H = 0$) and which has a Euclidean first-order osculating subspace at each point $x \in M^m$. Then M^m is totally geodesic.*

Proof. If one replaces (2.1.11) in the semiparallel condition (4.1.2), the latter becomes, due to (2.1.10), a system of algebraic equations on the components of the second fundamental form. Suppose that the tangent part of the frame adapted to M^m is orthonormal, so that $g_{ij} = \epsilon_i \delta_{ij}$, $g^{ij} = \epsilon_i \delta^{ij}$ (no sum!); then by means of the vector components this system can be written as

$$\begin{aligned} &\epsilon_p \left[\langle h_{i[k}, h_{l]p} \rangle h_{pj} + \langle h_{j[k}, h_{l]p} \rangle h_{ip} - \langle h_{ij}, h_{p[k} \rangle h_{l]p} \right] \\ &\quad + c \left[\epsilon_i \delta_{i[k} h_{l]j} + \epsilon_j \delta_{j[k} h_{l]i} \right] = 0 \end{aligned} \tag{4.1.4}$$

(summing over $p = 1, \ldots, m$). Contracting here with $g^{ik} = \epsilon_i \delta^{ik}$, using that $\frac{1}{m} \sum_{i=1}^m \epsilon_i h_{ii} = H$ is the mean curvature vector and denoting $\langle H, h_{lp} \rangle = h^H_{lp}$, one obtains

$$\begin{aligned} &\epsilon_p \Big[\left(m h^H_{lp} - \epsilon_i \langle h_{il}, h_{ip} \rangle \right) h_{pj} + \epsilon_i \left(\langle h_{ij}, h_{lp} \rangle - \langle h_{jl}, h_{ip} \rangle \right) h_{ip} \\ &\quad - \epsilon_i \left(\langle h_{ij}, h_{ip} \rangle h_{lp} - \langle h_{ij}, h_{lp} \rangle h_{ip} \right) \Big] + mc \left(h_{jl} - \epsilon_j \delta_{jl} H \right) = 0 \end{aligned}$$

(summing over i and p independently). If one sets $H = 0$, takes the scalar product with h_{lj}, and then sums over l and j independently, the result is

$$\begin{aligned} &mc \langle h_{lj}, h_{lj} \rangle - \epsilon_i \langle h_{lj}, h_{ip} \rangle \langle h_{lj}, h_{ip} \rangle \\ &\quad + 2\epsilon_i \left(\langle h_{ij}, h_{lp} \rangle \langle h_{lj}, h_{ip} \rangle - \langle h_{il}, h_{ip} \rangle \langle h_{lj}, h_{pj} \rangle \right) = 0. \end{aligned}$$

Now suppose the first-order osculating subspace is Euclidean. Then the term having coefficient 2 is the negative of a sum of squares. Indeed, then $\epsilon_i = 1$ for all $i \in \{1, \dots, m\}$; moreover, $\langle h_{ij}, h_{lp} \rangle = \sum_{\alpha=m+1}^{n} h_{ij}^{\alpha} h_{lp}^{\alpha}$ and $\langle h_{lj}, h_{ip} \rangle = \sum_{\beta=m+1}^{n} h_{lj}^{\beta} h_{ip}^{\beta}$. The coefficient 2 can be affected by repeating the term while interchanging the roles of the summation indices α and β. Then one gets $-\sum (h_{lp}^{\alpha} h_{ip}^{\beta} - h_{lp}^{\beta} h_{ip}^{\alpha})^2$, summed independently over i, l, p, α, β.

Then $c \leq 0$ implies that a sum of squares is ≤ 0, and therefore all these squares are zero, in particular, $h_{lj} = 0$. Hence M^m is totally geodesic.

Remark 4.1.8. In the special case of parallel submanifolds the assertion of Theorem 4.1.7 for $c = 0$ is proved in [Fe 74b], Lemma 4, and for $c \leq 0$ in [Tak 81], Lemma 1.6, using some known identities for minimal submanifolds. In the special case of submanifolds with parallel $\bar{\nabla} h$ this assertion is proved for $c \leq 0$ in [Mi 83b], Lemma 5, using an identity from [Ch 73b].

In general Theorem 4.1.7 is established for the case $c = 0$ and $s = 0$ in [De 85] using a result from [Ba 83]. In [Lu 2000a] this was extended, with a direct proof, to the case $c \leq 0$ and $s = 0$, and here it is now extended to hold for all s.

Remark 4.1.9. The assumptions of Theorem 4.1.7, that $c \leq 0$ and that the submanifold M^m has Euclidean first-order osculating subspace, are both essential. Indeed, it was shown in [Fe 74c] that an irreducible 1-parallel submanifold M^m in E^n belongs to a sphere (i.e., to an $N^{n-1}(c)$ with $c > 0$) and is minimal in the latter. On the other hand the classification of minimal semiparallel pseudo-Riemannian (timelike) surfaces M^2 in Lorentzian spacetime forms ${}_1N^n(c)$, given in [Lu 97b], showed that these surfaces, as a rule, are not totally geodesic.

Remark 4.1.10. In [DPV 97] it was established that a submanifold M^m in E^n with $m \geq 3$ is intrinsically semisymmetric and satisfies Chen's equality (see [Ch 2000]) if and only if M^m is either a minimal submanifold, or else a round hypercone in some totally geodesic subspace E^{m+1} of E^n. Together with Proposition 4.1.2 and Theorem 4.1.7, this implies that a semiparallel M^m in E^n with $m \geq 3$ satisfies Chen's equality if and only if M^m is either totally geodesic, or a round hypercone in some totally geodesic subspace E^{m+1} of E^n (see [DPV 97], Corollary 7).

4.2 The Semiparallel Condition from the Algebraic Viewpoint

Recall that while the parallel condition $\bar{\nabla} h = 0$ is a system of differential equations on the components of the second fundamental form h, the semiparallel condition (4.1.1) is a system of purely algebraic equations on these components. It is thus a pointwise condition (see, e.g., (4.1.4)), and therefore it can be treated in a purely algebraic way. This was already partly done in the course of investigations of parallel submanifolds, especially in [Fe 80], [Ba 83], [BR 83]. The possibility of using those concepts for the study of semiparallel submanifolds was mentioned first in [De 85].

Let V be a real (pseudo-)Euclidean vector space with scalar product $\langle\cdot,\cdot\rangle$, T a (pseudo-)Euclidean subspace of V, and $h : T \times T \to T^{\perp}$ a symmetric bilinear map, where $T^{\perp}$ is the orthogonal complement of T in V. Then (V, T, h) is called a *(pseudo-)Euclidean fundamental triplet* (for Euclidean V see [Lu 92b]; note that in [BR 83] the authors restricted themselves to the part (T, h), calling it the *initial data*).

Note that at every point x of a (pseudo-)Riemannian submanifold M^m in ${}_sN^n(c)$ there exists such a triplet, namely the one given by the tangent vector space $T = T_xM^m$ of M^m in the vector space $V = T_x({}_sN^n(c))$ and the second fundamental form $h = h_x$.

Having in mind this last context, one can introduce, for a (pseudo-)Euclidean fundamental triplet (V, T, h) and for a real constant c, a symmetric bilinear map $S : T \times T \to \operatorname{End} T$ by

$$\langle S(X,Y)Z, W\rangle = \langle h(X,Y), h(Z,W)\rangle + c\langle X,Y\rangle\langle Z,W\rangle.$$

Then the skew-symmetric bilinear map $R : T \times T \to \operatorname{End} T$, defined by

$$R(X,Y)Z = S(Y,Z)X - S(X,Z)Y, \tag{4.2.1}$$

is called the *curvature map* (with the constant c) and satisfies

$$\begin{aligned}\langle R(X,Y)Z, W\rangle &= \langle h(X,W), h(Y,Z)\rangle - \langle h(X,Z), h(Y,W)\rangle \\ &\quad + c(\langle Y,Z\rangle\langle X,W\rangle - \langle X,Z\rangle\langle Y,W\rangle)\end{aligned}$$

(the Gauss identity) and thus

$$\langle R(X,Y)Z, W\rangle = \langle R(Z,W)X, Y\rangle = -\langle Z, R(X,Y)W\rangle.$$

For $\xi \in T^{\perp}$ let $A_\xi : T \to T^{\perp}$ be the symmetric linear map defined by $\langle A_\xi X, Y\rangle = \langle h(X,Y), \xi\rangle$, called the *shape operator* for ξ (cf. Section 2.1). A skew-symmetric bilinear map $R^{\perp} : T \times T \to T^{\perp}$ can be introduced by

$$R^{\perp}(X,Y)\xi = h(X, A_\xi Y) - h(Y, A_\xi X),$$

or, equivalently, by

$$\langle R^{\perp}(X,Y)\xi, \eta\rangle = \langle [A_\xi, A_\eta]X, Y\rangle.$$

This $R^{\perp}$ is called the *normal curvature map* (cf. (2.1.9)).

The (pseudo-)Euclidean fundamental triplet (V, T, h) is said to be *semiparallel* (with constant c) if (cf. (4.1.1))

$$h(R(Z_1, Z_2)X, Y) + h(X, R(Z_1, Z_2)Y) - R^{\perp}(Z_1, Z_2)h(X,Y) = 0. \tag{4.2.2}$$

These concepts can also be considered in the context of *triple systems*, as follows.

A real vector space T (more generally, a unitary module over a ring K) together with a trilinear map $\{\} : T \times T \times T \to T$, where $(X, Y, Z) \mapsto \{XYZ\}$, is called a *Jordan triple system* (see [Mey 70], [Ne 86]) if

$$\{XYZ\} = \{ZYX\}, \tag{4.2.3}$$

$$\{W_1W_2\{XYZ\}\} - \{XY\{W_1W_2Z\}\} = \{\{W_1W_2X\}YZ\} - \{X\{W_2W_1Y\}Z\} \tag{4.2.4}$$

for all X, Y, Z, W_1, W_2 in T.

In case T is a (pseudo-)Euclidean vector space, if the conditions

$$\langle\{XYZ\}, W\rangle = \langle Z, \{YXW\}\rangle$$

and (4.2.3) are satisfied, then T is said to be a *(pseudo-)Euclidean triple system*; and if (4.2.4) is also satisfied, then it is a *(pseudo-)Euclidean Jordan triple system* (cf. [Ba 83], [BR 83]).

There is another way of introducing these triple sustems. Denoting $\{XYZ\} = L(X, Y)Z$ one obtains a bilinear map $L : T \times T \to \text{End}\, T$. The conditions for a Jordan triple system can then be written as

$$L(X, Y)Z = L(Z, Y)X,$$

$$[L(W_1, W_2), L(X, Y)] = L(L(W_1, W_2)X, Y) - L(X, L(W_2, W_1)Y),$$

and this is (pseudo-)Euclidean if

$$L(X, Y)^* = L(Y, X).$$

Proposition 4.2.1. *Let (V, T, h) be a (pseudo-)Euclidean fundamental triplet. Then $L = R + S$ turns the vector space T into a (pseudo-)Euclidean triple system. If (V, T, h) is a semiparallel fundamental triplet, then this triple system is a (pseudo-)Euclidean Jordan triple system.*

Proof. The first assertion follows directly from the definitions of S and R.

From the semiparallel condition it follows after some calculation, using the Gauss identity, that

$$[R(W_1, W_2), S(X, Y)] = S(R(W_1, W_2)X, Y) - S(X, R(W_2, W_1)Y)$$

(see [Fe 80], [Ba 83], [Lu 92b]), and hence

$$[R(W_1, W_2), R(X, Y)] = R(R(W_1, W_2)X, Y) - R(X, R(W_2, W_1)Y);$$

note that for a semiparallel submanifold the last equation is equivalent to its intrinsic semisymmetricity (cf. Proposition 4.1.2). One can see now that in the previous equation S can be replaced by L. To obtain the second assertion it remains to establish

$$[S(W_1, W_2), L(X, Y)] = L(S(W_1, W_2)X, Y) - L(X, S(W_2, W_1)Y).$$

This is done in [Fe 80] (for the Euclidean case $s = 0$, see p. 84; the argument works for $s > 0$), and in [Ba 83].

Remark 4.2.2. There is a known construction for producing a semisimple Lie algebra from a Jordan triple system (Koecher construction, see [Mey 70], also called the Kantor–Koecher–Tits construction [Ne 86]). Therefore, it is possible to introduce the theory of Lie groups into the study of semiparallel submanifolds, especially involving the parallel ones. For the latter, the investigations of D. Ferus are most extensive; they have been treated above in Section 3.6. The link with semiparallel submanifolds is the topic of a subsequent section below.

Remark 4.2.3. All this can also be given in the outer version. One must replace the ambient space V by the outer ambient space $V^* = V \oplus \mathbb{R}$ (orthogonal direct sum) and h by h^*, where

$$h^*(X, Y) = h(X, Y) - xc\langle X, Y\rangle;$$

according to (2.1.6), x denotes a generating element of $\mathbb{R}$ with norm squared equal to $c^{-1} = \text{const}$. Then

$$\langle S(X, Y)Z, W\rangle = \langle h^*(X, Y), h^*(Z, W)\rangle,$$

and all formulas from (4.2.1) and (4.2.2) remain in force after replacing h by h^*.

4.3 Decomposition of Semiparallel Fundamental Triplets

Let (V, T, h) and (V_1, T_1, h_1) be two pseudo-Euclidean fundamental triplets. The second one is said to be the *subtriplet* of the first one, if V_1 and T_1 are vector subspaces of V and T, respectively, with induced scalar product (nondegenerate symmetric bilinear form), and if $X, Y \in T_1$ then $h_1(X, Y) = h(X, Y) \in T_1^{\perp}$, where $T_1^{\perp}$ is the orthogonal complement of T_1 in V_1.

Let (V_ρ, T_ρ, h_ρ), $\rho \in \{1, \dots, r\}$, be r such subtriplets of such a (V, T, h). The latter is their *orthogonal direct sum*, if $V = V_1 \oplus \dots \oplus V_r$, $T = T_1 \oplus \dots \oplus T_r$ (orthogonal direct sums of pseudo-Euclidean vector spaces) and $h(X_1 + \dots + X_r, Y_1 + \dots + Y_r) = h_1(X_1, Y_1) + \dots + h_r(X_r, Y_r)$ for all $X_\rho, Y_\rho \in T_\rho$.

The *mean curvature vector* of (V, T, h) is a vector $H \in T^{\perp}$, defined by $\langle H, \xi\rangle = \frac{1}{m}(\text{trace } A_\xi)$, where $m = \dim T$ and ξ is an arbitrary vector of $T^{\perp}$; recall that the shape operator A_ξ was introduced in Section 4.2.

Taking $\xi = H =$ the mean curvature vector, one gets the *mean shape operator* A_H, which will play an important role below.

Theorem 4.3.1. *Let (V, T, h) be a (pseudo-)Euclidean semiparallel fundamental triplet, and let (V^*, T, h^*) be its corresponding outer triplet (see Remark 4.2.3). Suppose that T is a Euclidean vector subspace and that the mean shape operator A_H has r distinct eigenvalues. Then (V^*, T, h^*) is the orthogonal direct sum of its subtriplets $(V_\rho^*, T_\rho, h_\rho^*)$ for $\rho \in \{1, \dots, r\}$, where $T_1, \dots, T_r$ are the eigenspaces of A_H.*

Moreover, each of these subtriplets is a semiparallel triplet.

Proof. Let $\dim V = n$, and choose an orthonormal basis in V such that $e_i \in T$ and $e_\alpha \in T^\perp$, where $i, j, \dots \in \{1, \dots, m\}$ and $\alpha, \beta, \dots \in \{m+1, \dots, n\}$. Then $X = e_i X^i$, $Y = e_j Y^j$, and $h(X, Y) = h_{ij} X^i Y^j$, where $h_{ij} = h(e_i, e_j)$. The semiparallel condition (4.2.2), with h^* instead of h (see Remark 4.2.3) can be written as in (4.1.4) above as

$$\sum_p \left[\langle h^*_{i[k}, h^*_{l]p} \rangle h^*_{pj} + \langle h^*_{j[k}, h^*_{l]p} \rangle h^*_{ip} - \langle h^*_{ij}, h^*_{p[k} \rangle h^*_{l]p} \right] = 0, \qquad (4.3.1)$$

since all $\epsilon_p = 1$. Setting $i = j$ and summing, one obtains

$$\sum_{p,i} \left[\langle h^*_{i[k}, h^*_{l]p} \rangle h^*_{pi} + \langle h^*_{i[k}, h^*_{l]p} \rangle h^*_{ip} \right] - \sum_p \langle mH^*, h^*_{p[k} \rangle h^*_{l]p} = 0, \qquad (4.3.2)$$

where $H^* = \frac{1}{m} \sum h^*_{ii}$ is the *outer mean curvature vector*. The first sum easily reduces to zero. Consequently,

$$\sum_p \left[\langle H^*, h^*_{pk} \rangle h^*_{lp} - \langle H^*, h^*_{pl} \rangle h^*_{kp} \right] = 0,$$

or, equivalently,

$$\sum_p \left[(A_{H^*})_{pk} h^*_{lp} - (A_{H^*})_{pl} h^*_{kp} \right] = 0; \qquad (4.3.3)$$

here A_{H^*} is the *outer mean shape operator* defined by

$$A_{H^*}(X)Y = \langle H^*, h^*(X, Y) \rangle = A_H(X)Y + c\langle X, Y \rangle. \qquad (4.3.4)$$

Since A_H is a symmetric linear map on T, which was assumed to be Euclidean, there is an orthonormal basis of T for which the matrix of A_H is diagonal, i.e., $(A_H)_{ij} = \lambda_i \delta_{ij}$, and consequently $(A_{H^*})_{ij} = \lambda^*_i \delta_{ij}$, where $\lambda^*_i = \lambda_i + c$. As assumed, there are r distinct values among the eigenvalues λ; denote them by $\lambda_{(1)}, \dots, \lambda_{(r)}$, with multiplicities $p_1, \dots, p_r$. The basic vectors e_i can be renumbered so that the first p_1 vectors, denoted by e_{i_1}, refer to $\lambda_{(1)}, \dots$, the last p_r vectors, denoted by e_{i_r}, refer to $\lambda_{(r)}$. Then $(A_H)_{i_\rho j_\rho} = \lambda_{(\rho)} \delta_{i_\rho j_\rho}$, and $(A_H)_{i_\rho j_\sigma} = 0$ if $\rho \neq \sigma$.

Substituting $p = p_\rho, q = q_\sigma$ $(\rho \neq \sigma)$ in (4.3.2) gives

$$(\lambda^*_{(\rho)} - \lambda^*_{(\sigma)}) h^*_{p_\rho q_\sigma} = 0 \quad (\rho \neq \sigma).$$

Since here $\lambda^*_{(\rho)} \neq \lambda^*_{(\sigma)}$ $(\rho \neq \sigma)$, this implies that

$$h^*_{p_\rho q_\sigma} = 0 \quad (\rho \neq \sigma). \qquad (4.3.5)$$

The eigenspaces of A_H and A_{H^*} coincide; denote them by $T_\rho = \text{span}\{e_{i_\rho}\}$. Hence $h(X, Y) = h^*(X, Y) = 0$ for $X \in T_\rho$, $Y \in T_\sigma$ when $\rho \neq \sigma$.

Now let us return to (4.3.1) and take the scalar product with H^*. Then (4.3.2) implies that

$$\sum_p \Big[\langle h^*_{i[k}, h^*_{l]p}\rangle (A_{H^*})_{pj} + \langle h^*_{j[k}, h^*_{l]p}\rangle (A_{H^*})_{ip}\Big] = 0.$$

For the eigenbasis of A_{H^*}, this reduces to

$$\Big(\langle h^*_{ik}, h^*_{lj}\rangle - \langle h^*_{il}, h^*_{kj}\rangle\Big)(\lambda^*_i - \lambda^*_j) = 0.$$

Due to (4.3.5), these equations further reduce to $\langle h^*_{i_\rho k_\rho}, h^*_{j_\sigma l_\sigma}\rangle(\lambda^*_{(\rho)} - \lambda^*_{(\sigma)}) = 0$ and give the result

$$\langle h^*_{i_\rho k_\rho}, h^*_{j_\sigma l_\sigma}\rangle = 0 \quad (\rho \neq \sigma). \tag{4.3.6}$$

It follows that the subspaces $\mathrm{span}\{h^*_{i_1 j_1}\}, \dots, \mathrm{span}\{h^*_{i_r j_r}\}$ are mutually orthogonal in $T^{*\perp}$ and therefore can be extended, correspondingly, to the mutually orthogonal $T_1^{*\perp}, \dots, T_r^{*\perp}$, which are orthogonal complements of the eigenspaces $T_1, \dots, T_r$ in the corresponding subspaces $V_1^*, \dots, V_r^*$ of V^*. This completes the proof of the first assertion.

The last assertion now follows easily, by considering the semiparallel condition (4.3.1) for $i = i_\rho$ and $j = j_\rho$, and taking into account (4.3.6).

Remark 4.3.2. The assumption in Theorem 4.3.1 that T is Euclidean is essential to get the diagonal form of A_H in some orthonormal basis. The case of more general pseudo-Euclidean T is complicated because the diagonal form of A_H is not the only simplest form which can be obtained in a suitable orthonormal basis (see [Pe 66], Section 9). In particular, for the case of Lorentzian $T = {}_1E^m$ one has the following four possibilities over $\mathbb{R}$, explicitly given in [Mag 85]:

(1) the diagonal form $\mathrm{diag}\{\lambda_1, \dots, \lambda_m\}$;
(2) the forms that differ from the diagonal form by a special block only; for the latter there are three possible cases:

$$\text{I.}\ \begin{pmatrix} \lambda_0 & 0 \\ 1 & 0 \end{pmatrix}, \qquad \text{II.}\ \begin{pmatrix} \mu_0 & \nu_0 \\ -\nu_0 & \mu_0 \end{pmatrix}, \qquad \text{III.}\ \begin{pmatrix} \lambda_0 & 0 & 0 \\ 0 & \lambda_0 & 1 \\ -1 & 0 & \lambda_0 \end{pmatrix}.$$

The proof above works only for the possibility (1), if one takes $g_{11} = -1$, $g_{22} = \dots = g_{mm} = 1$. The other possibilities need a separate investigation, which will not be done here.

4.4 Triplets of Large Principal Codimension

Let (V, T, h) be a (pseudo-)Euclidean fundamental triplet. Then the span of $h^*(X, Y)$ over all $X, Y \in T$ is called the *outer principal normal subspace* of the outer ambient space (cf. with Section 2.4), and its dimension is called the *outer principal codimension*.

For an orthonormal basis $\{e_i\}$, $i \in \{1, \dots, m = \dim T\}$ of T, one has $h^*(X, Y) = h^*_{ij} X^i Y^j$, where $h^*_{ij} = h^*(e_i, e_j)$, and $h^*_{ij} = h^*_{ji}$. Hence the maximal value of the outer principal codimension is $\frac{1}{2}m(m+1)$, and this value is realized when all h^*_{ij}, $1 \le i \le j \le m$, are linearly independent vectors.

Let us consider semiparallel (pseudo-)Euclidean fundamental triplets with large outer principal codimension, starting first with the maximal case.

In general, the semiparallel condition is a system of algebraic equations on the components of the outer second fundamental form:

$$\sum_p \epsilon_p \left[\langle h^*_{i[k}, h^*_{l]p} \rangle h^*_{pj} + \langle h^*_{j[k}, h^*_{l]p} \rangle h^*_{ip} - \langle h^*_{ij}, h^*_{p[k} \rangle h^*_{l]p} \right] = 0, \qquad (4.4.1)$$

where $\epsilon_p = \langle e_p, e_p \rangle$ is $+1$ or -1 (cf. (4.3.1), where one had all $\epsilon_p = 1$). The scalar product $\langle h^*_{ij}, h^*_{kl} \rangle$ will be denoted below by $H^*_{ij,kl}$.

In the case of maximal outer principal codimension, the linear independence of h^*_{ij}, $1 \le i \le j \le m$, implies that the coefficients before these vectors in (4.4.1) must be zero.

If $i = j = k = a$ and $l = b$, where $a \ne b$, then the coefficients before h_{aa}, h_{ab}, h_{ad}, where a, b, d have distinct values, give, respectively,

$$H^*_{aa,ab} = 0, \qquad (4.4.2)$$

$$\epsilon_a H^*_{aa,aa} - \epsilon_b (3H^*_{aa,bb} - 2H^*_{ab,ab}) = 0, \qquad (4.4.3)$$

$$3H^*_{aa,bd} - H^*_{ab,ad} = 0. \qquad (4.4.4)$$

Similarly, if $i = k = b,\ j = l = a,\ a \ne b$, then the coefficients before h_{aa} and h_{bd} give

$$H^*_{aa,bb} - 2H^*_{ab,ab} = 0, \qquad (4.4.5)$$

$$2H^*_{ab,ad} - H^*_{aa,bd} = 0. \qquad (4.4.6)$$

But if $i = k = a,\ j = b$, and $l = d$, then before h_{bb} one obtains

$$H^*_{aa,bd} - H^*_{ab,ad} = 0. \qquad (4.4.7)$$

From (4.4.3) and (4.4.5) it follows that $H^*_{aa,aa} = 2\epsilon_a \epsilon_b H^*_{aa,bb}$. Here the right-hand side does not change if a and b are switched; therefore, $H^*_{aa,aa} = H^*_{bb,bb}$ for every distinct pair a, b. Denote their common value by

$$H^*_{aa,aa} = 4\kappa. \qquad (4.4.8)$$

Equations (4.4.3) and (4.4.5) now give

$$\epsilon_a \epsilon_b H^*_{aa,bb} = 2\kappa, \qquad (4.4.9)$$

$$\epsilon_a \epsilon_b H^*_{ab,ab} = \kappa. \qquad (4.4.10)$$

If $m = 3$, then (4.4.4) and (4.4.6) give, additionally,

$$H^*_{ab,ad} = H^*_{aa,ad} = 0; \tag{4.4.11}$$

but for $m \geq 3$ and four distinct a, b, d, f, then (4.4.1) gives, in addition,

$$H^*_{ab,df} = 0. \tag{4.4.12}$$

Proposition 4.4.1. *If a semiparallel (pseudo-)Euclidean fundamental triplet $(V.T, h)$ has outer principal normal subspace of maximal possible dimension, then*

$$\varepsilon_i \varepsilon_k H^*_{ij,kl} = (\delta_{ik}\delta_{jl} + \delta_{il}\delta_{jk} + 2\delta_{ij}\delta_{kl})\kappa, \tag{4.4.13}$$

with κ as defined in (4.4.8).

Proof. Indeed, equation (4.4.13) simply recapitulates equations (4.4.8) to (4.4.12).

Remark 4.4.2. Substituting $h^*_{ij} = h_{ij} - xc\delta_{ij}$ in (4.4.1), denoting $\langle h_{ij}, h_{kl}\rangle = H_{ij,kl}$, and replacing $H^*_{ij,kl}$ with $H_{ij,kl} + c\delta_{ij}\delta_{kl}$, one gets from (4.4.13)

$$\sum_p \epsilon_p \left[H_{i[k,l]p}h_{pj} + H_{j[k,l]p}h_{ip} - H_{ij,p[k}h_{l]p}\right]$$
$$+ c\left[\epsilon_l\left(\delta_{k[i}h_{j]l} + \delta_{ij}h_{kl}\right) - \epsilon_k\left(\delta_{l[i}h_{j]k} + \delta_{ij}h_{kl}\right)\right] = 0. \tag{4.4.14}$$

The following proposition shows that for semiparallel (pseudo-)Euclidean fundamental triplets with Euclidean T (i.e., where all $\epsilon_i = 1$), there is a lacuna in their possible principal codimensions below the maximal value.

Proposition 4.4.3. *There exist no semiparallel (pseudo-)Euclidean fundamental triplets (V, T, h) with Euclidean T of dimension $m \geq 3$ and with Euclidean principal normal subspace, whose principal codimension lies between the value $\frac{1}{2}m(m-1)+1$ and the maximal value $\frac{1}{2}m(m+1)$.*

Proof. Suppose that the principal codimension lies between these bounds and T is Euclidean. Then all h_{ij} must satisfy some linear equations, and one of these $h_{ij}\xi^{ij} = 0$ can be put in canonical form by a suitable orthogonal transformation of the basis $\{e_1, \ldots, e_m\}$. After suitable renumbering, this equation can be presented as

$$h_{mm} = \mu_1 h_{11} + \cdots + \mu_{m-1} h_{m-1\,m-1} = \sum_{a=1}^{m-1} \mu_a h_{aa}. \tag{4.4.15}$$

The lower bound for the principal codimension shows that all h_{ij} with $i \neq j$ must be linearly independent and $h_{11}, h_{22}, \ldots, h_{m-1\,m-1}$ cannot be mutually collinear; moreover none of the latter can be a linear combination of the former.

Recall that A_H is defined by $\langle A_H X, Y\rangle = \langle h(X, Y), H\rangle$ (see Section 4.2), so that for every orthonormal basis in T one has $(A_H)_{ij} = \langle h_{ij}, H\rangle$.

Let us first consider the case $r = 1$ in Theorem 4.3.1. Then $(A_H)_{ij} = \lambda_{(1)}\delta_{ij}$, or, equivalently, $\langle h_{ij}, H\rangle = \lambda_{(1)}\delta_{ij}$.

Hence, if one takes the scalar product of H with (4.4.15), one obtains

$$\lambda_{(1)} = (\mu_1 + \cdots + \mu_{m-1})\lambda_{(1)}.$$

Now if $\lambda_{(1)} = 0$, then $\langle h_{ij}, H\rangle = 0$, and therefore $\langle H, H\rangle = 0$. Since the principal normal subspace is assumed to be Euclidean, this implies $H = 0$, which implies $h_{ij} = 0$, due to the argument in the proof of Theorem 4.1.7. But this conclusion contradicts the assumption about the principal codimension. Therefore, $\lambda_{(1)} = 0$ is impossible, and hence $\mu_1 + \cdots + \mu_{m-1} = 1$ holds.

In (4.4.14), let us put $i = j = k = a, l = b \neq a$ (note that here $\epsilon_a = \epsilon_b = 1$). This gives a linear relation among h_{aa} and the h_{ij} with $i \neq j$; hence all coefficients must be zero. In particular, for the coefficient of h_{ab} this implies that

$$3H_{aa,bb} - 2H_{ab,ab} - H_{aa,aa} = 0,$$

for every pair of distinct a and b. Therefore, $H_{aa,aa} = H_{bb,bb}$, and consequently

$$\|h_{11}\|^2 = \|h_{22}\|^2 = \cdots = \|h_{m-1\,m-1}\|^2 = \sigma^2.$$

Now take (4.4.14) with $i = j = a$, $k = b \neq a$, and $l = m$; then the coefficient before h_{bm} is

$$\mu_a\sigma^2 + (\mu_b - 1)H_{aa,bb} + \mu_c H_{aa,cc} + \sum \mu_d H_{aa,dd} = 0;$$

here the value of index c is distinct from a and b (which is possible since $m \geq 3$), and summing is over all values d distinct from a, b, c (which, of course, is possible only if $m > 3$, for otherwise the range of d is empty). Here the roles of b and c can be interchanged, and hence $H_{aa,bb} = H_{aa,cc}$. As a result $H_{aa,bb} = \tau$ for all a, b. The last equation above now reduces to

$$\mu_a\sigma^2 + \left(\sum_{b\neq a} \mu_b - 1\right)\tau = 0,$$

which due to $\mu_1 + \mu_2 + \cdots + \mu_{m-1} = 1$ gives $\mu_a(\sigma^2 - \tau) = 0$ for every a. Therefore, $\sigma^2 - \tau = 0$ and thus $\langle h_{aa}, h_{aa} - h_{bb}\rangle = 0$, $\langle h_{bb}, h_{aa} - h_{bb}\rangle = 0$. This leads to $\|h_{aa} - h_{bb}\| = 0$, and hence to $h_{aa} = h_{bb}$ for every pair $a \neq b$. This is a contradiction to the above statement that $h_{11}, h_{22}, \ldots, h_{m-1\,m-1}$ are not mutually collinear. This finishes the proof for $r = 1$.

If $r > 1$, then by Theorem 4.3.1, (V, T, h) is the orthogonal direct sum of its subtriplets and $h_{ij} = h_{i_1 j_1} + h_{i' j'}$, where the terms on the right-hand side are orthogonal. Therefore, the maximal value of the principal normal codimension is

$$\frac{1}{2}m_1(m_1 + 1) + \frac{1}{2}(m - m_1)(m - m_1 + 1) = \frac{1}{2}m(m-1) + m(1 - m_1) + m_1^2,$$

where $1 \leq m_1 \leq \frac{m}{2}$; but $m \geq 3$ implies that this is less than $\frac{1}{2}m(m-1) + 2$. This completes the proof.

Remark 4.4.4. There is a direct connection between the result (4.4.13) of Proposition 4.4.1 and the result (3.3.4) obtained in the argument which led to Theorem 3.3.1. These results were first obtained in [Lu 89c]. Propositions 4.4.1 and 4.4.3 were derived in [Lu 92b] by purely algebraic arguments similar to those given above.

4.5 Semiparallel Submanifolds as Second-Order Envelopes of Parallel Submanifolds

As noted in Section 4.2, the semiparallel condition is simpler than the parallel condition: the first is algebraic, the second is a differential system. But from the geometrical viewpoint, on the other hand, the parallel submanifolds are simpler, because each (complete) parallel M^m is an orbit of a Lie group of isometries of ${}_sN^n(c)$, generated by the reflections in ${}_sN^n(c)$ with respect to the normal $(n–m)$-planes of M^m. Indeed, for every two points x and x' of such an M^m, there exists an isometry of ${}_sN^n(c)$ that maps M^m into itself and x into x', namely the reflection with respect to the $(n–m)$-plane normal to M^m at the midpoint x_0 of the geodesic between x and x'. (See Section 3.6, where this viewpoint is described in more detail.)

The aim of the present section is to show that semiparallel submanifolds can be characterized geometrically as second-order envelopes of families of parallel submanifolds. Second-order envelopes can be defined as follows (cf. [Je 77]).

Two curves λ and λ^* in ${}_sN^n(c)$ (the images of two smooth maps of some interval of the real line $\mathbb{R}$ into ${}_sN^n(c)$) are said to be first-order tangent at a common point x_0, corresponding to $t = 0$, if their tangent vectors at x_0 coincide. They are said to be second-order tangent at x_0 if, in addition, their curvature vectors $h(X^0, X^0)$ at x_0 coincide, where λ and λ^* are considered as one-dimensional submanifolds, and X^0 is their common unit tangent vector at x_0.

Two submanifolds M^m and M^{*m} with a common point x_0 in ${}_sN^n(c)$ are said to be *first-order* (or *second-order*) *tangent* at x_0, if for every curve λ through x_0 in M^m there is a curve λ^* through the same x_0 in M^{*m}, which is first-order (resp. second-order) tangent to λ at x_0.

Obviously, first-order tangency means that the tangent m-planes of these submanifolds at x_0 coincide. For second-order tangency, the following holds.

Proposition 4.5.1. *Two (pseudo-)Riemannian submanifolds M^m and M^{*m} of ${}_sN^n(c)$ are second-order tangent at a common point x_0 if and only if their fundamental triplets at x_0 coincide.*

Proof. Suppose these triplets coincide at x_0. Let λ be a curve in M^m through x_0 having arclength parameter s. The formulas in the proof of Theorem 3.5.1 show that $x(s)$ has unit tangent vector $\frac{d}{ds}x = e_iX^i$ of λ, and its curvature vector $\frac{d^2}{ds^2}x$ has normal (to M^m) component $h_{ij}X^iX^j$ (called the normal curvature vector of λ) and tangent component $\frac{1}{ds}e_i\nabla X^i$ (called the geodesic curvature vector of λ). It suffices now to consider in M^{*m} the curve λ^* through x_0 with the same unit tangent vector e_iX^i and with the same geodesic curvature vector at x_0 as λ above. Since by hypothesis the h_{ij}

are the same for M^m and M^{*m} at x_0 and X^i are also the same, the curvature vectors of λ and λ^* at x_0 coincide; thus these curves are second-order tangent at x_0. Since X^i can be chosen arbitrarily, the two submanifolds are second-order tangent at x_0.

The validity of the converse statement is clear.

Now the following definition can be given.

Suppose for every point x of a submanifold M^m of ${}_sN^n(c)$ there exists a submanifold M^{*m} in ${}_sN^n(c)$ that is second-order tangent to M^m at x. Then M^m is said to be the *second-order envelope* of the family of these submanifolds M^{*m}.

Example 4.5.2. Every curve M^1 with nonvanishing curvature in E^n is the second-order envelope of the family of its circles of curvature. Note that a circle is a one-dimensional parallel submanifold in some $E^2 \subset E^n$ (see Section 3.1). Here a certain degeneration is possible: if M^1 is a circle, then this family consists of the circle itself. But in general, second-order envelopes of circles are nontrivial, i.e., they do not reduce to a single circle.

Example 4.5.3. m-dimensional spheres in E^n are parallel submanifolds. If $m > 1$, there do not exist any nontrivial second-order envelopes of families of m-dimensional spheres in E^n. Indeed, if M^m and an m-sphere in E^n are second-order tangent at a common point, then this point is an umbilic point of M^m. But it is well known that if a submanifold in E^n consists only of umbilic points, then it is a sphere or a subset of a sphere (see Proposition 3.1.5).

Example 4.5.4. Let M^m be a normally flat submanifold in Euclidean space E^n, i.e., its normal connection $\nabla^\perp$ is flat, or equivalently, $\Omega_\alpha^\beta = 0$ at each point. From the last formulas of (2.1.9) and (2.1.10) for the bundle of adapted orthonormal frames, one can see that for every pair of distinct values α and β the matrices (h_{ij}^α) and (h_{kl}^β) commute and therefore can be simultaneously diagonalized by a suitable orthogonal transformation of the frame at every point. Then

$$h_{ij}^\alpha = k_i^\alpha \delta_{ij}$$

for every value of α. In general one obtains m mutually orthogonal basis vector fields whose integral curves are called the *lines of curvature* of the sumbmanifold M^m with flat normal connection $\nabla^\perp$.

Suppose, furthermore, that this M^m is also locally Euclidean, i.e., that $\Omega_i^j = 0$. Since here $c = 0$, the first formulas of (2.1.9) and (2.1.10) give $\langle k_i, k_j \rangle = 0$ for every pair of distinct values i, j, where $k_i = e_\alpha k_i^\alpha$. Suppose further that these m mutually orthogonal vectors, which are the curvature vectors of the lines of curvature, are nonzero; then $n \geq 2m$. It follows that this M^m is a second-order envelope of the tori $S^1(c_1) \times \cdots \times S^1(c_m)$, where the m circles $S^1(c_1), \ldots, S^1(c_m)$ are the circles of curvature at a point $x \in M^m$ of the lines of curvature of M^m. These circles lie on mutually orthogonal 2-planes.

In the particular case $m = 2$, these tori are Clifford tori. If one family of lines of curvature of this M^2 consists of circles (i.e., M^2 is a canal surface in E^n, $n \geq 4$), then the family of Clifford tori, each of which second-order envelops M^2, depends on one parameter. This family of Clifford tori reduces to M^2 itself if M^2 already is a Clifford torus.

These examples illustrate the following general assertion.

Theorem 4.5.5. *A submanifold M^m in ${}_sN^n(c)$ is semiparallel if and only if it is a second-order envelope of parallel submanifolds.*

Proof. Consider the pair consisting of a point $x \in {}_sN^n(c)$ and a fundamental triplet (V, T, h), where $V = T_x[{}_sN^n(c)]$ and T is (pseudo-)Euclidean T of $\dim T = m$. Let such a pair of x and (V, T, h) be called a *centered fundamental triplet* for ${}_sN^n(c)$. Denote by Φ the manifold formed by all centered triplets. For each such triplet, if one chooses an adapted frame, having the origin at x, the first m basis vectors e_i in T, and the next $n-m$ basis vectors e_α in $T^\perp$, one gets a *framed fundamental triplet* for ${}_sN^n(c)$. The manifold consisting of all these triplets will be denoted by Ψ. Local coordinates in Ψ are obtained by taking the local coordinates (x^I) of the point $x \in {}_sN^n(c)$, the elements of the nonsingular matrix (X_I^J) that transforms the natural basis of $\partial/\partial x^I$ into the basis adapted to (V, T, h) as above, and the components h_{ij}^α of h with respect to the basis just chosen.

The formulas (1.2.4), (1.4.2), and (1.4.3) hold, where now $g_{i\alpha} = 0$.

Let Ψ_S (resp. Φ_S) be the manifold of all framed (resp. centered) *semiparallel* fundamental triplets for ${}_sN^n(c)$. Consider the following differential system on Ψ_S:

$$\omega^\alpha = 0, \quad \omega_i^\alpha - h_{ij}^\alpha \omega^j = 0, \quad dh_{ij}^\alpha - h_{kj}^\alpha \omega_i^k - h_{ik}^\alpha \omega_j^k + h_{ij}^\beta \omega_\beta^\alpha = 0.$$

Taking exterior differentials of the left-hand sides of these equations, one sees that they vanish, due to the equations of the same system and the equations for the semiparallel condition. Thus the differential system is completely integrable (see the Frobenius' theorem, second variant, in [Ste 64], also [Li 55], Section 21).

Two framed fundamental triplets are said to be equivalent if $e_i' = e_j A_i^j$, $e_\alpha' = e_\beta A_\alpha^\beta$, with invertible basis change matrices, and ${}'h_{kl}^\beta = A_\alpha^\beta h_{ij}^\alpha A_k^i A_l^j$. This equivalence defines a mapping $\Psi_S \to \Phi_S$ that maps the above differential system into a well-defined completely integrable system on Φ_S, as can be easily seen.

It follows that for every fixed centered fundamental semiparallel triplet in Φ_S there is a unique integral submanifold of this differential system, that contains the fixed triplet and has maximal possible dimension, that is equal to the dimension of the involutive distribution on Φ_S corresponding to the differential system.

The first two groups of equations of this differential system show that the integral submanifold in Φ_S consists of centered fundamental triplets that belong to an m-dimensional submanifold M^{*m} in ${}_sN^n(c)$. The last group of equations shows that this M^{*m} is a parallel submanifold.

Now let M^m be a semiparallel submanifold in ${}_sN^n(c)$. Every fixed point x of M^m defines a centered semiparallel fundamental triplet $(T_x[{}_sN^n(c)], T_xM^m, h_x)$, and the fundamental triplet defines a parallel submanifold M^{*m} in ${}_sN^n(c)$, as is shown above. By Proposition 4.5.1, the submanifolds M^m and M^{*m} are second-order tangent at x, and hence M^m is the second-order envelope of the family of these M^{*m}, taken at all points x of M^m.

Remark 4.5.6. The fact that a centered semiparallel fundamental triplet defines a unique parallel submanifold was proved in [BR 83] (see Theorem 3) by using an algebraic approach (for the Riemannian case when $s = 0$). The proof is rather cumbersome, involving techniques from [Fe 80] and [Ba 83] developed for the Euclidean case when $c = 0$, and using certain constructions leading from Jordan triple systems to Lie groups. The proof given here above is more direct, following the Riemannian-case proof given in [Lu 90a] and generalizing it to the pseudo-Riemannian case considered here.

4.6 Second-Order Envelope of Segre Submanifolds

A new interesting example, in addition to Examples 4.5.2–4.5.4, is provided by the second-order envelope M^m of Segre submanifolds $S_{(p,q)}(k)$, $m = p+q$, with variable k in $N^n(c)$, $n > m$.

For such an envelope in its adapted orthonormal frame bundle $O(M^m, N^n(c))$ the expressions of dx and de_i must be the same as for $O(S_{(p,q)}(k), S^{pq+m}(k^2))$ (see the proof of Theorem 4.5.5). In the last frame bundle, one can use the same further frame adaption that was used in the proof of Theorem 3.2.1. Additionally, in the bundle $O(S_{(p,q)}(k), E^{pq+m+1})$, the unit frame vector e_{m+1} is specified as $e_{m+1} = kx$.

Then for the bundle of such adapted frames, there holds $dx = e_{i_1}\omega^{i_1} + e_{i_2}\omega^{i_2}$, and due to (2.1.5)

$$de_{i_1} = e_{j_1}\omega_{i_1}^{j_1} + e_{(i_1 j_2)}k\omega^{j_2} + e_{m+1}k\omega^{i_1} - cx\omega^{i_1},$$

$$de_{i_2} = e_{j_1}\omega_{i_2}^{j_2} + e_{(j_1 i_2)}k\omega^{j_1} + e_{m+1}k\omega^{i_2} - cx\omega^{i_2},$$

where x is now the radius vector of the point $x \in M^m$ in the outer ambient space E^{n+1} of $N^n(c)$. So for the submanifold geometry of M^m in $N^n(c)$, one has

$$\omega_{i_1}^{(j_1 k_2)} = \delta_{i_1}^{j_1}k\omega^{k_2}, \quad \omega_{i_2}^{(j_1 k_2)} = \delta_{i_2}^{k_2}k\omega^{j_1}, \quad \omega_i^{m+1} = k\omega^i, \quad \omega_i^{\xi} = 0, \qquad (4.6.1)$$

where $pq + m + 2 \leq \xi, \ldots \leq n + 1$.

After exterior differentiation, the last two equations above give

$$\omega^{i_1} \wedge \omega_{m+1}^{\xi} + \omega^{j_2} \wedge \omega_{(i_1 j_2)}^{\xi} = 0, \quad \omega^{i_1} \wedge \omega_{(i_1 j_2)}^{\xi} + \omega^{j_2} \wedge \omega_{m+1}^{\xi} = 0. \qquad (4.6.2)$$

Consider first the general case where $p > 1$ and $q > 1$. Then both i_1 and j_2 take more than one value. By Cartan's lemma ω_{m+1}^{ξ} is a linear combination of one ω^{i_1} and of all ω^{j_2} (from the first equation 4.6.2), but also of all ω^{i_1} and of one ω^{j_2} (from the second equation 4.6.2). Therefore, $\omega_{m+1}^{\xi} = 0$. The same argument also works for every $\omega_{(i_1 j_2)}^{\xi}$, and thus they are all zero, as well. Consequently, in this case the envelope M^m lies in an $N^{pq+m}(c)$.

After exterior differentiation, the third group of equations (4.6.1) give, separately for i_1 and i_2, that

$$d\ln k \wedge \omega^{i_1} + \sum_{j_2} \omega^{m+1}_{(i_1 j_2)} \wedge \omega^{j_2} = 0,$$

$$\sum_{i_1} \omega^{m+1}_{(i_1 j_2)} \wedge \omega^{i_1} + d\ln k \wedge \omega^{j_2} = 0.$$

The same argument just used above now gives

$$k = \text{const}, \quad \omega^{m+1}_{(i_1 j_2)} = 0. \tag{4.6.3}$$

The first group of equations (4.6.1) leads to

$$\sum_{l_1} (\delta^{j_1}_{i_1} \omega^{k_2}_{l_1} + \omega^{k_1}_{i_1} \delta^{j_1}_{l_1}) \wedge \omega^{l_1} + \sum_{l_2} \left(\delta^{j_1}_{i_1} \omega^{k_2}_{l_2} + \omega^{j_1}_{i_1} \delta^{k_2}_{l_2} - \omega^{(j_1 k_2)}_{(i_1 l_2)} \right) \wedge \omega^{l_2} = 0. \tag{4.6.4}$$

Now one shows via Cartan's lemma that $\omega^{k_2}_{i_1}$ is a linear combination of only ω^{l_2}. Indeed, taking $i_1 = j_1$ in (4.6.4), one obtains

$$2\omega^{k_2}_{i_1} \wedge \omega^{i_1} + \sum_{l_1 \neq i_1} \omega^{k_2}_{l_1} \wedge \omega^{l_1} + \sum_{l_2} (\omega^{k_2}_{l_2} - \omega^{(i_1 k_2)}_{(i_1 l_2)}) \wedge \omega^{l_2} = 0, \tag{4.6.5}$$

and thus

$$\omega^{k_2}_{i_1} = U^{k_2}_{i_1} \omega^{i_1} + \sum_{l_1 \neq i_1} V^{k_2}_{i_1 l_1} \omega^{l_1} + \sum_{l_2} W^{k_2}_{i_1 l_2} \omega^{l_2},$$

for an arbitrary fixed value i_1 and with $V^{k_2}_{i_1 l_1}$ symmetric with respect to the lower indices. Substituting this back into (4.6.5) gives

$$\sum_{l_1 \neq i_1} \left[2V^{k_2}_{i_1 l_1} \omega^{l_1} \wedge \omega^{i_1} + \sum_{k_1 \neq l_1} V^{k_2}_{l_1 k_1} \omega^{k_1} \wedge \omega^{l_1} \right] + (\ldots)^{k_2}_{l_2} \wedge \omega^{l_2} = 0.$$

This implies $V^{k_2}_{i_1 l_1} = 0$.

Taking $i_1 \neq j_1$ in (4.6.4) and substituting the above expression for $\omega^{k_2}_{i_1}$, one finds that

$$U^{k_2}_{i_1} \omega^{i_1} \wedge \omega^{j_1} + (\cdots)^{j_1 k_2}_{i_1 l_2} \wedge \omega^{l_2} = 0,$$

thus also $U^{k_2}_{i_1} = 0$, which proves the assertion about $\omega^{k_2}_{i_1}$.

Here the subindices 1 and 2 are in equivalent roles; therefore $\omega^{i_1}_{k_2}$ can be expressed in terms of ω^{l_1} only. Since $\omega^{k_2}_{i_1} + \omega^{i_1}_{k_2} = 0$, it follows that

$$\omega^{k_2}_{i_1} = 0. \tag{4.6.6}$$

Now putting $i_1 = j_1$ in (4.6.4) gives

$$\omega_{l_2}^{k_2} - \omega_{(i_1 l_2)}^{(i_1 k_2)} = Q_{l_2 j_2}^{k_2} \omega^{j_2},$$

where $Q_{l_2 j_2}^{k_2} = Q_{j_2 l_2}^{k_2}$. On the other hand, $Q_{l_2 j_2}^{k_2} = -Q_{k_2 j_2}^{l_2}$, and thus $Q_{l_2 j_2}^{k_2} = Q_{j_2 l_2}^{k_2} = -Q_{k_2 l_2}^{j_2} = -Q_{l_2 k_2}^{j_2} = Q_{j_2 k_2}^{l_2} = Q_{k_2 j_2}^{l_2} = -Q_{l_2 j_2}^{k_2}$; hence $Q_{l_2 j_2}^{k_2} = 0$, and so

$$\omega_{(i_1 l_2)}^{(i_1 k_2)} = \omega_{l_2}^{k_2} \tag{4.6.7}$$

for every fixed value of i_1. Here the roles of the subindices 1 and 2 can be exchanged, giving

$$\omega_{(l_1 i_2)}^{(k_1 i_2)} = \omega_{l_1}^{k_1}. \tag{4.6.8}$$

Now (4.6.4) with $i_1 \neq j_1$ reduces to

$$\sum_{l_2 \neq k_2} \omega_{(i_1 l_2)}^{(j_1 k_2)} \wedge \omega^{l_2} = 0;$$

therefore $\omega_{(i_1 l_2)}^{(j_1 k_2)}$ with $i_1 \neq j_1$ and $k_2 \neq l_2$ can be expressed in terms of ω^{h_2} only, and similarly, by ω^{h_1} only. Hence

$$\omega_{(i_1 l_2)}^{(j_1 k_2)} = 0 \quad (i_1 \neq j_1, k_2 \neq l_2). \tag{4.6.9}$$

Taking the exterior derivative of (4.6.6) one concludes, via (1.6.3) and (4.6.1), that $c\,\omega^{i_1} \wedge \omega^{k_2} = 0$, which implies $c = 0$. Thus the particular kind of second-order enveloping of M^m considered in this case can occur only in a locally Euclidean ambient space. The analysis of the system $\omega^{(i_1 j_2)} = \omega^{m+1} = 0$, (4.6.1), (4.6.3), and (4.6.6)–(4.6.9) shows that this system is completely integrable and determines a Segre submanifold $S_{(p,q)}(k)$ in $S^{pq+p+q}(k^2)$.

The result is a bit different if $p = 1, q > 1$. Then i_1 takes only one value 1; hence it is suitable here to use $a, b, \ldots$ instead of $i_2, j_2, \ldots$, and also $m + a$ instead of the index pair $(1a)$. Equations (4.6.1) now appear as

$$\omega_1^{m+a} = k\omega^a, \quad \omega_b^{m+a} = \delta_b^a k\omega^1, \quad \omega_1^{m+1} = k\omega^1, \quad \omega_a^{m+1} k\omega^a,$$
$$\omega_i^{\xi} = 0. \tag{4.6.10}$$

From (4.6.2) it now follows by Cartan's lemma that

$$\omega_{m+1}^{\xi} = \psi^{\xi}\omega^1, \quad \omega_{m+a}^{\xi} = \psi^{\xi}\omega^a. \tag{4.6.11}$$

Similarly, instead of (4.6.3) one now has

$$d \ln k = \kappa\omega^1, \quad \omega_{m+a}^{m+1} = \kappa\omega^a. \tag{4.6.12}$$

Exterior differentiation of the equations $\omega_1^{m+a} = k\omega^a$, $\omega_b^{m+a} = \delta_b^a k\omega^1$ gives

$$2(\omega_1^a + \kappa\omega^a) \wedge \omega^1 + \sum_b (\omega_b^a - \omega_{m+b}^{m+a}) \wedge \omega^b = 0, \tag{4.6.13}$$

$$(\omega_b^a - \omega_{m+b}^{m+a}) \wedge \omega^1 - (\omega_1^b + \kappa\omega^b) \wedge \omega^a - \delta_b^a \sum_c \omega_1^c \wedge \omega^c = 0. \tag{4.6.14}$$

As in the previous case, substituting first $a = b$ and then $a \neq b$ into the exterior equations (4.6.14) and (4.6.13), one gets

$$\omega_1^a = -\kappa\omega^a, \quad \omega_{m+b}^{m+a} = \omega_b^a. \tag{4.6.15}$$

Now, taking exterior derivatives of the second group of equations (4.6.12) gives $(d\kappa - \kappa^2\omega^1) \wedge \omega^a = 0$ for all values of a; hence $d\kappa = \kappa^2\omega^1$. The first equation (4.6.12) leads to an identity, but applying the same procedure to the first group of equations (4.6.15), and using (2.1.7) and (2.1.8), gives the result $c\,\omega^1 \wedge \omega^a = 0$. Hence $c = 0$.

Finally, taking exterior derivatives of the second group of equations (4.6.15), one obtains

$$\left[\sum_\xi (\psi^\xi)^2\right] \omega^a \wedge \omega^b = 0,$$

thus $\sum_\xi (\psi^\xi)^2 = 0$, so $\psi^\xi = 0$ and hence $\omega_{m+1}^\xi = \omega_{m+a}^\xi = 0$. This shows that the whole system is completely integrable. Therefore, in this case the second-order enveloping of M^m occurs only in a $2q+2$-dimensional locally Euclidean manifold, and it is determined up to some arbitrary real constants.

It remains to describe the geometrical construction of M^m in this case. Equations (4.6.15) imply that $d\omega^1 = 0$; therefore, at least locally, $\omega^1 = ds$ for some function s on M^m. For submanifolds with $s = \text{const}$, one has

$$dx = e_a\omega^a, \quad de_a = e_b\omega_a^b + (\kappa e_1 + k e_{m+1})\omega^a;$$

therefore they are $(m-1)$-dimensional spheres or their open subsets. The center of such a sphere has the radius vector $y = x + (\kappa^2 + k^2)^{-1}(\kappa e_1 + k e_{m+1})$. An easy calculation shows that $dy = 0$, and hence all these spheres have a common center.

The orthogonal trajectories of all these concentric spheres are defined by $\omega^a = 0$, and for each of them

$$dx = e_1 ds, \quad de_1 = k e_{m+1} ds, \quad de_{m+1} = -k e_1 ds.$$

It is easily seen that they are plane curves with curvature k. The previously derived equation $d\kappa = \kappa^2\omega^1$, where $\omega^1 = ds$, implies, for $\kappa \neq 0$, that $\kappa = -s^{-1}$. Now from the first equation (4.6.12) it follows that $k = Cs^{-1}$, where $C = \text{const}$. Hence all these orthogonal trajectories are congruent logarithmic spirals.

It is well known that a logarithmic spiral with polar equation $\rho = a^\varphi$ in the Euclidean plane has curvature $k = s^{-1}\tan\mu$, where s is the arclength from the pole, $\tan\mu = (\ln a)^{-1}$, and μ is the constant angle between the unit tangent vector and the radius vector x. In this notation $\rho = s\cos\mu$ and the pole has the radius vector $x + s\cos\mu(n\sin\mu - t\cos\mu)$, where t and n are the tangent and normal unit vectors of this spiral, respectively. If one writes this radius vector in the form

$x + s^2 \cos^2 \mu[e_{m+1}(s^{-1} \tan \mu) - s^{-1} e_1] = x + (k^2 + \kappa^2)^{-1}(e_{m+1}k + e_1\kappa)$, one sees that all these logarithmic spirals have their pole at the common center of the concentric generating $(m-1)$-dimensional spheres.

An M^m satisfying such a geometrical construction is suitably called a *logarithmic spiral tube*.

In the special case when $\kappa \equiv 0$, one has $k = \text{const}$, and the enveloping family of Segre submanifolds reduces to a single $S_{(1,q)}(k)$.

Finally, one notes that $S_{(1,1)}(k)$ is a torus $S^1(k) \times S^1(k)$ and hence the second-order envelope is a surface M^2 with flat $\bar{\nabla}$ (see Example 4.5.4).

The results of this investigation can be summarized as follows.

Theorem 4.6.1. *A semiparallel submanifold M^m in $N^n(c)$, which is a second-order envelope of Segre submanifolds $S_{(p,q)}(k)$ with $m = p + q$ and variable k, is either*

- *a single Segre submanifold with constant k, or its open subset, in a $(pq + m)$-dimensional submanifold of constant curvature k^2, embedded into $N^n(c)$, or*
- *for $p = q = 1$, a surface M^2 with flat $\bar{\nabla}$, or*
- *for $p = 1$, $q > 1$, a logarithmic spiral tube, or its open subset, in a $(2q + 2)$-dimensional locally Euclidean submanifold, embedded into $N^n(c)$.*

Remark 4.6.2. Observe that if $p > 1$ and $q > 1$, only the first case is possible: such an envelope is a single Segre submanifold, and the submanifold of constant curvature k^2 could be $N^n(c)$ itself (for example, when $n = pq + m$ and $c = k^2$, an $S^{pq+m}(k^2)$), or it could be a spherical submanifold of curvature k^2 in $N^n(c)$, $k^2 > c$.

In the third case, the ambient locally Euclidean submanifold can be $N^n(c)$ itself, when $n = 2q + 1$ and $c = 0$ (thus a E^{2q+1}), or such a submanifold $\tilde{N}^{2q+1}(0)$ in $N^n(c)$ (e.g., a horosphere in $H^{2q+2}(c)$).

Remark 4.6.3. For the case $c = 0$, Theorem 4.6.1 was previously established in [Lu 91a]. Now it is extended here to the case $c \neq 0$.

4.7 A New Approach to Veronese Submanifolds

An m-dimensional Veronese submanifold in ${}_sN^n(c) \subset {}_\sigma E^{n+1}$ was introduced in Section 3.3 as a not totally umbilic submanifold, which lies in its first-order outer osculating subspace of maximal possible dimension $\frac{1}{2}m(m+3)$, so that $n + 1 = \frac{1}{2}m(m+3)$, and, moreover, all its inner isometries are induced by the isometries of ${}_sN^n(c)$ (see Theorem 3.3.1). Such a submanifold has either a positive or negative definite inner metric and is characterized by the relation (3.3.4), where $\varepsilon\alpha$ is its constant curvature, and $\varepsilon = \pm 1$.

Note that (3.3.4) is a special case of (4.4.13). This leads one to consider semiparallel submanifolds M^m in ${}_sN^n(c) \subset {}_\sigma E^{n+1}$, whose first-order outer normal subspace is of maximal possible dimension $\frac{1}{2}m(m+1)$ at every point $x \in M^m$.

Proposition 4.4.1 implies that (4.4.13) holds for these submanifolds, where, as one recalls, $H^*_{ij,kl} = \langle h^*_{ij}, h^*_{kl} \rangle$. The first formulas (2.1.9) and (2.1.10) now imply $\Omega_i^j = \varepsilon_i \kappa \omega^i \wedge \omega^j$ (no sum!), where κ is a function on the submanifold.

From the first formula (2.1.8) it follows by exterior differentiation that

$$d\Omega_i^j = \omega_i^k \wedge \Omega_k^j - \Omega_i^k \wedge \omega_k^j$$

(the Bianchi identity; cf. (1.3.5)). Substituting into this the above expression for Ω_i^j, denoting $d\kappa = \kappa_k \omega^k$ and noting that due to (2.1.2) $\varepsilon_j \omega_i^j + \varepsilon_i \omega_j^i = 0$, one obtains $\sum_k \kappa_k \omega^k \wedge \omega^i \wedge \omega^j = 0$. If $m = 2$, i.e., if the submanifold is a surface, then this is an identity. But if $m > 2$, then the last equality implies $\kappa_k = 0$ and thus $\kappa = \text{const}$.

As a result, the following proposition can be stated.

Proposition 4.7.1. *If a semiparallel submanifold M^m of dimension $m \geq 3$ in ${}_sN^n(c)$ has first-order outer normal subspace of maximal possible dimension $\frac{1}{2}m(m+1)$, then it is intrinsically a (pseudo-)Riemannian space of constant curvature and* (4.4.13) *holds, where κ is a constant.*

Suppose now that this submanifold lies in its first-order outer osculating subspace, the dimension of which is, of course, $\frac{1}{2}m(m+3)$.

A consequence from (4.4.13) is that all $\langle h_{ij}^*, h_{kl}^* \rangle$ are constants on this submanifold M^m. Differentiating these expressions and using (3.1.2), where $e_\alpha h_{ijk}^\alpha = h_{ijk}$ belong now to the first-order outer normal subspace $\operatorname{span}\{h_{pq}^*\}$ at a fixed point $x \in M^m$, one obtains after some calculations

$$\langle h_{ijp}, h_{kl}^* \rangle + \langle h_{ij}^*, h_{klp} \rangle = 0.$$

Recall that due to (2.2.3), the h_{klp} are symmetric with respect to their indices. Therefore,

$$\langle h_{ijp}, h_{kl}^* \rangle = -\langle h_{klp}, h_{ij}^* \rangle = -\langle h_{kpl}, h_{ij}^* \rangle = \langle h_{ijl}, h_{kp}^* \rangle.$$

This shows that every index of the first triplet can be exchanged by every index of the second pair. Hence

$$\langle h_{ijp}, h_{kl}^* \rangle = \langle h_{klp}, h_{ij}^* \rangle = -\langle h_{ijp}, h_{kl}^* \rangle,$$

and so $\langle h_{ijp}, h_{kl}^* \rangle = 0$. Therefore, every h_{ijp} is orthogonal to all h_{kl}^*.

Suppose now that the first-order outer osculating subspace containing the submanifold M^m has a regular metric, i.e., is (pseudo-)Euclidean. Then the last assertion implies that $h_{ijp} = 0$ for all values of i, j, p. (Recall that a (pseudo-)Riemannian submanifold M^m in ${}_sN^n(c)$, whose first-order osculating subspaces have constant dimension and regular metric, is said to be *regular*; see Section 2.4.)

The result can be formulated as follows.

Theorem 4.7.2. *If a semiparallel regular (pseudo-)Riemannian submanifold M^m of dimension $m \geq 3$ in ${}_sN^n(c)$ lies in its first-order osculating subspace of maximal possible dimension $\frac{1}{2}m(m+3)$, then M^m is a parallel submanifold, has constant curvature, and satisfies equation* (4.4.13).

A parallel submanifold fitting this Theorem 4.7.2 is called (especially if it is complete) a (*general*) *Veronese submanifold*. The Veronese submanifolds of Theorem 3.3.1 are the special cases with positively or negatively definite inner metric. Recall that for the latter all of their inner isometries are induced by the isometries of the ambient space.

Remark 4.7.3. In algebraic geometry (see [Sha 88], Chapter I.4) the Veronese map v_2 between projective spaces means the following:

$$v_2 : P^m \longrightarrow P^{\frac{1}{2}m(m+3)}, \quad v_2(u_0, \dots, u_m) = (v_{ij}), \quad v_{ij} = u_i u_j,$$

where $0 \leq i, j \leq m$. If the real projective space P^m (resp. $P^{\frac{1}{2}m(m+3)}$) is considered as the manifold of one-dimensional vector subspaces of $\mathbb{R}^{m+1}$ (resp. of $\mathbb{R}^{\frac{1}{2}(m+1)(m+2)}$), then v_2 can be considered as a map $\mathbb{R}^{m+1} \longrightarrow \mathbb{R}^{\frac{1}{2}(m+1)(m+2)}$. After putting a (pseudo-)Euclidean metric on $\mathbb{R}^{m+1}$, this P^m can also be interpreted as a ${}_sS^m(c)$ or ${}_sH^m(c)$ with antipodal points identified (see Section 1.4). In this metric version, Veronese submanifolds were considered, e.g., in [Blo 86]. As mentioned above, the submanifolds of Theorem 3.3.1 are the special cases for $s = 0$ or $s = m$.

Note that for $s = 0$ and $m = 2$, Veronese surfaces were first studied in [Bor 28], and then for $m > 2$ in [Bla 53], [So 61] (see also [It 75]).

In [Lu 62] it was shown for a minimal surface in $N^4(c)$ that if either (i) the Gaussian curvature or (ii) the area of the normal curvature ellipse is constant, then this surface is a Veronese surface in $S^4(c)$. In [Br 85] it was shown that a minimal surface with Gaussian curvature $\frac{1}{3}$ in $S^n(1)$ is a Veronese surface in $S^4(1)$.

A Veronese submanifold can be also characterized by the second standard immersion of an m-dimensional sphere into a $\frac{1}{2}m(3m+1)$-dimensional sphere (see, e.g., [CW 71]).

Remark 4.7.4. Theorem 4.7.2 was first derived in [Lu 89c] for the case $s = c = 0$ and then in [Lu 96a] for the general case (see also [Lu 2000a], Chapter 19, for the case $s = 0$, $c \neq 0$). Meanwhile the same result was established for surfaces in a space form in [As 93] and [AM 94].

Remark 4.7.5. In connection with Proposition 4.7.1, the question arises whether there exist submanifolds fitting this proposition, but not Theorem 4.7.2 (i.e., not Veronese submanifolds). This question will be given a positive answer in Section 9.7: there exist such submanifolds, and they are second-order envelopes of families of congruent Veronese submanifolds (see Theorem 9.7.1 and Proposition 9.7.2).

5

Normally Flat Semiparallel Submanifolds

One of the main problems in the theory of semiparallel submanifolds M^m in $N^n(c)$ is how to classify all of them. Up to now, this classification problem has been solved only in certain particular cases. For an arbitrary dimension m, it has been done only for normally flat submanifolds: for the case $c = 0$, when $N^n(0) = E^n$ is a Euclidean space, in [Lu 89b], [Lu 90d, e], [Lu 91d, f] (see also [Ri 88], [Lu 2000a], [Li 2001]); and for the case $c \neq 0$, partly in [DN 93].

5.1 Principal Curvature Vectors and the Semiparallel Condition

A submanifold M^m in $N^n(c)$ is said to be *normally flat* if its normal connection $\nabla^\perp$ is flat, i.e., the curvature 2-forms Ω_α^β of $\nabla^\perp$ vanish: $\Omega_\alpha^\beta = 0$.

From (2.1.7) and (2.1.8) it then follows that every pair of matrices $\|h_{ij}^\alpha\|$ and $\|h_{ij}^\beta\|$ ($\alpha \neq \beta$) commute, and therefore they can all be simultaneously diagonalized by a suitable change of orthonormal basis in T_xM^m at an arbitrary point $x \in M^m$. Then $h_{ij}^\alpha = k_i^\alpha \delta_{ij}$, or, in vector form, $h_{ij} = k_i \delta_{ij}$. The vectors $k_1, \dots, k_m$ are called the *principal curvature vectors* of the normally flat submanifold M^m in $N^n(c)$; they are the natural generalizations of the principal curvatures of a hypersurface M^{n-1} in $N^n(c)$. The *outer principal curvature vectors* are defined as $k_i^* = k_i - xc$, and they are the principal curvature vectors of M^m as a submanifold of ${}_\sigma E^{n+1}$. Due to (1.6.2), (2.1.1)–(2.1.3), (2.4.1), (2.2.2), and (2.2.3) the following equations hold:

$$dx = \sum_{i=1}^m e_i \omega^i, \quad de_i = \sum_{j=1}^m e_j \omega_i^j + k_i^* \omega^i, \tag{5.1.1}$$

$$dk_i^* = -\sum_{j=1}^m e_j \langle k_i^*, k_j^* \rangle \omega^j + K_i \omega^i + \sum_{j \neq i} L_{ij} \omega^j, \tag{5.1.2}$$

$$(k_i^* - k_j^*) \omega_i^j = L_{ij} \omega^i + L_{ji} \omega^j + \sum_{l \neq i}^{l \neq j} E_{ijl} \omega^l, \quad i \neq j, \tag{5.1.3}$$

Ü. Lumiste, *Semiparallel Submanifolds in Space Forms*,
DOI 10.1007/978-0-387-49913-0_6, © Springer Science+Business Media, LLC 2009

where $K_i = h_{iii}$, $L_{ij} = h_{iij}$ $(i \neq j)$, $E_{ijl} = h_{ijl}$ $(i, j, l$ have three distinct values) and summing occurs only in terms containing the summation sign $\sum$; moreover, (5.1.2) and (5.1.3) remain valid when k_i, k_j are put in place of k_i^*, k_j^*, respectively.

Theorem 5.1.1. *A normally flat submanifold M^m in $N^n(c)$ is semiparallel if and only if its outer principal curvature vectors are either equal or pairwise orthogonal.*

Proof. In the above specialized frame bundle, (2.1.10) and (2.1.11) give

$$\Omega_i^j = -\langle k_i^*, k_j^* \rangle \omega^i \wedge \omega^j, \qquad \Omega_\alpha^\beta = 0; \tag{5.1.4}$$

thus the semiparallel condition (4.1.2) reduces to

$$(k_i^* - k_j^*)\langle k_i^*, k_j^* \rangle = 0. \tag{5.1.5}$$

The assertion now follows immediately.

Lemma 5.1.2. *For a normally flat semiparallel M^m in $N^n(c)$ the coefficients E_{ijl} in* (5.1.3) *are zero. If this M^m has r distinct principal curvature vectors $k_{(1)}, \dots, k_{(r)}$, and $k_{(\rho)}$ corresponds to the directions of the tangent basis vectors e_{i_ρ}, then*

$$L_{i_\rho j_\rho} = 0, \quad L_{i_\rho j_\tau} = \lambda_{(\rho) j_\tau}(k_{(\rho)} - k_{(\tau)}), \quad \omega_{i_\rho}^{j_\tau} = \lambda_{(\rho) j_\tau}\omega^{i_\rho} - \lambda_{(\tau) i_\rho}\omega^{j_\tau}, \tag{5.1.6}$$

$$dk_{(\rho)}^* = (k_{(\rho)}^*)^2 \sum_{j_\rho} e_{j_\rho}\omega^{j_\rho} + K_{i_\rho}\omega^{i_\rho} + \sum_{j \neq i_\rho} L_{i_\rho j}\omega^j, \tag{5.1.7}$$

$$\langle K_{i_\rho}, k_{(\tau)}^* \rangle = \lambda_{(\tau) i_\rho}(k_{(\rho)}^*)^2, \tag{5.1.8}$$

where $\rho \neq \tau$ are in $\{1, \dots, r\}$ and $(k_{(\rho)}^*)^2 = (k_{(\rho)})^2 + c$.
If $k_{(\rho)}$ is nonsimple, then $K_{i_\rho} = 0$.

Proof. Formulae (5.1.3) can contain E_{ijl} only if $m > 2$, and directly imply that $E_{i_\rho j_\rho l} = 0$. Due to symmetry, E_{ijl} is zero if any two of i, j, l lead to the same $k_{(\rho)}$. It follows that if $r = 1$ or $r = 2$, then all E_{ijl} are zero. For $r > 2$, consider $E_{i_\rho j_\tau l_\varphi}$ with three distinct ρ, τ, φ. From (5.1.3) it follows that $(k_{(\rho)}^* - k_{(\tau)}^*) \| E_{i_\rho j_\tau l_\varphi}$ and, similarly, $(k_{(\rho)}^* - k_{(\varphi)}^*) \| E_{i_\rho l_\varphi j_\tau} = E_{i_\rho j_\tau l_\varphi}$. Thus $E_{i_\rho j_\tau l_\varphi} = 0$ due to Theorem 5.1.1, and hence all E_{ijl} are zero.

Equations (5.1.7) follow from (5.1.2) by using the identity $\langle k_{(\rho)}^*, k_{(\tau)}^* \rangle = 0$ for $\rho \neq \tau$, which is a consequence of (5.1.5). To obtain (5.1.8) one has to differentiate this identity.

If $k_{(\rho)}$ is simple, then i_ρ takes only one value. If $k_{(\rho)}$ is nonsimple, then $k_{i_\rho} = k_{j_\rho} = k_{(\rho)}$, $i_\rho \neq j_\rho$, and from (5.1.2) it follows that $K_{i_\rho} = K_{j_\rho} = 0$ and $L_{i_\rho l} = L_{j_\rho l}$. Now (5.1.3) reduces to (5.1.6). This concludes the proof.

It is interesting that for the class of semiparallel submanifolds in a space form $N^n(c)$ there exist other conditions that imply normally flatness.

The first-order subspace $\operatorname{span}\{h_{ij}\} \subset T_x^\perp M^m$ of a submanifold M^m in $N^n(c)$, introduced in Section 2.4, is called the *principal normal subspace*; its dimension is called the *principal codimension* of such an M^m (cf. Section 4.4) and is often denoted by m_1.

Proposition 5.1.3. *If a semiparallel submanifold M^m in $N^n(c)$ has principal codimension $m_1 \leq 2$ and nonzero mean curvature vector $H \neq 0$, then this M^m is normally flat.*

Proof. If $m_1 = 0$, then $h = 0$ and M^m is totally geodesic, hence with flat $\nabla^\perp$. If $m_1 = 1$, the orthonormal frame bundle adapted to M^m can be adapted further so that $h_{ij} = h_{ij}^{m+1} e_{m+1}$ and thus $h_{ij}^\xi = 0$ for $\xi \in \{m+2, \dots, n\}$. Due to (2.1.9)–(2.1.11) all $\Omega_\alpha^\beta = 0$, thus $\nabla^\perp$ is again flat.

If $m_1 = 2$, the adaption can be chosen so that only H^{m+1} and H^{m+2} can be nonzero; moreover, $\Omega_{m+1}^{m+2} = -\Omega_{m+2}^{m+1}$. The consequence (4.3.2) of the semiparallel condition (4.1.2) reduces to

$$H^{m+2}\Omega_{m+2}^{m+1} = H^{m+1}\Omega_{m+1}^{m+2} = 0.$$

If $H \neq 0$ here, then $\Omega_{m+1}^{m+2} = 0$, and $\nabla^\perp$ is indeed flat, as was asserted.

Remark 5.1.4. If $H = 0$ for a semiparallel submanifold M^m in $N^n(c)$, then in the case $c \leq 0$ this M^m is totally geodesic, due to Theorem 4.1.7, and thus has flat $\nabla^\perp$.

But in the case $c > 0$ there exist semiparallel M^m in $N^n(c) \equiv S^n(c)$ with $H = 0$, nonflat $\nabla^\perp$, and $m_1 = 2$. An example is the Veronese surface M^2 in $S^4(1)$, which is a minimal surface of $S^4(1)$, has $m_1 = 2$, and nonflat $\nabla^\perp$ (see Sections 3.3 and 6.3).

Remark 5.1.5. If one considers the distribution defined for a fixed value of ρ, with nonsimple $k_{(\rho)}$, by $\omega^{j_\tau} = 0$ $(\tau \neq \rho)$, from (5.1.6) it follows that this distribution is a foliation (due to the Frobenius theorem, second version). The leaves of this foliation are called the *curvature surfaces* (see [Re 76, 79], [Ch 2000], Section 3.7); for a nonsimple $k_{(\rho)}$ this is of dimension ≥ 2. Then, due to Lemma 5.1.2, $K_{i_\rho} = 0$, and from (5.1.7) it follows that the corresponding principal curvature vector $k_{(\rho)}$ is parallel with respect to the normal connection along its curvature surface (cf. [DN 93], Lemma 2.2, where it is also stated that this curvature surface is totally umbilical and spherical).

5.2 Normally Flat Parallel Submanifolds

To study normally flat semiparallel submanifolds, let us first consider normally flat parallel submanifolds. According to Theorem 4.5.5, every semiparallel submanifold is a second-order envelope of parallel submanifolds, and obviously is normally flat if and only if the latter are normally flat.

Normally flat parallel submanifolds in Euclidean space have been investigated in [Wa 73] and [Sak 73], where it was shown that such a submanifold is either a product of several spheres and, possibly, a Euclidean subspace, or else an open part of such a product.

Generally, a submanifold M^m in ${}_\sigma E^n$ is said to be a *product of submanifolds* M^{m_ρ} in ${}_{\sigma_\rho}E^{n_\rho}$ $(\rho = 1, \dots, r)$ if

(i) $M^m = M^{m_1} \times \cdots \times M^{m_r}$,
(ii) ${}_\sigma E^n = {}_{\sigma_1} E^{n_1} \times \cdots \times {}_{\sigma_r} E^{n_r}$, where the product components on the right are pairwise totally orthogonal.

Recall that in Section 3.6 a submanifold in Euclidean space was called *reducible* if it is decomposable into a product; otherwise it was called *irreducible*. Now these terms can also be used for submanifolds in pseudo-Euclidean space, and also in space forms, considering their standard models in pseudo-Euclidean spaces.

The following gives a generalization of the results cited above to the case of normally flat submanifolds in space forms (see also [Lu 2000a]).

Let M^m be a normally flat parallel submanifold in $N^n(c)$. Then (5.1.1)–(5.1.3) hold, where $E_{ijl} = 0$ due to Lemma 5.1.2, and also $K_i = L_{ij} = 0$, because M^m is parallel. From (5.1.6) it follows that $\lambda_{(\rho)j_\tau} = 0$ and thus $\omega_{i_\rho}^{j_\tau} = 0\ (\rho \neq \tau)$. It follows that the distribution of subspaces $T_\rho = \text{span}\{e_{i_\rho}\}$ in the tangent vector spaces of M^m is a foliation for every fixed value of ρ. Indeed, this distribution is defined by the differential system $\omega^{j_\tau} = 0\ (\tau \neq \rho)$, but due to (1.4.3) and (2.1.3) $d\omega^j = \omega^i \wedge \omega_i^j$; thus now $d\omega^{j_\tau} = \omega^{i_\tau} \wedge \omega_{i_\tau}^{j_\tau}$; hence this system is totally integrable (see the Frobenius' theorem, second version, in [Ste 64]) and hence determines a foliation. For a leaf of this foliation, (5.1.1) and (5.1.2) imply

$$de_{i_\rho} = \sum_{j_\rho} e_{j_\rho} \omega_{i_\rho}^{j_\rho} + k^*_{(\rho)} \omega^{i_\rho}, \quad dk^*_{(\rho)} = -(k^*_{(\rho)})^2 \sum_{j_\rho} e_{j_\rho} \omega^{j_\rho}. \tag{5.2.1}$$

It follows that this leaf is a parallel submanifold $M_\rho^{m_\rho}$ and lies in ${}_{\sigma_\rho} E^{n_\rho}$, spanned by a point $x \in M_\rho^{m_\rho}$ and vectors $e_{i_\rho}, k^*_{(\rho)}$. For every pair of distinct ρ and σ these vectors are orthogonal; thus the above-considered normally flat parallel submanifold M^m is a product of parallel submanifolds $M_\rho^{m_\rho}$, $\rho \in \{1, \ldots, r\}$.

Here the following concept can be used. A submanifold M^m in $N^n(c)$ is said to be a *spherical submanifold*, if in (2.1.7) $h^*_{ij} = H^* g_{ij}$ and $\nabla^{*\perp} H^* = 0$, i.e., if M^m with $m > 1$ is totally umbilic or totally geodesic, and thus normally flat parallel (see Proposition 3.1.2), but for $m = 1$ this M^1 is a plane curve of constant curvature (see [DN 93], [Lu 2000a]).

If $c \geq 0$, i.e., $N^n(c)$ is E^n or $S^n(c)$, a spherical submanifold is a sphere or a circle (or their limit cases: a plane or a straight line). If $c < 0$ and thus $N^n(c)$ is $H^n(c)$, a spherical submanifold is a sphere or a horosphere, or an equidistant submanifold, including the one-dimensional cases: circle, horocircle, or equidistant curve (or their limit cases); cf. Remark 3.1.6.

Proposition 5.2.1. *A normally flat parallel submanifold M^m in $N^n(c)$ is a product of spherical submanifolds, or an open part of such a product.*

Such an M^m has flat ∇ (and thus flat $\bar{\nabla}$) if and only if every product component is one-dimensional, except possibly the plane (if $c = 0$) and the horosphere (if $c < 0$).

Proof. The first assertion summarizes the preceding analysis.

The second assertion follows from the fact that due to (5.1.4)

$$\Omega_{i_\rho}^{j_\tau} = 0 \quad (\rho \neq \tau), \quad \Omega_{i_\rho}^{j_\rho} = -(k^*_{(\rho)})^2 \omega^{i_\rho} \wedge \omega^{j_\rho}, \tag{5.2.2}$$

where $(k^*_{(\rho)})^2 = (k_{(\rho)})^2 + c$ (see Lemma 5.1.2). Therefore, ∇ is flat if and only if either $(k^*_{(\rho)})^2 \neq 0$ and i_ρ takes only one value, or $c = k_{(\rho)} = 0$, or $c < 0$ and $k^2_{(\rho)} + c = 0$.

Proposition 5.2.2. *An irreducible normally flat parallel submanifold M^m in $N^n(c)$ (i.e., a spherical submanifold) that is not a plane or straight line (if $c = 0$), or a horosphere or circle (if $c < 0$), is, in the geometry of ${}_\sigma E^{n+1}$, a sphere (in case $m = 1$, a circle) or its open part.*

Proof. For such an M^m the formulas above can be used, without the subscripts ρ. Therefore, $dx = e_i\omega^i$, $de_i = e_j\omega_i^j + k^*\omega^i$ and due to (5.2.1) $dk^* = -(k^*)^2 e_i\omega^i$. From here $d(k^*)^2 = 2\langle k^*, dk^*\rangle = 0$, thus $(k^*)^2$ is a constant, which is nonzero except for the two excluded submanifolds. It follows that M^m belongs to an $(m+1)$-dimensional plane in ${}_\sigma E^{n+1}$, which is spanned by the point x and vectors e_i, k^*. The point of this plane with radius vector $y = x + \|k^*\|^{-2}k^*$ is a fixed point because $dy = 0$; recall here that $\|k^*\|^2 = (k^*)^2$ is a nonzero constant, and therefore $\|y - x\|$ is also a nonzero constant.

The exceptional case, when $(k^*)^2 = k^2 + c = 0$ and either $k = c = 0$, or $c < 0$ and $k^2 = -c$, leads to the two excluded submanifolds.

Example 5.2.3. Consider the Plücker submanifold ${}_{lk}G^{2,p}(r)$ as an example. According to Theorem 3.2.3 this is a parallel submanifold. As is seen from (3.2.6), (3.2.7), and (3.2.9), its normal unit vectors are $\frac{1}{r}x = E_{[12]}$ and $E_{[uv]}$, and thus $\theta_u^{[12]} = -\frac{1}{r}\theta^u$, $\theta_{\bar{u}}^{[12]} = -\frac{1}{r}\theta^{\bar{u}}$,

$$\theta_{[uv]}^{w} = \frac{1}{r}(-\delta_u^w\theta^{\bar{v}} + \delta_v^w\theta^{\bar{u}}), \quad \theta_{[uv]}^{\bar{w}} = \frac{1}{r}(\delta_{\bar{u}}^{\bar{w}}\theta^{v} - \delta_{\bar{v}}^{\bar{w}}\theta^{u}). \tag{5.2.3}$$

Let us compute the normal curvature 2-forms. Obviously $\Theta_{[12]}^{[12]} = 0$. Furthermore,

$$\Theta_{[uv]}^{[12]} = \theta_{[uv]}^{w} \wedge \theta_{w}^{[12]} + \theta_{[uv]}^{\bar{w}} \wedge \theta_{\bar{w}}^{[12]} = 0,$$

as is easy to check. But $\Theta_{[uv]}^{[wy]}$ are not zero in general. Only for $p = 4$, where u, v, w, y can take only the values 3 and 4, does one have $\Theta_{[34]}^{[34]} = 0$. Thus ${}_{lk}G^{2,4}(r)$ is four-dimensional, parallel normally flat, and thus a product of two two-dimensional spherical submanifolds; if $l = k = 0$, then it is a product of two spheres.

Remark 5.2.4. Proposition 5.2.1 was first proved in the special case of normally flat parallel submanifolds of Euclidean space: for hypersurfaces in [SW 69], and more generally in [Wa 73] and [Sak 73] (see also [YI 71]). Here the spherical submanifolds are open parts of spheres or circles (except perhaps one, which can be a plane).

5.3 Adapted Frame Bundle for a Second-Order Envelope

A general normally flat semiparallel submanifold M^m in $N^n(c)$ is a second-order envelope of parallel submanifolds with flat $\nabla^\perp$, described in Theorem 5.2.1 as product submanifolds. This implies that such an M^m also has a generalized product structure, called the warped product structure. For Riemannian manifolds the concept of warped product was introduced in [Kr 57], [BiO'N 69] (see also [Hie 79], and [Nö 96]). In the investigation of normally flat semiparallel submanifolds this concept was used in [Lu 91d, f] and [DN 93].

As preparation, the orthonormal frame bundle is further adapted to the normally flat semiparallel M^m in $N^n(c)$, considering this M^m as a second-order envelope of the normally flat parallel submanifolds described in Proposition 5.2.1.

Here the notation for indices will be specified as follows. From now on, let ρ denote only those values for which $k^*_{(\rho)}$ is nonzero with non-zero $(k^*_{(\rho)})^2$ having the multiplicity $\nu_\rho > 1$; likewise for τ, etc.

These indices refer to the ν_ρ-dimensional components of the parallel product; let their range be $\{1, \dots, p\}$. Here e_{i_ρ} are tangent and $k^*_{(\rho)}$ is normal to such a component in ${}_\sigma E^{n+1}$.

The values of ρ (in the old sense of Lemma 5.1.2), for which $k^*_{(\rho)}$ is nonzero and has multiplicity 1, refer to the one-dimensional components of the parallel product. Let the number of such components be q and let them be denoted by subscripts a, $b, \dots$ in the range $\{\nu + 1, \dots, \nu + q\}$, where $\nu = \nu_1 + \dots + \nu_r$. Here e_a is tangent and k^*_a normal to one of these components in ${}_\sigma E^{n+1}$.

For the case $c = 0$, set $\rho = 0$ as the value of ρ for which $k^*_{(\rho)} = k_{(\rho)} = 0$. Due to (5.2.2), it refers to the plane component of the parallel product in E^n. Let the indices i_0, j_0 take values in $\{\nu + q + 1, \dots, \nu + q + \nu_0 = m\}$, where ν_0 is the dimension of this component.

Now due to Lemma 5.1.2, $K_{i_\rho} = K_{i_0} = 0$, and therefore (5.1.8) implies that

$$\lambda_{(0)i_\rho} = \lambda_{(a)i_\rho} = \lambda_{(\tau)i_\rho} = 0 \tag{5.3.1}$$

for every pair of distinct ρ and τ. Hence, due to (5.1.6),

$$L_{j_0 i_\rho} = L_{(a)i_\rho} = L_{j_\tau i_\rho} = 0, \tag{5.3.2}$$

$$\omega^{j_0}_{i_\rho} = \lambda_{(\rho)j_0}\omega^{i_\rho}, \quad \omega^a_{i_\rho} = \lambda_{(\rho)a}\omega^{i_\rho}, \quad \omega^{j_\tau}_{i_\rho} = 0. \tag{5.3.3}$$

The Euclidean case, when $c = 0$, has been analysed in [Lu 91d, f], where the ranges of indices $a, b, \dots$ and $i_0, j_0, \dots$ are joined by introducing the indices $u, v, \dots$ running through $\{\nu + 1, \dots, m\}$. Moreover, the parts of the orthonormal frames normal to M^m at arbitrary $x \in M^m$ are specified so that $e_{m+\rho}$ are collinear with $k_{(\rho)}$, and the succeeding e_{m+p+a^*}, with $a^* = a - \nu$ running through $\{1, \dots, q\}$, are collinear to $k_{(a)}$. Then $k_{(\rho)} = \kappa_\rho e_{m+\rho}$ and $k_{(a)} = \kappa_a e_{m+p+a^*}$. The remaining frame vectors normal to M^m are denoted by e_ξ.

The non-Euclidean case was treated in [DN 93] by a somewhat different method, not using the orthonormal frame bundle $O(M^m, N^n(c))$. The frame-bundle method will now be also extended to the case of $c \neq 0$, allowing more detail to be added to the results of [DN 93].

Suppose first that $c \geq 0$, i.e., consider a normally flat semiparallel submanifold M^m in E^n or in $S^n(c) \subset E^{n+1}$. Then this M^m, together with the adapted frame bundle introduced above, is defined by the following Pfaffian system (note that in $S^n(c)$ the range of i_0 is empty, and thus $\nu_0 = 0$):

$$\omega^{m+\rho} = \omega^{m+p+a^*} = \omega^{\xi} = 0, \tag{5.3.4}$$

$$\omega_{i_\rho}^{m+\tau} = \kappa_\rho \delta_\rho^\tau \omega^{i_\rho}, \quad \omega_u^{m+\tau} = 0, \quad \omega_{i_\rho}^{m+p+a^*} = 0, \quad \omega_u^{m+p+a^*} = \kappa_a \delta_u^a \omega^a, \tag{5.3.5}$$

$$\omega_{i_\rho}^{\xi} = \omega_u^{\xi} = 0, \tag{5.3.6}$$

where the left-hand sides contain coefficients from the formulas (1.4.2), and the ranges of indices are as indicated above.

Exterior differentiation and the use of (1.4.3) and Cartan's lemma lead to the following equations:

$$\omega_{i_\rho}^{j_\tau} = 0 \quad (\rho \neq \tau), \quad \omega_{i_\rho}^{u} = \lambda_{(\rho)u}\omega^{i_\rho}, \tag{5.3.7}$$

$$\omega_a^b = \kappa_a \gamma_a^b \omega^a - \kappa_b \gamma_b^a \omega^b, \quad \omega_a^{i_0} = \lambda_{(a)i_0}\omega^a, \tag{5.3.8}$$

$$\omega_{m+\rho}^{m+\tau} = 0 \quad (\rho \neq \tau), \quad \omega_{m+\rho}^{m+p+a^*} = -\kappa_a \kappa_\rho^{-1} \lambda_{(\rho)a}\omega^a, \quad \omega_{m+\rho}^{\xi} = 0, \tag{5.3.9}$$

$$\omega_{m+p+a^*}^{m+p+b^*} = \kappa_a \gamma_b^a \omega^a - \kappa_b \gamma_a^b \omega^b, \quad \omega_{m+p+a^*}^{\xi} = \varphi_a^{\xi}\omega^a, \tag{5.3.10}$$

$$d\kappa_\rho = \kappa_\rho \sum_u \lambda_{(\rho)u}\omega^u, \quad d\kappa_a = \psi_a \omega^a + \kappa_a \sum_{i_0} \lambda_{(a)i_0}\omega^{i_0}. \tag{5.3.11}$$

Note that here equations (5.3.7) coincide with (5.3.3), and those of (5.3.8) are the remaining part of (5.1.6), using the notation $\gamma_a^b = \kappa_a^{-1}\lambda_{(a)b}$; moreover, exterior differentiation of $\omega_{i_0}^{m+p+a} = 0$ gives $\lambda_{(0)a} = 0$. Equations (5.3.9)–(5.3.11) follow from (5.1.2) (with k_i instead of k_i^*) by replacing $k_{(\rho)} = \kappa_\rho e_{m+\rho}$ and $k_{(a)} = \kappa_a e_{m+p+a^*}$; moreover, $\varphi_a^{\xi} = \kappa_a^{-1} K_a^{\xi}$ and $\psi_a = K_a^{m+p+a^*}$.

In the case $c < 0$, when $N^n(c) = H^n(c)$, there is the possibility that $k_{(0)}^* \neq 0$. Since the other $p + q$ outer principal curvature vectors are orthogonal to this $k_{(0)}^*$, they all have positive scalar squares in ${}_1E^{n+1}$. Then, in addition to $k_{(\rho)} = \kappa_\rho e_{m+\rho}$, $k_{(a)} = \kappa_a e_{m+p+a}$, the next two normal frame vectors e_{m^*+1} and e_{m^*+2}, with $m^* = m + p + q$, can be chosen so that both are orthogonal to all $e_{m+\rho}$ and e_{m+p+a}, both have zero scalar squares, their scalar product is 1, and $k_{(0)}^* = \kappa_0 e_{m^*+1}$. The remaining frame vectors e_ξ normal to M^m in ${}_1E^{n+1}$ constitute an orthonormal set, and they span a subspace orthogonal to all the previous ones.

Here equations (5.3.4)–(5.3.6) remain the same, but we add

$$\omega^{m^*+1} = \omega^{m^*+2} = 0, \tag{5.3.12}$$

$$\omega_{i_\rho}^{m^*+1} = \omega_a^{m^*+1} = \omega_{i_\rho}^{m^*+2} = \omega_u^{m^*+2} = 0, \quad \omega_{i_0}^{m^*+1} = \kappa_0 \omega^{i_0}, \tag{5.3.13}$$

where the range of ξ is now $\{m^*+3, \dots, n+1\}$.

Moreover, exterior differentiation of the equations

$$\langle e_{I^*}, e_{m^*+1}\rangle = \langle e_{I^*}, e_{m^*+2}\rangle = \langle e_{m^*+1}, e_{m^*+1}\rangle = \langle e_{m^*+2}, e_{m^*+2}\rangle = 0,$$

and $\langle e_{m^*+1}, e_{m^*+2}\rangle = 1$, with I^* in the range $\{1, \dots, n+1\}$ of I, except for m^*+1 and m^*+2, yields

$$\omega_{I^*}^{m^*+1} + \omega_{m^*+2}^{I^*} = \omega_{I^*}^{m^*+2} + \omega_{m^*+1}^{I^*} = \omega_{m^*+1}^{m^*+2} = \omega_{m^*+2}^{m^*+1} = \omega_{m^*+1}^{m^*+1} + \omega_{m^*+2}^{m^*+2} = 0,$$

in addition to the usual $\omega_{I^*}^{J^*} + \omega_{J^*}^{I^*} = 0$.

Equations (5.3.7) also remain the same. In (5.3.8) the last group of equations becomes $\omega_a^{i_0} = \lambda_{(a)i_0}\omega^a - \lambda_{(0)a}\omega^{i_0}$, but in (5.3.9) the last equations have to be replaced by

$$\omega_{m+\rho}^{m^*+1} = -\kappa_0\kappa_\rho^{-1}\lambda_{(\rho)i_0}\omega^{i_0}, \quad \omega_{m+\rho}^{m^*+2} = \omega_{m+\rho}^{\xi} = 0. \tag{5.3.14}$$

In (5.3.10) the last group becomes

$$\omega_{m+p+a}^{m^*+1} = \kappa_a^{-1}K_a^{m^*+1}\omega^a, \qquad \omega_{m+p+a}^{m^*+2} = \kappa_a\kappa_0^{-1}\lambda_{(0)a}\omega^a, \qquad \omega_{m+p+a}^{\xi} = 0, \tag{5.3.15}$$

but (5.3.11) has to be complemented by

$$d\kappa_0 = \kappa_0(\omega_{m^*+1}^{m^*+1} + \lambda_{(0)a}\omega^a) + K_{i_0}^{m^*+1}\omega^{i_0}. \tag{5.3.16}$$

5.4 Second-Order Envelope as Warped Product

Now it is possible to realize the program announced at the beginning of the previous section.

Exterior differentiation applied to equations (5.3.7) gives

$$\sum_u \lambda_{(\rho)u}\lambda_{(\tau)u} = 0 \quad (\rho \neq \tau), \quad d\lambda_{(\rho)u} = \lambda_{(\rho)v}(\omega_u^v + \lambda_{(\rho)u}\omega^v). \tag{5.4.1}$$

Let the value ρ be fixed and consider the distribution defined on M^m by $\omega^{i_\tau} = 0$ $(\tau \neq \rho)$, $\omega^u = 0$. Due to (5.3.7) and (5.3.11) $d\omega^{i_\tau} = \omega^{j_\tau} \wedge (\omega_{j_\tau}^{i_\tau} + \delta_{j_\tau}^{i_\tau} d\ln\kappa_\tau)$, $d\omega^u = \omega^v \wedge \omega_v^u$, hence this distribution is a foliation. For each leaf $M_\rho^{\nu_\rho}$, one has $dx = e_{i_\rho}\omega^\rho$ (ρ fixed); and on M^m itself, equations (5.3.5)–(5.3.7) imply that

$$de_{i_\rho} = e_{j_\rho}\omega_{i_\rho}^{j_\rho} + \omega^{i_\rho}\left(\sum_u \lambda_{(\rho)u}e_u + k_{(\rho)}^*\right). \tag{5.4.2}$$

Hence each leaf $M_\rho^{\nu_\rho}$ is a totally umbilic submanifold with outer mean curvature vector

$$H_\rho^* = \sum_u \lambda_{(\rho)u} e_u + k_{(\rho)}^*. \tag{5.4.3}$$

Comparison of (5.4.2) with (2.1.7) shows that $h_{i_\rho j_\rho}^* = H_\rho^* \delta_{i_\rho j_\rho}$. Since $\nu_\rho > 1$, this implies that each leaf $M_\rho^{\nu_\rho}$ is a spherical submanifold (see Section 5.2). As follows from (5.2.2), it is intrinsically a Riemannian manifold of constant curvature $(H_\rho^*)^2$ (here $(H_\rho^*)^2$ is in the role of $(k_{(\rho)}^*)^2$).

A consequence from (5.4.1) is that H_ρ^* and H_τ^* are orthogonal for every pair of distinct ρ and τ. Therefore, the subspaces in ${}_\sigma E^{n+1}$ that contain the leaves $M_\rho^{\nu_\rho}$ and $M_\tau^{\nu_\tau}$ passing through a point $x \in M^m$ are totally orthogonal.

From (5.4.3) it follows, due to (2.1.3), (2.1.5), (5.4.1), and (5.3.11), that

$$dH_\rho^* = H_\rho^* \lambda_{(\rho)u} \omega^u - (H_\rho^*)^2 e_{i_\rho} \omega^{i_\rho}, \tag{5.4.4}$$

where $(H_\rho^*)^2 = \langle H_\rho^*, H_\rho^* \rangle = \sum_u \lambda_{(\rho)u}^2 + (k_{(\rho)}^*)^2$ and thus $d(H_\rho^*)^2 = 2(H_\rho^*)^2 \lambda_{(\rho)u} \omega^u$.

If $c \geq 0$, then each $(H_\rho^*)^2 > 0$; but if $c < 0$, this need not hold. Suppose now that $(H_\rho^*)^2 \neq 0$, for both cases of c. Note that in ${}_1E^{n+1}$, among mutually orthogonal vectors with nonzero scalar squares, only one of the squares is negative, while all others are positive.

Let us denote $r_\rho = \|H_\rho^*\|^{-1}$, where $\|H_\rho^*\| = \sqrt{(H_\rho^*)^2}$. From (5.4.2) and (5.4.3) it follows that every leaf $M_\rho^{\nu_\rho}$ is, in the geometry of ${}_\sigma E^{n+1}$, a sphere with radius r_ρ and normal unit vector $n_\rho = r_\rho H_\rho^*$, possibly an imaginary unit, for one value of ρ, in the case $c < 0$ (also see Remark 3.1.6 and Proposition 5.2.2).

It turns out that all these spheres which go through a point $x \in M^m$ lie on a $(\nu + p - 1)$-dimensional sphere in ${}_\sigma E^{n+1}$, whose radius is $r = \sqrt{r_1^2 + \cdots + r_p^2}$ and whose center has the radius vector $y = x + r_1 n_1 + \ldots r_p n_p$ (recall that here $\nu = \nu_1 + \cdots + \nu_p$).

Indeed, it can be deduced that

$$dr_\rho = r_{\rho u} \omega^u, \quad dn_\rho = -r_\rho^{-1} e_{i_\rho} \omega^{i_\rho}, \tag{5.4.5}$$

where $r_{\rho u} = -r_\rho \lambda_{(\rho)u}$, and therefore

$$dy = f_u \omega^u, \quad dr = r^{-1} \sum_\rho r_\rho r_{\rho u} \omega^u, \tag{5.4.6}$$

where $f_u = e_u + \sum_\rho r_{\rho u} n_\rho$. For the foliation on M^m defined by $\omega^u = 0$, one has $dy = 0$, $dr = 0$. This proves the assertion above.

Note also that (5.3.13) and (5.4.3) imply that $d(\kappa_\rho r_\rho) = 0$.

Now (5.4.2) is $de_{i_\rho} = e_{j_\rho} \omega_{i_\rho}^{j_\rho} + r_\rho^{-1} n_\rho \omega^{i_\rho}$. From this and from (5.4.5) it follows that the subspaces that contain the spherical leaves $M_\rho^{\nu_\rho}$ for a fixed value ρ are parallel in ${}_\sigma E^{n+1}$, because $\text{span}\{e_{i_\rho}, n_\rho\}$ is invariant on M^m, and these subspaces are orthogonal for different values of ρ, as was shown above.

Further, let us consider the submanifold in ${}_\sigma E^{n+1}$ that is described by the centers of the $(\nu + p - 1)$-dimensional spheres containing the leaves through an $x \in M^m$.

From (5.4.6) it follows that this is a $(q+\nu_0)$-dimensional submanifold $B^{m'}$, where $m' = q + \nu_0$, and f_u are its tangent vectors. A straightforward computation shows that $\langle f_u, n_\rho \rangle = 0$ and, using (5.3.5) and (5.3.13), that

$$df_u = f_v \omega_u^v + \sum_a \delta_u^a k_{(a)}^* \omega^a + \sum_{i_0} \delta_u^{i_0} k_{(0)}^* \omega^{i_0}. \tag{5.4.7}$$

The submanifold $B^{m'}$ lies in an $(n-\nu-p)$-dimensional subspace of ${}_\sigma E^{n+1}$ which is orthogonal to all the subspaces containing the leaves $M_\rho^{\nu_\rho}$. This can be deduced from (5.4.2)–(5.4.7) by starting with the identities $\langle dy, e_{i_\rho} \rangle = 0$, $\langle dy, n_\rho \rangle = 0$ and deriving successively $\langle d^2 y, e_{i_\rho} \rangle = 0$, $\langle d^2 y, n_\rho \rangle = 0$; $\langle d^3 y, e_{i_\rho} \rangle = 0$, $\langle d^3 y, n_\rho \rangle = 0, \ldots$.

A leaf $M_\rho^{\nu_\rho}$, which is a spherical submanifold of constant curvature $c_\rho = (H_\rho^*)^2 = r_\rho^{-2}$, will be denoted below by $Sph_\rho^{\nu_\rho}(c_\rho)$. For $c_\rho > 0$ it is a sphere $S_\rho^{\nu_\rho}(c_\rho)$, for $c_\rho < 0$ an equidistant submanifold $H_\rho^{\nu_\rho}(c_\rho)$.

The properties established above for a normally flat semiparallel M^m motivate the following general concept.

Consider a smooth fibre bundle immersed in a space ${}_\sigma E^{n+1}$ so that

(1) the base manifold is immersed as a smooth submanifold $B^{m'}$ in an n'-dimensional subspace of ${}_\sigma E^{n+1}$,
(2) every fibre is immersed as a product $Sph_1^{\nu_1}(c_1) \times \cdots \times Sph_p^{\nu_p}(c_p)$ of spherical submanifolds in an $(n-n')$-dimensional subspace, which is totally orthogonal to this n'-dimensional subspace and goes through the point $y \in B^{m'}$ which is y the center of the $(\nu^* + p - 1)$-dimensional sphere containing this product; here $n - n' = \nu + p$, $m - m' = \nu$,
(3) for every fixed value $\rho \in \{1, \ldots, p\}$ the subspaces which span $Sph_\rho^{\nu_\rho}(c_\rho)$ at different points of $B^{m'}$ are parallel to each other.

Then this immersed fibre bundle is said to be a *warped product submanifold*, denoted by

$$B^{m'} \times_{r_1} Sph_1^{\nu_1}(1) \times_{r_2} \cdots \times_{r_p} Sph_p^{\nu_p}(1),$$

which is, more specifically, either

$$B^{m'} \times_{r_1} S_1^{\nu_1}(1) \times_{r_2} \cdots \times_{r_p} S_p^{\nu_p}(1),$$

or, for $c < 0$ possibly

$$B^{m'} \times_{r_1} H_1^{\nu_1}(-1) \times_{r_2} S_2^{\nu_2}(1) \times_{r_3} \cdots \times_{r_p} S_p^{\nu_p}(1),$$

where r_1 is pure imaginary and $r_2, \ldots, r_p$ are real positive functions on $B^{m'}$.

Here $r_1, \ldots, r_p$ are called the *warping functions*; recall that $r_\rho = (\sqrt{c_\rho})^{-1}$ is the radius of the sphere in ${}_\sigma E^{n+1}$, which contains the spherical leaf $Sph_\rho^{\nu_\rho}(c_\rho)$.

It remains to consider the exceptional case, when $c < 0$ and $(H_\rho^*)^2 = 0$ for some value of ρ. There can only be one such value. Indeed, if also $(H_\tau^*)^2 = 0$ for $\tau \neq \rho$, then $\langle k_{(\rho)}^*, k_{(\tau)}^* \rangle = 0$ and $\sum_u \lambda_{(\rho)u} \lambda_{(\tau)u} = 0$ (see (5.4.1)) imply that $\langle H_\rho^*, H_\tau^* \rangle = 0$.

In ${}_1E^{n+1}$ two vectors with zero scalar square can be orthogonal only if they are collinear, but for $k^*_{(\rho)} = k_{(\rho)} - cx$ and $k^*_{(\tau)} = k_{(\tau)} - cx$ this is possible only if they are equal, which is excluded here.

For this single value ρ, the corresponding spherical leaf is intrinsically of constant curvature $(H^*_\rho)^2 = 0$, thus locally Euclidean, but extrinsically it is a horosphere of a hyperbolic space form realized as the standard model in ${}_1E^{n+1}$. The subspaces of ${}_1E^{n+1}$ which contain these horospheres are also parallel, and orthogonal to such subspaces for the other spherical leaves, as was shown above. Each of these horospheres is an orthogonal submanifold of straight lines in the direction of H^*_ρ with $(H^*_\rho)^2 = 0$, which by (5.4.4) are parallel in its subspace (in hyperbolic geometry and also in the geometry in ${}_1E^{n+1}$). The only difference from the general case above is that all horospheres are mutually congruent; therefore, no warping function is needed.

As a result, the submanifold M^m is simply the product of a warped product, described above, and a horosphere.

Recall that a submanifold in ${}_\sigma E^{n+1}$ is said to be irreducible if it is not a product of submanifolds of lower dimension (see Section 3.6).

From the preceding analysis above, the geometric structure of an irreducible normally flat semiparallel submanifold M^m in $N^n(c)$ can be described as follows.

Theorem 5.4.1. *Let M^m in $N^n(c)$ be an irreducible normally flat semiparallel submanifold with nonzero principal curvature vectors $k_{(1)}, \ldots, k_{p+q}$ of multiplicities $\nu_1 > 1, \ldots, \nu_p > 1$ and $\nu_{p+1} = \cdots = \nu_{p+q} = 1$. Then M^m is either*

- *a horosphere, if $c < 0$, $p = 1$, $q = 0$, and $(k^*_{(1)})^2 = 0$ for the only outer principal curvature vector $k^*_{(1)} \neq 0$, or*
- *a warped product submanifold, described above, with $p \leq m' = m - \nu_1 - \cdots - \nu_p$, satisfying the following conditions:*
  (1) *$B^{m'}$ is intrinsically locally Euclidean and with flat $\nabla^\perp$, i.e., with flat $\bar{\nabla}$.*
  (2) *The warping functions $r_1, \ldots, r_p$ are, with respect to some local affine coordinates in $B^{m'}$, nonconstant linear functions having mutually orthogonal gradients.*

Proof. The assertions concerning the horosphere and the warped product submanifold were obtained above, except for the inequality $p \leq m'$ and the conditions (1) and (2).

To prove (1), let us consider the formulas (5.4.7). It is seen that $B^{m'}$ is normally flat with outer curvature vectors $k^*_{(a)}$, which are mutually orthogonal. Therefore, due to (5.1.4), $B^{m'}$ is indeed locally Euclidean, and hence has flat $\bar{\nabla}$.

To prove (2), let us consider (5.4.5), where $r_{\rho u} = -r_\rho \lambda_{(\rho)u}$. Consider the covector field on the locally Euclidean $B^{m'}$ having the components $r_{\rho u}$ (ρ fixed) with respect to the moving frame $\{y; f_u\}$ on $B^{m'}$. Its covariant derivative is defined as $\nabla' r_{\rho u} = dr_{\rho u} - r_{\rho v}\omega^v_u$. An easy computation using (5.4.1) and (5.4.5) shows that $\nabla' r_{\rho u} = 0$. This verifies the assertion in (2) about the linearity of r_ρ. The assertion about the orthogonality of their gradients follows from the first equation (5.4.1).

Also the inequality $p \leq m'$ follows, because in the contrary case, there would be more than m' mutually orthogonal nonzero vectors $\sum_u \lambda_{(\rho)u} e_u$ in the m'-dimensional linear span of all the vectors e_u.

Finally, for an irreducible normally flat semiparallel M^m in E^n, none of the warping functions $r_1, \ldots, r_p$ can be constant. Indeed, if $r_\rho = \text{const}$, then $r_{\rho u} = 0$ and $\lambda_{(\rho)u} = 0$; hence (5.3.9) and (5.3.11) imply $\omega^u_{i_\rho} = \omega^{m+p+a}_{m+\rho} = 0$, but this implies that the submanifold M^m is the product of a sphere $S^{\nu_\rho}(c_\rho)$ and of the remaining warped product, i.e., it is not irreducible. This finishes the proof.

Remark 5.4.2. For the case $c = 0$, i.e., is when M^m is in E^n, the base submanifold $B^{m'}$ is extrinsically the envelope of a q-parameter family of m'-dimensional subspaces $E^{m'}$ in E^n.

Indeed, consider on $B^{m'}$ the distribution defined by $\omega^a = 0$, where a runs through $\{\nu+1, \ldots, \nu+q\}$. From (5.3.7) and (5.3.8) it follows that this distribution is a foliation whose leaves have, due to (5.4.6), the tangent vectors f_{i_0}. From (5.4.7) it follows in the case $c = 0$, when $k^*_{(0)} = 0$, that along each of these leaves, (5.3.8) implies that

$$df_a = 0, \quad df_{i_0} = f_{j_0}\omega^{j_0}_{i_0},$$

and hence these leaves are ν_0-dimensional planes in E^n and, moreover, the tangent subspace of $B^{m'}$ spanned by f_u is invariant along each leaf.

Conversely, if a warped product submanifold $B^{m'} \times_{r_1} S^{\nu_1}(1) \times_{r_2} \cdots \times_{r_p} S^{\nu_p}(1)$ in E^n satisfies the above conditions, then it is an irreducible normally flat semiparallel submanifold, as was shown in [Lu 91d], [Lu 91f].

Remark 5.4.3. In the special case when $p = 0$, the normally flat semiparallel M^m in $N^n(c)$ reduces to $B^{m'}$ with $m' = m$, i.e., has flat van der Waerden–Bortolotti connection $\bar{\nabla}$.

In the case $c = 0$ and $q \neq 0$, this M^m is the envelope of a q-parameter family of m-dimensional planes in E^n. This kind of M^m in E^n was studied in [Lu 87b].

In the case $c \neq 0$ (or $c = 0$ and $\nu_0 = 0$), this M^m has m distinct orthogonal nonzero outer principal curvature vector fields. They consist of the curvature vectors of the lines of curvature of M^m, which form an orthogonal holonomic conjugate net on M^m. Such types of submanifolds M^m were investigated by É. Cartan in [Ca 19] and they are called submanifolds of *Cartan type*, or, following [Che 47], *Cartan varieties* (see also [AG 93], Chapter 3).

Remark 5.4.4. Normally flat semiparallel submanifolds M^m in space forms $N^n(c)$ as warped product submanifolds were studied in [DN 93], mostly for $c \neq 0$. Theorem 5.4.1 contains some of the results obtained there.

5.5 Semiparallel Submanifolds of Principal Codimension 1

All semiparallel submanifolds M^m in $N^n(c)$ whose principal codimension is 1 are normally flat, as follows from the proof of Proposition 5.1.3. For such M^m, the orthonormal frame bundle $O(M^m, N^n(c))$ can be adapted further so that the first normal frame vector e_{m+1} at arbitrary $x \in M^m$ belongs to the one-dimensional principal normal subspace of M^m at x. Then $k_i = \kappa_i e_{m+1}$ and the semiparallel condition (5.1.5) reduces to $(\kappa_i - \kappa_j)(\kappa_i \kappa_j + c) = 0$, i.e., every two principal curvatures κ_i and κ_j are either equal or their product is $-c$.

The simplest case is a curve M^1, which is always semiparallel, namely with flat $\bar{\nabla}$, and there there is only one κ_1.

If $m > 1$ and there are two or more principal curvatures, it follows from the above that they can have only two distinct values. Indeed, if three among them, with indices i, j, k, are distinct, then $\kappa_i \kappa_j = \kappa_i \kappa_k = -c$, thus $\kappa_i(\kappa_j - \kappa_k) = 0$ and hence $\kappa_i = 0$. This is possible only if $c = 0$, but then $\kappa_j \kappa_k = 0$ leads to a contradiction. Therefore,

$$\kappa_1 = \cdots = \kappa_s = \kappa \neq 0, \quad \kappa_{s+1} = \cdots = \kappa_m = -c\kappa^{-1}$$

for some natural number s, $0 \leq s \leq m$.

Let us suppose first that $c = 0$, i.e., that $N^n(c)$ is the Euclidean space E^n. Here $s = 0$ is impossible, for principal codimension 1. Therefore, either

(1) $s = 1$, or
(2) $1 < s < m$, or
(3) $s = m$.

In the first subcase, when $s = 1$, M^m is defined by the differential system

$$\omega^{m+1} = \omega^\xi = 0, \quad \omega_1^{m+1} = \kappa\omega^1, \quad \omega_2^{m+1} = \cdots = \omega_m^{m+1} = \omega_i^\xi = 0,$$

where i runs through $\{1, \ldots, m\}$ and ξ through $\{m+2, \ldots, n\}$. By exterior differentiation, this system leads to

$$\omega_1^{i_0} = \lambda_{i_0}\omega^1, \quad \omega_{m+1}^\xi = \varphi_1^\xi \omega^1, \quad d\kappa = \psi\omega^1 + \kappa\lambda_{i_0}\omega^{i_0},$$

where i_0 runs through $\{2, \ldots, m\}$ and the notation in (5.3.8), (5.3.10), (5.3.11) has been simplified by writing $\lambda_{i_0} = \lambda_{(1)i_0}$. Here the distribution defined on M^m by $\omega^1 = 0$ is a foliation, for whose leaves there holds $de_{i_0} = e_{j_0}\omega_{i_0}^{j_0}$. Therefore, these leaves are $(m-1)$-dimensional planes, and thus M^m in E^n is generated by a one-parameter family of these generator planes. Since now

$$de_1 = e_{i_0}\omega_1^{i_0} + e_{m+1}\omega_1^{m+1} = \left(\sum_{i_0} e_{i_0}\lambda_{i_0} + \kappa e_{m+1}\right)\omega^1,$$

the tangent vector e_1 is constant along each of these plane generators. This means that the tangent m-dimensional plane of M^m depends only on one parameter and thus M^m is the envelope of the one-parameter family of these tangent planes.

In the second subcase, when $1 < s < m$, the submanifold M^m is defined by

$$\omega^{m+1} = \omega^\xi = 0, \quad \omega_{i_1}^{m+1} = \kappa\omega^{i_1}, \quad \omega_{i_0}^{m+1} = \omega_{i_1}^\xi = \omega_{i_0}^\xi = 0,$$

where i_1, etc., run through $\{1,\dots,s\}$ and i_0, etc., through $\{s+1,\dots,m\}$. The system above leads via exterior differentiation to

$$d\ln\kappa = \lambda_{j_0}\omega^{j_0}, \quad \omega_{i_1}^{j_0} = \lambda_{j_0}\omega^{i_1}, \quad \omega_{m+1}^{\xi} = 0,$$

and then to

$$d\lambda_{j_0} = \sum_{k_0} \lambda_{k_0}(\omega_{j_0}^{k_0} + \lambda_{j_0}\omega^{k_0}) \tag{5.5.1}$$

(cf. (5.3.11), (5.3.7), (5.3.9), (5.4.1)), where the notation has been simplified by putting $\lambda_{j_0} = \lambda_{(1)j_0}$. It follows that

$$dx = e_i\omega^i, \quad de_i = e_j\omega_i^j + e_{m+1}\kappa\delta_{ij_1}\omega^{j_1}, \quad de_{m+1} = -\kappa e_{i_1}\omega^{i_1}.$$

Hence, M^m lies in the $(m+1)$-dimensional subspace $E^{m+1} \subset E^n$, spanned at the point $x \in M^m$ by the vectors e_i, e_{m+1}, and thus M^m is a hypersurface.

Moreover, the vector $l = \sum_{j_0} \lambda_{j_0} e_{j_0}$ is invariant at each point $x \in M^m$, because $dl = l^2 e_{i_1}\omega^{i_1} + l(\lambda_{j_0}\omega^{j_0})$.

Let us suppose first that $l \neq 0$. Then the frame at each x can be specified so that e_{s+1} is collinear to l. Hence $\lambda_{s+1} \neq 0$, $\lambda_{i_0'} = 0$, where i_0', etc., run through $\{s+2,\dots,m\}$. This implies $\omega_{i_1}^{j_0'} = 0$, and due to (5.5.1), the equation $\omega_{j_0'}^{s+1} = 0$ holds.

Now considering the distributions on M^m defined by $\omega^{i_1} = \omega^{s+1} = 0$ and by $\omega^{i_0'} = 0$, one can see that both are foliations. For the leaves of the first of them, one has $de_{j_0'} = e_{k_0'}\omega_{j_0'}^{k_0'}$; thus the leaves are parallel $(m-s-1)$-dimensional planes in E^{m+1}.

For the leaves of the second foliation, orthogonal to the first one, one has

$$dx = e_{i_1}\omega^{i_1} + e_{s+1}\omega^{s+1}, \quad de_{i_1} = e_{j_1}\omega_{i_1}^{j_1} + (e_{s+1}\lambda_{s+1} + e_{m+1}\kappa)\omega^{i_1}, \tag{5.5.2}$$

$$de_{s+1} = -\lambda_{s+1}e_{i_1}\omega^{i_1}, \quad de_{m+1} = -\kappa e_{i_1}\omega^{i_1}. \tag{5.5.3}$$

This implies that every leaf is contained in the $(s+2)$-dimensional plane at the point x spanned by the vectors $e_1,\dots,e_{s+1}$ and e_{m+1}, and are thus totally orthogonal to the plane leaves of the first foliation. Moreover, $d(x + \lambda_{s+1}^{-1}e_{s+1}) = 0$. It follows that there is a fixed point with radius vector $z = x + \lambda_{s+1}^{-1}e_{s+1}$. Therefore, the leaf consists of straight lines with direction vectors e_{s+1} passing through this point z; hence the leaf is a cone. Orthogonal sections of the generator lines of this cone are the integral manifolds of the totally integrable equation $\omega^{s+1} = 0$. They are totally umbilic and therefore are s-dimensional spheres or their open subsets. Consequently, these leaves are round hypercones C^{s+1} in $(s+2)$-dimensional planes. Hence M^m is a product submanifold $C^{s+1} \times E^{m-s-1}$ in E^{m+1}.

The case $l = 0$ can be considered as the limit of cases where l tends to the zero vector, and thus λ_{s+1} tends to 0. Thus the limit case of the round cone C^{s+1} is the round cylinder $S^s \times E^1$, and $C^{s+1} \times E^{m-s-1}$ tends to $S^s \times E^{m-s}$.

Finally, in the third subcase when $s = m$, the submanifold M^m in E^n is totally umbilic, thus a sphere S^m or its open subset.

The following theorem summarizes the above results.

Theorem 5.5.1. *A semiparallel submanifold M^m with principal codimension* 1 *in Euclidean space E^n is either*

- *the envelope of a one-parameter family of m-dimensional planes, or*
- *a hypersurface which is a product $C^{s+1} \times E^{m-s-1}$ of an $(s+1)$-dimensional round cone and an $(m-s-1)$-dimensional plane, or a product $S^s \times E^{m-s}$, or*
- *a sphere S^m, or an open subset of one of the above.*

Remark 5.5.2. In the first subcase, Theorem 4.5.5 and the results presented in Remark 5.2.4 imply that such an M^m is a second-order envelope of round cylinders $S^1 \times E^{m-1}$. If $m = 1$ this is a curve.

In the second subcase, M^m is the second-order envelope of round cylinders $S^s \times E^{m-s}$.

Now let $c \neq 0$, i.e., let $N^n(c)$ be a non-Euclidean space form. Denoting here $\lambda = -c\kappa^{-1}$, the differential system which defines the above-considered semiparallel submanifold M^m in $N^n(c)$ is

$$\omega^{m+1} = \omega^{\xi} = 0, \quad \omega_{i_1}^{m+1} = \kappa\omega^{i_1}, \quad \omega_{i'}^{m+1} = \lambda\omega^{i'}, \quad \omega_i^{\xi} = 0, \tag{5.5.4}$$

where i' runs through $\{s+1, \ldots, m\}$, $\kappa \neq 0$, and $\kappa\lambda + c = 0$; hence also $\lambda \neq 0$.

Exterior differentiation of the last equations implies that $\omega_{m+1}^{\xi} = 0$. Hence M^m lies in an $N^{m+1}(c)$ which is spanned by x, all e_i, and e_{m+1}, and therefore is a hypersurface.

Here the simplest case is $s = 1$ and $m = 2$. This gives a semiparallel surface M^2 in $N^3(c)$ with principal curvatures κ and λ, and zero Gaussian curvature, due to $\kappa\lambda + c = 0$ (see, e.g., (5.1.4)).

Suppose now $s = m - 1 > 1$. Then i' takes only one value m, and exterior differentiation leads from (5.5.4) to

$$d\kappa = \psi\omega^m, \quad (\kappa - \lambda)\omega_{i_1}^m = \psi\omega^{i_1} + \chi_{i_1}\omega^m, \quad d\lambda = \chi_{i_1}\omega^{i_1} + \zeta\omega^m, \tag{5.5.5}$$

where $\kappa\lambda + c = 0$ gives by differentiation that $\chi_{i_1} = 0$, $\zeta = -\lambda\kappa^{-1}\psi$. Hence

$$d\kappa = \psi\omega^m, \quad \omega_{i_1}^m = \psi^*\omega^{i_1}, \tag{5.5.6}$$

where $\psi^* = \psi\kappa(\kappa^2 + c)^{-1}$; note that here $\kappa^2 + c \neq 0$, because otherwise $\kappa = \sqrt{-c}$ and then $\lambda = \sqrt{-c} = \kappa$, which is impossible here (it would imply $s = m$, which will be considered below).

The first equation (5.5.6) implies via exterior differentiation that $d\psi \wedge \omega^m = 0$, whence $d\psi = \varphi\omega^m$. The next two equations imply $d\psi^* = (\psi^*)^2\omega^m$, due to (1.3.9) (where now $g_{IK} = \delta_{IK}$).

Since due to (5.5.6), $d\omega^m = \omega^{i_1} \wedge \omega^m_{i_1} = 0$, there exists (at least locally) a function t on M^m such that $\omega^m = dt$. Now the equation for ψ^* can be integrated and gives $\psi^* = -t^{-1}$, by suitable choice of the initial value of t. Therefore, $\psi = -\frac{\kappa^2+c}{\kappa t}$ and so the first equation (5.5.6) is

$$\frac{d(\kappa^2 + c)}{\kappa^2 + c} = -2\frac{dt}{t}.$$

This implies

$$\kappa^2 = kt^{-2} - c, \tag{5.5.7}$$

where k is a constant.

As is seen from (5.5.4) and (5.5.6), the submanifolds of M^m defined by $t = \text{const}$ are all totally umbilic with mean curvature vector $e_m\psi^* + e_{m+1}\kappa$.

The curves of M^m defined by $\omega^{i_1} = 0$ $(i_1 = 1, \ldots, m-1)$ are orthogonal trajectories of the family of these totally umbilic submanifolds. For each of them,

$$dx = e_m dt, \quad de_m = -(c\kappa^{-1}e_{m+1} + cx)dt, \quad de_{m+1} = c\kappa^{-1}e_m dt.$$

Here t is the arclength parameter of this trajectory, whose principal normal in the geometry of $N^{m+1}(c)$ is the unit vector e_{m+1}, which is orthogonal to M^m. Hence the trajectory is a geodesic line of M^m. At the same time it is also a line of curvature of M^m and belongs to a plane $N^2(c)$, spanned at x by the vectors e_m (tangent) and e_{m+1} (normal). Considering an $(m-1)$-parameter family of the planes of these geodesic lines of curvature, if one wants to find the envelope of this family, one has to find the points with radius vectors $y = x + e_m y^m + e_{m+1} y^{m+1}$, having the property that dy belongs to the same plane. Since that component in the expression of dy which cannot belong to this plane, is $(1 - \psi^* y^m - \kappa y^{m+1})(e_1\omega^1 + \cdots + e_{m-1}\omega^{m-1})$, for the above points one must have $1 - \psi^* y^m - \kappa y^{m+1} = 0$. Hence, all planes of the geodesic lines of curvature intersect in the straight line whose equation was just obtained.

It is easy to see that the centers of the totally umbilic leaves satisfy this equation and therefore lie on this straight line. Therefore, in this case the hypersurface M^m is invariant with respect to the isometries of $N^{m+1}(c)$, leaving this straight line fixed, and hence is a rotation hypersurface in $N^{m+1}(c)$.

It remains to describe the profile curve, the geodesic line of curvature. From the above equations it follows that the geodesic curvature of this curve in M^m is $\kappa_g = -c\kappa^{-1}$. Then (5.5.7) implies that the natural equation of the profile curve in the geometry of $N^2(c)$ is

$$\kappa_g = \frac{ct}{\sqrt{k - ct^2}}. \tag{5.5.8}$$

Now let $s > 1$ and $m - s > 1$. Then it is better to denote the index i' in (5.5.4) by i_2, as in Section 5.3; note that here κ and λ play the roles of $\kappa_{(1)}$ and $\kappa_{(2)}$ of that section.

Equations (5.5.4) imply via exterior differentiation that

$$d\kappa \wedge \omega^{i_1} + (\lambda - \kappa)\omega^{i_1}_{j_2} \wedge \omega^{j_2} = 0, \quad (\lambda - \kappa)\sum_{i_1} \omega^{i_1}_{j_2} \wedge \omega^{i_1} + d\lambda \wedge \omega^{j_2} = 0.$$

Here the indices i_1 and j_2 both take more than one value. Therefore, due to Cartan's lemma, $d\kappa = \kappa_{j_2}\omega^{j_2}$, $d\lambda = \lambda_{i_1}\omega^{i_1}$, whence

$$(\lambda - \kappa)\omega^{i_1}_{j_2} = \kappa_{j_2}\omega^{i_1} + \lambda_{i_1}\omega^{j_2}.$$

But here $\kappa\lambda = -c$ leads to $\kappa_{j_2} = \lambda_{i_1} = 0$, so that κ and λ are constants, and $\omega^{i_1}_{j_2} = 0$ (cf. (5.3.7) and (5.3.11), where the range of u is now empty). Therefore,

$$de_{i_1} = e_{k_1}\omega^{k_1}_{i_1} + k^*_{(1)}\omega^{i_1}, \quad de_{j_2} = e_{k_2}\omega^{k_2}_{j_2} + k^*_{(2)}\omega^{j_2},$$

where $k^*_{(1)} = \kappa e_{m+1} - cx$ and $k^*_{(2)} = \lambda e_{m+1} - cx$. Here $\langle k^*_{(1)}, k^*_{(2)}\rangle = \kappa\lambda + c = 0$ and

$$dk^*_{(1)} = -(\kappa^2 + c)e_{i_1}\omega^{i_1}, \quad dk^*_{(2)} = -(\lambda^2 + c)e_{j_2}\omega^{j_2}.$$

Hence M^m is a product of an s-dimensional spherical submanifold and an $(m-s)$-dimensional spherical submanifold. Both are parallel submanifolds, and thus M^m is also parallel.

Finally, let $s = m$. Then M^m is totally umbilic, hence spherical and parallel.

The preceding results can be summarized as follows.

Theorem 5.5.3. *A semiparallel submanifold M^m with principal codimension* 1 *in a non-Euclidean space form $N^n(c)$ is either*

- *a curve M^1, or*
- *a hypersurface in $N^{m+1}(c) \subset N^n(c)$, which is either*
  - *a surface M^2 with zero Gaussian curvature in $N^3(c)$, or*
  - *a rotation hypersurface whose profile curve has natural equation* (5.5.8), *or*
  - *a parallel hypersurface, which is either spherical or a product of two spherical hypersurfaces of lower dimensions, or*
  - *an open subset of one of the above.*

Remark 5.5.4. The essential part of this theorem was proved for hypersurfaces in [Di 91b]. The proof given here is more direct and starts from a more general assumption. Also the case of a parallel hypersurface is studied here in more detail. The explicit parametric equations of the profile curves were derived at the same time in [Di 91b] for all particular cases.

5.6 Semiparallel Submanifolds of Principal Codimension 2 in Euclidean Space

It follows from Proposition 5.1.3 that a semiparallel submanifold M^m of principal codimension 2 and with nonzero mean curvature vector H in $N^n(c)$ is normally flat. Here the hypothesis about H is essential only if $c > 0$, because for $c \leq 0$ the identity $H \equiv 0$ leads by Theorem 4.1.7 to zero principal codimension, and thus is impossible in this case.

In the present section the case $c = 0$ will be considered, and thus $N^n(0) = E^n$ is a Euclidean space. A semiparallel M^m with principal codimension 2 is normally flat in this case. Since span$\{k_1, \dots, k_m\}$ now has dimension 2, only two principal curvature vectors can be nonzero and distinct, and therefore also orthogonal (see Section 5.1).

They are either (i) both with multiplicities > 1, or (ii) one of multiplicity > 1, the other simple, or (iii) both simple.

In subcase (i), the range of indices $a, b, \dots$ is empty, and ρ takes only two values 1 and 2, using the notation introduced in Section 5.3. Therefore, only equations (5.3.7), (5.3.9) and the first equations of (5.3.11) work here, where $u = j_0$. It follows that $d\omega^{i_\rho} = \omega^{j_\rho} \wedge \theta^{i_\rho}_{j_\rho}$, where $\theta^{i_\rho}_{j_\rho} = \omega^{i_\rho}_{j_\rho} + \delta^{i_\rho}_{j_\rho} \lambda_{(\rho)k_0}\omega^{k_0}$. Hence the distribution on M^m defined by $\omega^{i_1} = \omega^{i_2} = 0$ is a foliation, whose leaves have the tangent vectors e_{i_0}. It follows from (5.3.5)–(5.3.7) that along a leaf $de_{i_0} = e_{j_0}\omega^{j_0}_{i_0}$; hence every leaf is a plane in E^n.

On the other hand, it follows from (5.4.1) that the vectors $l_{(\rho)} = \sum_u \lambda_{(\rho)u} e_u$ are orthogonal and satisfy the equation $dl_{(\rho)} = l_{(\rho)}(\lambda_{(\rho)u}\omega^u) - l^2_{(\rho)}(e_{i_\rho}\omega^{i_\rho})$. Therefore, their directions are fixed on each of these plane leaves.

If both $l_{(\rho)}$ are zero, then $\lambda_{(\rho)u} = 0$. Due to (5.3.11), both κ_ρ are constants; and due to (5.3.7,) $\omega^u_{i_\rho} = 0$. Then it follows from (5.3.5) that M^m is a product submanifold $S^{\nu_1} \times S^{\nu_2} \times E^{m-\nu_1-\nu_2}$.

If both $l_{(\rho)}$ are nonzero, then the orthonormal frame bundle can be adapted further to M^m so that among the basis vectors e_{i_0} the last two e_{m-1} and e_m are collinear with $l_{(1)}$ and $l_{(2)}$, respectively. Then $\lambda_{(1)u'} = \lambda_{(1)m} = \lambda_{(2)u'} = \lambda_{(2)m-1} = 0$, $\lambda_{(1)m-1} = \lambda_1 \neq 0$, and $\lambda_{(2)m} = \lambda_2 \neq 0$, where $u', v', \dots$ run through $\{\nu_1 + \nu_2 + 1, \dots, m-2\}$. Now equations (5.3.7) imply

$$\omega^{j_2}_{i_1} = \omega^{u'}_{i_1} = \omega^{m}_{i_1} = \omega^{u'}_{j_2} = \omega^{m-1}_{j_2} = 0, \quad \omega^{m-1}_{i_1} = \lambda_1\omega^{i_1}, \quad \omega^m_{j_2} = \lambda_2\omega^{j_2};$$

moreover, equations (5.4.1) imply that

$$\omega^{m-1}_{u'} = \omega^m_{m-1} = \omega^m_{u'} = 0, \quad d\lambda_1 = \lambda_1^2\omega^{m-1}, \quad d\lambda_2 = \lambda_2^2\omega^m.$$

Therefore, the distributions defined on M^m by $\omega^{i_1} = \omega^{j_2} = \omega^{m-1} = \omega^m = 0$, then by $\omega^{j_2} = \omega^{u'} = \omega^m = 0$, and then by $\omega^{i_1} = \omega^{u'} = \omega^{m-1} = 0$ are foliations. Their leaves have the tangent vectors $e_{u'}$, then e_{i_1}, e_{m-1}, and then e_{j_2}, e_m, respectively. Here $de_{u'} = e_{v'}\omega^{v'}_{u'}$, whence

$$de_{i_1} = e_{k_1}\omega^{k_1}_{i_1} + \omega^{i_1}(\lambda_1 e_{m-1} + \kappa_1 e_{m+1}), \quad de_{m-1} = -\lambda_1 e_{i_1}\omega^{i_1},$$

and then

$$de_{j_2} = e_{k_2}\omega^{k_2}_{j_2} + \omega^{j_2}(\lambda_2 e_m + \kappa_2 e_{m+2}), \quad de_m = -\lambda_2 e_{j_2}\omega^{j_2};$$

moreover, (5.3.9) implies $de_{m+1} = -\kappa_1 e_{i_1}\omega^{i_1}$ and $de_{m+2} = -\kappa_2 e_{j_1}\omega^{j_2}$.

It follows that M^m is contained in an E^{m+2}, spanned at the point x by the vectors $e_1, \dots, e_m, e_{m+1}, e_{m+2}$, and the leaves of the three foliations above are parallel planes in E^{m+2} with totally orthogonal vector subspaces. The leaves of the first foliation are parallel $(m - \nu_1 - \nu_2 - 2)$-dimensional planes. And, for the leaves of the other two foliations, equations (5.5.2) and (5.5.3) can be used to obtain the conclusion that M^m is a product $C^{\nu_1+1} \times C^{\nu_2+1} \times E^{m-\nu_1-\nu_2-2}$ in E^{m+2}, where the first two components are round cones.

Finally, if one of $l_{(\rho)}$ is zero and the other nonzero, then one gets the limit case where one of these round cones tends to the round cylinder. This finishes the analysis of subcase (i).

In subcase (ii), the indices ρ and a each take only one value, namely 1 and $\nu_1 + 1$, respectively. Thus M^m is, according to the remark at the beginning of Section 5.3, a second-order envelope of the product submanifolds $S^{\nu_1} \times S^1 \times E^{m-\nu_1-1}$. Equations (5.3.7)–(5.3.11) reduce to

$$\omega_{i_1}^u = \lambda_{(1)u}\omega^{i_1}, \quad \omega_a^{i_0} = \lambda_{i_0}\omega^a, \tag{5.6.1}$$

$$\omega_{m+1}^{m+1+a} = -\kappa_a\kappa_1^{-1}\lambda_{(1)a}\omega^a, \quad \omega_{m+1}^{\xi} = 0, \quad \omega_{m+1+a}^{\xi} = \varphi_a^{\xi}\omega^a, \tag{5.6.2}$$

$$d\kappa_1 = \kappa_1 \sum_u \lambda_{(1)u}\omega^u, \quad d\kappa_a = \psi_a\omega^a + \kappa_a \sum_{i_0} \lambda_{i_0}\omega^{i_0}, \tag{5.6.3}$$

where the notation is simplified by introducing $\lambda_{i_0} = \lambda_{(a)i_0}$. In the argument which led to Theorem 5.4.1 above, one has in the current case (ii) $p = q = 1$, $\nu = \nu_1$, and $m' = \nu_0 + 1 = m - \nu$. Hence this theorem implies that M^m is a warped product submanifold $B^{m'} \times_r S^{\nu}(1)$ with locally Euclidean $B^{m'}$ and nonconstant linear warping function r; moreover, $B^{m'}$ is the envelope of a one-parameter family of m'-dimensional planes in E^n.

Such an M^m can be described in more detail as follows. Due to (5.3.5), (5.3.6), and (5.6.1), one has

$$de_{i_1} = e_{j_1}\omega_{i_1}^{j_1} + \left(\sum_u \lambda_{(1)u}e_u + \kappa_1 e_{m+1}\right)\omega^{i_1}, \tag{5.6.4}$$

$$de_a = -\lambda_{(1)a}e_{i_1}\omega^{i_1} + \left(\sum_{i_0} \lambda_{i_0}e_{i_0} + \kappa_a e_{m+1+a}\right)\omega^a, \tag{5.6.5}$$

$$de_{i_0} = -\lambda_{(1)i_0}e_{i_1}\omega^{i_1} - \lambda_{i_0}e_a\omega^a + e_{j_0}\omega_{i_0}^{j_0}. \tag{5.6.6}$$

Due to (5.6.1), the distribution defined on M^m by $\omega^{i_1} = 0$ is a foliation. Equations (5.6.5) and (5.6.6) imply that its leaves are envelopes of one-parameter families of m'-dimensional planes, like $B^{m'}$.

Also the distribution determined on M^m by $\omega^{i_0} = 0$ is a foliation. On each of its leaves, the system $\omega^a = 0$ defines a family of ν-dimensional spheres S^{ν}. Along each of these spheres, $d\lambda_{(1)a} = \lambda_{(1)a}\lambda_{(1)i_0}\omega^{i_0}$, due to (5.4.1) and (5.6.1); and if $\lambda_{(1)a} \neq 0$, then $d(x - \lambda_{(1)a}^{-1}e_a) = 0$, i.e., the point with radius vector $x - \lambda_{(1)a}^{-1}e_a$ is fixed. Therefore, the tangent lines to a leaf in the direction of e_a at all points of such a sphere S^ν generate a round cone with vertex at this fixed point. There exists a sphere $S^{\nu+1}$ which is tangent to this cone and to the above-considered leaf along this sphere S^ν. If $\lambda_{(1)a} = 0$, then instead of the cone there is a round cylinder, and a sphere $S^{\nu+1}$ with the properties mentioned above exists as before. Consequently, the leaf is the envelope of the one-parameter family of spheres $S^{\nu+1}$, and therefore it can be considered a *canal submanifold.* Moreover, the characteristic spheres S^ν in this leaf lie in parallel $(\nu+1)$-dimensional planes, as follows from (5.6.5), due to (5.4.1) and (5.6.1)–(5.6.3). Since the radius r of S^ν is a nonconstant linear function on $B^{m'}$, this canal submanifold can be considered a *warped cone.*

Finally, consider the distribution defined on M^m by the equation $\omega^a = 0$. Since (5.6.1) implies $d\omega^a = \omega^a \wedge \lambda_{i_0}\omega^{i_0}$, this distribution is also a foliation. For its leaves, equations (5.6.4) and (5.6.6) reduce to

$$de_{i_1} = e_{j_1}\omega_{i_1}^{j_1} + H_1^*\omega^{i_1}, \quad de_{i_0} = -\lambda_{(1)i_0}e_{i_1}\omega^{i_1} + e_{j_0}\omega_{i_0}^{j_0}.$$

For $H_1^* = \sum_u \lambda_{(1)u}e_u + \kappa_1 e_{m+1}$, it follows from (5.4.4) that $dH_1^* = H_1^*\lambda_{(1)i_0}\omega^{i_0} - (H_1^*)^2 e_{i_1}\omega^{i_1}$. These formulas show that every leaf is a hypersurface M^{m-1} of an m-dimensional plane, which is described in Theorem 5.5.1 as its middle case, i.e., is a product $C^{\nu+1} \times E^{m-\nu-2}$ (or its limiting case $S^\nu \times E^{m-\nu-1}$).

It remains to study subcase (iii) of the case $c = 0$. Then $p = 0$ and $q = 2$, and thus the range of indices i_ρ is empty and a, etc., run through $\{1, 2\}$. Equations (5.3.5)–(5.3.8) reduce to

$$\omega_1^{m+1} = \kappa_1\omega^1, \quad \omega_2^{m+2} = \kappa_2\omega^2,$$

$$\omega_2^{m+1} = \omega_{i_0}^{m+1} = \omega_1^{m+2} = \omega_{i_0}^{m+2} = \omega_a^\xi = \omega_{i_0}^\xi = 0,$$

$$\omega_1^2 = \kappa_1\gamma_1^2\omega^1 - \kappa_2\gamma_2^1\omega^2, \quad \omega_1^{i_0} = \lambda_{(1)i_0}\omega^1, \quad \omega_2^{i_0} = \lambda_{(2)i_0}\omega^2.$$

Therefore,

$$de_{i_0} = -\left(e_1\lambda_{(1)i_0}\omega^1 + e_2\lambda_{(2)i_0}\omega^2\right) + e_{j_0}\omega_{i_0}^{j_0},$$

$$de_1 = e_2\omega_1^2 + \left(\sum_{i_0} e_{i_0}\lambda_{(1)i_0} + \kappa_1 e_{m+1}\right)\omega^1,$$

$$de_2 = -e_1\omega_1^2 + \left(\sum_{i_0} e_{i_0}\lambda_{(2)i_0} + \kappa_2 e_{m+2}\right)\omega^2.$$

The equations $\omega^1 = \omega^2 = 0$ define on M^m a foliation whose leaves are $(m-2)$-dimensional planes in E^n, along which e_1 and e_2 are constant vectors. Hence M^m is the envelope of a two-parameter family of m-dimensional planes. Moreover, since now $\Omega_1^2 = \omega_1^\alpha \wedge \omega_\alpha^2 = 0$, this M^m has flat ∇ and hence is locally Euclidean. Recall also that $\nabla^\perp$ is flat, which, all together implies that $\bar\nabla$ is flat.

The above results can be summarized as follows.

Theorem 5.6.1. *A semiparallel submanifold M^m with principal codimension 2 in Euclidean space E^n is either*

- *a product submanifold $C^{\nu_1+1} \times C^{\nu_2+1} \times E^{m-\nu_1-\nu_2-2}$ in $E^{m+2} \subset E^n$, where the first two components are round cones of dimension > 2 with point-vertex (each of these two could also be a round cylinder), or*
- *a warped product submanifold $B^{m'} \times_r S^\nu(1)$, $\nu > 1$, $m' = m - \nu$, where $B^{m'}$ is the envelope of a one-parameter family of m'-dimensional planes in E^n, hence locally Euclidean, and r is a nonconstant linear function, or*
- *the envelope of a two-parameter family of m-dimensional planes, which has flat $\bar{\nabla}$, or*
- *an open subset of one of the above.*

Remark 5.6.2. Semiparallel submanifolds M^m in E^{m+2} were classified in [Lu 88b]. The preceding theorem generalizes those results to the case of semiparallel M^m with principal codimension 2 in E^n.

Note that for the submanifolds $B^{m'} \times_r S^\nu(1)$ of the middle class, more detailed information is given in the analysis above about the behavior of the fibre spheres $S^\nu(r^{-2})$ along the base submanifold $B^{m'}$: in every generating $(m'-1)$-plane, there is a direction in which they generate round cones, and in the other directions they generate round cylinders, and along the orthogonal trajectories of these plane generators, they generate canal submanifolds, which are warped cones.

5.7 Normally Flat Semiparallel Submanifolds of Principal Codimension 2 in Non-Euclidean Space Forms

As seen above, a semiparallel submanifold M^m of principal codimension 2 in a non-Euclidean space form $N^n(c)$ must be normally flat if $c \le 0$, but not if $c > 0$ (cf. Proposition 5.1.3 and Remark 5.1.4). Therefore, in order to include the case $c > 0$ in this section, normal flatness has to be assumed explicitly.

So let M^m be a normally flat semiparallel submanifold with principal codimension 2 in $N^n(c)$. Among its distinct principal curvature vectors $k_{(1)}, \dots, k_{(r)}$ at least one must be nonzero; and by renumbering if needed, $k_{(1)} \neq 0$. The orthonormal frame bundle $O(M^m, N^n(c))$ can be adapted so that

$$k_{(1)} = \kappa_1 e_{m+1}, \quad k_{(a)} = \kappa_a e_{m+1} + \mu_a e_{m+2},$$

where $a, b, \dots$ run through $\{2, \dots, r\}$ and $\kappa_1 \neq 0$.

Due to (5.1.5), for every value of a there holds $\langle k_{(1)}, k_{(a)} \rangle + c = \kappa_1 \kappa_a + c = 0$; thus $\kappa_1(\kappa_a - \kappa_b) = 0$, if $a \neq b$. It follows that $\kappa_2 = \dots = \kappa_r = \kappa$, where $\kappa \neq 0$, due to $c \neq 0$; and hence $\kappa_1 = -c\kappa^{-1}$. Moreover, $\langle k_{(a)}, k_{(b)} \rangle = \kappa^2 + \mu_a \mu_b + c = 0$, for $a \neq b$. Here at least one μ_a must be nonzero, because otherwise if all $\mu_a = 0$, one has $\kappa = \sqrt{-c}$ and thus also $\kappa_1 = \sqrt{-c}$, which implies $k_{(1)} = \dots = k_{(r)}$, which is impossible here. So, assume that $\mu_2 \neq 0$ (further renumbering if needed). Now $\mu_2(\mu_a - \mu_b) = 0$, if the indices 2, a, and b are distinct, which implies $\mu_3 = \dots =$

$\mu_r = \mu \neq 0$; but this leads to $k_3 = \cdots = k_r$, which is possible only for $r = 3$. Therefore, $\mu_3 = -\mu^{-1}(\kappa^2 + c)$ and thus

$$k_1 = -c\kappa^{-1}e_{m+1}, \quad k_{(2)} = \kappa e_{m+1} - \mu e_{m+2},$$
$$k_{(3)} = \kappa e_{m+1} + \mu^{-1}(\kappa^2 + c)e_{m+2}. \tag{5.7.1}$$

All this shows that the submanifold M^m satisfying the current hypotheses is defined by the following differential system:

$$\omega^{m+1} = \omega^{m+2} = \omega^{\xi} = 0,$$
$$\omega_{i_1}^{m+1} = -c\kappa^{-1}\omega^{i_1}, \quad \omega_{i_2}^{m+1} = \kappa\omega^{i_2}, \quad \omega_{i_3}^{m+1} = \kappa\omega^{i_3}, \tag{5.7.2}$$
$$\omega_{i_1}^{m+2} = 0, \quad \omega_{i_2}^{m+2} = \mu\omega^{i_2}, \quad \omega_{i_3}^{m+2} = -\mu^{-1}(\kappa^2 + c)\omega^{i_3}, \tag{5.7.3}$$
$$\omega_{i_1}^{\xi} = \omega_{i_2}^{\xi} = \omega_{i_3}^{\xi} = 0, \tag{5.7.4}$$

where ξ runs through $\{m+3, \ldots, n\}$.

By exterior differentiation of (5.7.4), it follows that

$$\omega_{m+1}^{\xi} \wedge \omega^{i_1} = 0, \quad [\kappa\omega_{m+1}^{\xi} + \mu\omega_{m+2}^{\xi}] \wedge \omega^{i_2} = 0,$$
$$[\kappa\omega_{m+1}^{\xi} - \mu^{-1}(\kappa^2 + c)\omega_{m+2}^{\xi}] \wedge \omega^{i_3} = 0.$$

Due to Cartan's lemma

$$\omega_{m+1}^{\xi} = U^{\xi}\omega^{i_1}, \quad \kappa\omega_{m+1}^{\xi} + \mu\omega_{m+2}^{\xi} = V^{\xi}\omega^{i_2},$$
$$\kappa\omega_{m+1}^{\xi} - \mu^{-1}(\kappa^2 + c)\omega_{m+2}^{\xi} = W^{\xi}\omega^{i_3}.$$

Eliminating ω_{m+2}^{ξ} and then substituting the expression of ω_{m+1}^{ξ}, one obtains

$$\kappa\mu^{-1}(\kappa^2 + \mu^2 + c)U^{\xi} = \mu^{-1}(\kappa^2 + c)V^{\xi} = \mu W^{\xi} = 0,$$

because ω^{i_1}, ω^{i_2}, and ω^{i_3} are linearly independent. Here $\kappa^2 + c \neq 0$ and $\kappa^2 + \mu^2 + c \neq 0$, because $\kappa^2 + c = 0$ would lead to $k_1 = k_2$ and $\kappa^2 + \mu^2 + c = 0$ to $k_2 = k_3$, which are both impossible here. Therefore, $U^{\xi} = V^{\xi} = W^{\xi} = 0$, and hence $\omega_{m+1}^{\xi} = \omega_{m+2}^{\xi} = 0$. Consequently, the submanifold M^m is contained in an $N^{m+2}(c) \subset N^n(c)$.

Further let us use the notation introduced in Section 5.3.

First let $p = 0$, i.e., let $k_{(1)}, k_{(2)}, k_{(3)}$ be simple k_1, k_2, k_3. Due to (5.1.4), $\Omega_i^j = -(\langle k_i, k_j\rangle + c)\omega^i \wedge \omega^j$, where now i, j are in $\{1, 2, 3\}$. From (5.7.1) one can easily deduce that all Ω_i^j are zero, thus the submanifold M^3 of $N^5(c)$, considered in this case, has flat van der Waerden–Bortolotti connection $\bar{\nabla}$.

An analysis of the above exterior system by means of Cartan's theory (see [Ca 45], [Fin 48], [Gri 83], [GJ 87], [BCGGG 91]) shows that this M^3 in $N^5(c)$ does exist and depends on six real holomorphic functions of one real argument. Indeed, in six exterior equations there are six secondary 1-forms: $d\kappa, d\mu, \omega_2^1, \omega_3^1, \omega_3^2, \omega_4^5$, and after applying Cartan's lemma, it turns out that Cartan's test is satisfied.

From (5.7.1) it is seen that $k_i^* = k_i - xc$ are three nonzero mutually orthogonal outer curvature vectors for this M^3 in ${}_\sigma E^6$; therefore the outer principal codimension of M^3 is 3. Hence this M^3 is a Cartan variety (see Remark 5.4.3).

If $k_{(1)}$ is nonsimple with multiplicity $\nu > 1$, but $k_{(2)}$ and $k_{(3)}$ are simple k_2, k_3, then in Theorem 5.4.1 $p = 1, m = \nu + 2, m' = 2$.

If $k_{(1)}$ and $k_{(2)}$ are nonsimple with multiplicities ν_1 and ν_2, but $k_{(3)}$ is simple k_3, then in this theorem $p = 2, m = \nu_1 + \nu_2 + 1$ and $m' = 1$.

If all $k_{(1)}, k_{(2)}, k_{(3)}$ are nonsimple, then $p = 3, m = \nu_1 + \nu_2 + \nu_3, m' = 0$, and M^m lies in an $(m+2)$-dimensional sphere of ${}_\sigma E^{m+3}$ as the product $Sph_1^{\nu_1}(c_1) \times Sph_2^{\nu_2}(c_2) \times Sph_3^{\nu_3}(c_2)$ in $N^{m+2}(c)$.

Recall that Theorem 5.4.1 describes only the irreducible normally flat M^m. For the complete classification of the normally flat semiparallel M^m of principal codimension 2, the requisite product submanifolds must also be added.

The results of all these considerations can be summarized as follows.

Theorem 5.7.1. *A normally flat semiparallel submanifold M^m with principal codimension 2 in a non-Euclidean space form $N^n(c)$ is either*

- *an M^3 with flat $\bar{\nabla}$ in an $N^5(c) \subset N^n(c)$, which is a Cartan variety, or*
- *a warped product submanifold $B^2 \times_r Sph^{m-2}(1)$ with $m > 3$ in an $N^{m+2}(c) \subset N^n(c)$, where B^2 is a surface with flat $\bar{\nabla}$ and r is a linear warping function on it, or*
- *a warped product submanifold $B^1 \times_{r_1} Sph^{\nu_1}(1) \times_{r_2} Sph^{\nu_2}(1)$, where $m = \nu_1 + \nu_2 + 1 > 4$, in an $N^{m+2}(c) \subset N^n(c)$; here B^1 is a curve, and r_1 and r_2 are some linear warping functions on it, or*
- *a product $Sph_1^{\nu_1}(c_1) \times Sph_2^{\nu_2}(c_2) \times Sph_3^{\nu_3}(c_2)$, where $m = \nu_1 + \nu_2 + \nu_3 > 5$, in an $N^{m+2}(c) \subset N^n(c)$, or*
- *a product of one of these $\tilde{m}$-dimensional submanifolds above and a totally geodesic $N^{m-\tilde{m}}(c)$, where $3 \leq \tilde{m} < m$, or*
- *a product of two semiparallel hypersurfaces, described in Theorem* 5.5.3.

6

Semiparallel Surfaces

In this chapter semiparallel submanifolds of dimension 2 (surfaces) in space and in spacetime forms ${}_sN^n(c)$, $s \geq 0$, will be considered. With respect to the dimension of the submanifold, this is the simplest nontrivial case, because all submanifolds of dimension 1 (curves) are trivially semiparallel, namely, with flat $\bar{\nabla}$; see Proposition 4.1.4.

The inner metric of M^2 is assumed to be regular, i.e., either Riemannian or pseudo-Riemannian. The surface is said to be, respectively, *spacelike* or *timelike*, using terms familiar from relativity theory. All semiparallel surfaces M^2 with regular metric in ${}_sN^n(c)$ will be classified and geometrically described.

6.1 Semiparallel Spacelike Surfaces

Deprez in his paper [De 85], where the concept of semiparallel submanifold M^m in Euclidean E^n was introduced, also gave a classification of these submanifolds for dimension $m = 2$, i.e., of semiparallel surfaces M^2 in E^n.

In this section this classification is extended to semiparallel spacelike surfaces M^2 (i.e., with Riemannian inner metric) in ${}_sN^n(c)$, under some supplementary regularity assumptions (see [Lu 99a][1]).

The tangent frame vectors in the orthonormal frame bundle adapted to such an M^2 transform according to

$$e'_1 = e_1 \cos\varphi + e_2 \sin\varphi, \quad e'_2 = -e_1 \sin\varphi + e_2 \cos\varphi. \tag{6.1.1}$$

Substitution into $dx = e_i\omega^i$ leads to

$$\omega^1 = \omega^{1'} \cos\varphi - \omega^{2'} \sin\varphi, \quad \omega^2 = \omega^{1'} \sin\varphi + \omega^{2'} \cos\varphi; \tag{6.1.2}$$

[1] A preliminary partial version of this extension was given in two master's theses [Rä 94] and [Fil 95] written under the author's supervision; the main results of the second thesis were published in [Saf 2001] (under a new name after marriage).

Ü. Lumiste, *Semiparallel Submanifolds in Space Forms*,
DOI 10.1007/978-0-387-49913-0_7,

hence for $h_{ij}\omega^i\omega^j$ the transformation formulas are

$$h_{1'1'} = \frac{1}{2}(h_{11} + h_{22}) + \frac{1}{2}(h_{11} - h_{22})\cos 2\varphi + h_{12}\sin 2\varphi, \tag{6.1.3}$$

$$h_{1'2'} = \frac{1}{2}(h_{22} - h_{11})\sin 2\varphi + h_{12}\cos 2\varphi, \tag{6.1.4}$$

$$h_{2'2'} = \frac{1}{2}(h_{11} + h_{22}) + \frac{1}{2}(h_{22} - h_{11})\cos 2\varphi - h_{12}\sin 2\varphi. \tag{6.1.5}$$

Setting $\frac{1}{2}(h_{11} - h_{22}) = A$, $h_{12} = B$ and recalling that $\frac{1}{2}(h_{11} + h_{22}) = H$ is the mean curvature vector (see Section 2.4), one obtains

$$A' = A\cos 2\varphi + B\sin 2\varphi, \quad B' = -A\sin 2\varphi + B\cos 2\varphi, \quad H' = H. \tag{6.1.6}$$

Hence span$\{A, B\}$ is an invariant subspace and H an invariant vector in the first-order normal subspace span$\{h_{11}, h_{12}, h_{22}\}$ = span$\{A, B, H\}$ at every point $x \in M^2$; moreover,

$$\langle A', B'\rangle = \langle A, B\rangle\cos 4\varphi + \frac{1}{2}(B^2 - A^2)\sin 4\varphi. \tag{6.1.7}$$

From here it follows that there exists a φ_0 such that $\langle A', B'\rangle = 0$; then $\varphi_0 \pm \frac{\pi}{2}$ gives the same result, but with the roles of A' and B' interchanged.

Suppose that this transformation has been done, i.e., suppose $\langle A, B\rangle = 0$.

Moreover, assume M^2 is regular, in the sense that its first-order normal subspaces span$\{A, B, H\}$ have the same dimension m_1 and also the subspaces span$\{A, B\}$ have the same dimension m_0; here, of course, $0 \leq m_1 \leq 3$ and $0 \leq m_0 \leq \min\{2, m_1\}$. In general, M^2 is a union of closures of its regular open subsets, and one can restrict attention to one of these sets.

In the most general case $m_1 = 3$ and $m_0 = 2$. Then the bundle of frames, orthonormal in its tangent part $\{e_1, e_2\}$, can be adapted further so that

$$A = ke_4, \quad B = le_5, \quad H = \alpha e_3 + \beta e_4 + \gamma e_5. \tag{6.1.8}$$

Then

$$h_{11} = \alpha e_3 + (\beta + k)e_4 + \gamma e_5, \quad h_{12} = le_5, \quad h_{22} = \alpha e_3 + (\beta - k)e_4 + \gamma e_5,$$

thus

$$\omega_1^3 = \alpha\omega^1, \quad \omega_1^4 = (\beta + k)\omega^1, \quad \omega_1^5 = \gamma\omega^1 + l\omega^2, \quad \omega_1^\xi = 0, \tag{6.1.9}$$

$$\omega_2^3 = \alpha\omega^2, \quad \omega_2^4 = (\beta - k)\omega^2, \quad \omega_2^5 = l\omega^1 + \gamma\omega^2, \quad \omega_2^\xi = 0, \tag{6.1.10}$$

where $\xi, \eta, \ldots$ run through $\{6, \ldots, n\}$.

In this general case $\alpha kl \neq 0$. But the formulas (6.1.8)–(6.1.10) also hold in the special degenerate cases, which are listed in rows (b)–(f) of Table 6.1.1, where in the last column the corresponding conditions are given, together with simplifications achievable by suitable adaption of the frame; only one possibility is shown, considering that A and B are interchangeable.

Table 6.1.1.

	m_1	m_0	Condition
(a)	3	2	$\alpha kl \neq 0$
(b)	2	2	$\alpha = 0, \quad kl \neq 0$
(c)	2	1	$l = 0, \quad \alpha k \neq 0$*
(d)	1	1	$\alpha = \gamma = l = 0, \quad k \neq 0$
(e)	1	0	$k = l = 0, \quad \alpha \neq 0$**
(f)	0	0	$k = l = H = 0$

*By further adaption of the frame, one can make $\gamma = 0$.
**One can make $\beta = \gamma = 0$.

For the last two subcases of the table, the surface M^2 can be characterized immediately. Indeed, in subcase (f) $h_{ij} = 0$, thus M^2 is totally geodesic, and in subcase (e) $h_{ij} = H\delta_{ij}$, thus M^2 is totally umbilic (see Proposition 3.1.2).

6.2 The Case of Regular Metrics

Assume now that span$\{A, B, H\}$ and span$\{A, B\}$ both have regular metrics. Then in (6.1.8), $\langle e_a, e_b \rangle = g_{ab} = \varepsilon_a \delta_{ab}$ $(a, b \cdots \in \{3, 4, 5\})$, where every ε_a is either 1 or -1. Moreover, $\langle e_a, e_\xi \rangle = g_{a\xi} = 0$, and due to (2.1.9),

$$\Omega_1^2 = -\Omega_2^1 = (\varepsilon_4 k^2 + \varepsilon_5 l^2 - H^2 - c)\omega^1 \wedge \omega^2, \tag{6.2.1}$$

$$\Omega_4^5 = -2\varepsilon_4 kl\omega^1 \wedge \omega^2, \quad \Omega_5^4 = 2\varepsilon_5 kl\omega^1 \wedge \omega^2; \tag{6.2.2}$$

all other Ω_i^j, Ω_α^β are zero.

The semiparallel condition (4.1.2) reduces to a system whose only essential equations are

$$\varepsilon_5 \gamma kl = 0, \tag{6.2.3}$$

$$l[\varepsilon_4 k(2k + \beta) + \varepsilon_5 l^2 - H^2 - c] = 0, \tag{6.2.4}$$

$$l[\varepsilon_4 k(2k - \beta) + \varepsilon_5 l^2 - H^2 - c] = 0, \tag{6.2.5}$$

$$k[\varepsilon_4 k^2 + 2\varepsilon_5 l^2 - H^2 - c] = 0. \tag{6.2.6}$$

In subcases (a) and (b) of the table, one has $kl \neq 0$. Then the first equation above yields $\gamma = 0$, the middle two imply $\beta = 0$ and $2\varepsilon_4 k^2 + \varepsilon_5 l^2 = H^2 + c$, and the last gives $\varepsilon_4 k^2 + 2\varepsilon_5 l^2 = H^2 + c$; hence $\varepsilon_4 k^2 = \varepsilon_5 l^2$, and thus $\varepsilon_4 = \varepsilon_5$.

This yields

$$\varepsilon_4 = \varepsilon_5 = \varepsilon = \pm 1, \quad k^2 = l^2, \quad H^2 = \varepsilon_3 \alpha^2 = 3\varepsilon k^2 - c; \tag{6.2.7}$$

moreover, one can make $k = l \neq 0$ by replacing e_5 by $-e_5$ if needed. Then due to (6.2.7),

$$\Omega_1^2 = -\varepsilon k^2 \omega^1 \wedge \omega^2, \quad \Omega_4^5 = -\Omega_5^4 = -2\varepsilon k^2 \omega^1 \wedge \omega^2, \tag{6.2.8}$$

i.e., the Gaussian curvature of M^2 is $K = \varepsilon k^2$.

In case (b) one has $H = 0$, i.e., M^2 is a minimal surface, and due to (6.2.7) this is possible only if $c \neq 0$ and $\varepsilon = \operatorname{sign} c$. Moreover, $c < 0$ is not possible here, because then due to Theorem 4.1.7 this minimal M^2 would be totally geodesic, which contradicts $k = l \neq 0$.

The normal curvature vector of M^2 is defined as $h(X, X)$, with $\|X\| = 1$ (see, e.g., [SchStr 35], B. II, Section 10). For $X = e_{1'}$, one has $h_{1'1'} = H + A\cos 2\varphi + B\sin 2\varphi$, due to (6.1.3). For subcases (a) and (b), it reduces to

$$h_{1'1'} = e_3\alpha + k(e_4\cos 2\varphi + e_5\sin 2\varphi), \tag{6.2.9}$$

and therefore the normal curvature vector has constant scalar square $4\varepsilon k^2 - c$ at every point $x \in M^2$, i.e., M^2 is an *isotropic surface* (cf. O'Neill [O'N 65]); moreover $H^2 = 3K - c$.

In particular, if $s = 0$, i.e., for a semiparallel M^2 in $N^n(c)$, one has $\varepsilon_3 = \varepsilon = 1$, and one can make $\alpha = \sqrt{3k^2 - c}$, by replacing e_3 by $-e_3$, if needed.

For subcases (c) and (d), $\varepsilon_4 k^2 - H^2 - c = 0$, and hence $\Omega_1^2 = 0$, $\Omega_4^5 = \Omega_4^5 = 0$; i.e., M^2 is locally Euclidean and has flat normal connection $\nabla^\perp$ and thus flat $\bar{\nabla}$.

Recall that in subcase (e), M^2 is totally umbilic, and in subcase (f), totally geodesic.

The above analysis can be summarized as follows.

Theorem 6.2.1. *Let M^2 be a semiparallel surface, obtained by an isometric immersion of a two-dimensional Riemannian manifold into ${}_sN^n(c)$ such that the regularity assumptions are satisfied. Then it can be characterized for subcases* (a)–(f) *of Table* 6.1.1 *as follows:*

(a) *M^2 is isotropic (cf. O'Neill* [O'N 65]*), $n \geq 5$, the first-order normal subspace of M^2 is three-dimensional and has definite metric (positive or negative), and moreover, $H^2 = 3K - c$ at every point $x \in M^2$;*
(b) *M^2 is a minimal isotropic surface of Gaussian curvature $K = \frac{1}{3}c > 0$;*
(c), (d) *M^2 has flat $\bar{\nabla}$;*
(e) *M^2 is totally umbilic;*
(f) *M^2 is totally geodesic.*

Note that due to Proposition 3.1.2, the totally umbilic M^2 of subcase (e) is a parallel surface, as is the totally geodesic M^2 of subcase (f).

For the geometric description of the semiparallel surfaces of subcases (c) and (d), Theorem 4.5.5 can be used, according to which every semiparallel submanifold is a second-order envelope of parallel submanifolds. It remains to describe the parallel surfaces of subcases (c) and (d), i.e., with flat $\bar{\nabla}$.

For these surfaces $l = \gamma = 0$; therefore, (6.1.9) and (6.1.10) give

$$h_{11}^3 = h_{22}^3 = \alpha, \quad h_{12}^3 = 0, \quad h_{11}^4 = \beta + k, \quad h_{22}^4 = \beta - k, \quad h_{12}^4 = 0, \quad h_{ij}^\rho = 0,$$

where $\rho \in \{5, \dots, n\}$. Now the parallel condition (3.1.1) reduces to the following essential equations:

$$d\alpha = d\beta = dk = \omega_1^2 = \omega_3^4 = \omega_3^\rho = \omega_4^\rho = 0.$$

Thus for these parallel surfaces one has

$$dx = e_1\omega^1 + e_2\omega^2, \quad de_1 = h_{11}^*\omega^1, \quad de_2 = h_{22}^*\omega^2,$$

where $h_{11}^* = \alpha e_3 + (\beta + k)e_4 - cx$, $h_{22}^* = \alpha e_3 + (\beta - k)e_4 - cx$. Since for subcases (c) and (d) $\varepsilon_4 k^2 - H^2 = 0$, it follows that $\varepsilon_4(k^2 - \beta^2) - \varepsilon\alpha^2 = 0$; hence

$$dh_{11}^* = -(h_{11}^*)^2 e_1\omega^1, \quad dh_{22}^* = -(h_{22}^*)^2 e_2\omega^2$$

with constant $(h_{11}^*)^2$ and $(h_{22}^*)^2$. Therefore, each such parallel surface is a product of two plane curves of constant curvature, defined by $\omega^1 = 0$ and $\omega^2 = 0$.

The result can be formulated as follows (cf. Theorem 3.6.1 and Examples 4.5.2 and 4.5.4).

Proposition 6.2.2. *The only parallel spacelike curves in ${}_sN^n(c)$ are the plane curves of constant curvature. Every surface of Theorem* 6.2.1 *which has flat* $\bar{\nabla}$ *but is not totally geodesic, is a second-order envelope of products of such curves.*

A more detailed description of the isotropic semiparallel surfaces of subcases (a) and (b) will be given in the next section.

Remark 6.2.3. The subspace span$\{A, B\}$ has a good geometric interpretation. At a point $x \in M^2$, consider the normal curvature vectors $h(X, X)$, for all tangent vectors X with $\|X\| = 1$; their endpoints describe the *normal curvature indicatrix* (see [SchStr 35], Vol. II, Section 12). Due to (6.1.8), the indicatrix is generally an ellipse whose plane has direction span$\{A, B\}$ and goes through the endpoint of the vector H, with startpoint placed at x; it could also be a degenerate form of such an ellipse (a line segment, if $m_1 = 1$, or a point, if $m_1 = 0$). The adaptation (6.1.8) means that e_4 and e_5 are directed along the axes of symmetry of this ellipse with semiaxes k and l (see [Ca 60]).

For subcases (a) and (b), this ellipse is a circle, and this characterizes the isotropic surfaces, if this holds at every point $x \in M^2$.

Remark 6.2.4. If $c = 0$ and $s = 0$, then Theorem 6.2.1 reduces to a result of Deprez [De 85]; the proof given above is more direct, and it also covers the general case of $c \neq 0$ and $s > 0$.

6.3 Veronese Surfaces

The most general and also most interesting case in Theorem 6.2.1 is, of course, subcase (a). Note that in all other subcases the semiparallel surface M^2 has inner metric of constant curvature.

This subcase (a) can be characterized by the property that the first-order osculating subspace of the surface M^2 has maximal possible dimension 5 at every point $x \in M^2$.
If $s = 0$, i.e., if $\varepsilon_3 = \varepsilon = 1$, the following assertion holds.

Proposition 6.3.1. *If a semiparallel surface M^2 in $N^n(c)$ belongs to subcase* (a) *of Theorem* 6.2.1 *and lies in an $N^5(c) \subset N^n(c)$, then it is a parallel surface and lies in a four-dimensional space form.*

Proof. For such an M^2, equations (6.1.9) and (6.1.10) reduce to

$$\omega_1^3 = \alpha\omega^1, \quad \omega_1^4 = k\omega^1, \quad \omega_1^5 = k\omega^2, \quad \omega_1^\xi = 0, \tag{6.3.1}$$

$$\omega_2^3 = \alpha\omega^2, \quad \omega_2^4 = -k\omega^2, \quad \omega_2^5 = k\omega^1, \quad \omega_2^\xi = 0, \tag{6.3.2}$$

where $\alpha = \sqrt{3k^2 - c}$. Taking exterior derivatives and applying the structure equations and Cartan's lemma, one gets

$$-\frac{1}{2}d\ln k = k_1\omega^1 + k_2\omega^2, \quad \frac{1}{5}(2\omega_1^2 - \omega_4^5) = -k_2\omega^1 + k_1\omega^2, \tag{6.3.3}$$

$$\frac{\alpha}{3k}\omega_3^4 = k_1\omega^1 - k_2\omega^2, \quad \frac{\alpha}{3k}\omega_3^5 = k_2\omega^1 + k_1\omega^2, \tag{6.3.4}$$

$$\frac{\alpha}{k}\omega_3^\xi + \omega_4^\xi = p_1^\xi\omega^1 + p_2^\xi\omega^2, \quad \frac{\alpha}{k}\omega_3^\xi - \omega_4^\xi = p_3^\xi\omega^1 + p_4^\xi\omega^2, \tag{6.3.5}$$

$$\omega_5^\xi = p_2^\xi\omega^1 + p_3^\xi\omega^2. \tag{6.3.6}$$

The same procedure applied to the equations in the first two rows above leads to

$$dk_1 = k_2\omega_1^2 + \left[\rho k_2^2 - \sigma k_1^2 + \frac{1}{30}(4\pi_1 - \pi_2)\right]\omega^1 - \left(\varphi k_1 k_2 + \frac{1}{12}\pi_3\right)\omega^2, \tag{6.3.7}$$

$$dk_2 = -k_1\omega_1^2 - \left(\varphi k_1 k_2 + \frac{1}{12}\pi_3\right)\omega^1 + \left[\rho k_1^2 - \sigma k_2^2 + \frac{1}{30}(4\pi_2 - \pi_1)\right]\omega^2, \tag{6.3.8}$$

where

$$\rho = \frac{3}{10}\alpha^{-2}(28k^2 - 5c), \quad \sigma = \frac{3}{10}\alpha^{-2}(22k^2 - 5c),$$

$$\varphi = 3\alpha^{-2}(5k^2 - c),$$

and

$$\pi_1 = \mathbf{p}_1\mathbf{p}_3 - \mathbf{p}_2^2, \quad \pi_2 = \mathbf{p}_2\mathbf{p}_4 - \mathbf{p}_3^2, \quad \pi_3 = \mathbf{p}_1\mathbf{p}_4 - \mathbf{p}_2\mathbf{p}_3; \tag{6.3.9}$$

here $\mathbf{p}_a = e_\xi p_a^\xi$ ($a \in \{1, 2, 3, 4\}$), and $\mathbf{p}_a \mathbf{p}_b = g_{\xi\eta} p_a^\xi p_b^\eta$ are scalar products.

If this M^2 lies in an $N^5(c)$, the set of values of ξ is empty. Then (6.3.7) and (6.3.8) can be used (in general they are useful also for $n > 5$, below) after exterior differentiation to obtain

$$k_i[50\alpha^4 + 3(252k^2 + 215c)(k_1^2 + k_2^2)] = 0; \quad i = 1, 2.$$

Here either $k_1 = k_2 = 0$, and thus $k = \text{const}$, or else $[\ldots] = 0$. It turns out that this last case also leads to $k = \text{const}$. Indeed, by direct calculation, $d(k_1^2 + k_2^2) = k\sigma(k_1^2 + k_2^2)dk$; applying the hypothesis $[\ldots] = 0$ and taking derivatives, one gets $P(k)dk = 0$, where $P(k)$ is a quartic polynomial in k with constant coefficients and leading term $2^3 \cdot 3^2 \cdot 511k^4$.

Hence (6.3.3), (6.3.4) reduce to

$$k = \text{const}, \quad \omega_3^4 = \omega_3^5 = 2\omega_1^2 - \omega_4^5 = 0. \tag{6.3.10}$$

The differential system consisting of $\omega^3 = \omega^4 = \omega^5 = 0$, (6.3.1), (6.3.2) (for $n = 5$), and (6.3.10) is totally integrable, since an easy computation shows that the covariant equations obtained by exterior differentiation are satisfied due to the equations of the same system. Hence the surface M^2 lies in $N^5(c)$ and depends on some arbitrary constants.

Since from (6.3.1), (6.3.2)

$$h_{ij}^3 = \alpha\delta_{ij}; \quad h_{11}^4 = -h_{22}^4 = k, \quad h_{12}^4 = 0;$$
$$h_{11}^5 = h_{22}^5 = 0, \quad h_{12}^5 = k, \tag{6.3.11}$$

substitution into (2.2.2) shows immediately that $\bar{\nabla} h_{ij}^\alpha = 0$, i.e., the surface M^2 is parallel.

For such an M^2, $dx = e_1\omega^1 + e_2\omega^2$,

$$de_1 = e_2\omega_1^2 + k(e_4\omega^1 + e_5\omega^2) + H^*\omega^1,$$
$$de_2 = -e_1\omega_1^2 + k(-e_4\omega^2 + e_5\omega^1) + H^*\omega^2, \tag{6.3.12}$$

where $H^* = -cx + \alpha e_3$, and

$$de_4 = -k(e_1\omega^1 - e_2\omega^2) + 2e_5\omega_1^2,$$
$$de_5 = -k(e_1\omega^2 + e_2\omega^1) - 2e_5\omega_1^2, \tag{6.3.13}$$
$$dH^* = -(H^*)^2(e_1\omega^1 + e_2\omega^2), \tag{6.3.14}$$

where $(H^*)^2 = c + \alpha^2 = \text{const}$. Now the point in ${}_\sigma E^5$ with radius vector $z = x + \frac{1}{(H^*)^2} H^*$ is a fixed point, because $dz = 0$; and $z - x$ has constant length $3k^2$. This ends the proof.

Remark 6.3.2. In the case $c = 0$ Proposition 6.3.1 reduces to the main result (for $m = 2$) of [Lu 89c], which was reproduced afterwards in [Lu 99a]. Here the hypothesis $c = 0$ implies considerable simplifications: $\rho = \frac{14}{5}$, $\sigma = \frac{11}{5}$, $\varphi = 5$; see the proof of Proposition 2 in [Lu 99a], where the case $s > 0$ was also treated. For the last case from (6.2.7) with $c = 0$, it follows that $\varepsilon_3 = \varepsilon$, so that only $s = 3$ is possible.

An extension to the case $c \neq 0$ (and $s = 0$) was then given in [Me 91], [AM 94], where the proof for subcase $c < 0$ is indirect and uses a classical result of Beltrami about differential parameters of a surface. The proof given here is taken from [Lu 2000a]; it is more direct and does not depend on the sign of c.

The converse to Proposition 6.3.1 also holds.

Proposition 6.3.3. *If a semiparallel surface M^2 in $N^n(c)$ belongs to subcase* (a) *of Theorem* 6.2.1 *and is parallel, then it lies in a four-dimensional space form.*

Proof. From (6.3.1, (6.3.2) it follows that (6.3.11) and $h^\xi_{ij} = 0$ hold. Substituting this into the parallel condition (3.1.1) gives $dk = \omega^4_3 = \omega^5_3 = \omega^5_4 - 2\omega^2_1 = 0$. Therefore, (6.3.12)–(6.3.14) hold and give the asserted conclusion, as in the proof of Proposition 6.3.1 above.

The situation is also similar for subcase (b).

Proposition 6.3.4. *If a semiparallel surface M^2 in $N^n(c)$ belongs to subcase* (b) *of Theorem* 6.2.1 *and lies in an $N^4(c) \subset N^n(c)$, then it is parallel.*

Proof. In subcase (b), $\alpha = 0$, and thus due to (6.2.7), $k^2 = l^2 = \frac{1}{3}c = \text{const} \neq 0$. Now (6.1.9) and (6.1.10) reduce to

$$\omega^3_1 = 0, \quad \omega^4_1 = k\omega^1, \quad \omega^5_1 = k\omega^2, \quad \omega^\xi_1 = 0,$$

$$\omega^3_2 = 0, \quad \omega^4_2 = -k\omega^2, \quad \omega^5_2 = k\omega^1, \quad \omega^\xi_2 = 0.$$

Here the outer equations can be joined together into $\omega^\phi_1 = \omega^\phi_2 = 0$, where $\phi \in \{3, 6, \ldots, n\}$.

If this M^2 lies in an $N^4(c)$, the set of values ϕ is empty and these outer equations disappear. The remaining equations lead after exterior differentiation to the single condition $2\omega^2_1 - \omega^5_4 = 0$, which by exterior differentiation gives an identity. Hence this surface M^2 which is contained in $N^4(c)$ depends on some arbitrary constants.

Now $h^4_{12} = h^5_{11} = h^5_{22} = 0$, $h^4_{11} = -h^4_{22} = h^5_{12} = k = \text{const}$, and substitution into (2.2.2) shows immediately that $\bar{\nabla} h^\alpha_{ij} = 0$, i.e., the surface M^2 is parallel, as asserted.

Remark 6.3.5. In subcase (b), for H^* one has $H^* = -cx$, where $c = 3k^2 > 0$. Now considering the surface M^2 in the outer version, and using in E^{n+1} the frame bundle introduced in Section 1.4 and adapted to M^2, then $x = -\frac{1}{\sqrt{c}}e_{n+1}$, and therefore $H^* = \sqrt{c}e_{n+1} = k\sqrt{3}e_{n+1}$.

For the surface M^2 of Proposition 6.3.4, the formulas (6.3.12) and (6.3.13) can be used. Here the first of them are

$$de_1 = e_2\omega_1^2 + k(e_4\omega^1 + e_5\omega^2) + k\sqrt{3}e_{n+1}\omega^1,$$
$$de_2 = -e_1\omega_1^2 + k(-e_4\omega^2 + e_5\omega^1) + k\sqrt{3}e_{n+1}\omega^2.$$

They show that the situation is the same as in Proposition 6.3.1 for $c = 0$, except that the subscript $n+1$ must be replaced by the subscript 3, which is unrestricted here.

The conclusion is that subcase (b) can be subsumed under (a) by replacing the ambient space form by its outer Euclidean space.

Note that the situation above was analysed already in [Lu 89c]; see Remark 6.3.2.

Using the results of [Blo 86], one can give a full characterization of the above-considered parallel surface M^2, when it is complete and connected. Namely, in [Blo 86] there is shown, in particular, that if a parallel complete connected Riemannian M^2 in E^n or ${}_sE^n$ $(s > 0)$ is planar geodesic (i.e., each of its geodesics lies in a 2-plane of E^n or ${}_sE^n$), then this M^2 is a Veronese surface, a surface homothetic to the image of the isometric Veronese immersion

$$x \in S^2\left(\frac{1}{3}\right) \subset \mathbb{R}^3 \to \frac{1}{\sqrt{6}}(x \cdot {}^tx - I_3) \in so(3),$$

where $so(3) = \{A \in \mathrm{sl}(3,\mathbb{R}) \mid {}^tA = A\}$ with $\langle A, B\rangle = \mathrm{tr}(A, B)$ in $\mathbb{R}^5$, or

$$x \in H^2\left(-\frac{1}{3}\right) \subset \mathbb{R}^3_1 \to \frac{1}{\sqrt{6}}(x \cdot {}^*x + I_3) \in so(1,2),$$

where ${}^*x = {}^t\mathrm{x}I_{1,2}$ and $so(1,2) = \{A \in \mathrm{sl}(3,\mathbb{R}) \mid I_{1,2}{}^tAI_{1,2} = A\}$ with $\langle A, B\rangle = -\mathrm{tr}\,AB$ in ${}_3\mathbb{R}^5$. The image of $S^2(\frac{1}{3})$ is minimal in $S^4(1)$ and the image of $H^2(-\frac{1}{3})$ is minimal in ${}_2H^4(-1)$.

This leads to the following.

Corollary 6.3.6. *A semiparallel complete connected surface M^2 which satisfies the conditions of Proposition* 6.3.1 *or* 6.3.4 *is a Veronese surface.*

Indeed, it remains to show that this M^2 is planar geodesic.

The differential equation $\omega_1^2 = \gamma_1\omega^1 + \gamma_2\omega^2$ determines, for every initial condition, a frame field on M^2 for which γ_1 is the geodesic curvature of the curve defined by $\omega^2 = 0$. Let this curve be a geodesic, i.e., let $\gamma_1 = 0$. Since for this curve $dx = e_1\omega^1$, $de_1 = (ke_4 + H^*)\omega^1$, $d(ke_4 + H^*) = -[k^2 + (H^*)^2]e_1\omega^1$, due to (6.3.12)–(6.3.14), the curve lies in a 2-plane spanned at the point x by vectors e_1, $ke_4 + H^*$.

The case of Proposition 6.3.4 is then already covered as a special case, as mentioned in Remark 6.3.5.

Remark 6.3.7. The geodesic curve just considered has curvature $\sqrt{k^2 + (H^*)^2}$, which is a constant $2k$. Hence it is a circle.

Note that these Veronese surfaces are the same as in case (2) of Theorem 3.3.1 for $m = 2$.

6.4 Second-Order Envelopes of Veronese Surfaces

The results of the preceding section make it possible to interpret the semiparallel surfaces of the most general subcase (a) of Theorem 6.2.1 as second-order envelopes of Veronese surfaces, via Proposition 4.5.1. Due to (6.2.8) the Gaussian curvature of Veronese surfaces is $K = \varepsilon k^2$. Here K is constant for each separate Veronese surface, but it could vary for the different members of the envelope. The question then arises, what is the nature of the function K on such an envelope?

This problem will be investigated here in the (pseudo-)Euclidean space ${}_sE^n$, i.e., supposing that $c = 0$. Recall that in E^5 the envelope reduces, due to Proposition 6.3.1, to a single Veronese surface. In E^6 there exist nonparallel second-order envelopes of congruent Veronese surfaces, as was shown in [Ri 91]. Afterwards in [Lu 96a], the same was shown for ${}_sE^6$ (and in ${}_sN^6(c)$), where s is 0, 3, or 4, and in [Lu 99a] also for such envelopes of noncongruent Veronese surfaces.

The most general result, established in [Lu 99a], is the following.

Theorem 6.4.1. *In ${}_sE^7$ there exist holomorphic semiparallel surfaces, depending on one holomorphic function of two real arguments, which are second-order envelopes of Veronese surfaces (here necessarily $s \in \{0, 3, 4, 5\}$). Each two-dimensional holomorphic Riemannian manifold M^2 can be immersed isometrically into ${}_sE^7$ as such a surface.*

Proof. To equations (6.3.7), (6.3.8) four equations are to be added, which follow from (6.3.5), (6.3.6) by exterior differentiation:

$$\delta p_1^\xi - 3p_2^\xi \omega_1^2 = q_1^\xi \omega^1 + q_2^\xi \omega^2 - (p_1^\xi + 6p_3^\xi)\vartheta, \tag{6.4.1}$$

$$\delta p_2^\xi + (p_1^\xi - 2p_3^\xi)\omega_1^2 = q_2^\xi \omega^1 + q_3^\xi \omega^2 + (3p_2^\xi - 2p_4^\xi)\vartheta, \tag{6.4.2}$$

$$\delta p_3^\xi - (p_4^\xi - 2p_2^\xi)\omega_1^2 = q_3^\xi \omega^1 + q_4^\xi \omega^2 + (3p_3^\xi - 2p_1^\xi)\vartheta, \tag{6.4.3}$$

$$\delta p_4^\xi + 3p_3^\xi \omega_1^2 = q_4^\xi \omega^1 + q_5^\xi \omega^2 - (p_4^\xi - 6p_2^\xi)\vartheta, \tag{6.4.4}$$

where $\delta p_a^\xi = dp_a^\xi + p_a^\eta \omega_\eta^\xi$ and $\vartheta = k_1\omega^1 + k_2\omega^2 = -\frac{1}{2} d \ln k$.

The differential system consisting of the equations $\omega^3 = \omega^4 = \omega^5 = \omega^\xi = 0$, (6.3.1)–(6.3.8), (6.4.1)–(6.4.4) is not (yet) involutive; therefore, exterior differentiation must be used once more. Most of these equations give identities, except (6.3.7), (6.3.8), (6.4.1)–(6.4.4).

The last four groups yield the following covariant exterior equations:

$$[\theta_1^\xi + (q_1^\xi + 6q_3^\xi)\vartheta] \wedge \omega^1 + [\theta_2^\xi + (q_2^\xi + 6q_4^\xi)\vartheta] \wedge \omega^2 = 0, \tag{6.4.5}$$

$$[\theta_2^\xi - (3q_2^\xi - 2q_4^\xi)\vartheta] \wedge \omega^1 + [\theta_3^\xi - (3q_3^\xi - 2q_5^\xi)\vartheta] \wedge \omega^2 = 0, \tag{6.4.6}$$

$$[\theta_3^\xi - (3q_3^\xi - 2q_1^\xi)\vartheta] \wedge \omega^1 + [\theta_4^\xi - (3q_4^\xi - 2q_2^\xi)\vartheta] \wedge \omega^2 = 0, \tag{6.4.7}$$

$$[\theta_4^\xi + (q_4^\xi + 6q_2^\xi)\vartheta] \wedge \omega^1 + [\theta_5^\xi + (q_5^\xi + 6q_3^\xi)\vartheta] \wedge \omega^2 = 0, \tag{6.4.8}$$

with secondary forms $\theta_b^\xi = dq_b^\xi + q_b^\eta \omega_\eta^\xi - r_b^\xi \omega_1^2$, where

$$r_1^\xi = 4q_2^\xi, \quad r_2^\xi = 3q_3^\xi - q_1^\xi, \quad r_3^\xi = 2(q_4^\xi - q_2^\xi), \quad r_4^\xi = q_5^\xi - 3q_3^\xi, \quad r_5^\xi = -4q_4^\xi.$$

Equations (6.3.7) and (6.3.8), if treated similarly, do not give exterior equations, but reduce to two algebraic equations

$$k_1\left[k^2 + \frac{42}{25}\epsilon(k_1^2 + k_2^2) + \frac{\epsilon}{50}(29\pi_1 + 24\pi_2)\right]$$

$$- \frac{1}{20}k_2\pi_3 - \frac{1}{30}\pi_{11} + \frac{2}{15}\pi_{21} - \frac{1}{12}\pi_{32} = 0, \tag{6.4.9}$$

$$k_2\left[k^2 + \frac{42}{25}\epsilon(k_1^2 + k_2^2) + \frac{\epsilon}{50}(24\pi_1 + 29\pi_2)\right]$$

$$- \frac{1}{20}k_1\pi_3 - \frac{1}{30}\pi_{22} + \frac{2}{15}\pi_{12} - \frac{1}{12}\pi_{31} = 0, \tag{6.4.10}$$

where k_1, k_2 and π_1, π_2, π_3 are given by (6.3.3) and (6.3.9), respectively, and

$$\pi_{11} = \mathbf{p}_2\mathbf{q}_4 + \mathbf{p}_4\mathbf{q}_2 - 2\mathbf{p}_3\mathbf{q}_3 - 3k_1Q_1, \quad \pi_{12} = \mathbf{p}_2\mathbf{q}_5 + \mathbf{p}_4\mathbf{q}_3 - 2\mathbf{p}_3\mathbf{q}_4 - 2k_2Q_1,$$

$$\pi_{21} = \mathbf{p}_3\mathbf{q}_1 + \mathbf{p}_1\mathbf{q}_3 - 2\mathbf{p}_2\mathbf{q}_2 - 2k_1Q_2, \quad \pi_{22} = \mathbf{p}_3\mathbf{q}_2 + \mathbf{p}_1\mathbf{q}_4 - 2\mathbf{p}_2\mathbf{q}_3 - 2k_2Q_2,$$

$$\pi_{31} = \mathbf{p}_4\mathbf{q}_1 + \mathbf{p}_1\mathbf{q}_4 - \mathbf{p}_3\mathbf{q}_2 - \mathbf{p}_2\mathbf{q}_3 - 2k_1Q_3,$$

$$\pi_{32} = \mathbf{p}_4\mathbf{q}_2 + \mathbf{p}_1\mathbf{q}_5 - \mathbf{p}_3\mathbf{q}_3 - \mathbf{p}_2\mathbf{q}_4 - 2k_2Q_3,$$

$$Q_1 = 2\mathbf{p}_1\mathbf{p}_3 + \mathbf{p}_2\mathbf{p}_4 - 3\mathbf{p}_3^2 - 3\mathbf{p}_2^2 - \mathbf{p}_3^2,$$

$$Q_2 = \mathbf{p}_1\mathbf{p}_3 + 2\mathbf{p}_2\mathbf{p}_4 - 3\mathbf{p}_2^2 - 3\mathbf{p}_3^2 - \mathbf{p}_1^2,$$

$$Q_3 = \mathbf{p}_1\mathbf{p}_4 + 2\mathbf{p}_3\mathbf{p}_4 + 2\mathbf{p}_1\mathbf{p}_2 + 3\mathbf{p}_2\mathbf{p}_3.$$

After differentiation, these algebraic relations (6.4.9), (6.4.10) lead to a system of two linear equations, which relate to each other the secondary forms θ_b^ξ of the exterior system (6.4.5)–(6.48), the form ω_1^2 and the basic forms ω^1, ω^2. The dependence on the latter two will be indicated indirectly below by (mod ω^1, ω^2). Writing the scalar products componentwise, using $e_\xi e_\eta = \epsilon_\xi \delta_{\xi\eta}$, these linear equations are as follows:

$$\sum_\xi \epsilon_\xi [8p_3^\xi\theta_1^\xi - (16p_2^\xi + 7p_4^\xi)\theta_2^\xi + (8p_1^\xi + 9p_3^\xi)\theta_3^\xi + 3p_2^\xi\theta_4^\xi - 5p_1^\xi\theta_5^\xi] + S_1\omega_1^2$$

$$= 0(\mathrm{mod}\,\omega^1, \omega^2), \tag{6.4.11}$$

$$\sum_\xi \epsilon_\xi [8p_2^\xi\theta_5^\xi - (16p_3^\xi + 7p_1^\xi)\theta_4^\xi + (8p_4^\xi + 9p_2^\xi)\theta_3^\xi + 3p_3^\xi\theta_2^\xi - 5p_4^\xi\theta_1^\xi] + S_2\omega_1^2$$

$$= 0(\mathrm{mod}\,\omega^1, \omega^2), \tag{6.4.12}$$

where S_1 and S_2 are some nonzero polynomials of k, k_1, k_2 and of the components of $\mathbf{p}_a$, $\mathbf{q}_b$.

The investigation of the Pfaffian system defining the above semiparallel surfaces M^2 is reduced now to the study of the system of covariant exterior equations (6.4.5)–(6.4.8) and of the last two linear equations for the secondary forms θ_a^ξ and the new secondary form ω_1^2.

Let $n = 7$. Then ξ takes values in $\{6, 7\}$. Provided $64 p_2^6 p_3^6 - 25 p_1^6 p_4^6 \neq 0$, one can solve the linear system (6.4.11), (6.4.12) with respect to θ_1^6 and θ_5^6, and obtain (mod ω^1, ω^2) their expressions as linear combinations of $\theta_2^6, \theta_3^6, \theta_4^6, \theta_1^7, \ldots, \theta_5^7$, and ω_1^2, where the coefficients are some rational functions of the coefficients on the right-hand sides of (6.3.3)–(6.3.6), (6.4.1)–(6.4.4).

These expressions are then substituted into the four exterior equations (6.4.5)–(6.4.8). After that, the latter contain nine secondary forms $\omega_1^2, \theta_2^6, \theta_3^6, \theta_4^6, \theta_1^7, \ldots, \theta_5^7$. The 8×9 matrix of the corresponding polar system (for some values u^1, u^2 of ω^1, ω^2) has an 8×8-determinant with a zero block and therefore equal to the product of two 4×4-determinants, both nonzero in general. For instance, one of them is

$$\begin{vmatrix} u^1 & u^2 & 0 & 0 \\ 0 & u^1 & u^2 & 0 \\ 0 & 0 & u^1 & u^2 \\ 0 & 0 & 0 & u^1 \end{vmatrix} = (u^1)^4.$$

Thus the first Cartan character is $s_1 = 8$. It follows that the second Cartan character is $s_2 = 1$, and hence $s_1 + 2s_2 = 10$.

It suffices to show that the general integral element of the above system with basis ω^1, ω^2 depends on 10 independent parameters. Then the Cartan criterion is satisfied (see [Ca 45], Section 85) or, equvalently, equality holds in Cartan's test inequality (see [BCGGG 91], p. 140). To show this, Cartan's lemma is applied to the exterior equations (6.4.5)–(6.4.8) for basis ω^1, ω^2. Since ξ takes the two values 6 and 7, this gives 12 new coefficients before ω^1, ω^2 in the expressions of $\theta_1^\xi, \ldots, \theta_5^\xi$. Substituting the latter into (6.4.11), (6.4.12) gives two expressions for ω_1^2 as linear combinations of ω^1, ω^2. These expressions must be equal, thus there are two linear dependencies among these new coefficients. Therefore, the number of independent coefficients is actually 10.

The first assertion of Theorem 6.4.1 is hence true.

To verify the second assertion of Theorem 6.4.1, one has to observe that the above system does not set any condition on the Gaussian curvature $K = \varepsilon k^2$ of the two-dimensional Riemannian space M^2 to be immersed. Indeed, this system involves conditions only on the quantities determining the immersion.

6.5 The Case of a Singular Metric

For a spacelike M^2 in ${}_sN^n(c)$ with $s > 0$ the metric regularity conditions of Section 6.2 may not necessarily be satisfied. In this section it is assumed that the first-order normal subspace span$\{h_{ij}\}$ = span$\{A, B, H\}$ has constant dimension m_1 on

M^2, $1 \le m_1 \le 3$, and at every point $x \in M^2$, it belongs to a three-dimensional subspace of the normal space $T_x^\perp M^2$, with Lorentz metric, and that the subspace span$\{A, B\}$ has constant dimension m_0, $1 \le m_0 \le 2$; moreover, at least one of the spaces span$\{A, B, H\}$ and span$\{A, B\}$ has singular metric.

For the subspaces of a Lorentz space, especially for the one-dimensional ones, terminology familiar from relativity theory will be used. If such a one-dimensional subspace is spanned by the vector $e \neq 0$ and $e^2 > 0$, or $e^2 < 0$, or $e^2 = 0$, then it is called, respectively, *spacelike*, or *timelike* (the regular case), or *lightlike* (the singular case).

6.5.1 The subcases where span$\{A, B\}$ has singular metric

Let us consider first the general case, where $m_1 = 3$, $m_0 = 2$, supposing here that the three-dimensional span$\{A, B, H\}$ is Lorentzian, thus regular, and that the two-dimensional span$\{A, B\}$ has singular metric (which does not vanish completely, due to $m_0 = 2$).

Recall that the tangent part $\{e_1, e_2\}$ of the orthonormal frame can be chosen so that $\langle A, B\rangle = 0$, and here A, B can be interchanged (see Section 6.1). Singularity of the metric on span$\{A, B\}$ leads to $\langle A, A\rangle\langle B, B\rangle = 0$, where only one factor can be zero, since this metric does not vanish completely.

One may assume $\langle B, B\rangle = 0$, $\langle A, A\rangle \neq 0$.

The part of the frame in the Lorentzian first-order normal subspace can be adapted so that (6.1.8) holds as before, only now this part is no longer orthonormal but must satisfy the conditions

$$\langle e_4, e_5\rangle = \langle e_5, e_5\rangle = 0; \tag{6.5.1}$$

moreover, e_3 and e_4 can be chosen so that

$$\langle e_3, e_4\rangle = \langle e_3, e_3\rangle = 0, \quad \langle e_4, e_4\rangle = \langle e_3, e_5\rangle = 1. \tag{6.5.2}$$

Note that e_3 and e_5 may be replaced by λe_3 and $\lambda^{-1}e_5$, respectively, for every $\lambda \neq 0$.

This leads to $H^2 = \beta^2 + 2\alpha\gamma$.

The formulas (6.1.9), (6.1.10) hold as well, but now due to (1.2.4),

$$\omega_1^1 = \omega_2^2 = 0, \quad \omega_1^2 = -\omega_2^1, \quad \omega_4^i = -\omega_i^4, \quad \omega_5^i = -\omega_i^3, \quad \omega_3^i = -\omega_i^5, \tag{6.5.3}$$

$$\omega_4^4 = \omega_5^3 = \omega_3^5 = 0, \quad \omega_4^3 = -\omega_5^4, \quad \omega_3^4 = -\omega_4^5, \quad \omega_5^5 = -\omega_3^3; \tag{6.5.4}$$

where $i \in \{1, 2\}$. Therefore, due to (6.1.9), (6.1.10), and (2.1.8),

$$-\Omega_2^1 = \Omega_1^2 = (k^2 - H^2 - c)\omega^1 \wedge \omega^2, \quad -\Omega_3^4 = \Omega_4^5 = -2kl\omega^1 \wedge \omega^2; \tag{6.5.5}$$

the remaining Ω_i^j, Ω_α^β are zero.

The semiparallel condition (4.1.2) reduces to a system whose only essential equations are

$$kl\alpha = 0, \tag{6.5.6}$$

$$\begin{aligned} k(k^2 - H^2 - c) &= 0, \\ l(2k^2 + k\beta - H^2 - c) &= 0, \\ l(2k^2 - k\beta - H^2 - c) &= 0, \end{aligned} \tag{6.5.7}$$

where the last equation can be replaced by

$$kl\beta = 0. \tag{6.5.8}$$

In the general subcase above, with $m_1 = 3$, $m_0 = 2$, one has $\alpha kl \neq 0$ in (6.1.8) (see (a) in Table 6.1.1). Comparing with (6.5.6) gives the following.

Proposition 6.5.1. *There exists no semiparallel spacelike surface M^2 in ${}_sN^n(c)$ with three-dimensional Lorentzian first-order normal subspace, whose normal curvature indicatrix spans a two-dimensional plane with singular metric.*

Next, let span$\{A, B\}$ be two-dimensional and have singular metric, as before, but let span$\{A, B, H\}$ coincide with it. This is subcase (b) of Table 6.1.1, thus $\alpha = 0$, $kl \neq 0$. Here equation (6.5.8) gives $\beta = 0$, and the previous equations lead to $k^2 - H^2 - c = 2k^2 - H^2 - c = 0$, which are impossible for $k \neq 0$. This implies the following.

Proposition 6.5.2. *There exists no semiparallel spacelike surface M^2 in ${}_sN^n(c)$ whose normal curvature indicatrix spans a two-dimensional plane which coincides with the two-dimensional first-order normal subspace and has singular metric.*

Now let span$\{A, B\}$ be one dimensional, i.e., $m_0 = 1$, and have singular metric. Then $\langle A, B\rangle = A^2 = B^2 = 0$, and (6.1.7) shows that $\langle A', B'\rangle = 0$ independently of φ; thus the tangent part of the adapted frame is free, and it can be used to make $A' = 0$ according to the first equation (6.1.6). Having done this, one can use the above frame with (6.1.8) and $k = 0$. This implies $B \neq 0$, since $m_0 = 1$, and thus $l \neq 0$. Therefore, the semiparallel conditions (6.5.6)–(6.5.7) reduce here to a single condition $H^2 + c = 0$ (which is equivalent to $\beta^2 + 2\alpha\gamma + c = 0$). All this implies, via (6.5.5), that all Ω_i^j, Ω_α^β are zero, i.e., the surface M^2 has flat $\bar{\nabla}$.

More detailed geometrical description of these M^2 can be given by considering the corresponding parallel surfaces. This will be done for the special case $c = 0$, i.e., in (pseudo-)Euclidean space ${}_sE^n$. Then $H^2 = 0$ and hence both B and H are lightlike.

Let span$\{A, B, H\}$ = span$\{B, H\}$ be two dimensional. Then $m_1 = 2$ and one gets subcase (c) of Table 6.1.1. The lightlike B and H are noncollinear here, and hence span$\{B, H\}$ has regular metric. Since $A = 0$, the frame vector e_4 is now free, together with e_3, which can be taken collinear to nonzero H. This gives $\alpha \neq 0$, $\beta = \gamma = 0$.

The parallel condition (4.1.2) reduces to the following essential equations:

$$\omega_1^2 = \omega_3^4 = \omega_4^3 = \omega_4^\xi = \omega_3^\xi = 0, \tag{6.5.9}$$

$$dl = l\omega_3^3, \quad d\alpha = -\alpha\omega_3^3; \tag{6.5.10}$$

therefore $d(\alpha l) = 0$. Due to these equations, and also due to (6.5.3), (6.5.4),

$$\begin{aligned}
dx &= e_1\omega^1 + e_2\omega^2,\\
de_1 &= \alpha e_3\omega^1 + le_5\omega^2 = H\omega^1 + B\omega^2,\\
de_2 &= \alpha e_3\omega^2 + le_5\omega^1 = H\omega^2 + B\omega^1,\\
dH &= -(\alpha l)(e_1\omega^2 + e_2\omega^1), \quad dB = -(\alpha l)dx;
\end{aligned}$$

therefore the parallel surface M^2 is contained in a four-dimensional Lorentz space ${}_1E^4 \subset {}_sE^n$, spanned at the point x by vectors e_1, e_2, H, and B. Moreover, for $z_0 = x + (\alpha l)^{-1}B$, one has $dz_0 = 0$; hence the point with radius vector z_0 is a fixed point. Due to $x = z_0 - (\alpha l)^{-1}B$ and $\langle B, B\rangle = 0$, the surface M^2 lies on the light cone $C^3 \subset {}_1E^4$ with vertex at this fixed point. Note that span$\{A, B\}$ = span B is directed along a generator line through x of this cone C^3 and the mean curvature vector H is parallel to a different generator line.

Lastly, let span$\{B, H\}$ be one-dimensional. Then $m_1 = 1$ and one gets subcase (d) of Table 6.1.1. Hence one has $\alpha = \beta = 0$ in (6.1.8).

The parallel condition (4.1.2) now reduces to the following essential equations:

$$\begin{aligned}
\omega_1^2 &= \omega_5^3 = \omega_5^4 = \omega_5^\xi = 0,\\
d\gamma &= -\gamma\omega_5^5, \quad dl = -l\omega_5^5;
\end{aligned}$$

hence for $\gamma l^{-1} = h$ one has $dh = 0$. Due to these equations, and also due to (6.5.3), (6.5.4),

$$\begin{aligned}
dx &= e_1\omega^1 + e_2\omega^2,\\
de_1 &= l(h\omega^1 + \omega^2)e_5, \quad de_2 = l(\omega^1 + h\omega^2)e_5, \quad d(le_5) = 0.
\end{aligned}$$

Therefore, the parallel surface M^2 is contained in a three-dimensional semi-Euclidean space, spanned at the point $x \in M^2$ by vectors e_1, e_2, and le_5, the latter being constant with zero length squared.

Here $d\omega^1 = d\omega^2 = 0$, and at least locally $\omega^1 = du^1$, $\omega^2 = du^2$. Consequently, $e_1 = x_1$, $e_2 = x_2$, where subscripts on the vector x denote partial derivatives with respect to the variables u^1, u^2. Further, $x_{11} = x_{22} = \gamma e_5 = H$, $x_{12} = le_5 = B$, $x_{111} = x_{112} = x_{122} = x_{222} = 0$, where $H = hB$, h = const, and B is a constant vector with zero length squared. Therefore, this parallel M^2 can be represented by the equation

$$x = \frac{1}{2}[h(u^1)^2 + h(u^2)^2 + 2u^1u^2]B + c_1u^1 + c_2u^2,$$

where c_1 and c_2 are constant vectors, noncollinear like x_1 and x_2. (Note that here the absolute term has been made zero by exchanging the initial point).

Thus with respect to affine coordinates u^1, u^2, v in the three-dimensional semi-Euclidean space with basis vectors c_1, c_2 and $\frac{1}{2}B$, the parallel surface M^2 has the equation

$$v = h[(u^1)^2 + (u^2)^2] + 2u^1u^2$$

and therefore is a *paraboloid*, elliptic or hyperbolic. Note that the basis vector along the v-axis has zero length squared and therefore has singular direction of the semi-regular metric of the semi-Euclidean space containing the paraboloid. The same direction is singular also for the paraboloid as a quadric surface.

6.5.2 The subcases where span$\{A, B\}$ has regular metric

In this case, span$\{A, B, H\}$ must have singular metric. Here, too, only the case $c = 0$ will be considered.

Consider first the subcase with $m_0 = 2$. Then m_1 must be 3. Therefore, subcase (b) is impossible, and only (a) can occur. Taking the frame vectors e_3, e_4, and e_5 so that (6.1.8) holds, it follows that $\langle e_3, e_3\rangle = \langle e_3, e_4\rangle = \langle e_3, e_5\rangle = 0$, and, as before, $\langle e_4, e_4\rangle = \varepsilon_4$, $\langle e_5, e_5\rangle = \varepsilon_5$, $\langle e_4, e_5\rangle = 0$. Here one needs to choose one of the next frame vectors, e.g., e_6, so that $\langle e_4, e_6\rangle = \langle e_5, e_6\rangle = \langle e_6, e_6\rangle = 0$. It can also be assumed that $\langle e_3, e_6\rangle = 1$, which requires that $n \geq 6$. If $n > 6$ then the remaining frame vectors e_ξ can be chosen to make an orthonormal basis.

After differentiation, these conditions on the frame vectors give

$$\omega_3^i = -\,\omega_i^6 = 0, \quad \omega_4^i = -\varepsilon_4\omega_i^4, \quad \omega_5^i = -\varepsilon_5\omega_i^5, \quad \omega_6^i = -\omega_i^3,$$
$$\omega_4^4 = \omega_5^5 = 0, \quad \varepsilon_4\omega_5^4 = -\varepsilon_5\omega_4^5, \quad \omega_6^6 = -\omega_3^3,$$
$$\omega_4^3 = -\,\varepsilon_4\omega_6^4, \quad \omega_5^3 = -\varepsilon_5\omega_6^5, \quad \omega_4^6 = -\varepsilon_4\omega_3^4, \quad \omega_5^6 = -\varepsilon_5\omega_3^5.$$

An easy calculation shows that for the curvature 2-forms, the formulas (6.2.1), (6.2.2) hold with only the difference that now, due to $\varepsilon_3 = 0$, one has $H^2 = \varepsilon_4\beta^2 + \varepsilon_5\gamma^2$.

The semiparallel condition (4.1.2) reduces to the same system as in Section 6.2, and yields similarly $\beta = \gamma = 0$; thus $H^2 = 0$, which implies $2\varepsilon_4k^2 + \varepsilon_5l^2 = \varepsilon_4k^2 + 2\varepsilon_5l^2 = 0$. This is in contradiction with $kl \neq 0$ and $\varepsilon_4\varepsilon_5 \neq 0$. Therefore, subcases (a) and (b) are impossible here.

Now consider subcase (c) of Table 6.1.1, where $m_0 = 1$ and $m_1 = 2$. It is assumed here in subsection 6.5.2 that the one-dimensional span$\{A, B\}$ has regular metric (i.e., space- or timelike) and that the two-dimensional span$\{A, B, H\}$ has singular metric. Thus the one-dimensional regular span$\{A, B\}$ is spacelike, where A and B are collinear vectors. Making, as before, $\langle A, B\rangle = 0$ with interchangeable A and B, one can get $B = 0$ and use the above frame with (6.1.8) and $l = 0$, $\alpha \neq 0$; but now $\varepsilon_3 = 0$, $\varepsilon_4 = 1$, and thus $H^2 = \beta^2$. Moreover, one can make $\gamma = 0$.

Now the semiparallel conditions of Section 6.1 can be used, and they give a single essential equation $k^2 - H^2 = 0$, which implies $\beta^2 = k^2$ and $\Omega_i^j = \Omega_\alpha^\beta = 0$, i.e., $\bar{\nabla}$ is flat and either $\beta = k$ or $\beta = -k$.

For a geometrical description of a semiparallel M^2 in this subcase, the corresponding parallel surfaces can be considered. By Theorem 4.5.5, a semiparallel surface M^2 is the second-order envelope of these surfaces.

First set $\beta = k$. Then the parallel condition (4.1.2) reduces to the following essential equations:

$$\omega_1^2 = \omega_3^4 = \omega_4^3 = \omega_4^\xi = \omega_3^\xi = 0,$$

$$dk = 0, \quad d\alpha = -\alpha\omega_3^3.$$

These equations imply via (6.5.3), (6.5.4) that

$$dx = e_1\omega^1 + e_2\omega^2,$$

$$de_1 = (H + A)\omega^1, \quad de_2 = (H - A)\omega^2,$$

$$d(H + A) = -(2k)^2 e_1\omega^1, \quad d(H - A) = 0.$$

Here $H + A = \alpha e_3 + 2k e_4$ has a constant length $2k$, and $H - A = \alpha e_3$ is a constant lightlike vector. It is seen that the integral curves of $\omega^2 = 0$ are the congruent circles $S^1(2k)$ of curvature $2k$ on parallel Euclidean 2-planes, spanned by the spacelike vectors e_1 and $H + A$. For the integral curves of $\omega^1 = 0$, at least locally $\omega^2 = ds$ and $\dot{x} = e_2, \ddot{x} = \alpha e_3 =$ const, and hence these curves are congruent parabolas on parallel singular (i.e., semi-Euclidean) 2-planes, spanned by the spacelike e_2 and lightlike constant αe_3; the axes of these parabolas go in direction of this constant lightlike vector. The parallel surface is a translation surface of these circles and parabolas.

The other possibility is that $k = -\beta$. The result is the same, only the roles of the circles and parabolas are interchanged.

Subcase (d) of Table 6.1.1, where $m_1 = m_0 = 1$, is impossible here 6.5.2, because here in Section 6.5, it is assumed that span$\{A, B\}$ and span$\{A, B, H\}$ do not both have regular metric.

Summarizing the investigations in Section 6.5, the following theorem can be stated.

Theorem 6.5.1. *If a semiparallel Riemannian surface M^2 in ${}_sN^n(c)$, $s > 0$, has its first-order normal subspaces* span$\{A, B, H\}$ *contained at every point $x \in M^2$ in a three-dimensional subspace with Lorentz metric in the normal space $T_x^\perp M^2$, and at least one of the subspaces* span$\{A, B, H\}$ *and* span$\{A, B\}$ *(the latter spanned by the normal curvature indicatrix) has singular metric, then M^2*

- *is either totally umbilic or totally geodesic, or*
- *has flat $\bar{\nabla}$ and is a second-order envelope of parallel surfaces, each of which is, in the case $c = 0$, either*
  (1) *a surface on a light cone C^3 in Minkowski space, such that its normal curvature indicatrix is a segment of a generator of C^3 and its mean curvature vector is directed along another generator, or*
  (2) *a paraboloid in three-dimensional semi-Euclidean space, such that the singular direction of this paraboloid (as a quadric surface) is singular also for the metric of the space, or*

(3) *a translation surface of circles and parabolas, the latter on parallel semi-Euclidean planes, whose singular direction coincides with the direction of diameters of the parabolas.*

Remark 6.5.2. Here the question arises: Do there exist nontrivial second-order envelopes of the parallel surfaces (1), (2), or (3), which do not reduce to a single parallel surface? This problem was treated in [Saf 2001] and solved affirmatively.

6.6 Semiparallel Timelike Surfaces in Lorentz Spacetime Forms

Let M^2 be a semiparallel timelike surface in a Lorentz spacetime form ${}_1N^n(c)$. The aim is now to classify and describe geometrically all such M^2 (see [Lu 97b]).

The frame bundle adapted to such an M^2 (see Section 2.1) will be used here, so that the tangent part $\{e_1, e_2\}$ of the frame is orthonormal,with $g_{11} = -1$, $g_{22} = 1$, $g_{12} = 0$, and therefore due to (1.2.4) one has

$$\omega_1^1 = \omega_2^2 = 0, \quad \omega_2^1 = \omega_1^2.$$

This part can be transformed according to

$$e_1' = \varepsilon_1(e_1 \cosh\varphi + e_2 \sinh\varphi), \quad e_2' = \varepsilon_2(e_1 \sinh\varphi + e_2 \cosh\varphi),$$

where $\varepsilon_1^2 = \varepsilon_2^2 = 1$.

Similar to Section 6.1, this leads to $\omega_1^{\prime 2} = \varepsilon_1\varepsilon_2(\omega_1^2 + d\varphi)$ and

$$h_{11}' = \frac{1}{2}(h_{11} - h_{22}) + \frac{1}{2}(h_{11} + h_{22})\cosh 2\varphi + h_{12}\sinh 2\varphi,$$

$$h_{12}' = \varepsilon_1\varepsilon_2\left[\frac{1}{2}(h_{11} + h_{22})\sinh 2\varphi + h_{12}\cosh 2\varphi\right],$$

$$h_{22}' = \frac{1}{2}(-h_{11} + h_{22}) + \frac{1}{2}(h_{11} + h_{22})\cosh 2\varphi + h_{12}\sin 2\varphi.$$

Denoting $\frac{1}{2}(h_{11} + h_{22}) = A$, $h_{12} = B$, and $\frac{1}{2}(-h_{11} + h_{22}) = H$, one obtains

$$A' = A\cosh 2\varphi + B\sinh 2\varphi, \quad B' = \varepsilon_1\varepsilon_2[A\sinh 2\varphi + B\cosh 2\varphi], \quad H' = H.$$

Hence, as in Section 6.1 at each point $x \in M^2$, span$\{A, B\}$ is an invariant subspace and H is an invariant vector (the mean curvature vector) in the first-order normal subspace span$\{h_{11}, h_{12}, h_{22}\} = $ span$\{A, B, H\}$; moreover,

$$\langle A', B'\rangle = \varepsilon_1\varepsilon_2\left[\langle A, B\rangle \cosh 4\varphi + \frac{1}{2}(A^2 + B^2)\sinh 4\varphi\right]. \tag{6.6.1}$$

Here a *special case* can be distinguished, where $A = B = 0$. Then $-h_{11} = h_{22}$, $h_{12} = 0$, i.e., $h_{ij} = Hg_{ij}$, and the surface M^2 is totally umbilic (if $H \neq 0$) or totally geodesic (if $H = 0$) (see Section 2.4), thus it is parallel (see Proposition 3.1.2).

Recall that in the general situation of Section 6.1 there always existed a transformation giving $\langle A', B' \rangle = 0$. In this section, this is not the case for timelike surfaces, since $-1 < \tanh 4\varphi < 1$. Therefore, one has the *principal case*, when this can be done, and the *exceptional case*, when this is impossible, i.e., when the right-hand side of (6.6.1) is never zero.

The corresponding criterion follows from the inequality

$$\langle A, B \rangle^2 \leq \frac{1}{4}(A^2 + B^2)^2, \tag{6.6.2}$$

which can be deduced in the following way. Since span$\{A, B\}$ now lies in a Euclidean vector subspace, normal to timelike M^2 in ${}_1N^n(c)$ with Lorentz metric, one has $\langle A, B \rangle^2 \leq A^2 \cdot B^2$. On the other hand, since $(a^2 - b^2)^2 \geq 0$ for any two real numbers a and b, one also has $4a^2b^2 \leq (a^2 + b^2)^2$.

The exceptional case is where equality holds in (6.6.2), and both sides are nonzero. It turns out that this is equivalent to $A = \varepsilon B \neq 0$, where ε is either 1 or -1. Indeed, let $\langle A, B \rangle^2 = \frac{i}{4}(A^2 + B^2)^2 \neq 0$. Since $\langle A, B \rangle = \|A\| \cdot \|B\| \cdot \cos\alpha$, this gives $\cos^2\alpha = \frac{1}{4}(\lambda + \lambda^{-1} + 2)$, where $\lambda = \|A\|^2 \cdot \|B\|^{-2}$, which in turn implies that $\lambda^2 - 2(2\cos^2\alpha - 1)\lambda + 1 = 0$ has a real root, whence $\cos\alpha = \pm 1$ and $\lambda = 1$. The converse is obvious.

6.6.1 The principal case

In this case $A \neq \varepsilon B$, and by a suitable transformation one can get $\langle A, B \rangle = 0$.

Let m_1 and m_0 be defined as in Section 6.1. In general the orthonormal frame bundle can be adapted so that (6.1.8) holds, except that now

$$A = \frac{1}{2}(h_{11} + h_{22}), \quad H = \frac{1}{2}(-h_{11} + h_{22}).$$

Thus instead of (6.1.9), (6.1.10), one now has

$$\omega_1^3 = -\alpha\omega^1, \quad \omega_1^4 = (k - \beta)\omega^1, \quad \omega_1^5 = -\gamma\omega^1 + l\omega^2,$$

$$\omega_2^3 = \alpha\omega^2, \quad \omega_2^4 = (k + \beta)\omega^2, \quad \omega_2^5 = l\omega^1 + \gamma\omega^2,$$

therefore instead of (6.2.1) and (6.2.2) now

$$\Omega_1^2 = \Omega_2^1 = (c - k^2 + l^2 + H^2)\omega^1 \wedge \omega^2, \quad -\Omega_5^4 = \Omega_4^5 = 2kl\omega^1 \wedge \omega^2; \tag{6.6.3}$$

here $H^2 = \alpha^2 + \beta^2 + \gamma^2$ and all other Ω_i^j, Ω_α^β are zero.

The particular subcases of the table are the same as in Section 6.1, with the only difference that here the orthogonal vectors A and B are not interchangable. Therefore, (c) must be further divided into subsubcases (c_1) with $k > 0$, $l = 0$, where one can get $\alpha = 0$, $\gamma \neq 0$; and (c_2) with $k = 0$, $l > 0$, where one can make $\alpha = 0$ $\beta \neq 0$. Likewise, subcase (d) is to be divided into (d_1) and (d_2), which differ from the previous ones only by $\gamma = 0$ and $\beta = 0$, respectively.

The semiparallel condition (4.1.2) reduces to a system whose only essential equations are

$$kl\gamma = 0,$$
$$k(c - k^2 + 2l^2 + H^2) = 0,$$
$$l(c - 2k^2 + l^2 + k\beta + H^2) = 0,$$
$$l(c - 2k^2 + l^2 - k\beta + H^2) = 0,$$

where the last one can be replaced by

$$kl\beta = 0.$$

In general, when $m_1 = 3$, $m_0 = 2$ and thus $kl > 0$, this system leads to contradiction, because then $\beta = \gamma = 0$, $c - k^2 + 2l^2 + \alpha^2 = c - 2k^2 + l^2 + \alpha^2 = 0$ and these last two equations give $k^2 + l^2 = 0$.

In the particular case (b) $m_1 = m_0 = 2$, when $\alpha = 0$, the same contradiction occurs.

In conclusion, the following assertion can be formulated.

Proposition 6.6.1. *In subcases* (a) *and* (b) *of Table* 6.1.1, *there are no semiparallel timelike surfaces* M^2 *in* ${}_1N^n(c)$.

In the particular cases (c) and (d), when either $k > 0$ and $l = 0$, or else $k = 0$ and $l > 0$, the above system gives either $c - k^2 + H^2 = 0$, or $c + l^2 + H^2 = 0$, respectively.

Due to (6.6.3), $\Omega_1^2 = 0$, $\Omega_3^4 = 0$ for both these subcases, i.e., $\bar{\nabla}$ is flat.

In the first subcase one can make $\alpha = 0$, and thus $k^2 - \beta^2 = c + \gamma^2$. Then

$$h_{11} = (k - \beta)e_4 - \gamma e_5, \quad h_{12} = 0, \quad h_{22} = (k + \beta)e_4 + \gamma e_5. \tag{6.6.4}$$

Let us consider the corresponding parallel surface. Then the condition (3.1.1) must be satisfied and leads to

$$dk = d\beta = d\gamma = \omega_1^2 = \omega_4^5 = \omega_4^3 = \omega_5^3 = \omega_4^\xi = \omega_5^\xi = 0, \quad \xi \in \{6, \dots, n\}.$$

Hence for this parallel surface

$$dx = e_1\omega^1 + e_2\omega^2,$$
$$de_1 = [ke_4 + (cx - H)]\omega^1, \quad de_2 = [ke_4 - (cx - H)]\omega^2,$$
$$d[ke_4 + (cx - H)] = 2k(k - \beta)e_1\omega^1, \quad d[ke_4 - (cx - H)] = -2k(k + \beta)e_2\omega^2,$$

where the vectors in square brackets are orthogonal due to $k^2 - (c + H^2) = 0$ (see above). It follows that this parallel surface either lies in a ${}_1E^4$, ${}_2E^4$, or in a semipseudo-Euclidean 4-space, spanned at the point x by the mutually orthogonal vectors $e_1, e_2, e_4, cx - \gamma e_5$, where $(cx - \gamma e_5)^2 = c + \gamma^2 = k^2 - \beta^2$ is either (1) positive, (2) negative, or (3) zero, respectively; here $d(cx - \gamma e_5) = (c + \gamma^2)dx$.

In cases (1) and (2), it can be seen that the surface is a product of two plane curves of constant curvature. Moreover, the point with radius vector

$$z = x - (c + \gamma^2)^{-1}(cx - \gamma e_5)$$

is a fixed point z_0 in ${}_1E^4$ or ${}_2E^4$, since $dz = 0$. Since $(x - z_0)^2 = (c + \gamma^2)^{-1}$, the distance between x and z_0 is a constant, c_0. Hence this parallel surface belongs to an ${}_1S^3(c_0) \subset {}_1E^4$ or to an ${}_1H^3(c_0) \subset {}_2E^4$.

In case (3), where $c + \gamma^2 = k^2 - \beta^2 = 0$, and thus $c < 0$, the vector $cx - \gamma e_5$ is a constant vector with zero length squared, orthogonal to other vectors of the 4-space. Thus the latter is semipseudo-Euclidean. Here either (3_1) $k - \beta = 0$, or (3_2) $k + \beta = 0$.

In subcase (3_1), when $k - \beta = 0$, one has

$$de_1 = (cx - \gamma e_5)\omega^1, \quad de_2 = [-(cx - \gamma e_5) + 2ke_4]\omega^2,$$

where $cx - \gamma e_5 = p_1$ is, in this case, a constant vector with zero length squared. Here $\omega_1^2 = 0$ implies that, at least locally, $\omega^1 = du^1$, $\omega^2 = du^2$. For the u^1-curves $\frac{dx}{du^1} = e_1$, $\frac{d^2x}{(du^1)^2} = p_1$; therefore, $x = \frac{1}{2}p_1(u^1)^2 + p_2u^1 + p_3$, where p_2, p_3 are also constant vectors, and hence these curves are congruent parabolas on parallel semipseudo-Euclidean 2-planes which differ only by translations. Since $d[-(cx-\gamma e_5)+2ke_4] = -4k^2e_2du^2$, the u^2-curves are congruent circles of curvature $2k$ on the parallel planes E^2 spanned at x, by $e_2, -(cx - \gamma e_5) + 2ke_4$, and are thus orthogonal to the 2-planes of these parabolas. Hence the parallel surface here is a product of such a parabola and such a circle.

In the second subcase (3_2), where $k + \beta = 0$, the roles of the parabola and circle are interchanged.

The semiparallel surface M^2 of the first subcase with $k > 0$, $l = \alpha = 0$, $k^2 - \beta^2 = c + \gamma^2$, is a second-order envelope of these parallel products or translation surfaces. Its h is diagonalizable, thus its set of lines of curvature is enveloped by the generating plane curves of these parallel surfaces. Using the Cartan–Kähler theory of Pfaffian systems, it can be shown that these semiparallel surfaces (c_1) exist, depend on $2(n-2)$ arbitrary real functions of one argument, and are nonparallel in general (see [Lu 97b]).

In the other subcase with $k = 0, l > 0$, one can also make $\alpha = 0$, and thus $c + l^2 + \beta^2 + \gamma^2 = 0$, whence $c < 0$. Then

$$-h_{11} = h_{22} = \beta e_4 + \gamma e_5, \quad h_{12} = le_5. \tag{6.6.5}$$

The parallel condition (3.1.1) now reduces to

$$dl = d\beta = d\gamma = \omega_1^2 = \omega_4^5 = \omega_4^\rho = \omega_5^\rho = 0.$$

Hence for this parallel surface

$$dx = e_1\omega^1 + e_2\omega^2,$$

$$de_1 = (cx - \beta e_4)\omega^1 + (-\gamma\omega^1 + l\omega^2)e_5,$$
$$de_2 = -(cx - \beta e_4)\omega^2 + (l\omega^1 + \gamma\omega^2)e_5,$$
$$d(cx - \beta e_4) = (c + \beta^2)dx, \quad de_5 = (-\gamma\omega^1 + l\omega^2)e_1 - (l\omega^1 + \gamma\omega^2)e_2.$$

It is seen that this parallel surface lies in a ${}_2E^4$, spanned at the point x by mutually orthogonal vectors $e_1, e_2, cx - \beta e_4, e_5$ with $e_1^2 = -1$, $(cx - \beta e_4)^2 = -(l^2 + \gamma^2) < 0$.

The point with radius vector $z = x - (c + \beta^2)^{-1}(cx - \beta e_4)$ is a fixed point z_0, since $dz = 0$. Since $(x - z)^2 = -(l^2 + \gamma^2)^{-1}$, the distance between x and z_0 is an imaginary constant $i(l^2 + \gamma^2)^{-\frac{1}{2}}$. Hence the parallel surface here belongs to an ${}_1H^3(c^*) \subset {}_2E^4$.

The asymptotic lines of this surface, with respect to the projective structure in ${}_1H^3(c^*)$, are defined by $\omega^1(-\gamma\omega^1 + l\omega^2) + \omega^2(l\omega^1 + \gamma\omega^2) = 0$ or, for general $\gamma \neq 0$, by

$$[(l - \sqrt{l^2 + \gamma^2})\omega^1 + \gamma\omega^2][l + \sqrt{l^2 + \gamma^2})\omega^1 + \gamma\omega^2] = 0.$$

The tangent vectors of these curves are

$$e_1^* = \gamma e_1 - (l - \sqrt{l^2 + \gamma^2})e_2,$$
$$e_2^* = -\gamma e_1 + (i + \sqrt{l^2 + \gamma^2})e_2$$

and they are linearly independent if $\gamma \neq 0$. For these vectors

$$de_1^* = c^*x^*[\gamma\omega^1 + (l - \sqrt{l^2 + \gamma^2})\omega^2] + e_5\sqrt{l^2 + \gamma^2}[(l - \sqrt{l^2 + \gamma^2})\omega^1 + \gamma\omega^2],$$
$$de_2^* = -c^*x^*[\gamma\omega^1 + (l + \sqrt{l^2 + \gamma^2})\omega^2] + e_5\sqrt{l^2 + \gamma^2}[(l + \sqrt{l^2 + \gamma^2})\omega^1 + \gamma\omega^2],$$

where $x^* = x - z$. On the asymptotic lines of the first family,

$$dx^* = \gamma^{-1}e_1^*\omega^1, \quad de_1^* = -\gamma^{-1}(e_1^*)^2 \cdot c^*x^*\omega^1,$$

and on the asymptotic lines of the second family,

$$dx^* = \gamma^{-1}e_2^*\omega^1, \quad de_2^* = -\gamma^{-1}(e_2^*)^2 \cdot c^*x^*\omega^1,$$

where $(e_1^*)^2 = 2l(l - \sqrt{l^2 + \gamma^2}) < 0$, $(e_2^*)^2 = 2l(l + \sqrt{l^2 + \gamma^2}) > 0$. It is seen that these asymptotic lines are geodesics of ${}_1H^3(c^*)$, i.e., straight lines of the projective structure of ${}_1H^3(c^*)$. Hence the parallel surface here is projectively a quadric. From point of view of the metric in ${}_1H^3(c^*)$, it is an orbit of a 2-parametric subgroup of isometries of ${}_1H^3(c^*)$.

In case (d) of Table 6.1.1, when $\gamma = 0$, the asymptotic lines are defined by $2l\omega^1\omega^2 = 0$; they are geodesics of ${}_1H^3(c^*)$ too; note that here $c^* = c$. Therefore, the result is the same: the parallel surface is projectively a quadric. Moreover, since these asymptotic lines are orthogonal at each point of the surface, the latter is a minimal surface of ${}_1H^3(c)$.

For all values of γ, the semiparallel surface M^2 here is a second-order envelope of these parallel surfaces. It is unique up to $2(n-2)$ arbitrary real functions of one argument, and in general is nonparallel (see [Lu 97b]).

The results of this analysis can be summarized as follows.

Theorem 6.6.2. *A semiparallel timelike surface M^2 of principal case in a Lorentz space-time form ${}_1N^n(c)$ is the union of the closures of its open subsets, each of which is either*

(1) *a special subcase, i.e., a totally geodesic or umbilic surface, or*
(2) *has flat $\bar{\nabla}$ and either*
  (2a) *is a second-order envelope of parallel surfaces in ${}_1E^n$, ${}_1S^n(c)$, or ${}_1H^n(c)$, and each such surface either*
    (2a′) *lies in an ${}_1S^3(c_0) \subset {}_1E^4$ or in an ${}_1H^3(c_0) \subset {}_2E^4$, and is a product of two plane curves of constant curvature in ${}_1E^4$ or ${}_2E^4$, respectively, or*
    (2a″) *lies in a semipseudo-Euclidean 4-space and is the product surface of a parabola with imaginary arclength and diameters in null direction, and of a circle $S^1(c)$ in a plane E^2, orthogonal to the 2-plane of this parabola, or*
  (2b) *is a second-order envelope in ${}_1H^n(c)$ of parallel surfaces, each of which is projectively a ruled quadric in some ${}_1H^3(c^*)$, generated by geodesics of ${}_1H^3(c^*)$, one family having real arclength, and the other family, imaginary arclength.*

Remark 6.6.3. There is a subcase of (2*b*) satisfying (6.6.5), where on each of the parallel ruled quadrics, two generating families of geodesics are orthogonal to each other at each point. It follows that M^2 is a minimal surface of ${}_1H^n(c)$.

Note that the surfaces of (2a) satisfying (6.6.4) are minimal, i.e., $\beta = \gamma = 0$, iff $h_{11} = h_{22} = \sqrt{c}e_4$, $h_{12} = 0$. Then the surface is totally umbilic in ${}_1S^n(c)$.

6.6.2 The exceptional case

The exceptional case is characterized by $A = \varepsilon B \neq 0$. Then $A' = \varepsilon\varepsilon_1\varepsilon_2 B'$ for every φ and by a suitable choice of ε_1 and ε_2 (replacing e_1 by $-e_1$ and e_2 by $-e_2$, if needed) one can make $A' = B'$. So, one may suppose that $A = B \neq 0$.

Here always $m_0 = 1$. In general, $m_1 = 2$ and the normal basis vectors e_3 and e_4 can be chosen so that

$$A = B = \frac{1}{2}(h_{11}+h_{22}) = h_{12} = ke_4, \quad a > 0, \quad H = \frac{1}{2}(-h_{11}+h_{22}) = \alpha e_3+\beta e_4;$$

then

$$h_{11} = -\alpha e_3 + (k-\beta)e_4, \quad h_{12} = ke_4, \quad h_{22} = \alpha e_3 + (k+\beta)e_4$$

and

$$\omega_1^3 = -\alpha\omega^1, \quad \omega_1^4 = (k-\beta)\omega^1 + k\omega^2, \quad \omega_1^\rho = 0, \tag{6.6.6}$$

$$\omega_2^3 = \alpha\omega^2, \quad \omega_2^4 = k\omega^1 + (k+\beta)\omega^2, \quad \omega_2^\rho = 0, \tag{6.6.7}$$

where $\rho \in \{5, \dots, n\}$. Hence $\Omega_1^2 = (c + \alpha^2 + \beta^2)\omega^1 \wedge \omega^2$, $\Omega_3^4 = 0$, and all other $\Omega_\alpha^\beta = 0$, thus $\nabla^\perp$ is flat.

The semiparallel condition (4.1.2) reduces to a single equation $k(c+\alpha^2+\beta^2) = 0$, thus to $c + \alpha^2 + \beta^2 = 0$. Consequently, ∇ is also flat, and it follows that $c \leq 0$.

In general $c < 0$ and the semiparallel surface in this case lies in ${}_1H^n(c)$.

Due to the Pfaffian system above,

$$dx = e_1\omega^1 + e_2\omega^2,$$

$$de_1 = cx\omega^1 + e_2\omega_1^2 - \alpha e_3\omega^1 + e_4[(k-\beta)\omega^1 + k\omega^2],$$

$$de_2 = -cx\omega^2 + e_1\omega_1^2 + \alpha e_3\omega^2 + e_4[k\omega^1 + (k+\beta)\omega^2];$$

thus the curves with $\omega^2 = -\omega^1$, considered in ${}_2E^{n+1}$, are straight lines in null (or lightlike) directions, since for these curves $dx = (e_1 - e_2)\omega^1$, $d(e_1 - e_2) = -(e_1 - e_2)\omega_1^2$, $(e_1 - e_2)^2 = 0$. Hence this semiparallel surface in ${}_2E^{n+1}$ is a ruled surface with generators in null directions, and it lies in ${}_1H^n(c)$. Using the Cartan–Kähler theory of Pfaffian systems, it can be proved that this surface exists uniquely up to the choice of $2n - 5$ real functions of one argument (see [Lu 97b]).

For the corresponding parallel surface, the following equations have to be added:

$$d\alpha = d\beta = \omega_3^4 = dk - 2k\omega_1^2 = \omega_3^\rho = \omega_4^\rho = 0.$$

Let $\beta \neq 0$. This parallel surface belongs to a ${}_2E^4$, spanned at x by mutually orthogonal vectors $e_1, e_2, e_4, cx - \alpha e_3$, with $e_1^2 = -1$, $e_2^2 = e_3^2 = 1$, $(cx - \alpha e_3)^2 = c + \alpha^2 = -\beta^2 < 0$.

Since $d(cx - \alpha e_3) = (c + \alpha^2)dx = -\beta^2 dx$, the point with radius vector $z = x + \beta^{-2}(cx - \alpha e_3)$ is a fixed point z_0 for this parallel surface, at the constant imaginary distance $i|\beta|^{-1}$ from x. Hence this surface lies in a ${}_1H^3(c^*)$, where $c^* = -\beta^2$.

Its asymptotic lines with respect to the projective structure of ${}_1H^3(c^*)$ are defined by $h_{ij}\omega^i\omega^j = 0$, i.e., by

$$(\omega^1 + \omega^2)[(k-\beta)\omega^1 + (k+\beta)\omega^2] = 0.$$

One family consists of straight lines in null directions of $e_1 - e_2$, considered above. A curve of the other family has tangent vector $(k+\beta)e_1 - (k-\beta)e_2$ with length squared $-4k\beta \neq 0$, and is noncollinear to $e_1 - e_2$. Moreover,

$$d[(k+\beta)e_1 - (k-\beta)e_2] = -\beta^2(x - z)\theta + [(k+\beta)e_1 - (k-\beta)e_2]\omega_1^2,$$

where $\theta = 2\alpha(\omega^1 - \omega^2)$. Hence these curves are geodesics of ${}_1H^3(c^*)$, i.e., straight lines of the projective structure of ${}_1H^3(c^*)$, and so the parallel surface is projectively a quadric.

The semiparallel surface M^2 of the exceptional case with $m_1 = 2$, $m_0 = 1$, and $\beta \neq 0$ in ${}_1H^n(c)$ is a second-order envelope of these parallel surfaces.

There is another subcase with $m_1 = 2$, $m_0 = 1$, namely the subcase where $\beta = 0$, $\alpha \neq 0$. Then $c + \alpha^2 = 0$, thus $c = -\alpha^2 < 0$, and so $d\alpha = 0$.

A semiparallel surface belonging to this subcase is defined by equations (6.6.6) and (6.6.7) with $\beta = 0$. Here the treatment can be simplified considerably.

Recall that $\Omega_1^2 = 0$, thus $d\omega_1^2 = 0$, and therefore, at least locally, $\omega_1^2 = d\psi$. Now the transformation of the tangent part $\{e_1, e_2\}$ of the frame in the adapted frame bundle (see the introduction of Section 6.6) will be applied. This transformation leads to $\omega_1'^2 = \omega_1^2 + d\varphi$, and then to $\omega_1'^2 = d(\psi + \varphi)$. Taking $\varphi = -\psi$, one gets $\omega_1'^2 = 0$, and it may thus be assumed that $\omega_1^2 = 0$.

For this surface M^2 one has $dx = e_1\omega^1 + e_2\omega^2$,

$$de_1 = (cx - \alpha e_3)\omega^1 + ke_4(\omega^1 + \omega^2), \quad de_2 = -(cx - \alpha e_3)\omega^2 + ke_4(\omega^1 + \omega^2),$$

where $d\omega^1 = d\omega^2 = 0$, due to $\omega_1^2 = 0$, and at least locally $\omega^1 = du^1$, $\omega^2 = du^2$.

Denoting $e_1' = e_1 + e_2$, $e_2' = e_1 - e_2$, $2du = du^1 + du^2$, $2dv = du^1 - du^2$ one obtains

$$\begin{gathered} dx = e_1' du + e_2' dv, \quad de_1' = 4ke_4 du + 2(cx - \alpha e_3)dv, \\ de_2' = 2(cx - \alpha e_3)du. \end{gathered} \tag{6.6.8}$$

It is seen that for the curves with $u = \text{const}$ one has $dx = e_2' dv$, $de_2' = 0$, $(e_2')^2 = 0$; thus these curves are straight lines, and hence the semiparallel M^2 is a ruled surface with generators in null directions. Using the Cartan–Kähler theory, it can be proved that an M^2 of this subcase exists, and is given uniquely up to $2n - 3$ real functions of one variable (see [Lu 97b]).

Here equations (6.6.8) can be partly integrated. Indeed, here $e_1' = x_u$ and $e_2' = x_v$ are the partial derivatives of x. From the last equation it is seen that $x_{vv} = 0$. Thus $x_v = y(u)$; hence $x = y(u)v + z(u)$. Further, $x_u = y'(u)v + z'(u)$ and

$$x_{uv} = y'(u) = 2(cx - \alpha e_3). \tag{6.6.9}$$

The integration can be continued for the corresponding parallel surfaces. In fact, (3.1.1) in addition to (6.6.6) and (6.6.7) implies

$$dk = 0, \quad \omega_3^4 = -\omega_4^3 = \omega_3^\rho = \omega_4^\rho = 0 \quad (\rho \in \{5, \ldots, n\});$$

thus $d(cx - \alpha e_3) = 0$, since $c + \alpha^2 = 0$, and $de_4 = 2ke_2' du$. Now from (6.6.7), $x_{uuv} = y''(u) = 0$; hence $y(u) = p_1 u + p_2$, where p_1 and p_2 are some constant vectors.

Further, $x_{uu} = 4ke_4$, $x_{uuu} = 8k^2 x_v = 8k^2 y(u)$. This implies that $z''' = 8k^2(p_1 u + p_2)$; therefore,

$$x = (p_1 u + p_2)v + 8k^2\left(\frac{1}{24}p_1 u^4 + \frac{1}{6}p_2 u^3 + \frac{1}{2}p_3 u^2 + p_4 u + p_5\right), \tag{6.6.10}$$

where p_3, p_4 and p_5 are additional constant vectors.

From here $x - 8k^2 p_5 = \xi_1 p_1 + \xi_2 p_2 + \xi_3 p_3 + \xi_4 p_4$, where $\xi_1 = uv + \frac{1}{3}k^2u^4$, $\xi_2 = v + \frac{4}{3}k^2u^3$, $\xi_3 = 4k^2u^2$, and $\xi_4 = 8k^2u$ satisfy the equations $2\xi_2\xi_4 - (\xi_3)^2 - 16k^2\xi_1 = 0$, $(\xi_4)^2 - 16k^2\xi_3 = 0$. Hence the surface is contained in $H_1^n(r)$, thus $\langle x, x\rangle = \frac{1}{c}$ is satisfied identically with respect to u and v. It follows that the matrix of $\langle p_\varphi, p_\psi\rangle$, $\varphi, \psi \in \{1, \cdots, 5\}$, is

$$\begin{pmatrix} 0 & 0 & 0 & 0 & \kappa \\ 0 & 0 & 0 & -\kappa & 0 \\ 0 & 0 & \kappa & 0 & 0 \\ 0 & -\kappa & 0 & 0 & 0 \\ \kappa & 0 & 0 & 0 & \lambda, \end{pmatrix}$$

where $\lambda = (64ck^2)^{-1}$ and κ is a nonzero real number. Therefore, $p_1, \ldots, p_5$ is a basis of ${}_2E^5$, which contains ${}_1H^4(c)$ as a standard model. The equations above show that the parallel M^2 is an algebraic surface in this case, namely, the intersection of two quadrics.

The semiparallel surface M^2 of the present subcase is a second-order envelope of these parallel algebraic ruled surfaces.

Now consider the particular subcase with $m_1 = m_0 = 1$, which is characterized by $\alpha = 0$.

First suppose $\beta \neq 0$. The geometry of the corresponding semiparallel surface M^2 is the same as above, its codimension and degree of arbitrariness are reduced substantially. Namely, in its Pfaffian system now $\omega_1^{\rho'} = \omega_2^{\rho'} = 0$, $\rho' \in \{3, 5, \ldots, n\}$, which gives by exterior differentiation that $\omega_4^{\rho'} \wedge \omega_1^4 = \omega_4^{\rho'} \wedge \omega_2^4 = 0$, and hence $\omega_4^{\rho'} = 0$, because ω_1^4 and ω_2^4 are linearly independent since $(k - \beta)(k + \beta) - k^2 = -\beta^2 \neq 0$. It follows that this M^2 lies in an ${}_1H^3(c)$; and it exists, uniquely up to one real function of one variable (see [Lu 97b]). Furthermore, it is a second-order envelope of the corresponding parallel surfaces, which are projective quadrics in ${}_1H^3(c)$, and it is a ruled surface with generators in null directions.

Finally, let $\alpha = \beta = 0$. Then $c = 0$, and hence this timelike semiparallel surface M^2 of the exceptional case lies in ${}_1E^n$; and since now $H = 0$, it is a minimal surface. Equations (6.6.8) now reduce to

$$dx = e_1'du + e_2'dv, \quad de_1' = 4ke_4du, \quad de_2' = 0.$$

Here $x_v = e_2' = p_0$ is a constant vector; hence the curves with $u =$ const on this M^2 are parallel straight lines of null direction, so M^2 is a cylinder. Moreover, now $x = p_0v + q(u)$, thus $x_u = e_1' = q'(u)$, and hence $x_{uu} = 4ke_4 = q''$.

For a corresponding parallel surface, $de_4 = 2ke_2'du$ (see above), which implies $q''' = 8k^2p_0$, and thus $q = \frac{4}{3}k^2p_0u^3 + \frac{1}{2}p_1u^2 + p_2u + p_3$, where p_1, p_2, p_3 are some constant vectors. Consequently, this parallel cylindrical surface is given by

$$x = p_0\left(v + \frac{4}{3}k^2u^3\right) + \frac{1}{2}p_1u^2 + p_2u + p_3.$$

Here $v + \frac{4}{3}k^2u^3 = a = \text{const}$ gives a family of congruent parabolas, and hence the parallel surface is a parabolic cylinder with generators in a null direction.

The semiparallel M^2 of this subcase is a second-order envelope of such cylinders. The results of the analysis for this case can be summarized as follows.

Theorem 6.6.4. *A semiparallel timelike surface M^2 of exceptional case in a Lorentz spacetime form ${}_1N^n(c)$ is the union of the closures of its open subsets, each of which is*

(i) *a ruled surface with flat $\bar{\nabla}$ and with generators in null directions, lying in ${}_1H^n(c)$, and is a second-order envelope of parallel surfaces, each of which is either*
- *a quadric in some ${}_1H^3(c^*)$, or*
- *an algebraic surface, defined by* (6.6.10) (*intersection of two quadrics*) *in some ${}_1H^4(c)$, or*

(ii) *a cylindrical surface in ${}_1E^n$ with generators in null direction, which is a second-order envelope of parabolic cylinders, each contained in some ${}_1E^3$.*

Remark 6.6.5. As shown for the parallel parabolic cylinder above, each of them contains a family of congruent parabolas defined by $v + \frac{4}{3}k^2u^3 = a = \text{const}$. Similarly, on the parallel algebraic ruled surfaces defined by (6.6.8), a simple algebraic curve can be indicated. Namely, adding to (6.6.8) the equation $v + \frac{1}{3}u^3 = 0$ leads to $x = k^2(p_2u^3 + 4p_3u^2 + 8p_4u + 8p_5)$, and thus to a cubic curve.

Remark 6.6.6. As is noted above, the semiparallel timelike cylindrical surfaces M^2 in ${}_1E^n$ of Theorem 6.6.4 are minimal surfaces. They are not the only semiparallel minimal timelike surfaces in ${}_1N^n(c)$ which are not totally geodesic, as seen from Remark 6.6.3.

These minimal semiparallel timelike surfaces can be considered as objects of geometrical string theory, which plays an important role in theoretical particle physics and cosmology.

The fact that a timelike surface with flat $\bar{\nabla}$ in ${}_1E^3$ is minimal iff it is a cylinder with lightlike generators, was proved in [Va 91] (see also [VdW 90]). Corresponding surfaces in ${}_1E^n$ were considered in [Mag 84]. (In some of those references, minimal timelike surfaces are called maximal or extremal.)

6.7 Spacelike 2-Parallel Surfaces

Among semiparallel surfaces M^2 in a space form $N^n(c)$, a special class consists of 2-parallel surfaces (see Proposition 4.1.5), which will be studied in this section.

First, consider 2-parallel curves; their product surfaces are special cases of spacelike 2-parallel surfaces.

A plane curve in a space form $N^n(c)$ is called a *clothoid*, or *Cornu spiral*, if its natural equation is $k = as$, for some constant a.

Proposition 6.7.1. *A 2-parallel curve in $N^n(c)$ is a Cornu spiral, either on a plane $N^2(c) \subset N^n(c)$ or on a totally umbilic surface $N^2(c^*)$ of $N^n(c)$.*

Proof. For a curve $dx = e_1\omega^1$; if e_1 is a unit vector, then $\omega^1 = ds$, where s is the arclength parameter. Now (2.4.1) reduces to $de_1 = h^*_{11}ds$, since $\omega_1^1 = 0$; then $h^*_{11} = h_{11} - cx$, and for a 2-parallel curve $h_{11} \neq 0$. Taking the unit vector e_2 to lie in the normal subspace, so that $h_{11} = k_1 e_2$, one obtains $de_1 = (k_1 e_2 - cx)ds$, where k_1 is the curvature of the curve, moreover $\omega_1^2 = k_1 ds$. Further, (3.1.2) reduces to $dh_{11} = -k_1^2 e_1 ds + h_{111} ds$, where $dh_{11} = d(k_1 e_2) = e_2 dk_1 + k_1(-\omega_1^2 e_1 + \omega_2^\alpha e_\alpha)$, and for a 2-parallel curve, $h_{111} \neq 0$. Therefore,

$$h^2_{111} = \frac{dk_1}{ds}, \quad h^\rho_{111} ds e_\rho = \omega_2^\rho e_\rho,$$

where $\rho \in \{3, \dots, n\}$. In general the part $e_3, \dots, e_n$ of the orthonormal frame can be chosen so that $h^\rho_{111} e_\rho = k_1 k_2 e_3$ for some k_2. Then

$$h^3_{111} = k_1 k_2, \quad h^4_{111} = \dots = h^n_{111} = 0.$$

The 2-parallel condition $\bar{\nabla} h^\alpha_{ijk} = 0$, where

$$\bar{\nabla} h^\alpha_{ijk} = h^\alpha_{ljk}\omega_i^l + h^\alpha_{ilk}\omega_j^l + h^\alpha_{ijl}\omega_k^l - h^\beta_{ijk}\omega_\beta^\alpha,$$

then reduces to

$$\frac{d^2 k_1}{ds^2} - k_1 k_2^2 = 0, \quad \frac{d(k_1 k_2)}{ds} + \frac{dk_1}{ds} k_2 = 0, \quad k_1 k_2 k_3 = 0, \tag{6.7.1}$$

where at least one of $\frac{dk_1}{ds}$ and $k_1 k_2$ is nonzero, because $\bar{\nabla} h \neq 0$.

If $k_1 \neq 0$, $\frac{dk_1}{ds} \neq 0$, $k_2 = 0$, then $\frac{d^2 k_1}{ds^2} = 0$, and thus $k_1 = as + b$ with constant $a \neq 0$ and b. The origin of s can be chosen so that $b = 0$. Hence the curve has the natural equations $k_1 = as$, $k_2 = 0$ and is a Cornu spiral on a plane of $N^n(c)$.

If $k_1 k_2 \neq 0$, then $k_3 = 0$. Hence the curve lies in a three-dimensional subspace $N^3(c) \subset N^n(c)$. The middle equation (6.7.1) is $\frac{dk_2}{ds} k_2^{-1} = -2\frac{dk_1}{ds} k_1^{-1}$ and yields $k_2 = q k_1^{-2}$, where $q = \text{const} \neq 0$. Substitution into the first equation (6.7.1) gives $\frac{d^2 k_1}{ds^2} = q^2 k_1^{-3}$; thus $k_1 = \sqrt{As^2 + 2Bs + C}$ with constants A, B, C satisfying $AC - B^2 = q^2$. A suitable choice of the origin of s makes $B = 0$, so that

$$k_1 = \sqrt{As^2 + C}, \quad k_2 = q k_1^{-2}, \quad q^2 = AC.$$

Hence $q = \varepsilon\sqrt{AC}$, where ε is 1 or -1. Now a straightforward computation shows that in ${}_\sigma E^4$ the point with radius vector

$$x_0 = x + k_1^{-1}\left(e_2 + \varepsilon\sqrt{\frac{A}{C}} s e_3\right)$$

is a fixed point, since $dx_0 = 0$. Moreover $(x_0 - x)^2 = C^{-1} = \text{const}$. Therefore, this curve lies on a sphere, whose unit normal vector is $n = k_1^{-1}(\sqrt{C} e_2 + \varepsilon\sqrt{A} s e_3)$. Thus the geodesic unit normal vector of this curve on the sphere is $n_g = k_1^{-1}(-\varepsilon\sqrt{A}se_2 + \sqrt{C}e_3)$. Now $\frac{de_1}{ds} = k_1 e_2$ has the geodesic normal component $-\varepsilon\sqrt{A}sn_g$. Thus the geodesic curvature of the curve is $k_g = as$, where $a = -\varepsilon\sqrt{A}$ is a constant. Hence the curve is a Cornu spiral on a sphere in ${}_\sigma E^4$, which intersects $N^3(c)$ along a totally umbilic surface $N^2(c^*)$. This finishes the proof.

Now consider 2-parallel surfaces.

Theorem 6.7.2. *Each 2-parallel surface M^2 in $N^n(c)$ has flat $\bar{\nabla}$ and is either*

- *a product of 2-parallel or parallel curves, at least one of which is 2-parallel, or*
- *a surface in a three-dimensional totally umbilic $N^3(c^*)$, generated by the geodesics of $N^3(c^*)$ which go in directions of the binormals of a curve in $N^3(c^*)$ with geodesic curvature $k_g = as$, and with constant geodesic torsion $\kappa_g = \sqrt{c}$.*

The last surface is often called the *B-scroll* of this curve, following [DaN 81].

Proof. Since 2-parallel surfaces are semiparallel, Theorem 6.2.1 can be used, but with subcases (e) and (f) omitted, because totally umbilic surfaces and totally geodesic surfaces are parallel and therefore not 2-parallel.

It can be shown that in subcases (a) and (b), the surface also reduces to a parallel one, namely to a Veronese surface.

Due to Remark 6.3.5 it suffices to analyse subcase (a) since it also encompasses (b).

Then (6.3.11) holds, and $h_{ij}^{\xi} = 0$. Substitution into (2.2.3), considering (2.2.2), gives the following system of nine equations for superscripts $\alpha = 3, 4, 5$:

$$
\begin{aligned}
d\alpha &= k\omega_3^4 + h_{111}^3\omega^1 + h_{112}^3\omega^2,\\
0 &= k\omega_3^5 + h_{112}^3\omega^1 + h_{122}^3\omega^2,\\
d\alpha &= -\omega_3^4 + h_{122}^3\omega^1 + h_{222}^3\omega^2,\\
dk &= -\alpha\omega_3^4 + h_{111}^4\omega^1 + h_{112}^4\omega^2,\\
0 &= -k(2\omega_1^2 - \omega_4^5) + h_{112}^4\omega^1 + h_{122}^4\omega^2,\\
-dk &= -\alpha\omega_3^4 + h_{122}^4\omega^1 + h_{222}^4\omega^2,\\
0 &= k(2\omega_1^2 - \omega_4^5) - \alpha\omega_3^5 + h_{111}^5\omega^1 + h_{112}^5\omega^2,\\
0 &= -k(2\omega_1^2 - \omega_4^5) - \alpha\omega_3^5 + h_{122}^5\omega^1 + h_{222}^5\omega^2,\\
dk &= h_{112}^5\omega^1 + h_{122}^5\omega^2.
\end{aligned}
$$

Here dk, ω_3^4, ω_3^5, and $2\omega_1^2 - \omega_3^4$ can be eliminated and, since ω^1 and ω^2 are linearly independent, this leads to the following 10 relations on the components of $\bar{\nabla}h$:

$$
h_{111}^4 - h_{122}^4 = 2h_{112}^5, \quad h_{112}^4 - h_{222}^4 = h_{122}^5,
$$
$$
2h_{112}^4 = h_{122}^5 - h_{111}^5, \quad 2h_{122}^4 = h_{222}^5 - h_{112}^5,
$$

$$k(h^4_{111}+h^4_{122}) = -\alpha(h^3_{111}-h^3_{122}), \quad k(h^4_{112}+h^4_{222}) = -\alpha(h^3_{112}-h^3_{222}),$$
$$k(h^5_{111}+h^5_{122}) = -2\alpha h^3_{112}, \quad k(h^5_{112}+h^5_{222}) = -2\alpha h^3_{122},$$
$$3kh^5_{112} = \alpha(h^3_{111}+h^3_{122}), \quad 3kh^5_{122} = \alpha(h^3_{112}+h^3_{222}).$$

The 2-parallel condition implies via (2.2.6) that $h^\alpha_{ijkl} = 0$, and thus (2.2.7) gives $\bar{\Omega} \circ h^\alpha_{ijk} = 0$, or, more explicitly,

$$h^\alpha_{ljk}\Omega^l_i + h^\alpha_{ilk}\Omega^l_j + h^\alpha_{ijl}\Omega^l_k - h^\beta_{ijk}\Omega^\alpha_\beta = 0. \tag{6.7.2}$$

Then (2.1.9), together with (6.3.1) and (6.3.2), imply $\Omega^2_1 = -k^2\omega^1 \wedge \omega^2 \neq 0$, $\Omega^3_\alpha = 0$, $\Omega^\xi_4 = \Omega^\xi_5 = 0$ (where $\xi \in \{6, \ldots, n\}$); therefore from (6.7.2) $h^3_{ijk} = 0$. Now the above 10 relations imply that $h^4_{ijk} = h^5_{ijk} = 0$. Moreover, since $\Omega^\xi_3 = \Omega^\xi_4 = \Omega^\xi_5 = 0$, then (6.7.2) implies $h^\xi_{ijk} = 0$. As a result, all $h^\alpha_{ijk} = 0$, thus $\bar{\nabla} h = 0$. Hence the surface is parallel; due to Proposition 3.1.7, it lies in an $N^5(c)$; and due to Proposition 6.3.1 and Corollary 6.3.6, it is a Veronese surface, i.e., not 2-parallel.

So it remains to study the surfaces of subcases (c) and (d), with the additional 2-parallel condition. Recall that for these, $\bar{\nabla}$ is flat, i.e., $\Omega^2_1 = \Omega^\beta_\alpha = 0$; that proves the first assertion.

For subcase (c), $l = 0$, $\alpha k \neq 0$, and one can make $\gamma = 0$. Therefore,

$$h^3_{11} = h^3_{22} = \alpha, \ \ h^5_{12} = 0, \ \ h^4_{11} = \beta + k, \ \ h^4_{12} = 0, \ \ h^4_{22} = \beta - k, \ \ h^5_{ij} = h^\xi_{ij} = 0,$$

where $\xi \in \{6, \ldots, n\}$. Substitution into (2.2.3), using (2.2.2), gives

$$h^3_{112} = h^3_{122} = h^5_{112} = h^5_{122} = h^\xi_{112} = h^\xi_{122} = 0,$$
$$2k\omega^2_1 = h^4_{112}\omega^1 + h^4_{122}\omega^2, \tag{6.7.3}$$
$$d\alpha = (\beta + k)\omega^4_3 + h^3_{111}\omega^1, \quad d\alpha = (\beta - k)\omega^4_3 + h^3_{222}\omega^2,$$
$$d(\beta + k) = -\alpha\omega^4_3 + h^4_{111}\omega^1 + h^4_{112}\omega^2,$$
$$d(\beta - k) = -\alpha\omega^4_3 + h^4_{122}\omega^1 + h^4_{222}\omega^2,$$
$$\alpha\omega^\rho_3 + (\beta + k)\omega^\rho_4 = h^\rho_{111}\omega^1, \quad \alpha\omega^\rho_3 + (\beta - k)\omega^\rho_4 = h^\rho_{222}\omega^2,$$

where $\rho \in \{5, 6, \ldots, n\}$. From here, in particular,

$$2k\omega^4_3 = -h^3_{111}\omega^1 + h^3_{222}\omega^2, \quad 2k\omega^\rho_4 = h^\rho_{111}\omega^1 - h^\rho_{222}\omega^2. \tag{6.7.4}$$

Recall that for a semiparallel M^2 in E^n, in subcase (c) it follows from (6.2.6) that $k^2 - H^2 - c = 0$, i.e., $(\beta + k)(\beta - k) + \alpha^2 + c = 0$. After differentiation, this gives

$$(\beta - k)h^4_{111} + (\beta + k)h^4_{122} = -\alpha h^3_{111}, \tag{6.7.5}$$
$$(\beta - k)h^4_{112} + (\beta + k)h^4_{222} = -\alpha h^3_{222}. \tag{6.7.6}$$

Now apply the 2-parallel condition $\bar{\nabla} h^{\alpha}_{ijk} = 0$. In particular, $\bar{\nabla} h^{3}_{112} = 0$ and $\bar{\nabla} h^{3}_{122} = 0$ reduce to

$$h^3_{111}h^4_{112} = 0, \quad h^3_{222}h^4_{122} = 0, \quad h^3_{111}h^4_{122} = h^3_{222}h^4_{112},$$

which by (6.7.3) and (6.7.4) is equivalent to

$$\omega^2_1 \omega^4_3 = 0.$$

Similarly $\bar{\nabla} h^{\rho}_{112} = 0$ and $\bar{\nabla} h^{\rho}_{122} = 0$ imply $\omega^2_1 \omega^{\rho}_4 = 0$, which all together means that

$$\omega^2_1 \omega^{\alpha}_4 = 0.$$

Here the case $\omega^2_1 = 0$ gives the first case of Theorem 6.7.2. Indeed, here $h_{12} = 0$, and by (6.7.3), $h_{112} = h_{122} = 0$. Hence for this surface

$$dx = e_1\omega^1 + e_2\omega^2, \quad de_1 = h^*_{11}\omega^1, \quad de_2 = h^*_{22}\omega^2,$$

where the vectors

$$h^*_{11} = \alpha e_3 + (\beta + k)e_4 - cx, \quad h^*_{22} = \alpha e_3 + (\beta - k)e_4 - cx$$

are orthogonal since $\langle h^*_{11}, h^*_{22}\rangle = \alpha^2 + \beta^2 - k^2 + c = 0$.

For these vectors, (3.1.2) and (3.1.3) imply

$$dh^*_{11} = (h_{111} - e_1\langle h^*_{11}, h^*_{11}\rangle)\omega^1, \quad dh^*_{22} = (h_{222} - e_2\langle h^*_{22}, h^*_{22}\rangle)\omega^2.$$

Differentiation of the last relation gives $\langle h_{111}, h^*_{22}\rangle = \langle h^*_{11}, h_{222}\rangle = 0$.

For h_{ijk}, a formula anologous to (3.1.2) holds, but with an additional lower index. Here the 2-parallel condition means that $e_\alpha h^{\alpha}_{ijkl} = 0$, and therefore

$$dh_{111} = -e_1\langle h_{111}, h^*_{11}\rangle\omega^1, \quad dh_{222} = -e_2\langle h_{222}, h^*_{22}\rangle\omega^2.$$

This implies that the surface is a product of curves defined by $\omega^2 = 0$ and $\omega^1 = 0$, and therefore is at least 2-parallel. This proves the first part of Theorem 6.7.2.

It remains to study the case $\omega^{\alpha}_4 = 0$, where, in particular, $\omega^3_4 = 0$. Then $h^3_{ijk} = h^{\rho}_{ijk} = 0$, due to (6.7.3) and (6.7.4). This must be substituted into all the formulas above. In particular, $d\alpha = 0$, so that $\alpha = \text{const}$.

It turns out that if $\beta^2 - k^2 = 0$, then the situation is the same as before. Indeed, if $\beta + k = 0$, this implies $h^4_{111} = h^4_{112} = 0$, and by (6.7.6) also $h^4_{222} = 0$; hence $\omega^2_1 = 0$. If $\beta - k = 0$, one just has to interchange the roles of the lower indices 1 and 2.

So, assume now that $\beta^2 - k^2 \neq 0$; then $\alpha^2 + c \neq 0$. From (6.7.5) and (6.7.6) it follows that there exist some functions λ and μ such that

$$h^4_{111} = \lambda(\beta + k), \quad h^4_{122} = -\lambda(\beta - k), \quad h^4_{112} = \mu(\beta + k), \quad h^4_{222} = -\mu(\beta - k).$$

The above expressions for $d(\beta + k)$, $d(\beta - k)$ and $2k\omega^2_1$ now imply

$$\frac{d(\beta+k)}{\beta+k} = \frac{d(\beta-k)}{\beta-k} = \lambda\omega^1 + \mu\omega^2,$$

$$2k\omega_1^2 = \mu(\beta+k)\omega^1 - \lambda(\beta-k)\omega^2.$$

Substitution into $\bar{\nabla} h_{ijk}^4 = 0$ gives

$$\lambda^2(\beta-k) + \mu^2(\beta+k) = 0,$$

$$d\lambda = \frac{k-3\beta}{2k}\lambda(\lambda\omega^1 + \mu\omega^2), \quad d\mu = -\frac{k+3\beta}{2k}\mu(\lambda\omega^1 + \mu\omega^2). \quad (6.7.7)$$

Multiplying the first of the preceding equations by $\beta + k = B$ gives $\lambda^2 C^2 = \mu^2 B^2$, where $C^2 = \alpha^2 + c$. Thus there exists a function ψ such that

$$\lambda = 2kBC^{-1}\psi, \quad \mu = 2k\varepsilon\psi, \quad \varepsilon = \pm 1;$$

hence

$$\omega_1^2 = \psi(\varepsilon B\omega^1 + C\omega^2).$$

Substitution into (6.7.7) gives

$$d\psi = \varepsilon\psi\left(\frac{2C}{B} - \frac{3B}{c}\right)\omega_1^2;$$

moreover

$$dB = \varepsilon C^{-1}(B^2 + C^2)\omega_1^2.$$

Recall that the M^2 in $N^n(c)$ with the specialized orthonormal frame field chosen above satisfies the Pfaffian equations

$$\omega^3 = \omega^4 = \omega^\rho = 0 \quad (\rho = 5, 6 \ldots, n),$$

$$\omega_1^3 = \alpha\omega^1, \quad \omega_2^3 = \alpha\omega^2, \quad \omega_1^4 = B\omega^1, \quad \omega_2^4 = -\frac{C^2}{b}\omega^2,$$

$$\omega_1^\rho = \omega_2^\rho = 0, \quad \omega_3^4 = \omega_3^\rho = \omega_4^\rho = 0,$$

where, as one recalls, α is a constant.

Adding the three previous equations to these equations, one obtains a Pfaffian system which is totally integrable, as is easy to check via Frobenius' theorem (e.g., its second version in [Ste 64]).

To obtain a geometrical description of the surface M^2 defined by this system, one starts with the case $\alpha = 0$. Then $C^2 = c$, and M^2 belongs to an $S^3(c)$. Its asymptotic lines are defined by $h_{ij}^4\omega^i\omega^j = 0$, i.e., by $(B\omega^1)^2 - (C\omega^2)^2 = 0$. The frame in the tangent plane can be rotated so as to obtain new basis vectors

$$f_1 = B_c^{-1}(\sqrt{c}e_1 - \varepsilon Be_2), \quad f_2 = B_c^{-1}(\varepsilon B + \sqrt{c}e_2),$$

where $B_c = \sqrt{B^2 + c}$. Then $dx = f_1\theta^1 + f_2\theta^2$, where

$$\theta^1 = B_c^{-1}(\sqrt{c}\omega^1 - \varepsilon B\omega^2), \quad \theta^2 = B_c^{-1}(\varepsilon B\omega^1 + \sqrt{c}\omega^2).$$

Since $dB_c = \varepsilon B(\sqrt{c})^{-1} B_c \omega_1^2$, an easy computation shows that

$$df_1 = -cx\theta^1 + \varepsilon\sqrt{c}\, e_4\theta^2, \quad df_2 = -cx\,\theta^2 + (\varepsilon\sqrt{c}\,\theta^1 + 2H_c\,\theta^2)e_5,$$

where $H_c = \frac{1}{2}\left(B - \frac{c}{B}\right)$ is the mean curvature in spherical geometry.

The asymptotic lines, defined by $\theta^2 = 0$, are geodesics of $S^3(c)$, due to $dx = f_1\theta^1$, $df_1 = -cx\theta^1$. Their orthogonal trajectories are defined by $\theta^1 = 0$ and for them

$$dx = f_2\theta^2, \quad df_2 = -cx\theta^2 + 2H_c e_3\theta^2, \quad de_3 = -(\varepsilon\sqrt{c} f_1 + 2H_c f_2)\theta^2.$$

It is seen that in spherical geometry, the trajectory has unit tangent vector f_2, unit principal normal and binormal vectors e_3 and f_1, respectively, curvature $k_c = 2H_c$, and torsion $\kappa_c = -\varepsilon\sqrt{c}$, which is a constant.

For the curvature of this trajectory one has that $dk_c = p\theta^2$, where $p = \varepsilon(\sqrt{c})^{-1}\psi B^{-2}B_c^5$. Due to the above equations, $dp = 0$, so that $p =$ const. Since θ^2 is the differential of the arclength parameter s_c of a trajectory, it follows that $\frac{dk_c}{ds_c} = p =$ const. Therefore, $k_c = ps_c + q$, where q is a constant, and by suitably choosing the origin of s_c, one gets $k_c = ps_c$. This proves the theorem in the special case $\alpha = 0$.

The general case $\alpha \neq 0$ can be reduced to the previous one. Recall that α is a constant; moreover $\omega_4^\rho = 0$. Therefore, $z = x + \alpha^{-1}e_4$ is a constant vector, since $dz = 0$.

Let $c \neq 0$. The point in ${}_\sigma E^{n+1}$ with radius vector z is a fixed point. Hence the surface M^2 lies in the intersection of the standard model of $N^n(c)$ with the n-dimensional hypersphere of real radius $(\sqrt{\alpha})^{-1}$ around this point. Here $\langle z, z\rangle = C^2(c\alpha^2)^{-1}$ is nonzero since $C^2 \neq 0$, and has the same sign as c. It follows that the above intersection is an $(n-1)$-dimensional sphere in $N^n(c)$ containing M^2; this gives the desired reduction.

Let $c = 0$, i.e., $N^n(c) = E^n$. Then M^2 lies in a hypersphere of E^n around a fixed point, as above, with the same radius as above, again giving the desired reduction. This finishes the proof.

Remark 6.7.3. The 2-parallel surfaces in E^n were first investigated in [LM 84], but with an important case omitted. The omitted case was studied in [Lu 86], where the above Theorem 6.7.2 was proved for the special case of E^n (see also [Lu 87c], Proposition 2). This result was then generalized in [Lu 2000a], Theorem 18.2, to 2-parallel M^2 in $N^n(c)$.

Spacelike or timelike 2-parallel surfaces M^2 in pseudo-Euclidean spaces ${}_sE^n$ or pseudo-Riemannian spacetime forms ${}_sN^n(c)$ have not yet been investigated completely, as far as the author knows.

Remark 6.7.4. In a Riemannian manifold, the Laplace operator is $\Delta = g^{ij}\nabla_i\nabla_j$. For a submanifold M^m in E^n the well-known Beltrami equation states that $\Delta x = mH$, where x is the radius vector of an arbitrary point of M^m and H is the mean curvature vector at this point. It follows that $\Delta^2 x = \Delta(\Delta x) = m\Delta H$. Consequently, x is biharmonic (i.e., $\Delta^2 x = 0$) if and only if H is harmonic (i.e., $\Delta H = 0$); see [Ch 2000].

For the normal connection $\nabla^\perp$ of M^m in E^n, one can introduce the operator $\Delta^\perp = g^{ij}\nabla_i^\perp\nabla_j^\perp$. If an M^m in E^n satisfies $\Delta^\perp H = 0$, then M^m is said to be *weak biharmonic*.

Weak biharmonic curves M^1 in E^n were investigated in [BG 95] and surprisingly we found that they coincide with 2-parallel curves, which are classified in Proposition 6.7.1 above. As a generalization, it was shown in [KALM 03] that every locally Euclidean (i.e., with flat ∇) 2-parallel submanifold M^m in E^n is weak biharmonic. Conversely, it was also shown in [KALM 03] that if a surface M^2 with flat $\bar\nabla$ in E^4 is weak biharmonic, and if one family of its lines of curvature consists of geodesics, then this M^2 is 2-parallel.

6.8 q-Parallel Surfaces as Semiparallel Surfaces

Theorem 6.7.2, which classifies the 2-parallel surfaces M^2 in $N^n(c)$, also states that all such M^2 have flat $\bar\nabla$. This can be generalized to q-parallel surfaces with $q \geq 2$, and so it follows that they are all special cases of semiparallel surfaces.

Theorem 6.8.1. *Every q-parallel surface M^2 in $N^n(c)$ with $q \geq 2$ has flat $\bar\nabla$, and thus is semiparallel* (*see Proposition* 4.1.4).

Proof. The flatness of $\bar\nabla$ for such submanifolds can be shown by using an idea of V. Mirzoyan [Mi 91d], for the case $c = 0$. In the special case of surfaces this was done in a more simple realization in [Lu 91c]. Here it will be generalized to $c \neq 0$.

The connection ∇ of a surface M^2 is flat if and only if $\Omega_1^2 = -K\omega^1 \wedge \omega^2$ is zero, where K is the Gaussian curvature.

Suppose it is not so, i.e., suppose $\Omega_1^2 \neq 0$ and M^2 is q-parallel with $q \geq 2$, so that $\bar\nabla^{q-1}h \neq 0$. Introduce the notation

$$F_{ij}^{(s)} = h_{ip_2\ldots p_s}^{\alpha} h_{\alpha j}^{p_2\ldots p_s}.$$

The square of $\bar\nabla^{s-2}h$ is the trace $F^{(s)} = F_i^{(s)i}$. Applying the Laplace–Beltrami operator $\Delta = \nabla_p\nabla^p$ gives

$$\frac{1}{2}\Delta F^{(s)} = (\bar\nabla_p\bar\nabla^p h_{ip_2\ldots p_s}^{\alpha})h_{\alpha}^{ip_2\ldots p_s} + F^{(s+1)}. \tag{6.8.1}$$

Similar to (2.2.3) and (2.2.6), the following sequence of implications can be verified:

$$\nabla F_{ij}^{(s)} = F_{ijk}^{(s)}\omega^k \Rightarrow \nabla F_{ijk}^{(s)} \wedge \omega^k = \Omega \circ F_{ij}^{(s)}$$
$$\Rightarrow \nabla F_{ijk}^{(s)} = F_{ijkl}^{(s)}\omega^l \Rightarrow \nabla F_{ijkl}^{(s)} \wedge \omega^l = \Omega \circ F_{ijk}^{(s)},$$

where the operator Ω acts as in (2.2.5).

If M^2 is q-parallel, then $\bar{\nabla} h_{ijp_1\dots p_{q-1}}$ is zero; this yields $F^{(q-1)}_{ijkl} = 0$ and thus $\Omega \circ F^{(q-1)}_{ijk} = 0$. The last equations form a 6×6 linear homogeneous system with unknowns

$$F^{(q-1)}_{111}, \quad F^{(q-1)}_{112}, \dots, F^{(q-1)}_{222},$$

which has a nonvanishing determinant, due to $\Omega_1^2 \neq 0$. Consequently, $F^{(q-1)}_{ijk} = 0$ and hence $\nabla F^{(q-1)}_{ij} = 0$. Substituting all this into (6.8.1) with $s = q-1$, the result is $F^{(q)} = 0$ and thus $h_{ijp_1\dots p_q} = 0$; but this contradicts the assumption $\bar{\nabla}^{q-1} h \neq 0$.

The second part of the proof, concerning the flatness of $\nabla^\perp$, can be given in an indirect way as follows. The subbundle $O(M^2, N^n(c))$ of adapted orthonormal frames has a canonical section for which (6.1.8) hold. Then, similarly to (6.2.1) and (6.2.2) for M^2 in ${}_sE^n$, one now gets

$$\Omega_1^2 = (-c + k^2 + l^2 - H^2)\omega^1 \wedge \omega^2, \quad \Omega_4^5 = -2kl\omega^1 \wedge \omega^2,$$

with all other Ω_i^j, Ω_α^β being zero.

Suppose M^2 is q-parallel, $q \geq 2$, and $\Omega_4^5 \neq 0$; it is known already that $\Omega_1^2 = 0$. The first assumption gives $h^\alpha_{ijp_1\dots p_q} = 0$ and from (2.2.7) it follows that $\Omega \circ h^\alpha_{ijp_1\dots p_{q-1}} = 0$. The only nontrivial equations here are

$$h^4_{ijp_1\dots p_{q-1}} \Omega_4^5 = 0, \quad h^5_{ijp_1\dots p_{q-1}} \Omega_4^5 = 0,$$

thus $h^a_{ijp_1\dots p_{q-1}} = 0$, where $a, b, \dots \in \{4, 5\}$. From $\bar{\nabla}(\bar{\nabla}^{q-1}h) = \bar{\nabla}^q h = 0$ it follows that $\bar{\nabla} h^a_{ijp_1\dots p_{q-1}} = 0$ and thus $h^b_{ijp_1 p_{q-2}} \Omega^a_b = 0$ due to (2.2.7); hence $h^b_{ijp_1 p_{q-2}} = 0$. This process can be repeated, and after $q-2$ steps it gives $h^4_{ij} = h^5_{ij} = 0$; which, due to (2.1.9) and (2.1.10), contradicts the assumption $\Omega_4^5 \neq 0$. This finishes the proof.

Now a method can be used from [Lu 91c], where it was called the *polynomial map method*, and derived for all q-parallel submanifolds M^m with flat $\bar{\nabla}$ in E^n.

The flatness of $\bar{\nabla}$ is equivalent to the existence of a section in the adapted orthonormal frame bundle $O(M^m, E^n)$ which is parallel with respect to $\bar{\nabla}$, and thus is characterized by $\omega_i^j = \omega_\alpha^\beta = 0$ (see, e.g., [Ch 73b], Chapter 4, Section 1). For such a section, which is determined up to a constant orthogonal transformation in TM^m and $T^\perp M^m$, there exists an atlas of local coordinate systems $\{u^1, \dots, u^m\}$ such that $\omega^i = du^i$, and therefore from (2.2.3)

$$h^\alpha_{ijk} = \partial h^\alpha_{ij}/\partial u^k = \partial h^\alpha_{ik}/\partial u^j.$$

The second equality shows that $h^\alpha_{ij} du^j$ is an exact differential, $d\chi_i^\alpha$, so that $h^\alpha_{ij} = \partial \chi_i^\alpha / \partial u^j$. Now $h^\alpha_{ij} = h^\alpha_{ji}$ implies that $\chi_i^\alpha du^i = d\chi^\alpha$, thus

$$h^\alpha_{ij} = \frac{\partial^2 \chi^\alpha}{\partial u^i \partial u^j}, \dots, h_{ijp_1\dots p_s} = \frac{\partial^{s+2}\chi^\alpha}{\partial u^i \partial u^j \partial u^{p_1} \dots \partial u^{p_s}}.$$

The q-parallel condition $\bar{\nabla}^q h = 0$ requires that the last quantities above must be zero for $s = q$, but this means that χ^α are polynomials of degree $\leq q + 1$, with at least one of them having degree exactly $q + 1$.

It follows that $\chi = e_\alpha \chi^\alpha$ is a vector valued polynomial of degree $q + 1$ in $u^1, \dots, u^m$; consequently, the h_{ij} are the same, but of degree $q - 1$. When these polynomial expressions are substituted into the conditions $\Omega_i^j = \Omega_\alpha^\beta = 0$ expressing the flatness of $\bar{\nabla}$, they lead via (2.1.9) and (2.1.10) to some relations connecting these polynomials. Consequently, one obtains a system of equations for the vectorial coefficients, whose analysis permits one to classify and describe the q-parallel normally flat submanifolds.

This polynomial map method has been used in [Lu 91c] for surfaces M^2 in E^n. In particular, the following theorem for 3-parallel surfaces was established.

Theorem 6.8.2. *A 3-parallel surface M^2 in E^n is either*

(i) *a product of two curves, or*
(ii) *a B-scroll in a three-dimensional sphere $S^3(c) \subset E^n$ of a curve whose spherical curvature is a second-degree polynomial in the arclength parameter and whose spherical torsion τ satisfies $\tau^2 = c$.*

Proof. The proof is given in [Lu 91c].

Remark 6.8.3. The new curves which are to be added to get all surfaces of case (i) were also characterized in [Lu 91c].

Note that B-scroll here means the same as in Theorem 6.7.2. Note also that the proof of (ii) by the polynomial map method is rather laborious and uses the analysis of the above-mentioned system of vectorial equations.

Remark 6.8.4. The following result was stated in [Di 90a] and then proved in [Di 91a], [Di 92]:

A q-parallel surface M^2 in $S^3(c)$ ($q \geq 2$) is a B-scroll of a curve whose spherical curvature is a polynomial of degree $q - 1$ of the arclength parameter and whose spherical torsion τ satisfies $\tau^2 = c$.

Theorems 6.7.2 and 6.8.2 show that, at least for $q = 2$ or 3, a q-parallel surface M^2 in E^n is either a product of curves or a B-scroll in an $S^3(c) \subset E^n$. It can be conjectured that perhaps this also happens in E^n if $q > 3$.

The polynomial map method gives a way of solving this problem, but it is clear that it would be technically rather complicated.

Remark 6.8.5. In [DV 90] it is shown that in the context of complex geometry the case of a B-scroll is impossible, independently from the dimension and codimension. Namely,

- *if M^m is a complex submanifold in a complex space form $N^n(c)$ then M^m is k-parallel if and only if $k \leq 1$.*

In the context of affine differential geometry there are several papers on k-parallel surfaces, normalized by the affine normal vector field in affine 3-space. Let M^2 be a nondegenerate surface in such a space, let ∇ be the canonical affine connection, and let h be the second fundamental form. As is shown in [NP 89], if $\nabla^2 h = 0$ but $\nabla h \neq 0$, then M^2 is congruent to an open part of the Cayley surface $z = xy + y^3$ by an equiaffine transformation. Here $\nabla^2 h = 0$ is equivalent to $\nabla C = 0$, where C is called the cubic form.

This result is generalized in [Vr 88] and in [DV 90], where it is shown that if $\nabla^q h = 0$ for $q \in \{2, 3, 4, 5\}$ then M^2 is locally equivalent to the generalized Cayley surface $z = xy + x^3 P(x)$, where P is a polynomial in x of degree at most $q - 2$. (See also [NP 89], [NM 89], [Vr 91], [DV 91], and the survey with references in [LMSS 96].)

7

Semiparallel Three-Dimensional Submanifolds

A complete classification has been given up to now not only for semiparallel surfaces but also for semiparallel three-dimensional submanifolds M^3, at least for positive definite metrics. For M^3 in Euclidean space E^n the classification is found in [Ri 86], [LR 90], [Lu 90b]. In this chapter the classification is extended to the case of M^3 in non-Euclidean space forms $N^n(c)$.

7.1 Semiparallel Submanifolds M^3 of Principal Codimension $m_1 \leq 2$

Semiparallel submanifolds M^3 with principal codimension 1 in $N^n(c)$ are a special case of the submanifolds M^m described in Theorems 5.5.1 and 5.5.3. Setting $m = 3$ in those theorems, one obtains the following.

Theorem 7.1.1. *A semiparallel submanifold M^3 with principal codimension* 1 *in $N^n(c)$ is*

- *for $c = 0$, i.e., in Euclidean space $N^n(0) = E^n$, either*
  - *the envelope of a one-parameter family of three-dimensional planes, or*
  - *a hypersurface in $E^4 \subset E^n$ which is either the product $C^{s+1} \times E^{2-s}$ of an $(s+1)$-dimensional round cone and an $(2-s)$-dimensional plane (with $s = 1, 2$), or the product $S^s \times E^{3-s}$ of a sphere S^s and a plane E^{3-s} (with $s = 2, 3$),*
- *for $c \neq 0$, i.e., in non-Euclidean space form $N^n(c)$, either*
  - *a hypersurface of rotation whose profile curve has the natural equation (5.5.8), or*
  - *a parallel hypersurface which is either spherical, or is the product of a spherical surface and a spherical curve, one of which is geodesic,*
- *or an open subset of one of the above.*

Ü. Lumiste, *Semiparallel Submanifolds in Space Forms*,
DOI 10.1007/978-0-387-49913-0_8, © Springer Science+Business Media, LLC 2009

Semiparallel submanifolds M^3 with principal codimension 2 and flat $\nabla^\perp$ in $N^n(c)$ can be considered as special cases for $m = 3$ of the submanifolds M^m described in Theorems 5.6.1 and 5.7.1.

If $c \leq 0$, then the flatness of $\nabla^\perp$ follows from the assumption $m_1 = 2$, but if $c > 0$, then the assumption $H \neq 0$ must be added (see Proposition 5.1.4 and Remark 5.1.4). Therefore, one needs to investigate only the following case, which will be considered here separately.

Let M^3 be a semiparallel submanifold with principal codimension $m_1 = 2$, nonflat $\nabla^\perp$, and zero mean curvature vector $H \equiv 0$ in $S^n(c)$, i.e., in $N^n(c)$ with $c > 0$.

Since $H = \frac{1}{3}(h_{11} + h_{22} + h_{33})$, the condition $H \equiv 0$ means that

$$h_{11} + h_{22} + h_{33} = 0. \tag{7.1.1}$$

Moreover, the principal normal subspace span$\{h_{ij}\}$ must be of dimension $m_1 = 2$ at every point $x \in M^3$. Therefore, the orthonormal frame bundle $O(M^3, S^n(c))$ can be adapted so that the frame vectors e_4 and e_5 belong to this subspace span$\{h_{ij}\}$. Since the matrix of h^4_{ij} is symmetric, the frame vectors e_1, e_2, e_3 tangent to M^3 can be chosen at any x so that $h^4_{ij} = 0$, for $i \neq j$. Denote $h^4_{ii} = \lambda_i$; not all of them are zero, because otherwise it would contradict the assumption $m_1 = 2$. Here $\lambda_1 + \lambda_2 + \lambda_3 = 0$ due to (7.1.1). Since $\nabla^\perp$ is assumed to be nonflat, it follows from (2.1.9) and (2.1.10) that

$$0 \neq \Omega^5_4 = (\lambda_2 - \lambda_1)h^5_{12}\omega^1 \wedge \omega^2 + (\lambda_3 - \lambda_2)h^5_{23}\omega^2 \wedge \omega^3 + (\lambda_1 - \lambda_3)h^5_{31}\omega^3 \wedge \omega^1.$$

Hence at least one of the coefficients is nonzero. By renumbering if needed, one can obtain $(\lambda_1 - \lambda_2)h^5_{12} \neq 0$, $\lambda_1 \neq 0$.

The semiparallel condition (4.1.2) for $\alpha = 4$ reduces to

$$(\lambda_j - \lambda_i)\Omega^j_i = -h^5_{ij}\Omega^5_4; \tag{7.1.2}$$

therefore,

$$h^5_{11} = h^5_{22} = h^5_{33} = 0, \quad (\lambda_1 - \lambda_2)\Omega^2_1 = h^5_{12}\Omega^5_4 \neq 0, \tag{7.1.3}$$

and thus $\Omega^2_1 \neq 0$.

Further, it follows from (4.1.2) with $\alpha = 5$ and either $i = 1$, $j = 3$, or $i = 2$, $j = 3$, that

$$h^5_{23}\Omega^2_1 + h^5_{12}\Omega^2_3 = 0, \quad h^5_{13}\Omega^1_2 + h^5_{21}\Omega^1_3 = 0, \tag{7.1.4}$$

whence (7.1.2) leads to

$$(\lambda_3 - \lambda_1)\Omega^3_1 = -h^5_{13}\Omega^5_4, \quad (\lambda_3 - \lambda_2)\Omega^3_2 = -h^5_{23}\Omega^5_4.$$

After multiplying by $h^5_{12} \neq 0$, equations (7.1.3), (7.1.4), and $\lambda_1 + \lambda_2 + \lambda_3 = 0$ imply that

$$h^5_{13}\lambda_1 = h^5_{23}\lambda_2 = 0. \tag{7.1.5}$$

From here $h^5_{13} = 0$ and thus the second relation (7.1.4) gives $\Omega^3_1 = -\Omega^1_3 = 0$.

But here $h^5_{23} = 0$ is not possible. Indeed if it were, then the first relation (7.1.4) would give $\Omega^3_2 = 0$. Hence, (4.1.2) with $\alpha = 5$, $i = j = 1$ and $i = j = 2$ implies

$$2h^5_{12}\Omega^2_1 - \lambda_1\Omega^5_4 = 0, \quad -2h^5_{12}\Omega^2_1 - \lambda_2\Omega^5_4 = 0, \tag{7.1.6}$$

thus $(\lambda_1 + \lambda_2)\Omega^5_4 = 0$ and so $\lambda_2 = -\lambda_1$, $\lambda_3 = 0$. This implies

$$\omega^4_1 = \lambda_1\omega^1, \quad \omega^4_2 = -\lambda_1\omega^2, \quad \omega^4_3 = 0, \quad \omega^5_1 = h^5_{12}\omega^2, \quad \omega^5_2 = h^5_{12}\omega^1, \quad \omega^5_3 = 0,$$

which implies via (2.1.8) that $\Omega^3_1 = c\omega^3 \wedge \omega^1 \neq 0$, since $c > 0$. This contradicts $\Omega^3_1 = 0$.

Hence the only possibility here is that $h^5_{23} \neq 0$, $\lambda_2 = 0$, and thus $-\lambda_3 = \lambda_1 = \lambda \neq 0$, whence (7.1.4) gives, in addition to $\Omega^3_1 = 0$, that $\Omega^3_2 \neq 0$.

The semiparallel condition (4.1.2) with $\alpha = 5$ and $i = j = 1$, and with $\alpha = 4$ and $i = 1$, $j = 2$ yields, respectively,

$$2h^5_{12}\Omega^2_1 - \lambda\Omega^5_4 = 0, \quad \lambda\Omega^2_1 - h^5_{12}\Omega^5_4 = 0.$$

Hence $2(h^5_{12})^2 = \lambda^2$. Again, (4.1.2) with $\alpha = 5$ and $i = j = 2$, or $i = 1$, $j = 3$ gives, respectively,

$$h^5_{12}\Omega^2_1 - h^5_{23}\Omega^3_2 = 0, \quad h^5_{23}\Omega^2_1 - h^5_{12}\Omega^3_2 = 0;$$

hence $(h^5_{23})^2 = (h^5_{12})^2$. Replacing e_5 by $-e_5$ and e_3 by $-e_3$ if needed, one obtains $h^5_{23} = h^5_{12} = \frac{1}{\sqrt{2}}\lambda$; thus

$$\omega^4_1 = \lambda\omega^1, \quad \omega^4_2 = 0, \quad \omega^4_3 = -\lambda\omega^3,$$

$$\omega^5_1 = \frac{1}{\sqrt{2}}\lambda\omega^2, \quad \omega^5_2 = \frac{1}{\sqrt{2}}\lambda(\omega^1 + \omega^3), \quad \omega^5_3 = \frac{1}{\sqrt{2}}\lambda\omega^2.$$

Now rotate e_2 and e_3 in their plane by the angle $\frac{\pi}{4}$ and take $-e_4$ instead of e_4. Then the differential system defining M^3 above is

$$\omega^4 = \omega^5 = \omega^\rho = 0, \tag{7.1.7}$$

$$\omega^4_1 = \lambda\omega^3, \quad \omega^4_2 = 0, \quad \omega^4_3 = \lambda\omega^1, \tag{7.1.8}$$

$$\omega^5_1 = \lambda\omega^2, \quad \omega^5_2 = \lambda\omega^1, \quad \omega^5_3 = 0, \tag{7.1.9}$$

$$\omega^\rho_1 = \omega^\rho_2 = \omega^\rho_3 = 0, \tag{7.1.10}$$

where $\rho = 6, \ldots, n$. Now $0 = \Omega^3_1 = (\lambda^2 - c)\omega^1 \wedge \omega^3$ which implies $\lambda = \pm\sqrt{c}$. Reversing e_1, e_2, e_3 if needed, one finally gets $\lambda = \sqrt{c}$. Then exterior differentiation and Cartan's lemma (i.e., differential prolongation) lead to

$$\omega^2_1 = 0, \quad \omega^3_1 = 0, \quad \omega^3_2 - \omega^5_4 = 0, \quad \omega^\rho_4 = \omega^\rho_5 = 0.$$

The whole resulting system is totally integrable and determines the submanifold M^3 up to constants. For this M^3,

$$dx = e_1\omega^1 + e_2\omega^2 + e_3\omega^3, \quad de_1 = \sqrt{c}(e_4\omega^2 + e_5\omega^3) - cx\omega^1,$$

$$de_2 = e_3\omega_2^3 + \sqrt{c}e_4\omega^1 - cx\omega^2, \quad de_3 = -e_2\omega_2^3 + \sqrt{c}e_5\omega^1 - cx\omega^3,$$

$$de_4 = -\sqrt{c}(e_1\omega^2 + e_2\omega^1) + e_5\omega_2^3, \quad de_5 = -\sqrt{c}(e_1\omega^3 + e_3\omega^1) - e_4\omega_2^3.$$

It is seen that such an M^3 lies in an $S^5(c) \subset S^n(c)$, which is intersected by the E^6 spanned at the point o by the vectors $x, e_1, e_2, e_3, e_4, e_5$. The equation $\omega^1 = 0$ defines a foliation on M^3 whose leaves are totally geodesic in $S^5(c)$ and hence are its great 2-spheres. By $\omega^2 = \omega^3 = 0$ a family of curves in M^3 is defined, which are great circles of $S^5(c)$ and orthogonal to the leaves of the foliation. All this shows that the submanifold M^3 is a Segre submanifold $S_{(1,2)}(a)$ in $S^5(a^2)$, where now $a^2 = c$ (see Section 3.2).

The result, together with Theorems 5.6.1 and 5.7.1 for $m = 3$, can be summarized as follows.

Theorem 7.1.2. *A semiparallel submanifold M^3 with principal codimension* 2 *is*

- *in a Euclidean space E^n, either*
  - *a warped product $B^1 \times_r S^2(1)$, where B^1 is a curve and r is a nonconstant linear function on it, or*
  - *the envelope with flat $\bar{\nabla}$ of a two-parameter family of three-dimensional planes, or*
  - *the product of a semiparallel surface with principal codimension* 1 *(see Theorem* 5.5.1 *for the case of $m = 2$) and a curve;*
- *in a non-Euclidean $N^n(c)$, either*
  - *a Cartan variety, thus with flat $\bar{\nabla}$, in an $N^5(c) \subset N^n(c)$, or*
  - *a warped product $B^1 \times_r Sph^2(1)$, where B^1 is a curve and r is a nonconstant linear function on it, or*
  - *a Segre submanifold $S_{(1,2)}(a)$ in an $S^5(a^2)) \subset N^n(c)$, or*
  - *the product of a semiparallel surface with principal codimension* 1 *(see Theorem* 5.5.3 *for the case of $m = 2$) and a curve;*
- *or an open part of one of the above.*

Remark 7.1.3. For Euclidean space E^n, this classification result was proved for the most part in [LR 90]. The warped products $B^1 \times_r S^2(1)$ are described there as orthogonal type canal submanifolds, meaning that each of them is the envelope of a one-parameter family of 3-spheres such that the curvature vector of each orthogonal trajectory of characteristic 2-spheres is orthogonal to the four-dimensional space of the 3-sphere of the family.

7.2 Nonminimal Semiparallel M^3 of Principal Codimension $m_1 = 3$

Here $\nabla^\perp$ may or may not be flat.

Let us start with the first possibility, i.e., that $\nabla^\perp$ is flat. Then the three principal curvature vectors k_1, k_2, k_3 are distinct, due to $m_1 = 3$, and from Theorem 5.1.2 it follows that the corresponding outer principal curvature vectors k_1^*, k_2^*, k_3^* are mutually orthogonal. Now (5.1.4) implies $\Omega_i^j = 0$, thus M^3 has flat $\bar{\nabla}$.

In Euclidean space E^n (i.e., for $c = 0$) such an M^m was considered in Example 4.5.4. For $m = 3$ this means that the submanifold M^3 is a second-order envelope of the tori $S^1(c_1) \times S^1(c_2) \times S^1(c_3)$, composed of the curvature circles of the lines of curvature of this M^3. Each of these tori is a three-dimensional parallel submanifold, and their envelope M^3 has an orthogonal holonomic net of lines of curvature, so it is a Cartan variety (see Remark 5.4.3).

If $c \neq 0$, a similar interpretation can be made in ${}_\sigma E^{n+1}$, but here in the case $c < 0$ the curvature circles of the tori must be considered in a more general sense, namely horocycles and equidistant curves could also be involved (see, e.g., [Nu 61]).

Now let $\nabla^\perp$ be nonflat. Here M^3 is assumed to be nonminimal; therefore, $H \neq 0$.

The orthonormal frame bundle adapted to M^3 can be adapted further so that e_4, e_5, e_6 belong to the three-dimensional principal normal subspace and, moreover, e_6 is collinear with H. Then $h_{ij}^\xi = 0$ ($\xi \in \{7, \dots, n\}$) and $H^4 = H^5 = H^\xi = 0$, but $H^6 \neq 0$. Now (4.3.2) implies that $\Omega_6^\alpha = 0$ in addition to $\Omega_\xi^\alpha = 0$, which follows immediately from (2.1.8). In particular, $\Omega_6^4 = 0$, but this implies that the matrices $\|h_{ij}^6\|$ and $\|h_{ij}^4\|$ commute and hence can be simultaneously diagonalized by a suitable choice of e_1, e_2, e_3. That implies then

$$h_{ij}^6 = \kappa_i \delta_{ij}, \quad h_{ij}^4 = \lambda_i \delta_{ij}.$$

Here $\lambda_1 + \lambda_2 + \lambda_3 = 0$ due to $H^4 = 0$. Since $\nabla^\perp$ is nonflat, one has $\Omega_4^5 \neq 0$; therefore, (2.1.9), (2.1.10) give

$$0 \neq \Omega_4^5 = (\lambda_2 - \lambda_1) h_{12}^5 \omega^1 \wedge \omega^2 + (\lambda_3 - \lambda_2) h_{23}^5 \omega^2 \wedge \omega^3 + (\lambda_1 - \lambda_3) h_{31}^5 \omega^3 \wedge \omega^1.$$

Hence at least one of the coefficients is nonzero. After a renumbering if needed, one obtains $(\lambda_1 - \lambda_2) h_{12}^5 \neq 0$, $\lambda_1 \neq 0$.

The situation is now similar to that of Section 7.1. The argument that led to (7.1.3) and (7.1.4) also holds here and also yields the conclusions $h_{13}^5 = 0$ and $\Omega_1^3 = 0$. Moreover, (7.1.5) shows that here

$$\text{either (a)} \lambda_2 = 0, \text{ or } \quad \text{(b) } h_{23}^5 = 0.$$

For case (a) further arguments from Section 7.1 can be used, leading to the system (7.1.7)–(7.1.10), which must now be completed by the equations

$$\omega^6 = 0, \quad \omega_1^6 = \kappa_1 \omega^1, \quad \omega_2^6 = \kappa_2 \omega^2, \quad \omega_3^6 = \kappa_3 \omega^3,$$

and instead of ρ one has $\xi = 7, \dots, n$. The semiparallel condition (4.1.2) for $\alpha = 6$ gives $(\kappa_j - \kappa_i)\Omega_i^j = 0$. Therefore, $\kappa_1 = \kappa_2 = \kappa_3 = \kappa \neq 0$, because $m_1 = 3$, $\Omega_1^2 \neq 0$, $\Omega_3^2 \neq 0$, and thus $h_{ij}^6 = \kappa \delta_{ij} \omega^i$. Now (2.1.8) and the equations of the above system imply

$$0 = \Omega_1^3 = (\lambda^2 - \kappa^2 - c)\omega^1 \wedge \omega^3.$$

From here $\lambda = \pm\sqrt{\kappa^2 + c}$, and $\kappa^2 + c > 0$ if $c < 0$. The system can now be transformed to

$$\omega^4 = \omega^5 = \omega^6 = \omega^\xi = 0,$$

$$\omega_1^4 = \sqrt{\kappa^2 + c}\omega^2, \quad \omega_2^4 = \sqrt{\kappa^2 + c}\omega^1, \quad \omega_3^4 = 0,$$

$$\omega_1^5 = \sqrt{\kappa^2 + c}\omega^3, \quad \omega_2^5 = 0 \quad , \quad \omega_3^5 = \sqrt{\kappa^2 + c}\omega^1,$$

$$\omega_1^6 = \kappa\omega^1, \quad \omega_2^6 = \kappa\omega^2, \quad \omega_3^6 = \kappa\omega^3,$$

$$\omega_1^\xi = \omega_2^\xi = \omega_3^\xi = 0.$$

By exterior differentiation and Cartan's lemma (i.e., differential prolongation) the equations of the last row give

$$\omega_4^\xi = \frac{1}{\sqrt{\kappa^2 + c}} p^\xi \omega^2, \quad \omega_5^\xi = \frac{1}{\sqrt{\kappa^2 + c}} p^\xi \omega^3, \quad \omega_6^\xi = \frac{1}{\kappa} p^\xi \omega^1, \tag{7.2.1}$$

and the equations of the penultimate row give

$$d \ln \kappa = a\omega^1, \quad \omega_4^6 = \frac{\kappa}{\sqrt{\kappa^2 + c}} a\omega^2, \quad \omega_5^6 = \frac{\kappa}{\sqrt{\kappa^2 + c}} a\omega^3. \tag{7.2.2}$$

The remaining equations lead to

$$\omega_1^2 = -\frac{\kappa^2}{\kappa^2 + c} a\omega^2, \quad \omega_1^3 = -\frac{\kappa^2}{\kappa^2 + c} a\omega^3, \quad \omega_2^3 = \omega_4^5. \tag{7.2.3}$$

Now (7.2.1) yields $dp^\xi = -p^\eta \omega_\eta^\xi - p^\xi \omega^1$, and from the first equation of (7.2.2) $da = A\omega^1$. The first two equations of (7.2.3) imply via exterior differentiation that $A = \frac{a^2(\kappa^2 - 2c)}{\kappa^2 + c} - \frac{c(\kappa^2 + c)}{\kappa^2}$. The remaining equations (7.2.2), (7.2.3) imply

$$\sum_\xi (p^\xi)^2 = -c\frac{a^2\kappa^2 + (\kappa^2 + c)^2}{\kappa^2 + c}. \tag{7.2.4}$$

The last result shows that $c > 0$ is not possible here. Indeed, $\kappa^2 + c > 0$, as shown above, and therefore $c > 0$ implies that the right-hand side is negative.

Also $c < 0$ is not possible, as can be shown by a technically more complicated argument. Namely, differentiating (7.2.4), using that $d\sum_\xi (p^\xi)^2 = 2\sum_\xi (p^\xi)^2\omega^1$ and $d\kappa = a\kappa\omega^1$, $da = A\omega^1$, as well as the expression for A, one obtains a rather complicated relation between a and κ, which contains a contradiction. A simpler indirect proof can be given via Theorem 4.5.5, i.e., the fact that a submanifold M^m in $N^n(c)$ is semiparallel if and only if it is the second-order envelope of parallel submanifolds. It turns out that for the case considered here, namely for semiparallel M^3 of case (a) in $H^n(c)$ with $c < 0$, the corresponding parallel submanifolds do not exist. Indeed, from the above differential system it follows that the parallel condition, i.e., $\bar{\nabla} h = 0$, leads to $a \equiv 0$, which gives $A \equiv 0$ and thus $0 = \kappa^2 + c = \lambda^2$, which is impossible here.

Therefore, $c = 0$ is the only possible case. Then (7.2.4) leads to $p^\xi = 0$ and so to the semiparallel M^3 of case (a) in E^6. Then the above expression for A gives $A = a^2$ and thus $da = a^2 ds$, since from (7.2.3) it follows that $d\omega^1 = 0$, so that $\omega^1 = ds$, at least locally.

For the corresponding parallel submanifold $a \equiv 0$, so κ is a constant; hence

$$de_1 = e_4 \kappa \omega^2 + e_5 \kappa \omega^3 + e_6 \kappa \omega^1,$$

$$de_2 = e_3 \omega_2^3 + e_4 \kappa \omega^1 + e_6 \kappa \omega^2, \quad de_3 = -e_2 \omega_2^3 + e_5 \kappa \omega^1 + e_6 \kappa \omega^3.$$

It is seen that this submanifold belongs to the sphere $S^5(\kappa^{-2})$ in E^6 whose center has the radius vector $x + \kappa^{-1} e_6$, and it is a Segre submanifold $S_{(1,2)}(\kappa^{-1})$ (see Sections 3.2 and 4.6).

A general semiparallel M^3 of case (a) is the second-order envelope of these Segre submanifolds $S_{(1,2)}(\kappa^{-1})$ with variable κ. It follows from Theorem 4.6.1 that this envelope is, in general, a logarithmic spiral tube (or its open subset in E^6), because in that theorem now $p = 1, q = 2$.

The analysis above gives the following.

Proposition 7.2.1. *A nonminimal semiparallel M^3 in $N^n(c)$ with principal codimension $m_1 = 3$ and nonflat $\nabla^\perp$ can be of case* (a) *only for $c = 0$. Such an M^3 lies in $N^6(0) = E^6$, is a second-order envelope of Segre submanifolds $S_{(1,2)}$, and thus is, in general, a logarithmic spiral tube, or its open subset.*

Now consider the other possible case (b) above: $h_{23}^5 = 0$. Here the consequences (7.1.2)–(7.1.5) derived from the semiparallel condition (4.1.2) remain valid, as well as the equalities $h_{13}^5 = 0$ and $\Omega_1^3 = 0$. Moreover, $h_{23}^5 = 0$ and (4.1.2) imply $-\lambda_2 = \lambda_1 = \lambda \neq 0$, $\lambda_3 = 0$, as was shown after (7.1.5), and (7.1.4) now implies $\Omega_2^3 = 0$.

The system (7.1.7)–(7.1.10) must be completed, as above, by the following equations:

$$\omega_1^6 = \kappa_1 \omega^1, \quad \omega_2^6 = \kappa_2 \omega^2, \quad \omega_3^6 = \kappa_3 \omega^3, \quad \omega_1^\xi = \omega_2^\xi = \omega_3^\xi = 0.$$

Therefore, (2.1.8) implies $\Omega_1^3 = -(\kappa_1\kappa_3 + c)\omega^1 \wedge \omega^3$ and $\Omega_2^3 = -(\kappa_2\kappa_3 + c)\omega^2 \wedge \omega^3$, thus $\kappa_1\kappa_3 + c = \kappa_2\kappa_3 + c = 0$. Using (4.1.2) with $\alpha = 6$ and $i = 1, j = 2$ one gets $(\kappa_1 - \kappa_2)\Omega_1^2 = 0$; hence $\kappa_1 = \kappa_2 = \kappa$, and so $\kappa_3 = -c\kappa^{-1}$. Now $\Omega_1^2 = [\lambda^2 + (h_{12}^5)^2 - \kappa^2 - c]\omega^1 \wedge \omega^2$ and $\Omega_4^5 = -2\lambda h_{12}^5 \omega^1 \wedge \omega^2$; therefore, (7.1.3) and (7.1.6) reduce to $\lambda^2 + 2(h_{12}^5)^2 - \kappa^2 - c = 0$ and $2\lambda^2 + (h_{12}^5)^2 - \kappa^2 - c = 0$. It follows that $(h_{12}^5)^2 = \lambda^2$. Replacing e_5 by $-e_5$ if needed, one gets $h_{12}^5 = \lambda$ and so $3\lambda^2 = \kappa^2 + c$.

Summing up, the only nonzero coefficients of the second fundamental form are $h^4_{11} = -h^4_{22} = h^5_{12} = \lambda$, $h^6_{11} = h^6_{22} = \kappa$, $h^6_{33} = -c\kappa^{-1}$, where $\kappa = \sqrt{3\lambda^2 - c}$ (reversing e_6 if needed); all other coefficients h^α_{ij} are zero. Therefore, the semiparallel M^3 of case (b) in $N^n(c)$ is defined by the Pfaffian system

$$\omega^4 = \omega^5 = \omega^6 = \omega^\xi = 0, \tag{7.2.5}$$

$$\omega^4_1 = \lambda\omega^1, \quad \omega^4_2 = -\lambda\omega^2, \quad \omega^4_3 = 0, \tag{7.2.6}$$

$$\omega^5_1 = \lambda\omega^2, \quad \omega^5_2 = \lambda\omega^1, \quad \omega^5_3 = 0, \tag{7.2.7}$$

$$\omega^6_1 = \kappa\omega^1, \quad \omega^6_2 = \kappa\omega^2, \quad \omega^6_3 = -c\kappa^{-1}\omega^3, \tag{7.2.8}$$

$$\omega^\xi_1 = \omega^\xi_2 = \omega^\xi_3 = 0, \tag{7.2.9}$$

where $\kappa = \sqrt{3\lambda^2 - c}$ and $\xi = 7, \dots, n$.

Now it is easy to find the corresponding parallel M^3. Substituting the values of the coefficients of the second fundamental form h obtained above into the parallel condition (3.1.1), one finds that

$$d\lambda = \omega^3_1 = \omega^3_2 = \omega^6_4 = \omega^6_5 = \omega^5_4 - 2\omega^2_1 = \omega^\xi_4 = \omega^\xi_5 = \omega^\xi_6 = 0; \tag{7.2.10}$$

hence $\lambda =$ const, thus also $\kappa =$ const. It is easy to check that if these new equations are added to equations (7.2.5)–(7.2.9), one gets a totally integrable system, thus this M^3 exists, uniquely up to some real constants.

For its geometric characterization, note first that now $d\omega^3 = 0$, and therefore $\omega^3 = ds$, at least locally. Considering the surfaces in M^3 defined by $s =$ const, i.e., by $\omega^3 = 0$, and comparing the equations resulting from (7.2.6)–(7.2.9) with equations (6.3.1) and (6.3.2), one finds that these surfaces are Veronese surfaces, which are moreover mutually congruent (in (6.3.1), (6.3.2), replace k with λ and superscript 3 with 6).

Since for each of them, $dx = e_1\omega^1 + e_2\omega^2$,

$$de_1 = e_2\omega^2_1 + f_1\omega^1 + e_5\lambda\omega^2, \quad de_2 = -e_1\omega^2_1 + f_2\omega^2 + e_5\lambda\omega^1,$$

where $f_1 = \lambda e_4 + \kappa e_6 - cx$, $f_2 = -\lambda e_4 + \kappa e_6 - cx$, and

$$df_1 = -2\lambda^2(2e_1\omega^1 + e_2\omega^2) + 2\lambda\omega^2_1 e_5, \quad df_2 = -2\lambda^2(e_1\omega^1 + 2e_2\omega^2) - 2\lambda\omega^2_1,$$

$$de_5 = -\lambda(e_1\omega^2 + e_2\omega^1) - \omega^2_1\lambda^{-1}(f_1 - f_2),$$

each such Veronese surface lies in the five-dimensional space ${}_\sigma E^5$, spanned at the point $x \in M^3$ by the vectors e_1, e_2, f_1, f_2, e_5.

The orthogonal trajectories of these surfaces are defined by $\omega^1 = \omega^2 = 0$ and for each of them

$$dx = e_3 ds, \quad de_3 = -c\kappa^{-1}(e_6 + \kappa x)ds, \quad d(e_6 + \kappa x) = \kappa^{-1}(c + \kappa^2)e_3 ds.$$

This trajectory has constant curvature $\sqrt{3c}\lambda\kappa^{-1}$ and lies in the plane spanned at the point $x \in M^3$ by the vectors $e_3, e_6 + \kappa x$. The latter are orthogonal to the five-dimensional space above, thus the above parallel M^3 is the product of a Veronese surface and a plane curve of this constant curvature.

According to Theorem 4.5.5, every semiparallel submanifold is the second-order envelope of the corresponding parallel submanifolds. Therefore, the following proposition can be stated.

Proposition 7.2.2. *If a nonminimal semiparallel M^3 in $N^n(c)$ with principal codimension $m_1 = 3$ and nonflat $\nabla^\perp$ is of case* (b), *then it is the second-order envelope of products of Veronese surfaces and plane curves. If $c = 0$, then each of these curves is a straight line, but if $c \neq 0$, then its curvature is equal to $\sqrt{3c}\lambda(\sqrt{3\lambda^2 - c})^{-1}$, where the quantity λ characterizes the Veronese surface.*

For a more detailed investigation of the submanifolds M^3 of Proposition 7.2.2, which are characterized by the Pfaffian system (7.2.5)–(7.2.9), exterior differentiation must again be used.

The last equations $\omega_3^4 = \omega_3^5 = 0$, $\omega_3^6 = -c\kappa^{-1}\omega^3$ in (7.2.6)–(7.2.8) give, respectively, the exterior equations

$$\lambda(\omega_1^3 \wedge \omega^1 - \omega_2^3 \wedge \omega^2) + c\kappa^{-1}\omega_4^6 \wedge \omega^3 = 0,$$

$$\lambda(\omega_1^3 \wedge \omega^2 + \omega_2^3 \wedge \omega^1) + c\kappa^{-1}\omega_5^6 \wedge \omega^3 = 0,$$

$$3\lambda^2(\omega_1^3 \wedge \omega^1 + \omega_2^3 \wedge \omega^2) + c\kappa^{-1}d\kappa \wedge \omega^3 = 0.$$

Using Cartan's lemma, one obtains

$$\omega_1^3 = \tau\omega^1 + \varphi_1\omega^3, \quad \omega_2^3 = \tau\omega^2 - \varphi_2\omega^3, \tag{7.2.11}$$

$$c\kappa^{-1}\lambda^{-1}\omega_4^6 = \varphi_1\omega^1 + \varphi_2\omega^2 + \psi\omega^3, \tag{7.2.12}$$

$$c\kappa^{-1}\lambda^{-1}\omega_5^6 = -\varphi_2\omega^1 + \varphi_1\omega^2 + \phi\omega^3, \tag{7.2.13}$$

$$c\kappa^{-1}d\kappa = 3\lambda^2(\varphi_1\omega^1 - \varphi_2\omega^2) + \chi\omega^3. \tag{7.2.14}$$

Then the first two equations in (7.2.6) and (7.2.7) give, respectively,

$$(d\lambda - \lambda\tau\omega^3 - \kappa\omega_4^6) \wedge \omega^1 + \lambda(2\omega_1^2 - \omega_4^5) \wedge \omega^2 = 0, \tag{7.2.15}$$

$$\lambda(2\omega_1^2 - \omega_4^5) \wedge \omega^1 - (d\lambda - \lambda\tau\omega^3 + \kappa\omega_4^6) \wedge \omega^2 = 0, \tag{7.2.16}$$

and

$$-[\lambda(2\omega_1^2 - \omega_4^5) + \kappa\omega_5^6] \wedge \omega^1 + (d\lambda - \lambda\tau\omega^3) \wedge \omega^2 = 0, \tag{7.2.17}$$

$$(d\lambda - \lambda\tau\omega^3) \wedge \omega^1 + [\lambda(2\omega_1^2 - \omega_4^5) - \kappa\omega_5^6] \wedge \omega^2 = 0; \tag{7.2.18}$$

and from the first two equations in (7.2.8) it follows that

$$(d\kappa - 3\kappa^{-1}\lambda^2\tau\omega^3 + \lambda\omega_4^6) \wedge \omega^1 + \lambda\omega_5^6 \wedge \omega^2 = 0, \qquad (7.2.19)$$

$$\lambda\omega_5^6 \wedge \omega^1 + (d\kappa - 3\kappa^{-1}\lambda^2\tau\omega^3 - \lambda\omega_4^6) \wedge \omega^2 = 0. \qquad (7.2.20)$$

Here $d\kappa = d\sqrt{3\lambda^2 - c} = 3\kappa^{-1}\lambda d\lambda$.

From (7.2.15)–(7.2.18) one obtains via Cartan's lemma that

$$d \ln \lambda - \tau\omega^3 = \frac{1}{2}[(P - 5A)\omega^1 + (Q - 5B)\omega^2],$$

$$2\omega_1^2 - \omega_4^5 = -5B\omega^1 + 5A\omega^2,$$

$$\lambda^{-1}\kappa\omega_4^6 = \frac{1}{2}[-(P + 5A)\omega^1 + (Q + 5B)\omega^2],$$

$$\lambda^{-1}\kappa\omega_5^6 = -\frac{1}{2}[(Q + 5B)\omega^1 + (P + 5A)\omega^2].$$

Now substitution into (7.2.19), (7.2.20) gives $P = A$, $Q = B$, so that

$$d \ln \lambda = -2(A\omega^1 + B\omega^2) + \tau\omega^3, \quad 2\omega_1^2 - \omega_4^5 = -5(B\omega^1 - A\omega^2), \qquad (7.2.21)$$

$$\omega_4^6 = 3\kappa^{-1}\lambda(A\omega^1 - B\omega^2), \quad \omega_5^6 = -3\kappa^{-1}\lambda(B\omega^1 + A\omega^2). \qquad (7.2.22)$$

Comparing these results with (7.2.12) and (7.2.13), one finds that

$$cA = cB = \varphi_1 = \varphi_2 = 0. \qquad (7.2.23)$$

Finally, exterior differentiation of equations (7.2.9) gives

$$(\kappa\omega_6^\xi + \lambda\omega_4^\xi) \wedge \omega^1 + \lambda\omega_5^\xi \wedge \omega^2 = 0, \qquad (7.2.24)$$

$$\lambda\omega_5^\xi \wedge \omega^1 + (\kappa\omega_6^\xi - \lambda\omega_4^\xi) \wedge \omega^2 = 0,$$

$$c\kappa^{-1}\omega_6^\xi \wedge \omega^3 = 0.$$

Further analysis now proceeds separately for the cases $c \neq 0$ and $c = 0$. Suppose first that $c \neq 0$.

Theorem 7.2.3. *Every semiparallel M^3 in $N^n(c)$, $c \neq 0$, $n \geq 6$, satisfying the hypotheses of Proposition* 7.2.2 *(i.e., nonflat $\nabla^\perp$, $H \neq 0$, and of case* (b)*) reduces to a parallel M^3 in $N^6(c)$, which is the product of a Veronese surface and a plane curve of constant curvature.*

Proof. Here (7.2.23) implies $A = B = \varphi_1 = \varphi_2 = 0$, and so equations (7.2.11), (7.2.21), and (7.2.22) reduce to

$$\omega_1^3 = \tau\omega^1, \quad \omega_2^3 = \tau\omega^2, \qquad (7.2.25)$$

$$d \ln \lambda = \tau\omega^3, \quad 2\omega_1^2 - \omega_4^5 = 0, \quad \omega_4^6 = 0, \quad \omega_5^6 = 0. \qquad (7.2.26)$$

From (7.2.24) one gets via Cartan's lemma that

$$\lambda\omega_4^\xi = p^\xi\omega^1 + q^\xi\omega^2, \quad \lambda\omega_5^\xi = q^\xi\omega^1 - p^\xi\omega^2, \quad \omega_6^\xi = 0.$$

Now the second equation (7.2.26) leads by exterior differentiation to

$$\left[2\tau^2 + \lambda^{-2}\left(\sum_\xi (p^\xi)^2 + \sum_\xi (q^\xi)^2\right)\right]\omega^1 \wedge \omega^2 = 0.$$

This is possible only if $\tau = p^\xi = q^\xi = 0$, i.e., if M^3 is as asserted in the theorem.

If $c = 0$, then M^3 is by Proposition 7.2.2 the second-order envelope of products of Veronese surfaces and straight lines (they can be called *Veronese cylinders*). The dimension of the ambient space E^n plays a decisive role here.

Theorem 7.2.4. *Let M^3 be a nonminimal semiparallel submanifold in E^n with principal codimension $m_1 = 3$ and nonflat $\nabla^\perp$, and of case* (b) *(thus a second-order envelope of Veronese cylinders). If $n = 6$, then M^3 is a single Veronese cylinder, and hence a parallel M^3 in E^6.*

Proof. Here $\kappa = \sqrt{3}\lambda$ due to $c = 0$; hence in (7.2.21) and (7.2.22) the coefficient $3\kappa^{-1}\lambda$ is equal to $\sqrt{3}$, but the equalities (7.2.23) do not imply $A = B = 0$ in this case. Now exterior differentiation and Cartan's lemma give

$$dA = B\omega_1^2 + \frac{1}{5}(14B^2 - 11A^2 - \tau^2)\omega^1 - 5AB\omega^2 + A\tau\omega^3, \tag{7.2.27}$$

$$dB = -A\omega_1^2 - 5AB\omega^1 + \frac{1}{5}(14A^2 - 11B^2 - \tau^2)\omega^2 + B\tau\omega^3. \tag{7.2.28}$$

From here exterior differentiation leads to $AL = BL = 0$, where $L = 25\lambda^2 + 42(A^2 + B^2) - 28\tau^2$. If $L \neq 0$, then $A = B = 0$, as before. If $L = 0$, then by differentiation one gets $AP = BP = 0$, where $P = 115\lambda^2 + 14 \cdot 15(A^2 + B^2) \neq 0$, due to $\lambda \neq 0$, and thus $A = B = 0$ again. Now it follows from (7.2.27) and (7.2.28) that $\tau = 0$. This finishes the proof.

If $n > 6$ the situation is a bit different.

Theorem 7.2.5. *Every second-order envelope M^3 of Veronese cylinders in E^n, $n > 6$, is a cone with a point-vertex (or its limiting case, a cylinder), whose director surface is a second-order envelope of Veronese surfaces.*

Proof. From (7.2.12) and (7.2.13) it follows that $\varphi_1 = \varphi_2 = 0$, because now $c = 0$; thus (7.2.11) reduce to

$$\omega_1^3 = \tau\omega^1, \quad \omega_2^3 = \tau\omega^2. \tag{7.2.29}$$

Adding these equations to the system (7.2.5)–(7.2.9), and applying exterior differentiation and Cartan's lemma, one gets

$$d\tau = \tau^2\omega^3. \tag{7.2.30}$$

If $\tau \neq 0$, then the vector $z = x + \tau^{-1} e_3$ satisfies

$$dz = dx - \tau^{-2}(\tau^2 \omega^3) e_3 + \tau^{-1}[-e_1 \tau \omega^1 - e_2 \tau \omega^2] = 0.$$

Thus the point with radius vector z belonging to the straight line generators of M^3 is a fixed point for the whole M^3 in E^n. It follows that M^3 is a cone with a point-vertex.

If $\tau = 0$, then $de_3 = 0$, and thus M^3 is a cylinder whose straight line generators have the direction $e_3 = \text{const}$.

The surfaces in M^3 orthogonal to the straight line generators are defined by the Pfaffian equation $\omega^3 = 0$. Substituting this into (7.2.21) and (7.2.22), and comparing the results with (6.3.3) and (6.3.4), one sees that they coincide when k, α, k_1, k_2 are identified with λ, κ, A, B, respectively, and the indices 3, 4, 5 are replaced by 4, 5, 6. In the same way one sees that (6.3.1) and (6.3.2) coincide with the first two equations in (7.2.5)–(7.2.9). This finishes the proof.

The question of existence of M^3 satisfying Theorem 7.2.5 in the limiting case, i.e., of a cylinder on a second-order envelope of Veronese surfaces, reduces to the analogous problem for such an envelope which was solved in Section 6.4 (see Theorem 6.4.1 or [Lu 99a]).

Considering the existence problem of the cones of Theorem 7.2.5, one could also be guided by the argument in the proof of Theorem 6.4.1, used now with $c = 0$.

Since $c = 0$, one now has $\kappa = \sqrt{3}\lambda$ in the exterior equations (7.2.24), and therefore they reduce to

$$(\sqrt{3}\omega_6^\xi + \omega_4^\xi) \wedge \omega^1 + \omega_5^\xi \wedge \omega^2 = 0, \quad \omega_5^\xi \wedge \omega^1 + (\sqrt{3}\omega_6^\xi - \omega_4^\xi)\wedge^2 = 0$$

(note that the exterior equation that follows them disappears in this case). Cartan's lemma then implies

$$\sqrt{3}\omega_6^\xi + \omega_4^\xi = p_1^\xi \omega^1 + p_2^\xi \omega^2, \quad \omega_5^\xi = p_2^\xi \omega^1 + p_3^\xi \omega^2,$$

$$\sqrt{3}\omega_6^\xi - \omega_4^\xi = p_3^\xi \omega^1 + p_4^\xi \omega^2.$$

It is convenient to introduce the vectors $p_a = e_\xi p_a^\xi$, where $a, b, \dots \in \{1, 2, 3, 4\}$, and denote $p_a^2 = \sum_\xi (p_a^\xi)^2$, $\langle p_a, p_b \rangle = \sum_\xi p_a^\xi p_b^\xi$. Instead of (7.2.27) and (7.2.28) one now has

$$dA = B\omega_1^2 + \left[\frac{1}{5}(14B^2 - 11A^2 - \tau^2) + P\right]\omega^1 - (5AB + Q)\omega^2 + A\tau\omega^3, \tag{7.2.31}$$

$$dB = -A\omega_1^2 - (5AB + Q)\omega^1 + \left[\frac{1}{5}(14A^2 - 11B^2 - \tau^2) + R\right]\omega^2 + B\tau\omega^3, \tag{7.2.32}$$

where

$$P = \frac{1}{30}[4(\langle p_2, p_4\rangle - p_3^2) - (\langle p_1, p_3\rangle - p_2^2)],$$

$$Q = \frac{1}{12}(\langle p_1, p_4\rangle - \langle p_2, p_3\rangle),$$

$$R = \frac{1}{30}[4(\langle p_1, p_3\rangle - p_2^2) - (\langle p_2, p_4\rangle - p_3^2)].$$

Here (7.2.31) and (7.2.32) can be considered as (6.3.7) and (6.3.8), completed by the terms containing τ, with different notation; some of the differences were already noted in the proof of Theorem 7.2.5.

Moreover, the equations obtained by differential prolongation (i.e., exterior differentiation and Cartan's lemma) differ very slightly from the corresponding equations in the proof of Theorem 6.4.1. Namely, on the right sides of (6.4.1)–(6.4.4), one must add the terms $p_a^{\xi}\tau\omega^3$. The argument then proceeds as before, with only slight changes. The result is that the cones M^3 of Theorem 7.2.5 do exist in E^7, and depend on one real holomorphic function of two real arguments.

Remark 7.2.6. Theorem 7.2.5 was established in [Lu 91d], but without the existence argument. Instead of that, some geometric properties were indicated which cannot hold for the cones of Theorem 7.2.5.

7.3 Semiparallel M^3 of Principal Codimension $m_1 = 4$

The preceding Section 7.2 omitted discussion of minimal semiparallel M^3 with principal codimension $m_1 = 3$ in $N^n(c)$. These can occur only for $c > 0$, because if $c \leq 0$ then Theorem 4.1.7 implies that a minimal semiparallel submanifold is totally geodesic and thus has $m_1 = 0$. Since the standard model of $N^n(c)$ with $c > 0$ is a sphere in E^{n+1}, and the outer principal codimension of the above-considered M^3 in E^{n+1} is $m_1^* = 4$, the investigation of such M^3 can be done in the framework of the present section.

So let M^3 be a semiparallel submanifold with $m_1 = 4$ in $N^n(c)$. Then among the six vectors $h_{ij} = e_\alpha h_{ij}^\alpha$ there must be two independent linear relations $h_{ij}\xi^{ij} = 0$ and $h_{ij}\eta^{ij} = 0$, which define a one-parameter family $\rho(h_{ij}\xi^{ij}) + \sigma(h_{ij}\eta^{ij}) = 0$ of such relations. In this family, the singular case corresponds to a root of the cubic equation $\det|\rho\xi^{ij} + \sigma\eta^{ij}| = 0$ with respect to ρ/σ or σ/ρ. There is at least one real root, and thus one basic relation can be presented in the form $h_{ij}\xi_1^{(i}\xi_2^{j)} = 0$, where $\xi_1 = \xi_1^i e_i$ and $\xi_2 = \xi_2^j e_j$ are two vectors in $T_x M^3$.

Two cases can occur here.

(A) Let the two vectors be distinct. After normalization of ξ_1 and ξ_2, the frame vector e_2 can be taken orthogonal to them and e_1 and e_3 chosen collinear to $\xi_1 + \xi_2$ and $\xi_1 - \xi_2$, respectively. Then the special basic relation above is $h_{11}(\xi_1^1)^2 - h_{33}(\xi_1^3)^2 = 0$. Here $\xi_1^1\xi_1^3 \neq 0$; the roles of e_1 and e_3 can be interchanged by taking $-\xi_2$ instead of ξ_2. Hence the following subcases occur:

(A$_1$) $(\eta^{12})^2 + (\eta^{23})^2 \neq 0$. Here it can be supposed that $\eta^{23} \neq 0$ and the basic relations are $h_{33} = \mu h_{11}$, $h_{23} = \nu_1 h_{11} + \nu_2 h_{22} + \nu_3 h_{12} + \nu_4 h_{13}$ with $\mu \neq 0$.

(A$_2$) $\eta^{12} = \eta^{23} = 0$, $\eta^{13} \neq 0$. Then $h_{33} = \mu h_{11}$, $h_{13} = \nu_1 h_{11} + \nu_2 h_{22}$ with $\mu \neq 0$.
(A$_3$) $\eta^{12} = \eta^{23} = \eta^{13} = 0$. Here either
(A$_3'$) $\eta^{22} \neq 0$ and $h_{33} = \mu h_{11}$, $h_{22} = \nu h_{11}$ with $\mu \neq 0$, or
(A$_3''$) $\eta^{22} = 0$, then $\eta^{11} \neq 0$ and $h_{33} = h_{11} = 0$.

(B) Let ξ_1 and ξ_2 have the same direction, which one can assume to be that of e_3. Then the special basic relation above is $h_{33} = 0$ and the roles of e_1 and e_2 could be interchanged by taking $-e_3$ instead of e_3. Here the following subcases occur:
(B$_1$) $(\eta^{13})^2 + (\eta^{23})^2 \neq 0$. One can suppose $\eta^{23} \neq 0$, and this leads to the limiting case of (A$_1$), where $\mu = 0$.
(B$_2$) $\eta^{13} = \eta^{23} = 0$, $\eta^{12} \neq 0$. Then $h_{33} = 0$, $h_{12} = \lambda_1 h_{11} + \lambda_2 h_{22}$.
(B$_3$) $\eta^{13} = \eta^{23} = \eta^{12} = 0$. This leads either to the limiting case of (A$_3'$), where $\mu = 0$, or to the case (A$_3''$).

So one must consider three subcases (A$_2$), (A$_3''$), (B$_2$), and two subcases (A$_1$), (A$_3'$) with their limiting cases when $\mu = 0$.

In each case, the semiparallel condition (4.1.2), or equivalently (4.4.1), must hold:

$$\sum_k \{h_{,kj}\, H^*_{i[p,q]k} + h_{ik} H^*_{j[p,q]k} - H^*_{ij,k[p} h_{q]k}\} = 0, \tag{7.3.1}$$

where, as one recalls, $H^*_{ij,kl} = \langle h^*_{ij}, h^*_{kl}\rangle = H_{ij,kl} + c$ with $H_{ij,kl} = \langle h_{ij}, h_{kl}\rangle$ due to (2.1.6), and all $\epsilon_k = 1$ here. Each equation of this system is a linear dependence between vectors h_{ij} with different pairs $\{ij\}$. In the current case $m_1 = 4$ there must always be four linearly independent vectors among them; therefore the coefficient of each of the latter must be zero.

In what follows, equation (7.3.1) will be referred to as $[ij, pq]$ and, if the coefficient of a vector h_{rs} is set equal to zero, then this condition will be referred to as $[ij, pq|rs]$.

First consider subcase (B$_2$). Here the vectors h_{11}, h_{22}, h_{13}, h_{23} are linearly independent, and $h_{33} = 0, h_{12} = \lambda_1 h_{11} + \lambda_2 h_{22}$. Now $[12, 13|13] : H_{13,13} + H_{12,12} = 0$, where $H_{12,12} = \lambda_1^2 H_{11,11} + 2\lambda_1\lambda_2 H_{11,22} + \lambda_2^2 H_{22,22}$ turns out to be zero, as follows easily from $[11, 13|23] : H_{11,12} = 0$ and $[22, 23|13] : H_{12,22} = 0$. Therefore, $H_{13,13} = 0$, which is a contradiction. Hence (B$_2$) is impossible for a semiparallel M^3 in $N^n(c)$. The same argument shows that (A$_3''$) is impossible also.

Next consider subcase (A$_3'$). Here the vectors h_{11}, h_{12}, h_{13}, h_{23} are linearly independent, and $h_{33} = \mu h_{11}, h_{22} = \nu h_{11}$. Then

$$[11, 12|12] : \; 3H^*_{11,22} - 2H^*_{12,12} - H^*_{11,11} = 0, \tag{7.3.2}$$

$$[22, 12|12] : \; 3H^*_{11,22} - 2H^*_{12,12} - H^*_{22,22} = 0, \tag{7.3.3}$$

thus $H^*_{22,22} = H^*_{11,11}$ and therefore $\nu^2 = 1$. The case $\nu = -1$ is impossible because (7.3.2) would then give the contradiction $2H^*_{11,11} + H^*_{12,12} = 0$. Hence $\nu = 1$ and $H^*_{11,22} = H^*_{12,12}$. Now

$$[13, 12|23] : \; H^*_{11,22} - H^*_{12,12} - H^*_{13,13} = 0$$

gives the contradiction $H^*_{13,13} = 0$ and so (A'_3) is impossible for a semiparallel M^3 in $N^n(c)$.

The same can be shown for (A_2), where the vectors $h_{11}, h_{22}, h_{12}, h_{23}$ are linearly independent, and $h_{33} = \mu h_{11}, h_{13} = \nu_1 h_{11} + \nu_2 h_{22}$ with $\mu \neq 0$. Here (7.3.2), (7.3.3) hold as before and thus $H^*_{11,11} = H^*_{22,22}$. On the other hand, [33, 12|12] gives $\mu(H^*_{11,22} - H^*_{11,11}) = 0$; hence $H^*_{11,11} = H^*_{22,22} = H^*_{11,22}$. This is impossible because h_{11}, h_{22} are linearly independent.

It turns out that the semiparallel condition (7.3.1) can be satisfied only in subcase (A_1), where the vectors $h_{11}, h_{22}, h_{12}, h_{13}$ are linearly independent and, as one recalls,

$$h_{33} = \mu h_{11}, \quad h_{23} = \nu_1 h_{11} + \nu_2 h_{22} + \nu_3 h_{12} + \nu_4 h_{13}. \tag{7.3.4}$$

Here the following conditions will be used.

$$\begin{aligned}
&[11, 12|11]: && H^*_{11,12} - \nu_1 H^*_{11,13} = 0,\\
&[11, 12|22]: && -H^*_{11,12} - \nu_2 H^*_{11,13} = 0,\\
&[22, 12|11]: && H^*_{12,22} + \nu_1(2H^*_{12,23} - 3H^*_{22,13}) = 0,\\
&[22, 12|22]: && -H^*_{12,22} + \nu_2(2H^*_{12,23} - 3H^*_{22,13}) = 0,\\
&[12, 12|11]: && 2H^*_{12,12} - H^*_{11,22} + \nu_1(H^*_{11,23} - H^*_{12,13}) = 0,\\
&[12, 12|22]: && -2H^*_{12,12} + H^*_{11,22} + \nu_2(H^*_{11,23} - H^*_{12,13}) = 0,\\
&[33, 12|11]: && H^*_{33,12} + \nu_1(2H^*_{22,13} - 2H^*_{12,23} - H^*_{13,33}) = 0,\\
&[33, 12|22]: && -H^*_{33,12} + \nu_2(2H^*_{22,13} - 2H^*_{12,23} - H^*_{13,33}) = 0.
\end{aligned}$$

Suppose $\nu_1 + \nu_2 \neq 0$. From the first three pairs of these conditions

$$\begin{aligned}
&H^*_{11,12} = H^*_{11,13} = 0,\\
&H^*_{12,22} = 2H^*_{12,23} - 3H^*_{22,13} = 0,\\
&2H^*_{12,12} - H^*_{11,22} = H^*_{11,23} - H^*_{12,13} = 0.
\end{aligned}$$

Therefore, due to (7.3.4), $H^*_{33,12} = \mu H^*_{11,12} = 0$, $H^*_{33,13} = \mu H^*_{11,13} = 0$. Now the third pair of these conditions reduces to $H^*_{22,13} = H^*_{12,23}$ and together with the relation above gives $H^*_{22,13} = H^*_{12,23} = 0$. Thus by (7.3.4), $\nu_3 H^*_{12,12} + \nu_4 H^*_{12,13} = 0$.

Further,

$$\begin{aligned}
&[13, 12|11]: && \nu_1(H^*_{11,22} - H^*_{12,12} - H^*_{13,13}) + H^*_{12,13} = 0,\\
&[13, 12|22]: && \nu_2(H^*_{11,22} - H^*_{12,12} - H^*_{13,13}) - H^*_{12,13} = 0;
\end{aligned}$$

hence $H^*_{12,13} = H^*_{11,22} - H^*_{12,12} - H^*_{13,13} = 0$; consequently, $H^*_{11,23} = 0$, $H^*_{12,12} = H^*_{13,13} = \chi^2 \neq 0$, and thus $\nu_3 = 0$, $H^*_{11,22} = 2\chi^2$. Now from

$$[11, 12|12]: \quad 3H^*_{11,22} - 2H^*_{12,12} - H^*_{11,11} = 0,$$

$$[22, 12|12]: \quad 3H^*_{11,22} - 2H^*_{12,12} - H^*_{22,22} = 0$$

one obtains $H^*_{11,11} = H^*_{22,22} = 4\chi^2$, where $\chi \neq 0$, because otherwise h^*_{11}, h^*_{22} would be orthogonal vectors with zero length squared, which could occur only in ${}_1E^{n+1}$, and would imply the collinearity of these vectors, which is impossible.

On the other hand, $[13, 12|13]: \quad \nu_4(H^*_{11,22} - H^*_{12,12}) = 0$ gives $\nu_4\chi^2 = 0$, thus $\nu_4 = 0$; and now $[22, 12|13]$ yields $H^*_{22,23} = 0$. But $H^*_{11,23} = 0$ and $H^*_{22,23} = 0$ together give a contradiction:

$$4\chi^2\nu_1 + 2\chi^2\nu_2 = 0, \quad 2\chi^2\nu_1 + 4\chi^2\nu_2 = 0, \quad \nu_1 + \nu_2 \neq 0.$$

Consequently, $\nu_1 = -\nu_2 = \nu$ and

$$H^*_{11,12} = \nu H^*_{11,13}, \quad H^*_{22,12} = \nu(3H^*_{22,13} - 2H^*_{12,23}), \tag{7.3.5}$$

$$2H^*_{12,12} - H^*_{11,22} = \nu(H^*_{12,13} - H^*_{11,23}), \tag{7.3.6}$$

$$\mu H^*_{11,12} = \nu(2H^*_{12,23} - 2H^*_{22,13} + \mu H^*_{11,13}). \tag{7.3.7}$$

The following relations will be used next:

$$[11, 13|11]: \quad (1-\mu)H^*_{11,13} - \nu H^*_{11,12} = 0,$$

$$[11, 13|22]: \quad \nu H^*_{11,12} = 0,$$

$$[22, 13|11]: \quad (1-\mu)H^*_{22,13} + \nu(2\mu H^*_{11,12} - 2H^*_{22,13} - H^*_{22,12}) = 0,$$

$$[22, 13|22]: \quad \nu(2\mu H^*_{11,12} - 2H^*_{22,13} - H^*_{22,12}) = 0,$$

$$[12, 13|11]: \quad 2H^*_{12,13} - H^*_{11,23} - \mu H^*_{12,13} + \nu(\mu H^*_{11,11} - H^*_{12,12} - H^*_{13,13}) = 0,$$

$$[12, 13|22]: \quad H^*_{11,23} - H^*_{12,13} - \nu(\mu H^*_{11,11} - H^*_{12,12} - H^*_{13,13}) = 0,$$

$$[13, 12|11]: \quad \nu(H^*_{11,22} - H^*_{12,12} - H^*_{13,13}) + H^*_{12,13}$$
$$+ (1-\mu)(H^*_{12,13} - H^*_{11,23}) = 0,$$

$$[13, 12|22]: \quad \nu(H^*_{11,22} - H^*_{12,12} - H_{13,13})^* + H^*_{12,13} = 0,$$

$$[23, 12|11]: \quad H^*_{12,23} + \mu(H^*_{12,23} - H^*_{13,22}) - \nu H^*_{13.23} = 0,$$

$$[23, 12|22]: \quad H^*_{13,22} - 2H^*_{12,23} + \nu H^*_{13,23} = 0,$$

$$[23, 13|11]: \quad H^*_{13,23} + \mu(\mu H^*_{11,12} - 2H^*_{13,23}) - \nu H^*_{12,23} = 0,$$

$$[23, 13|22]: \quad H^*_{13,23} - \mu H^*_{11,12} + \nu H^*_{12,23} = 0.$$

Suppose $\mu \neq 1$. Then from the first three pairs of relations $H^*_{11,13} = H^*_{22,13} = H^*_{12,13} = 0$, and (7.3.5) gives $H^*_{11,12} = 0$. The next three pairs of relations give

$H^*_{11,23} = H^*_{12,23} = H^*_{13,23} = 0$, and from (7.3.5)–(7.3.7) it follows that $H^*_{22,12} = 2H^*_{12,12} - H^*_{11,22} = 0$. Now $H^*_{12,23} = H^*_{13,23} = 0$ and (7.3.4) imply $\nu_3 H^*_{12,12} = 0$, $\nu_4 H^*_{13,13} = 0$, thus $\nu_3 = \nu_4 = 0$. Adding

$$[22, 23|11]: \quad -\mu H^*_{22,23} + \nu(3\mu H^*_{11,22} - 2H^*_{23,23} - H^*_{22,22}) = 0,$$

$$[22, 23|22]: \quad H^*_{22,23} - \nu(3\mu H^*_{11,22} - 2H^*_{22,23} - H^*_{22,22}) = 0$$

one obtains $(1 - \mu)H^*_{22,23} = 0$, and therefore $H^*_{22,23} = 0$. This together with $H^*_{11,23} = 0$ gives

$$\nu(H^*_{11,11} - H^*_{11,22}) = 0, \quad \nu(H^*_{11,22} - H^*_{22,22}) = 0.$$

Here $\nu \neq 0$ yields a contradiction: $H^*_{11,11} = H^*_{11,22} = H^*_{22,22}$ and h_{11}, h_{22} could not be linearly independent. But $\nu = 0$ leads to a contradiction, too. Indeed, then $h_{23} = 0$ and $[23, 12|13]: \quad H^*_{12,12} - H^*_{11,22} = 0$ contradicts $2H^*_{12,12} - H^*_{11,22} = 0$, which was obtained above.

So, we must have $\mu = 1$, and thus

$$h_{33} = h_{11}, \quad h_{23} = \nu(h_{11} - h_{22}) + \nu_3 h_{12} + \nu h_{13}.$$

Suppose $\nu \neq 0$. Then $H^*_{11,12} = H^*_{11,13} = 0$ due to (7.3.5) and to the relation [11, 13|11] obtained above. The relations [33, 13|22] and [33, 13|12] yield $H^*_{13,23} = H^*_{11,23} = 0$, but [11, 13|12] and [11, 13|13] give $H^*_{12,13} = H^*_{11,11} - H^*_{13,13} = 0$. Substitution into the relation [12, 13|22] obtained above leads to a contradiction $\nu H^*_{12,12} = 0$.

Hence $\nu = 0$ and thus $h_{33} = h_{11}$, $h_{23} = \nu_3 h_{12} + \nu_4 h_{13}$. From (7.3.5) and other relations above it follows that

$$H^*_{11,12} = H^*_{22,12} = H^*_{11,22} - 2H^*_{12,12} = H^*_{12,13} = H^*_{13,22} - 2H^*_{12,23} = 0,$$
$$H^*_{11,23} = H^*_{13,23} = 0.$$

Thus $\nu_4 H^*_{13,13} = 0$, so $\nu_4 = 0$ and $h_{23} = \nu_3 h_{12}$. Now

$$[12, 23|13]: \quad H^*_{11,22} - H^*_{12,12} - \nu_3^2 H^*_{12,12} = 0$$

gives $(1 - \nu_3^2)H^*_{12,12} = 0$, and hence $\nu_3 = \pm 1$. Here the case $\nu_3 = 1$, where $h_{33} = h_{11}$, $h_{23} = h_{12}$, can be reduced to the case $\nu_3 = -1$ by taking $-e_1$ instead of e_1.

Now in the case $\nu_3 = -1$, where $h_{33} = h_{11}$, $h_{23} = -h_{12}$, one applies the transformation $e'_3 = \frac{1}{\sqrt{2}}(e_1 + e_3)$, $e'_2 = \frac{1}{\sqrt{2}}(-e_1 + e_3)$, $e'_1 = e_2$ to get

$$h'_{13} = h'_{23} = 0.$$

A straightforward verification shows that all semiparallel conditions for $h_{13} = h_{23} = 0$ and linearly independent h_{11}, h_{22}, h_{12}, h_{33} reduce to

$$H^*_{11,12} = H^*_{11,33} = H^*_{22,12} = H^*_{22,33} = H^*_{12,33} = 0,$$

$$H^*_{11,11} = H^*_{22,22} = 2H^*_{11,22} = 4H^*_{12,12}.$$

Denoting now $H_{12,12} = \lambda^2$, $H_{33,33} = \mu^2$ and taking e_4, e_5, e_6, e_7 collinear with $h_{11}-h_{22}, h_{12}, h_{11}+h_{22}, h_{33}$, respectively, it follows that the only nonzero coefficients of the second fundamental form are $h^4_{11} = -h^4_{22} = h^5_{12} = \lambda$, $h^6_{11} = h^6_{22} = \kappa$, $h^7_{33} = \mu$. Hence the above-considered M^3 is defined by the differential system

$$\omega^4 = \omega^5 = \omega^6 = \omega^7 = \omega^\rho = 0, \tag{7.3.8}$$

$$\omega^4_1 = \lambda\omega^1, \quad \omega^4_2 = -\lambda\omega^2, \quad \omega^4_3 = 0, \tag{7.3.9}$$

$$\omega^5_1 = \lambda\omega^2, \quad \omega^5_2 = \lambda\omega^1, \quad \omega^5_3 = 0, \tag{7.3.10}$$

$$\omega^6_1 = \kappa\omega^1, \quad \omega^6_2 = \kappa\omega^2, \quad \omega^6_3 = 0, \tag{7.3.11}$$

$$\omega^7_1 = 0, \quad \omega^7_2 = 0 \quad \omega^7_3 = \mu\omega^3, \tag{7.3.12}$$

$$\omega^\rho_1 = 0, \quad \omega^\rho_2 = 0, \quad \omega^\rho_3 = 0, \tag{7.3.13}$$

where $\rho = 8, \dots, n$ and $\lambda > 0$, $\kappa = \sqrt{3\lambda^2 - c}$, $\mu > 0$.

It is important to find the corresponding parallel M^3. The parallel condition (3.1.1) now reduces to

$$d\lambda = d\mu = \omega^3_1 = \omega^3_2 = 2\omega^2_1 - \omega^5_4 = \omega^6_4 = \omega^6_5 = 0, \tag{7.3.14}$$

$$\omega^7_4 = \omega^7_5 = \omega^7_6 = \omega^\rho_4 = \omega^\rho_5 = \omega^\rho_6 = \omega^\rho_7 = 0. \tag{7.3.15}$$

The equations $\omega^3_1 = \omega^3_2 = 0$ give by exterior differentiation that $c\omega^3 \wedge \omega^1 = c\omega^3 \wedge \omega^2 = 0$. This is possible only for $c = 0$, i.e., only for parallel M^3 in E^7, where $\kappa = \sqrt{3}\lambda$. Consequently, their second-order envelopes—the corresponding semiparallel M^3—can also exist only in Euclidean spaces E^n.

There is an analogy here with the system (7.2.5)–(7.2.10), and the geometric consequences are also similar. Moreover, $\omega^3 = ds$, at least locally, and the surfaces defined by s = const are Veronese surfaces whose orthogonal trajectories, defined by $\omega^1 = \omega^2 = 0$, satisfy the following equations:

$$dx = e_3 ds, \quad de_3 = \mu e_7 ds, \quad de_7 = -\mu e_3 ds,$$

where μ = const. Hence these trajectories are congruent plane curves of constant curvature $\mu > 0$, thus circles; and hence the parallel M^3 is the product of a Veronese surface and a circle in E^7.

Now one can answer the question about the minimal semiparallel M^3 of principal codimension $m_1 = 3$ in $N^n(c)$ with $c > 0$, which was posed at the beginning of this section.

From (7.3.9)–(7.3.12) one can see that the mean curvature vector $H = \frac{1}{3}(h_{11} + h_{22} + h_{33})$ of the above-considered M^3 is equal to $H = \frac{1}{3}(\sqrt{3}\lambda e_6 + \mu e_7)$. Thus for the corresponding parallel M^3, where λ and μ are constants, one gets $dH = -\frac{1}{3}[3\lambda^2(e_1\omega^1 + e_2\omega^2) + \mu^2 e_3\omega^3]$.

For the special case where $\mu^2 = 3\lambda^2$, one obtains $dH = -\lambda^2 dx$, and therefore $d(x + \lambda^{-2}H) = 0$; thus the point in E^7 with radius vector $z = x + \lambda^{-2}H$ is a fixed point for this parallel M^3. Hence this M^3 lies in a sphere $S^6(\lambda^4) \subset E^7$. With respect to this sphere, its mean curvature vector is zero and its principal codimension is 3.

This special case can be characterized by a simple relationship between the geometric invariants of the product components. Namely, the mean curvature vector of a Veronese surface (with $\omega^3 = 0$) is $\sqrt{3}\lambda e_6$, and since $d[x + (\sqrt{3}\lambda)^{-1}e_6] = 0$, this surface belongs to a four-dimensional sphere of radius $(\sqrt{3}\lambda)^{-1}$. In this case, the circle of constant curvature has the same radius μ.

The results of this analysis can be summarized as follows.

Theorem 7.3.1. *A semiparallel M^3 with principal codimension $m_1 = 4$ in $N^n(c)$ can exist only if $c = 0$, i.e., in Euclidean E^n, and is a second-order envelope of products of Veronese surfaces and circles.*

Among such products there exists a parallel M^3 which is minimal in a 6-dimensional sphere S^6 and has principal codimension $m_1 = 3$ in this S^6. In this case, the radius of the four-dimensional sphere containing the Veronese surface is equal to the radius of the circle.

In order to investigate in more detail the envelopes M^3 of Theorem 7.3.1, exterior differentiation must be applied to equations (7.3.8)–(7.3.13), where now $\kappa = \lambda\sqrt{3}$.

The equations $\omega_3^4 = \omega_3^5 = \omega_3^6 = \omega_3^7 - \mu\omega^3 = 0$ then give the exterior equations

$$\lambda\omega_3^1 \wedge \omega^1 - \lambda\omega_3^2 \wedge \omega^2 + \mu\omega_4^7 \wedge \omega^3 = 0,$$

$$\lambda\omega_3^1 \wedge \omega^2 + \lambda\omega_3^2 \wedge \omega^1 + \mu\omega_5^7 \wedge \omega^3 = 0,$$

$$\sqrt{3}(\lambda\omega_3^1 \wedge \omega^1 + \lambda\omega_3^2 \wedge \omega^2) + \mu\omega_6^7 \wedge \omega^3,$$

$$\omega_3^1 \wedge \omega^1 + \omega_3^2 \wedge \omega^2 + d\ln\mu \wedge \omega^3 = 0.$$

Using Cartan's lemma, one gets

$$\lambda\omega_3^1 = A\omega^1 + F\omega^3, \quad \lambda\omega_3^2 = A\omega^2 + G\omega^3,$$

$$\mu\omega_4^7 = F\omega^1 - G\omega^2 + H\omega^3,$$

$$\mu\omega_5^7 = G\omega^1 + F\omega^2 + I\omega^3, \quad \mu\omega_6^7 = \sqrt{3}(F\omega^1 + G\omega^2 + J\omega^3),$$

$$d\ln\mu = \lambda^{-1}(F\omega^1 + G\omega^2) + K\omega^3.$$

The other equations (7.3.12) give

$$\lambda(\omega_4^7 + \sqrt{3}\omega_6^7) \wedge \omega^1 + \lambda\omega_5^7 \wedge \omega^2 + \mu\omega_3^1 \wedge \omega^3 = 0,$$

$$\lambda\omega_5^7 \wedge \omega^1 + \lambda(-\omega_4^7 + \sqrt{3}\omega_6^7) \wedge \omega^2 + \mu\omega_3^2 \wedge \omega^3 = 0,$$

and thus $F = G = H = I = 0$, $J = \kappa^{-2}\mu^2 A$. Denoting $-\lambda^{-1}A = \tau$, $K = \nu$ one obtains

$$\omega_1^3 = \tau\omega^1, \quad \omega_2^3 = \tau\omega^2 \tag{7.3.16}$$

$$\omega_4^7 = 0, \quad \omega_5^7 = 0, \quad \omega_6^7 = -\mu\kappa^{-1}\tau\omega^3, \quad d\ln\mu = \nu\omega^3. \tag{7.3.17}$$

Exterior differentiation of the remaining equations results in the exterior equations (7.2.15)–(7.2.20), where now $\kappa = \lambda\sqrt{3}$. Therefore, (7.2.21) and (7.2.22) also hold here, now with $\kappa^{-1}\lambda = (\sqrt{3})^{-1}$.

Finally, from (7.3.13)

$$(\omega_4^\rho + \sqrt{3}\omega_6^\rho) \wedge \omega^1 + \omega_5^\rho \wedge \omega^2 = 0,$$

$$\omega_5^\rho \wedge \omega^1 + (-\omega_4^\rho + \sqrt{3}\omega_6^\rho) \wedge \omega^2 = 0, \quad \omega_7^\rho \wedge \omega^3 = 0,$$

which gives

$$\omega_4^\rho + \sqrt{3}\omega_6^\rho = p_1^\rho\omega^1 + p_2^\rho\omega^2, \quad -\omega_4^\rho + \sqrt{3}\omega_6^\rho = p_3^\rho\omega^1 + p_4^\rho\omega^2, \tag{7.3.18}$$

$$\omega_5^\rho = p_2^\rho\omega^1 + p_3^\rho\omega^2, \tag{7.3.19}$$

and $\omega_7^\rho = q^\rho\omega^3$.

One can see that (7.2.21), (7.2.22) coincide with equations (6.3.3), (6.3.4) for the second-order envelope of Veronese surfaces, and similarly (7.3.18), (7.3.19) coincide with (6.3.5), (6.3.6), where some notations are a bit different. Therefore, the further investigation of semiparallel M^3 of Theorem 7.3.1 repeats in most details the analysis given in Section 6.3, complemented now with the analysis of the remaining equations which describe the role of the other component. The full argument is rather complicated technically, and will be omitted here.

7.4 Higher Principal Codimensions: Conclusions

Principal codimension $m_1 = 5$ is impossible for a semiparallel M^3 in $N^n(c)$, due to Proposition 4.4.3. Indeed, if $m = 3$, then $\frac{1}{2}m(m+1) = 6$ and $\frac{1}{2}m(m-1)+1 = 4$, and $m_1 = 5$ lies exactly between these bounds.

Principal codimension $m_1 = 6$ is the maximal possible value of m_1 for a submanifold M^3. If a semiparallel M^3 with $m_1 = 6$ lies in an $N^9(c)$, then by Theorem 4.7.2 it is parallel and is a Veronese submanifold. Therefore, a semiparallel M^3 with $m_1 = 6$ in $N^n(c)$, $n > 9$, is the second-order envelope of three-dimensional Veronese submanifolds. Such envelopes deserve a special investigation, which will be done in Section 9.7 below. It is proved there, that if a submanifold M^m with $m > 2$ is the second-order envelope of Veronese submanifolds, then the latter must be congruent (Proposition 9.7.2) and this M^m cannot be parallel, i.e., cannot be a single Veronese submanifold, but intrinsically is a Riemannian manifold of the same constant curvature (Theorem 9.7.1).

As a summary of the results in the present chapter, the following two theorems can be formulated.

Theorem 7.4.1. *A semiparallel M^3 with principal codimension m_1 in Euclidean space E^n is*

- *for $m_1 = 1$:*
  - *the envelope of a one-parameter family of three-dimensional planes, or*
  - *a hypersurface in $E^4 \subset E^n$, which is a product of an $(s+1)$-dimensional round cone and a $(2-s)$-dimensional plane (s is 1 or 2); the cone could degenerate into a cylinder;*
- *for $m_1 = 2$:*
  - *a warped product $B^1 \times_r S^2(1)$ with nonconstant linear function r, or*
  - *the envelope with flat $\bar{\nabla}$ of a two-parameter family of three-dimensional planes, or*
  - *the product of a semiparallel surface with principal codimension 1 and a curve;*
- *for $m_1 = 3$:*
  - *a cone with a point-vertex (or its limiting case, a cylinder), whose director surface is the second-order envelope of Veronese surfaces, or*
  - *a logarithmic spiral tube in $E^6 \subset E^n$;*
- *for $m_1 = 4$:*
  - *the second-order envelope of products of the Veronese surfaces and circles;*
- *for $m_1 = 6$:*
  - *the second-order envelope of three-dimensional Veronese submanifolds.*

There exists no semiparallel M^3 with $m_1 = 5$.

Theorem 7.4.2. *A semiparallel M^3 with principal codimension m_1 in a non-Euclidean space form $N^n(c)$ is*

- *for $m_1 = 1$:*
  - *a hypersurface of rotation, whose profile curve has natural equation* (5.5.8), *or*
  - *a parallel hypersurface, which is either spherical, or a product of a spherical surface and a spherical curve;*
- *for $m_1 = 2$:*
  - *a Cartan variety, thus with flat $\bar{\nabla}$, in an $N^5(c) \subset N^n(c)$, or*
  - *a warped product $B^1 \times_r Sph^2(1)$ with a nonconstant linear function r, or*
  - *a Segre submanifold $S_{(1,2)}(a)$ in an $S^5(a^2) \subset N^n(c)$, or*
  - *the product of a semiparallel surface with principal codimension 1 and a curve;*
- *for $m_1 = 3$:*
  - *the product of a Veronese surface and plane curve of constant curvature in $N^6(c) \subset N^n(c)$, such that the curvature of this curve is connected to the quantity λ characterizing the Veronese surface, and is either $\sqrt{3c}\lambda(\sqrt{3\lambda^2 - c})^{-1}$, or $\sqrt{3}\lambda$ in the special case $c > 0$;*
- *for $m_1 = 6$:*
  - *the second-order envelope of three-dimensional Veronese submanifolds.*

There exists no semiparallel M^3 with $m_1 = 4$ or $m_1 = 5$.

Remark 7.4.3. In both theorems above, one could in some cases just have an open subset instead of the whole M^3 itself. Among the second-order envelopes of parallel submanifolds, single parallel manifolds are also included as special cases.

Remark 7.4.4. For higher dimensions $m > 3$, a complete classification of semiparallel submanifolds M^m is still missing. Some initial results for $m = 4$ and Euclidean space E^n were established in [Ri 97], [Ri 99], [Ri 2000].

8

Decomposition Theorems

As an algebraic preparation for decomposition theorems of semiparallel submanifolds, Theorem 4.3.1 of Section 4.3 on the decomposition of semiparallel fundamental triplets will be considered first.

8.1 Decomposition of Semiparallel Submanifolds

Recall that in Section 5.2 a submanifold M^m in ${}_\sigma E^n$ is said to be decomposable into a *product of submanifolds* M^{m_ρ} in ${}_{\sigma_\rho}E^{n_\rho}$ $(\rho = 1, \ldots, r)$ if

(i) $M^m = M^{m_1} \times \cdots \times M^{m_r}$,
(ii) ${}_\sigma E^n = {}_{\sigma_1}E^{n_1} \times \cdots \times {}_{\sigma_r}E^{n_r}$, where the factors on the right side are pairwise totally orthogonal.

A submanifold which is decomposable into a product is called *reducible*, otherwise it is *irreducible*.

Proposition 8.1.1. *Let a submanifold M^m in ${}_\sigma E^n$ be decomposable into a product of submanifolds as above. Then this M^m is semiparallel (resp. parallel, $\bar{\nabla}$-flat, or 2-parallel) if and only if every M^{m_ρ} in ${}_{\sigma_\rho}E^{n_\rho}$ is semiparallel (resp. parallel, $\bar{\nabla}_\rho$-flat, or with parallel $\bar{\nabla}_\rho h_\rho$, where at least one of them is 2-parallel).*

Proof. Choose an adapted moving frame for M^m such that $\{x; e_{i_\rho}, e_{\alpha_\rho}\}$ is adapted to M^{m_ρ} for every $\rho = 1, \ldots, r$. Then $\omega_{i_\rho}^{j_\tau} = 0$, $\omega_{\alpha_\rho}^{\beta_\tau} = 0$ if $\rho \neq \tau$, and among h_{ij}^α only $h_{i_\rho j_\rho}^{\alpha_\rho}$ can be nonzero. It is seen that among $\bar{\nabla} h_{ij}^\alpha$ only components with the same lower subindex ρ can be nonzero and they are exactly the $\bar{\nabla}_\rho h_{i_\rho j_\rho}^{\alpha_\rho}$ for the factor M^{m_ρ}. The assertion concerning parallel submanifolds now follows immediately.

The assertion for 2-parallel submanifolds follows from the fact that among h_{ijk}^α, which are symmetric in all three lower indices, only components with the same lower subindex ρ can be nonzero. The same also holds for $\bar{\nabla} h_{ijk}^\alpha$.

Ü. Lumiste, *Semiparallel Submanifolds in Space Forms*,
DOI 10.1007/978-0-387-49913-0_9, © Springer Science+Business Media, LLC 2009

It then follows that among Ω_i^j and Ω_α^β, only $\Omega_{i_\rho}^{j_\rho}$ and $\Omega_{\alpha_\rho}^{\beta_\rho}$ can be nonzero. The assertions for semiparallel submanifolds and flat $\bar{\nabla}$ (i.e., with $\Omega_i^j = \Omega_\alpha^\beta = 0$) follow now in the same way.

To investigate the decomposition of semiparallel submanifolds M^m in $N^n(c) \subset {}_\sigma E^{n+1}$ the algebraic setup in Theorem 4.3.1 can be used. Namely, after comparing (4.1.1) and (4.2.2) it is obvious that a submanifold M^m in $N^n(c)$ is semiparallel iff at any point $x \in M^m$ the fundamental triplet (V, T, h) with $V = T_x N^n(c)$, $T = T_x M^m$, and $h = h_x$ is semiparallel, or equivalently, (V^*, T, h^*) with $V^* = T_x({}_\sigma E^{n+1})$ and $h^* = h_x^*$ is semiparallel. Therefore, considering a submanifold M^m, let T denote its tangent vector bundle, h its second fundamental form, A_H its mean shape operator, etc. The eigensubspaces $T_1, \dots, T_r$ of A_H, introduced in Theorem 4.3.1, form corresponding distributions T_ρ on every open subset of M^m, on which the number r of different eigenvalues of A_H and their multiplicities $m_1, \dots, m_r$ are constant. In general, M^m is the union of the closures of all these open subsets. In the investigations below, M^m will be considered only on the domain of one of these subsets.

It was shown in the proof of Theorem 4.3.1 that for every pair of different eigensubspaces T_ρ and T_σ of A_H, one has $h(X, Y) = 0$ for all $X \in T_\rho$, $Y \in T_\sigma$ with $\rho \neq \sigma$. In the differential geometry of submanifolds, it is said that in this case the subspaces T_ρ and T_σ are *conjugate* (with respect to the second fundamental form h; see, e.g., [AG 93], 3.5).

Theorem 8.1.2. *Let M^m be a Riemannian semiparallel submanifold in $N^n(c) \subset {}_\sigma E^{n+1}$ on which the mean shape operator A_H has r distinct eigenvalues of constant multiplicities $m_1, \dots, m_r$. Then the r eigendistributions $T_1, \dots, T_r$ of A_H define foliations whose leaves are mutually orthogonal conjugate semiparallel submanifolds $M_1^{m_1}, \dots, M_r^{m_r}$ with orthogonal first-order outer normal subspaces at any point $x \in M^m$.*

Proof. It is shown first that the eigendistributions are foliations. From the expression for the components of the mean curvature vector H (see Section 2.4) it follows that $\nabla^\perp H^\alpha = H_k^\alpha \omega^k$, where $H_k^\alpha = \frac{1}{m}\sum_{i=1}^m h_{iik}^\alpha$. Therefore, for the components $h_{ij}^H = \langle H, h_{ij}\rangle = g_{\alpha\beta} H^\alpha h_{ij}^\beta$ of A_H with respect to a frame with orthonormal tangent part the following hold:

$$\nabla h_{ij}^H = g_{\alpha\beta}(H_k^\alpha h_{ij}^\beta + H^\alpha h_{ijk}^\beta)\omega^k. \tag{8.1.1}$$

Writing this for the canonically adapted frame bundle of A_H, one obtains $h_{i_\rho j_\rho}^H = \lambda_{(\rho)}\delta_{i_\rho j_\rho}$, $h_{i_\rho j_\sigma}^H = 0$ $(\rho \neq \sigma)$, and therefore, in particular,

$$(\lambda_{(\rho)} - \lambda_{(\sigma)})\omega_{i_\rho}^{j_\sigma} = g_{\alpha\beta} H^\alpha h_{i_\rho j_\sigma k}^\beta \omega^k, \quad \rho \neq \sigma. \tag{8.1.2}$$

The eigenspace distribution of A_H for $\lambda_{(\rho)}$ is defined by the differential system

$$\omega^{j_1} = \dots = \omega^{j_{\rho-1}} = \omega^{j_{\rho+1}} = \dots = \omega^{j_r} = 0, \tag{8.1.3}$$

where the superscripts run all possible values. By (1.4.3),

$$d\omega^{j_\sigma} = \omega^{i_\rho} \wedge \omega_{i_\rho}^{j_\sigma} + \sum_{\tau \neq \rho} \omega^{i_\tau} \wedge \omega_{i_\tau}^{j_\sigma} \quad (\sigma \neq \rho).$$

The first group of terms on the right side is $\omega^{i_\rho} \wedge (\lambda_{(\rho)} - \lambda_{(\sigma)})^{-1} h^H_{i_\rho j_\sigma k}\omega^k$, and due to the symmetry of h^H_{ijk} it reduces to

$$(\lambda_{(\rho)} - \lambda_{(\sigma)})^{-1} \sum_{\tau \neq \rho} h^H_{i_\rho j_\sigma k_\tau} \omega^{i_\rho} \wedge \omega^{k_\tau}.$$

This shows that for each σ distinct from ρ, the exterior differential $d\omega^{j_\sigma}$ becomes zero as an algebraic consequence of the system (8.1.3). Hence the distribution defined by this system is a foliation T_ρ; this is true for $\rho = 1, \ldots, r$.

The other assertions now follow immediately from the Theorem 4.3.1 and from the result contained in its proof that $h(X, Y) = 0$ for all $X \in T_\rho$, $Y \in T_\sigma$ with $\rho \neq \sigma$.

This finishes the proof of Theorem 8.1.2.

A submanifold M^m in $N^n(c)$ is said to be *pseudoumbilic* if its mean curvature shape operator A_H is proportional to the identity operator (see [Ch 73b], [Ch 2000], 5.4), i.e., if for the orthonormal frame bundle $O(M^m, N^n(c))$ there holds $\langle H, h_{ij}\rangle = \lambda\delta_{ij}$, or, equivalently by (2.1.6),

$$\langle H^*, h^*_{ij}\rangle = \lambda^*\delta_{ij}, \tag{8.1.4}$$

where $\lambda^* = \lambda + c$.

Proposition 8.1.3. *The leaves of the foliations $T_1, \ldots, T_r$ in Theorem* 8.1.2 *are pseudoumbilic submanifolds in $N^n(c)$.*

Proof. The mean curvature vector of a leaf of T_ρ is $H_\rho = \frac{1}{m_\rho}\sum_{i_\rho} h_{i_\rho j_\rho}$, and for M^m, therefore, $H = \frac{1}{m}\sum_\rho \sum_{i_\rho} h_{i_\rho i_\rho} = \frac{1}{m}\sum_\rho m_\rho H_\rho$. Hence, $h^H_{ij} = \langle H, h_{ij}\rangle = \frac{1}{m}\langle \sum_\sigma m_\sigma H_\sigma, h_{ij}\rangle$.

On the other hand, in Section 4.3 it was shown that $(A_H)^i_j = H^\beta h^i_{\beta j} = \langle H, h_{ij}\rangle$; therefore, the relation $(A_H)^{i_\rho}_{j_\rho} = \lambda_{(\rho)}\delta^{i_\rho}_{j_\rho}$ used in the proof of Theorem 4.3.1 is equivalent to $\langle H, h_{i_\rho j_\rho}\rangle = \lambda_{(\rho)}\delta_{i_\rho j_\rho}$.

All this holds also in the outer version, with sign *, where the last equality is $\langle H^*, h^*_{i_\rho j_\rho}\rangle = \lambda^*_{(\rho)}\delta_{i_\rho j_\rho}$ with $H^* = \frac{1}{m}\sum_\rho m_\rho H^*_\rho$ and $\lambda^*_{(\rho)} = \lambda_{(\rho)} + c$. Substituting this expression of H^* and using that by (4.3.7) $\langle H^*_\sigma, h^*_{i_\rho j_\rho}\rangle = 0$ $(\rho \neq \sigma)$, one obtains $\langle H^*_\rho, h^*_{i_\rho j_\rho}\rangle = \frac{m}{m_\rho}\lambda^*_{(\rho)}\delta_{i_\rho j_\rho}$. Therefore, every leaf of T_ρ is indeed pseudoumbilic as asserted, since (8.1.4) holds, with $\lambda^* = \frac{m}{m_\rho}\lambda^*_{(\rho)}$.

Remark 8.1.4. The notion of pseudoumbilic submanifold M^m in $N^n(c)$ has importance only when $m > 1$, because if $m = 1$, then every curve M^1 in $N^n(c)$ is pseudoumbilic, since then (8.1.4) is satisfied trivially: $H^* = h^*_{11}$ is then the outer curvature vector of the curve and $\lambda^* = \langle H^*, H^* \rangle$.

Therefore, Proposition 8.1.3 has meaning only for foliations whose dimension, i.e., the multiplicity of the corresponding eigenvalue of A_H, is greater than 1. Otherwise, the leaves are simply a set of curves. Of course, the assertions of Theorem 8.1.2 are also valid for these curves.

In order to obtain the product decomposition of a semiparallel submanifold M^m in $N^n(c) \subset {}_\sigma E^{n+1}$, some additional assumptions must be added to those of Theorem 8.1.2.

Theorem 8.1.5. *Let M^m be as in Theorem 8.1.2 and let $T'_1, \dots, T'_s$ be direct sums of $T_1, \dots, T_r$ which at every point $x \in U$ satisfy $T'_1 \oplus \cdots \oplus T'_s = T_x M^m$, so that every T'_φ, $1 \le \varphi \le s$, is parallel for the Riemannian connection ∇ induced on M^m by immersion. Then around each of its points the submanifold M^m coincides locally with a product of semiparallel submanifolds $U^{m'_1}, \dots, U^{m'_s}$, where every $U^{m'_\varphi}$ is a leaf of T'_φ.*

Proof. Since the foliations $T_1, \dots, T_r$ are mutually orthogonal and conjugate, the $T'_1, \dots, T'_s$ are also mutually orthogonal and conjugate, i.e.,

$$\langle T'_\varphi, T'_\psi \rangle = 0, \quad h(T'_\varphi, T'_\psi) = 0 \quad (\varphi \neq \psi).$$

Moreover, due to (4.3.6)

$$\langle h^*_{i'_\varphi k'_\varphi}, h^*_{j'_\psi l'_\psi} \rangle = 0 \quad (\varphi \neq \psi). \tag{8.1.5}$$

If every T'_φ is parallel in ∇, then $\omega_{i'_\varphi}^{j'_\psi} = 0$ for every pair $\varphi \neq \psi$, thus T'_φ is a foliation. Now (2.2.3), (2.2.2) imply that among h_{ijk} only $h_{i'_\varphi j'_\varphi k'_\varphi}$ $(\varphi = 1, \dots, s)$ can be nonzero.

Therefore, differentiation of (8.1.5) yields

$$\langle h_{i'_\varphi k'_\varphi p'_\varphi}, h^*_{j'_\psi l'_\psi} \rangle = 0 \quad (\varphi \neq \psi). \tag{8.1.6}$$

For semiparallel M^m one has $\nabla h^\alpha_{ijk} = h^\alpha_{ijkl}\omega^l$ (see Section 3.4), where h^α_{ijkl} are symmetric in the lower indices. It follows that among h_{ijkl} only $h_{i'_\varphi j'_\varphi k'_\varphi l'_\varphi}$ can be nonzero.

From (8.1.6) one can deduce by differentiation that

$$\langle h_{i'_\varphi k'_\varphi p'_\varphi q'_\varphi}, h^*_{j'_\psi l'_\psi} \rangle = 0, \quad \langle h_{i'_\varphi k'_\varphi p'_\varphi}, h_{j'_\psi l'_\psi q'_\psi} \rangle = 0, \tag{8.1.7}$$

if $\varphi \neq \psi$. Repeating this procedure, one sees that in general

$$\langle h_{i'_\varphi k'_\varphi \dots u'_\varphi}, h_{j'_\psi l'_\psi \dots v'_\psi} \rangle = 0 \quad (\varphi \neq \psi) \tag{8.1.8}$$

for every admissible choice of indices; this can be proved by induction.

The procedure terminates when all vectors $h_{i'_\varphi \dots u'_\varphi}$ with $p+1$ indices are contained in the linear span of these vectors with 2, 3,...,p indices. Such a p exists because ${}_\sigma E^{n+1}$ is finite dimensional.

From (1.6.2), (2.2.1), (2.4.1), (3.1.2) and from their differential prolongations it follows that

$$\nabla e_{i'_\varphi} = h^*_{i'_\varphi j'_\varphi} \omega^{j'_\varphi},$$

$$\nabla h^*_{i'_\varphi j'_\varphi} = \sum_{l'_\varphi} e_{l'_\varphi} \langle h^*_{i'_\varphi j'_\varphi}, h^*_{k'_\varphi l'_\varphi} \rangle \omega^{k'_\varphi} + h_{i'_\varphi j'_\varphi k'_\varphi} \omega^{k'_\varphi},$$

$$\cdots\cdots\cdots\cdots\cdots\cdots$$

$$\nabla h_{i'_\varphi \ldots u'_\varphi} = \sum_{l'_\varphi} e_{l'_\varphi} \langle h_{i'_\varphi \ldots u'_\varphi}, h^*_{p'_\varphi l'_\varphi} \rangle \omega^{p'_\varphi} + h_{i'_\varphi \ldots u'_\varphi p'_\varphi} \omega^{p'_\varphi},$$

where

$$\nabla e_{i'_\varphi} = de_{i'_\varphi} - e_{j'_\varphi} \omega^{j'_\varphi}_{i'_\varphi}, \quad \nabla h^*_{i'_\varphi j'_\varphi} = dh^*_{i'_\varphi j'_\varphi} - h^*_{k'_\varphi j'_\varphi} \omega^{k'_\varphi}_{i'_\varphi} - h^*_{i'_\varphi k'_\varphi} \omega^{k'_\varphi}_{j'_\varphi},$$

and so on; here $\nabla h_{i'_\varphi \ldots u'_\varphi}$ must have p indices.

It is seen that a leaf $U^{m'_\varphi}$ of the foliation T'_φ, containing $x \in U^m$, lies around x in the n_φ-plane defined by x and the linear span of the vectors $e_{i'_\varphi}, h^*_{i'_\varphi j'_\varphi}, \ldots, h_{i'_\varphi \ldots u'_\varphi}$.

All such pairs of n_φ-planes and n_ψ-planes ($\varphi \neq \psi$) are totally orthogonal due to (8.1.5)–(8.1.8). Therefore, around x the submanifold M^m is actually a product of submanifolds $U^{m'_\varphi}$ ($\varphi = 1, \ldots, s$), which are semiparallel, as follows from Proposition 8.1.1 and Theorem 8.1.2.

This finishes the proof of Theorem 8.1.5.

Remark 8.1.6. In the special case $c = 0$, i.e., for submanifolds in E^n, Theorem 8.1.5 was announced in [Lu 87a], with a reference to [Mo 71]. A direct proof for this case was given in [Lu 88a]; the proof just given above extends it now to the case $c \neq 0$.

For the 2-parallel submanifolds, which due to Proposition 4.1.5 constitute a special class of semiparallel ones, Theorem 8.1.5 with $c = 0$ was proved previously in [Lu 87c].

An interesting modification of Theorem 8.1.5 was given by V. Mirzoyan in [Mi 91c], where the assumption that every T'_φ is parallel with respect to ∇ is replaced by another assumption also involving ∇ and referring directly to the foliations T_ρ.

Following [ChK 52] the dimension of the tangent subspace $\{Z \in T_x M^m : R(X, Y)Z = 0 \text{ for all } X, Y \in T_x M^m\}$ of a Riemannian manifold M^m is called the *index of nullity* of this M^m at x.

If the index of nullity of M^m is zero, there exist $X, Y \in T_x M^m$ such that $R(X, Y)Z = 0$ implies $Z = 0$. In other words, the matrix of $R_{ij}(X, Y) = R_{ij,kl} X^k Y^l$ has maximal rank m and hence for every $W \in T_x M^m$ there exists $Z \in T_x M^m$ such that

$$W = R(X, Y)Z.$$

Theorem 8.1.7. *Let M^m be as in Theorem 8.1.2 and let all eigenvalues of A_H be nonzero and all leaves of the foliations T_ρ have zero index of nullity at every point. Then M^m is a product of these leaves.*

Proof. One has to establish that for arbitrary $X \in T_x M^m$ and $W_\rho \in T_\rho$ the covariant derivative $\nabla_X W_\rho$ belongs to T_ρ. This would imply that the foliation T_ρ is parallel for ∇ and then Theorem 8.1.5 gives the desired conclusion.

Due to the assumption about the nullity index, there exist X_ρ, Y_ρ, Z_ρ in T_ρ such that $W_\rho = R(X_\rho, Y_\rho)Z_\rho$. Thus

$$\begin{aligned}\nabla_{V_\sigma} W_\rho &= \nabla_{V_\sigma}[R(X_\rho, Y_\rho)Z_\rho] \\ &= [(\nabla_{V_\sigma} R)(X_\rho, Y_\rho)]Z_\rho + R(\nabla_{V_\sigma} X_\rho, Y_\rho)Z_\rho \\ &\quad + R(X_\rho, \nabla_{V_\sigma} Y_\rho)Z_\rho + R(X_\rho, Y_\rho)\nabla_{V_\sigma} Z_\rho. \end{aligned} \tag{8.1.9}$$

From (2.1.10) and from $h^*_{i_\rho p_\sigma} = 0$ $(\rho \neq \sigma)$ it follows that $R_{i_\rho j_\rho, p_\sigma q_\sigma} = 0$, if $\rho \neq \sigma$. For semiparallel M^m, (4.1.1) holds in its outer version and this yields $R_{i_\rho j_\sigma, pq} = 0$ $(\rho \neq \sigma)$. Thus only $R_{i_\rho j_\rho, p_\rho q_\rho}$ can be nonzero and hence the last three summands on the right side of (8.1.9) belong to T_ρ.

It remains to show that the same is true for the first summand. The Bianchi identity (1.3.6) implies that

$$(\nabla_{V_\sigma} R)(X_\rho, Y_\rho) = -(\nabla_{X_\rho} R)(Y_\rho, V_\sigma) - (\nabla_{Y_\rho} R)(V_\sigma, X_\rho). \tag{8.1.10}$$

Here, as in (8.1.9),

$$\begin{aligned}[(\nabla_{X_\rho} R)(Y_\rho, V_\sigma)]Z_\rho &= \nabla_{X_\rho}[R(Y_\rho, V_\sigma)Z_\rho] - R(\nabla_{X_\rho} Y_\rho, V_\sigma)Z_\rho \\ &\quad - R(Y_\rho, \nabla_{X_\rho} V_\sigma)Z_\rho - R(Y_\rho, V_\sigma)\nabla_{X_\rho} Z_\rho.\end{aligned}$$

For the case $\rho \neq \sigma$, only $-R(Y_\rho, \nabla_{X_\rho} V_\sigma)Z_\rho$ can be nonzero on the right side, and it belongs indeed to T_ρ (by the same argument as above).

So the result is as follows: If $\rho \neq \sigma$ then $\nabla_{V_\sigma} W_\rho$ belongs to T_ρ.

Further, $\langle V_\rho, W_\sigma \rangle = 0$ implies

$$\langle \nabla_{X_\rho} V_\rho, W_\sigma \rangle + \langle V_\rho, \nabla_{X_\rho} W_\sigma \rangle = 0,$$

where $\nabla_{X_\rho} W_\sigma$ belongs to T_σ and thus the second term above is zero. Since W_σ is arbitrary for every $\sigma \neq \rho$, it follows that $\nabla_{X_\rho} V_\rho$ belongs to $T\rho$.

Therefore, $\nabla_X W_\rho$ belongs to T_ρ for any vector field X on M^m. Hence every foliation T_ρ is parallel with respect to ∇, and now Theorem 8.1.5 can be applied, for the situation where every T'_φ is a single T_ρ. This verifies the assertion.

Remark 8.1.8. Theorem 8.1.7 was proved by V. Mirzoyan in [Mi 91c] in an indirect way, based on some properties of the so-called V-decomposition of a semisymmetric Riemannian manifold which was introduced and investigated by Z. I. Szabó [Sza 82]. The proof given above uses the same idea but is more direct.

8.2 Decomposition of Parallel Submanifolds

A normally flat parallel submanifold is decomposed in Section 5.2 as a product of spherical submanifolds. This was done as preparation for the decomposition of normally flat semiparallel submanifolds. In this section, general normally nonflat parallel submanifolds will be decomposed.

Every parallel submanifold M^m in $N^n(c) \subset {}_\sigma E^{n+1}$ is semiparallel (see Proposition 4.1.3); therefore, all results of the previous Section 8.1 can be used for them. By Proposition 8.1.3, the leaves of the eigenfoliations of A_H are pseudoumbilic submanifolds. Now suppose the parallel condition also holds.

Proposition 8.2.1. *A pseudoumbilic parallel submanifold M^m in $N^n(c) \subset {}_\sigma E^{n+1}$ is a minimal submanifold,*

- *in general, if $\lambda^* \neq 0$ in* (8.1.4)*, in a hypersphere of ${}_\sigma E^{n+1}$, or,*
- *in particular, if $\lambda^* = 0$ in* (8.1.4)*, in an $(m-1)$-dimensional horosphere of $H^n(c) \subset {}_1En+1$, which could reduce to an m-dimensional plane or to its open subset.*

Proof. The parallel condition is $\bar{\nabla}^* h^*_{ij} + \sum_k e_k \langle h^*_{ij}, h^*_{kl}\rangle \omega^l = 0$ (see (3.1.2) with $h_{ijk} = 0$). Summing this over $i = j$ one obtains $dH^* + \sum_k e_k \langle H^*, h^*_{kl}\rangle \omega^l = 0$. On the other hand the pseudoumbilic condition (8.1.4) gives $\langle H^*, h^*_{kl}\rangle = \lambda^* \delta_{kl}$, so $dH^* + \lambda^*(e_k \omega^k) = 0$, and thus $dH^* = -\lambda^* dx$.

Further, summing in (8.1.4) over $i = j$ leads to $\langle H^*, H^*\rangle = \lambda^*$ and differentiation then gives $d\lambda^* = -\lambda^* 2\langle dx, H^*\rangle = 0$.

This shows that if $\lambda^* \neq 0$ then $d[x + (\lambda^*)^{-1} H^*] = 0$, so that the point in ${}_\sigma E^{n+1}$ with radius vector $z = x + (\lambda^*)^{-1} H^*$ is a fixed point. Therefore, M^m is contained in a hypersphere with center at this point, whose radius is the reciprocal of the length of H^*. Since H^* points along the radius of this hypersphere, the mean curvature vector of M^m with respect to this hypersphere is zero; hence M^m is minimal in the latter.

If $\lambda^* = 0$ then $0 = \langle H^*, H^*\rangle = \langle H, H\rangle + c$ shows that only $c \leq 0$ is possible here. If $c = 0$, then $H = 0$ and by Theorem 4.1.7, M^m is totally geodesic in E^n, thus an m-dimensional plane or its open subset. If $c < 0$ then M^m is contained in an $(n-1)$-dimensional horosphere of $H^n(c)$, orthogonal to the straight lines of $H^n(c)$ with direction vector H, parallel in the Lobachevsky sense, i.e., intersected from $H^n(c) \subset {}_1E^{n+1}$ by the two-dimensional planes of ${}_1E^{n+1}$ containing the origin point o and the lightlike vector H^*.

Theorem 8.2.2. *Let M^m be a parallel submanifold in $N^n(c)$ and let $\lambda_{(1)}, \ldots, \lambda_{(r)}$ be distinct eigenvalues of A_H with constant multiplicities $m_1, \ldots, m_r$ on some open $U \subset M^m$. Then U is the product of pseudoumbilic parallel submanifolds $U^{m_1}, \ldots, U^{m_r}$, described in Proposition* 8.2.1.

Proof. Since a parallel M^m is also semiparallel, one can use Theorem 8.1.2, under some simplifying circumstances. Namely, for a parallel M^m characterized by $\bar{\nabla} h^\alpha_{ij} = 0$, (2.2.3) implies $h^\alpha_{ijk} = 0$. Therefore, in (8.1.1) and (8.1.2) the right sides are equal to zero, because also $H^\alpha_k = 0$, thus $\nabla h^H_{ij} = 0$ and $\omega^{j_\sigma}_{i_\rho} = 0$ $(\rho \neq \sigma)$. Hence all T_ρ are themselves parallel for the Riemannian connection ∇ induced on M^m by immersion. By Theorem 8.1.2, this parallel M^m coincides, locally around every point of U, with a product $U^{m_1}, \ldots, U^{m_r}$, where every U^{m_ρ} is a leaf of T_ρ, and is by Proposition 8.1.1 a parallel submanifold, and by Proposition 8.1.3 it is also pseudoumbilic.

Now, indeed, Proposition 8.2.1 can be applied, which finishes the proof.

Remark 8.2.3. Theorem 8.2.2 generalizes Proposition 5.2.1, which describes normally flat parallel submanifolds M^m in $N^n(c)$ as products of spherical submanifolds. But in those parts which concern the possible one-dimensional product components, they coincide. Indeed, such components in Theorem 8.2.2 are obviously normally flat parallel ones and therefore are spherical curves by Proposition 5.2.1, and thus they are the plane curves of constant curvature in $N^n(c)$ described in Section 5.2.

Remark 8.2.4. For general parallel submanifolds of Euclidean space E^n, Theorem 8.2.2 was proved by Ferus [Fe 74a, b]. Later in [Fe 80] he showed that a complete parallel M^m in E^n is a *symmetric orbit*, i.e., the orbit of a Lie group acting by isometries of E^n, and symmetric with respect to each of its normal subspaces.

Ferus' main result in [Fe 80] is as follows.

Theorem 8.2.5. *A general symmetric orbit M^m in Euclidean space E^n is a product submanifold*

$$M^{m_1} \times \cdots \times M^{m_s} \times S^1(c_{s+1}) \times \cdots \times S^1(c_{s+q}) \times E^{m_0},$$

where $m_1 > 1, \ldots, m_s > 1$, and the factors lie fully in their own subspaces, which are mutually totally orthogonal in E^n. Moreover, each M^{m_σ} is a standardly imbedded symmetric R-space, and in particular, minimal in an $S^{n_\sigma}(c_\sigma) \subset E^{n_\sigma+1}$.

Generalizations of this theorem to the case of symmetric orbits M^m in $N^n(c)$ were given in [Tak 81] and [BR 83]. Note that Theorem 8.2.2 above can be considered as a preparation for such a generalization. Indeed, the multiplicities $m_1, \ldots, m_r$ must be separated into those which are > 1, and those which are 1. Replacing the local approach by a global one, only the assertion about the standardly imbedded symmetric R-spaces needs justification, and Theorem 3.6.1 can be used for that.

A factor M^{m_σ} in Theorem 8.2.5, i.e., an irreducible symmetric orbit, which is of dimension $m_\sigma > 1$ and not totally geodesic, will be called a *main symmetric orbit* (cf. [Lu 96c, d]).

Recall that the semiparallel submanifolds of Theorem 8.1.2 are second-order envelopes of the corresponding parallel ones (see Theorem 4.5.5), which are the symmetric orbits of Theorem 8.2.5. The leaves of the eigenfoliations $T_1, \ldots, T_r$ of A_H, which have dimension > 1 and are not totally geodesic, have the property that each of them is a second-order envelope of main symmetric orbits. They will be called the *main leaves* of the semisymmetric submanifold.

For instance, the spherical factors of an irreducible normally flat semiparallel submanifold considered as a warped product (see Section 5.4) are examples of such main leaves.

8.3 Decomposition of Normally Flat 2-Parallel Submanifolds

Two important classes of semiparallel submanifolds, besides the class of parallel ones, consist of 2-parallel submanifolds and of submanifolds with flat van der Waerden–Bortolotti connection $\bar{\nabla}$ (see Propositions 4.1.4 and 4.1.5).

Among 2-parallel submanifolds, the normally flat ones deserve special attention. In this section it will be shown that each of them is a product submanifold whose nonparallel components can only have dimension 1 or 2.

Theorem 8.3.1. *A normally flat 2-parallel submanifold M^m in $N^n(c) \subset_\sigma E^{n+1}$ is a product of 2-parallel curves or surfaces and possibly a normally flat parallel submanifold.*

The proof will result from the lemma and two propositions that follow.

Lemma 8.3.2. *A normally flat 2-parallel submanifold M^m in $N^n(c)$ satisfies, in addition to* (5.1.1)–(5.1.3) *and the assertions of Lemma* 5.1.2, *also*

$$dK_i = -\sum_{l=1}^{m} \langle K_i, k_l \rangle e_l \omega^l + 3 \sum_{j \neq i} L_{ij} \omega_i^j, \tag{8.3.1}$$

$$dL_{ij} = -\sum_{l=1}^{m} \langle L_{ij}, k_l \rangle e_l \omega^l + (2L_{ji} - K_i)\omega_i^j + \sum_{l \neq i}^{l \neq j} L_{il} \omega_j^l, \tag{8.3.2}$$

and if $r \geq 3$, then the coefficients in (5.1.6) *satisfy*

$$\lambda_{(\rho) j_\tau} \lambda_{(\rho) l_\varphi} = 0, \quad \lambda_{(\rho) j_\tau} \lambda_{(\tau) l_\varphi} = \lambda_{(\rho) l_\varphi} \lambda_{(\varphi) j_\tau}, \tag{8.3.3}$$

for every distinct triple of indices ρ, τ and φ.

Proof. The relations (8.3.1) and (8.3.2) are immediate consequences from the 2-parallel condition $\bar{\nabla} h^\alpha_{ijl} = 0$. Substituting into it $i = i_\rho$, $j = j_\tau$, $l = l_\varphi$ with distinct ρ, τ, φ, one gets via $E_{ijl} = 0$ that

$$(L_{i_\rho l_\varphi} - L_{j_\tau l_\varphi})\omega_{i_\rho}^{j_\tau} + (L_{j_\tau i_\rho} - L_{l_\varphi i_\rho})\omega_{j_\tau}^{l_\varphi} + (L_{l_\varphi j_\tau} - L_{i_\rho j_\tau})\omega_{l_\varphi}^{i_\rho} = 0.$$

After substitution from (5.1.6), the result is (8.3.3).

Proposition 8.3.3. *The field of subspaces* $\operatorname{span}\{e_{i_\rho}\}$ *of Lemma* 5.1.2 *coincides with the eigenfoliation T_ρ of Theorem* 8.1.2 *for every value of ρ. If this field corresponds to a principal curvature vector $k_{(\rho)}$ which is either zero or nonzero nonsimple, then this eigenfoliation is parallel for ∇ and its leaves are parallel submanifolds. The one-dimensional foliations corresponding to nonzero simple principal curvature vectors k_i span a foliation T' which is also parallel for ∇, and M^m is a product of its leaves and of the parallel submanifolds above.*

Proof. Since $H = \frac{1}{m}(k_1 + \cdots + k_m)$, one has

$$\langle H, h_{i_\rho j_\rho} \rangle = \frac{1}{m} \langle k_1 + \cdots + k_m, k_{(\rho)} \delta_{i_\rho j_\rho} \rangle = \frac{m_\rho}{m} k_{(\rho)}^2 \delta_{i_\rho j_\rho}, \quad \langle H, h_{i_\rho j_\tau} \rangle = 0$$

for $\rho \neq \tau$. Thus $\lambda_\rho = \frac{m_\rho}{m} k_{(\rho)}^2$; this proves the first assertion.

Let $k_{(1)} = 0$ and so $\lambda_1 = 0$. Then $K_{i_1} = L_{i_1 j} = 0$ due to (5.1.2). Now (5.1.6) and (8.3.2) give, respectively, $\omega_{i_1}^{j_\tau} = -\lambda_{(\tau)i_1}\omega^{j_\tau}$ and $2L_{j_\tau i_1}\omega_{i_1}^{j_\tau} = 0$, both for $\tau \neq 1$. Again by (5.1.6), $L_{j_\tau i_1} = \lambda_{(\tau)i_1}k_{(\tau)}$ and so, after substitution, $\lambda_{(\tau)i_1}k_{(\tau)}(-\lambda_{(\tau)i_1}\omega^{j_\tau}) = 0$; thus $\lambda_{(\tau)i_1} = 0$ and hence $\omega_{i_1}^{j_\tau} = 0$. Therefore, T_1 is parallel for ∇.

Let $k_{(\rho)}$ be nonzero nonsimple. Then Lemma 5.1.2 gives $K_{i_\rho} = 0$ and now substitution into (8.3.1) implies

$$0 = \sum_{\tau\neq\rho} L_{i_\rho j_\tau}\omega_{i_\rho}^{j_\tau} = \sum_{\tau\neq\rho} \lambda_{(\rho)j_\tau}(k_{(\rho)} - k_{(\tau)})(\lambda_{(\rho)j_\tau}\omega^{i_\rho} - \lambda_{(\tau)i_\rho}\omega^{j_\tau})$$

(summing by j_τ), and thus $\lambda_{(\rho)j_\tau} = 0$, i.e., $L_{i_\rho j_\tau} = 0$. Now substitution into (8.3.2) gives

$$0 = 2L_{j_\tau i_\rho}\omega_{i_\rho}^{j_\tau} = 2\lambda_{(\tau)i_\rho}(k_{(\rho)} - k_{(\tau)})\lambda_{(\tau)i_\rho}\omega^{j_\tau},$$

thus $\lambda_{(\tau)i_\rho} = 0$. Consequently, $\omega_{i_\rho}^{j_\tau} = 0$ for $\tau \neq \rho$ and T_ρ is parallel in ∇.

The field of tangent subspaces spanned by the one-dimensional eigenspaces corresponding to nonzero simple principal curvature vectors k_i is totally orthogonal to the parallel eigenfoliations considered in this proof, and therefore it is also parallel in ∇. This field is defined by the system $\omega^{i_1} = \omega^{i_\rho} = 0$, and since all $d\omega^{i_1}$ and $d\omega^{i_\rho}$ reduce to zero due to the equations of this system, as follows from the results above $\omega_{j_\tau}^{i_1} = -\omega_{i_1}^{j_\tau} = 0$, $\omega_{j_\tau}^{i_\rho} = -\omega_{i_\rho}^{j_\tau} = 0$, this field is also a foliation T'. The submanifold M^m is thus a product of the leaves of all these totally orthogonal foliations.

Moreover, as is seen from the proof, the components of the third fundamental form are zero (see, e.g., (5.1.6)) for all leaves of these foliations except T'; hence these leaves are parallel submanifolds, as asserted.

Proposition 8.3.4. *The one-dimensional foliations which span T' can be joined into pairs so that every pair spans a two-dimensional foliation parallel in ∇, and every leaf of T' is a product of leaves of all these one- or two-dimensional foliations.*

Proof. Let a leaf of T' be denoted simply by M^m. Now (5.1.2) and (5.1.6) take the form

$$dk_i = (-k_i^2 e_i + K_i)\omega^i + \sum_{j\neq i} L_{ij}\omega^j, \tag{8.3.4}$$

$$L_{ij} = \lambda_{ij}(k_i - k_j), \quad \omega_i^j = \lambda_{ij}\omega^i - \lambda_{ji}\omega^j. \tag{8.3.5}$$

By differentiating $\langle k_i, k_j\rangle = 0$ $(i \neq j)$, one obtains

$$\langle K_i, k_j\rangle = k_i^2\lambda_{ji}, \quad (i \neq j). \tag{8.3.6}$$

After exterior differentiation of (8.3.4), one can see that all tangential terms cancel, as well as the normal terms with $\omega^i \wedge \omega^j$, and the result is

$$0 = \sum_{j \neq i}^{l \neq i} [\lambda_{il}\lambda_{lj}(k_i - k_l) + \lambda_{ij}\lambda_{jl}(k_i - k_j)]\omega^j \wedge \omega^l.$$

So in addition to (8.3.3), in which the first equations take the form $\lambda_{ij}\lambda_{il} = 0$, one obtains $\lambda_{ij}\lambda_{jl} = 0$, where i, j, l have any three distinct values; moreover, the second equations of (8.3.3) reduce to identities.

Let us consider for some three distinct values i, j, k the matrix

$$\begin{pmatrix} 0 & \lambda_{ij} & \lambda_{ik} \\ \lambda_{ji} & 0 & \lambda_{jk} \\ \lambda_{ki} & \lambda_{kj} & 0 \end{pmatrix}.$$

The last equalities say that in every row and in every side of Sarrus's "+"-triangle, there can be only one nonzero entry. Up to permutations there are only two possibilities: the nonzero entries are either (1) only λ_{jk} and λ_{kj} or (2) only λ_{jk} and λ_{ik}. In the second case $\lambda_{ij} = \lambda_{ji} = \lambda_{ki} = \lambda_{kj} = 0$ and therefore, in particular, $L_{ij} = L_{ji} = 0$. Now (8.3.2) and (8.3.5) imply that

$$\sum_{l \neq i}^{l \neq j} \lambda_{il}(k_i - k_l)(\lambda_{jl}\omega^j - \lambda_{lj}\omega^l) = 0,$$

whence $\lambda_{ik}\lambda_{jk} = 0$. Therefore, this second case is impossible.

As a final result, in every principal (3×3)-matrix of the $(m \times m)$-matrix

$$\begin{pmatrix} 0 & \lambda_{12} & \lambda_{13} & \dots & \lambda_{1m} \\ \lambda_{21} & 0 & \lambda_{23} & \dots & \lambda_{2m} \\ \lambda_{31} & \lambda_{32} & 0 & \dots & \lambda_{3m} \\ \vdots & \vdots & \vdots & \ddots & \vdots \\ \lambda_{m1} & \lambda_{m2} & \lambda_{m3} & \dots & 0 \end{pmatrix}$$

nonzero entries can occur only in pairs of entries symmetric across the principal diagonal. Moreover the indices in two such pairs cannot have a common value. Without loss of generality one may assume that these pairs are $(\lambda_{12}, \lambda_{21})$, $(\lambda_{34}, \lambda_{43})$, $(\lambda_{56}, \lambda_{65})$, etc. Thus, due to (8.3.5), among ω_i^j only $\omega_1^2, \omega_3^4, \omega_5^6$ can (but need not) be nonzero. In view of Theorem 8.1.2, this proves the assertion.

Now, to prove Theorem 8.3.1 it suffices to combine the results obtained above.

Remark 8.3.5. Theorem 8.3.1 for the case $c = 0$ was given in [Lu 89a], where descriptions of the nonparallel components of the submanifold were also added; these descriptions are given here in Section 6.8.

Note that the normally flat parallel submanifolds mentioned in Theorem 8.3.1 were described in Section 5.2 above.

Remark 8.3.6. General (i.e., normally nonflat) 2-parallel submanifolds M^m in $N^n(c)$ have been investigated up to now only for dimensions $m = 2$ and $m = 3$. These surfaces ($m = 2$) are considered above in Section 6.7 (see Theorem 6.7.2), and such three-dimensional submanifolds were studied for $c = 0$ in [Lu 90c] and in general in [Lu 2000a], where Theorem 22.1 states the following:

Every three-dimensional 2-parallel submanifold M^3 in $N^n(c)$ is reducible in the sense that it is a product of curves and surfaces with $\bar{\nabla} h = 0$ or $\bar{\nabla}^2 h = 0$, at least one of which has $\bar{\nabla} h \neq 0$.

Note that 2-parallel curves are also described in Section 6.7 (see Proposition 6.7.1).

8.4 Structure of Submanifolds with Flat van der Waerden–Bortolotti Connection

An important class of semiparallel submanifolds, besides the classes of parallel and 2-parallel ones, consists of submanifolds with flat van der Waerden–Bortolotti connection $\bar{\nabla}$ (see Proposition 4.1.4). The most general among them were introduced by É. Cartan [Ca 19] in a projective treatment as those m-dimensional submanifolds which carry a holonomic net of conjugate curves having osculating subspace of dimension $2m$ at each point; they were later called Cartan varieties (see Remark 5.4.3). They emerged again in investigations by R. Mullari [Mu 61, 62a] on submanifolds with absolute principal directions and were then studied in [Lu 87b].

Recall that two directions of tangent vectors $X,\ Y \in T_x M^m$ of a submanifold M^m are called *conjugate* (see [SchStr 35] Vol. II, Section 11; [AG 93], Sections 3.1–3.4), if $h(X, Y) = h_{ij} X^i Y^j = 0$ (see also Sect 8.1). If m linearly independent tangent vector fields have mutually conjugate directions at every point $x \in M^m$, then their integral curves are said to form a *conjugate net* on M^m. Both of these concepts arose first in projective differential geometry and have given rise to several investigations (see [AG 93], Chapter 3; [AG 96], Chapters 2 and 3).

Absolute principal directions were introduced for the special case of surfaces in Euclidean 4-space by Wong [Won 52] and then in general by Mullari [Mu 61], [Mu 62a].

It is known that if two vector subspaces T and T^* are given in an n-dimensional Euclidean vector space V, then we have the following.

Lemma 8.4.1. *The stationary values of the angle φ between directions of unit vectors $t \in T$ and $u \in T^*$ occur so that their sides are mutually orthogonal in T as well in T^*; moreover, the 2-subspaces of V containing these angles with stationary values are mutually totally orthogonal, and orthogonal to T and T^*.*

Proof. Let orthonormal bases in T and T^* be chosen as follows: e_i in T, and e_a in T^*, where $i, j, \ldots$ are in $\{1, \ldots, m = \dim T\}$, and $a, b, \ldots$ are in $\{1, \ldots, m^* = \dim T^*\}$. Let $t = e_i t^i$, $u = e_a u^a$. Denoting

$$\Phi = \langle t, u\rangle + \lambda(\langle t, t\rangle - 1) + \mu(\langle u, u\rangle - 1)$$

one obtains these stationary angles and their sides from the system $\partial\Phi/\partial t^i = 0$, $\partial\Phi/\partial u^a = 0$, which leads to

$$\langle e_i, u\rangle + 2\lambda\langle e_i, t\rangle = 0, \quad \langle t, e_a\rangle + 2\mu\langle u, e_a\rangle = 0; \tag{8.4.1}$$

this yields $\cos\varphi = \langle t, u\rangle = -2\lambda = -2\mu$ and then

$$\langle e_i, e_a\rangle u^a - t^i \cos\varphi = 0, \quad \langle e_i, e_a\rangle t^i - u^a \cos\varphi = 0. \tag{8.4.2}$$

Eliminating u^a, one obtains

$$\sum_{a=1}^{m^*} \langle e_i, e_a\rangle\langle e_j, e_a\rangle t^j = t^i \cos\varphi;$$

hence the sides of stationary angles go in the eigendirections of a symmetric matrix, which are, as is known, mutually orthogonal in T; the eigenvalues are the squares of cosines of these angles (the same holds of course in T^*).

For two different eigenvalues $\cos\varphi = \langle t, u\rangle$ and $\cos\varphi^* = \langle t^*, u^*\rangle$, the system (8.4.2) implies $t\cos\varphi = \sum_{i=1}^{m} e_i\langle e_i, u\rangle$, $u^*\cos\varphi^* = \sum_{a=1}^{m^*} e_a\langle t^*, e_a\rangle$; therefore, $\langle t, u^*\rangle\cos\varphi\cos\varphi^* = \langle t^*, u\rangle\cos^2\varphi$ (for T) $=\langle t^*, u\rangle\cos^2\varphi^*$ (for T^*); thus $\langle t^*, u\rangle(\cos^2\varphi - \cos^2\varphi^*) = 0$, and so $\langle t^*, u\rangle = 0$. In the same way, $\langle u^*, t\rangle = 0$. This, together with the conclusion $\langle e_i, u - \langle t, u\rangle t\rangle = 0$ from (8.4.1), verifies the last assertion of the lemma.

Corollary 8.4.2. *The problem of finding the sides of nonzero stationary angles, in the sense of Lemma 8.4.1, is equivalent to finding the directions of unit vectors $t \in T$ with the following properties: there exists $v \in V$, orthogonal to T, such that $t + v \in T^*$, and the length squared of v is stationary.*

Indeed, then $\langle v, v\rangle = \tan^2\varphi$.

Let M^m be a submanifold in E^n and $\lambda : \mathbb{R} \supset [0, 1] \to M^m$ be a smooth path with arclength parameter s.

Lemma 8.4.3. *The limit position of the system of sides in $T = T_{\lambda(0)}M^m$ of the stationary angles between T and $T^* = T_{\lambda(s)}M^m$, as $s \to 0$, depends only on the point $x = \lambda(0)$ and on the direction of the tangent vector X of λ at x.*

Proof. A neighborhood U of x in M^m can be chosen so that it contains the above-considered path between $x = \lambda(0)$ and $\lambda(s)$, and has a section of the tangent frame bundle defined on it, whose integral curves are the coordinate curves of a map, geodesic at x (see, e.g., [Ra 53], Section 91). For these coordinates x^i, one has $e_i = \partial/\partial x^i$, $\omega^i = dx^i$, $\omega_i^j = \Gamma_{ik}^j dx^k$, $(\Gamma_{ik}^j)_{\lambda(0)} = 0$. Along the above path, $\omega^i = X^i ds$, $ds = s - 0 = s$, and

$$(e_i)_{\lambda(s)} = (e_i)_{\lambda(0)} + (de_i)_{\lambda(0)} + o_i,$$

where $de_i = e_j\omega_i^j + h_{ij}X^i s$, $\lim\limits_{s\to 0} \frac{1}{s} o_i = 0$.

Since for $t = t^i e_i$ with arbitrary constant t^i, there holds $dt = t^j\omega_j^i e_i + t^i h_{ij} X^j s + t^i o_i$ with $(\omega_j^i)_{\lambda(0)} = 0$, it follows that for $t + v$ in Corollary 8.4.2 one can take $(t + dt)_{\lambda(0)} = t + t^i (hij)_{\lambda(0)} X^j s + t^i o_i^{norm}$, which yields $v = t^i h_{ij} X^j s + t^i o_i^{norm}$.

Hence, the system of sides of Lemma 8.4.1 is determined from $\partial\Psi/\partial t^i = 0$, where $\Psi = \langle v, v\rangle + (\langle t, t\rangle - 1)$. Then dividing by s^2 and taking the limit as $s \to 0$ one gets

$$t^i(\langle h_{ij}, h_{kl}\rangle X^j X^l - \lambda g_{ik}) = 0, \tag{8.4.3}$$

and thus

$$\det |\langle h_{ij}, h_{kl}\rangle X^j X^l - \lambda g_{ik}| = 0.$$

Hence the solutions do indeed depend only on x and X.

The directions defined by (8.4.3), and thus by Lemma 8.4.3, are called the *principal directions with respect to $X \in T_x M^m$*.

Eliminating λ from two arbitrary equations of (8.4.3), it follows that for each of these directions

$$t^i t^p \langle h_{ij}, \quad g_{pq} h_{kl} - g_{pk} h_{ql}\rangle X^j X^l = 0. \tag{8.4.4}$$

Remark 8.4.4. The concept of a system of mutually orthogonal principal directions with respect to a given tangent direction was introduced by Wong in [Won 52] in the special case $m = 2$, $n = 4$; it was then generalized to the situation of Lemma 8.4.3 by Mullari in [Mu 61], [Mu 62a], where equations (8.4.4) were also deduced.

These same authors also introduced in these articles the concept of a system of *absolute principal directions* of M^m in E^n (in [Won 52] for the case $m = 2, n = 4$ only), which they defined as the system of mutually orthogonal directions formed by the principal directions with respect to *all* tangent directions of M^m at a point $x \in M^m$, if such a system does exist.

It is seen from (8.4.4) that these directions are defined by the tangent vectors $t = e_i t^i$ satisfying

$$t^i t^p [\langle h_{ij}, \quad g_{pq} h_{kl} - g_{pk} h_{ql}\rangle + \langle h_{il}, \quad g_{pq} h_{kj} - g_{pk} h_{qj}\rangle] = 0. \tag{8.4.5}$$

Due to Lemmas 8.4.1 and 8.4.3, the absolute principal directions are mutually orthogonal if they exist; therefore the basis vectors e_a $(a = 1, \dots, m)$ of the orthonormal frame adapted to M^m can be taken in these directions. Then equations (8.4.5) are satisfied by $t^i = \delta^i_a$, $g_{pq} = \delta_{pq}$ and reduce to

$$\langle h_{aj}, \quad \delta_{aq} h_{kl} - \delta_{ak} h_{ql}\rangle + \langle h_{al}, \quad \delta_{aq} h_{kj} - \delta_{ak} h_{qj}\rangle = 0,$$

and thus to trivial identities, except for $a = q \neq k$:

$$\langle h_{aj}, \quad h_{kl}\rangle + \langle h_{al}, \quad h_{kj}\rangle = 0 \quad (a \neq k); \tag{8.4.6}$$

here $a = k \neq q$ gives the same, only k is to be replaced by q.

Lemma 8.4.5. *If M^m in E^n has a field of absolute principal directions which form a conjugate net, then M^m has flat connection $\bar\nabla$. Conversely, if M^m has flat connection $\bar\nabla$, then its principal basis at an arbitrary point $x \in M^m$ consists of vectors with absolute principal directions of M^m.*

Proof. For the adapted frame above one has (8.4.6), and since the net is conjugate, one has $h_{aj} = k_a \delta_{aj}$, etc. Hence $\langle k_a, k_k \rangle = 0$ $(a \neq k)$. Then in (2.1.9) (with $c = 0$) $h^\alpha_{ij} = k^\alpha_i \delta_{ij}$ and thus $R^j_{i,pq} = R^\beta_{\alpha,pq} = 0$; hence this M^m has flat $\bar{\nabla}$.

The argument works in reverse, and hence the converse is also true.

The last lemma states, in other words, that the class of those submanifolds M^m in E^n which carry a conjugate net of lines with absolute principal directions coincides with the class of M^m having flat $\bar{\nabla}$, i.e., with a subclass of semiparallel submanifolds in E^n.

The rest of the present section presents some results obtained in [Mu 61] and [Lu 87b] about the structure of submanifolds belonging to this subclass.

Recall that a normally flat submanifold M^m having m distinct orthogonal nonzero outer principal curvature vectors at each point is said to be of *Cartan type* (see Remark 5.4.3). Due to (5.1.4) they have flat $\bar{\nabla}$.

Theorem 8.4.6. *Let a Riemannian submanifold M^m in ${}_sE^n$ with flat van der Waerden–Bortolotti connection $\bar{\nabla}$ be not of Cartan type, and suppose that on an open set $U \subset M^m$ there are exactly $m^* = m - q$ nonzero vectors, $0 < m^* < m$, among its principal curvature vectors. Then there exists a normally flat Riemannian submanifold $M^{m^*} \subset U$ of Cartan type, with its normal field ν^q of q-dimensional Euclidean subspaces which is parallel with respect to $\nabla^{*\perp}$, so that U consists of open subsets of q-planes through points $x \in M^{m^*}$, in the direction of ν^q_x.*

Conversely, if one has a normally flat Riemannian M^{m^*} in ${}_sE^n$ with a field ν^q of q-dimensional Euclidean subspaces normal to M^{m^*} and parallel with respect to $\nabla^{*\perp}$, then the submanifold M^m, $m = m^* + q$, in ${}_sE^n$, formed by all points of q-planes through arbitrary $x \in M^{m^*}$ in direction of ν^q_x, is a Riemannian submanifold M^m in ${}_sE^n$ with flat $\nabla^\perp$. This M^m has flat $\bar{\nabla}$ if and only if the components of the principal curvature vectors of M^{m^*} at every point $x \in M^m$, orthogonal to the q-plane at this x, are mutually orthogonal.

Proof. Let the principal curvature vectors $k_1, \dots, k_m$ of M^m be renumbered so that the first m^* of them are nonzero and the next q are zero: $k_{i^*} \neq 0$ for $1 \leq i^* \leq m^*$ and $k_{(0)} = k_u = 0$ for $m^* + 1 \leq u \leq m$. Since M^m with flat $\bar{\nabla}$ is assumed to be semiparallel, the nonzero $k_1, \dots, k_{m^*}$ are mutually orthogonal, due to Theorem 5.1.2. One can use (5.1.2), which for $i = u$ yields $L_{ui^*} = 0$, and thus from (5.1.6) $\lambda_{(0)i^*} = 0, \quad \omega^u_{i^*} = \lambda_{i^*u}\omega^{i^*}$.

The last relation together with $\omega^{i^*}_u = -\omega^u_{i^*}$ shows that the differential system $\omega^{i^*} = 0$ on M^m is completely integrable. The leaves of the corresponding foliation are q-planes, because for them $dx = \omega^u e_u$, $de_u = \omega^v_u e_v$.

The differential system $\omega^u = 0$ on M^m is completely integrable as well. Indeed, $d\omega^u = \omega^{i^*} \wedge \omega^u_{i^*} + \omega^v \wedge \omega^u_v = \omega^v \wedge \omega^u_v$. The leaves of this foliation are m^*-dimensional submanifolds M^{m^*}. Consider one of them. On it one has

$$de_{i^*} = e_{j^*}\omega^{j^*}_{i^*} + e_u\omega^u_{i^*} + k_{i^*}\omega^{i^*};$$

hence

$$h_{i^*j^*} = \left(k_{i^*} + \sum_{u=m^*+1}^{m} e_u \lambda_{i^*u}\right) \delta_{i^*j^*}.$$

Now it follows easily that this M^{m^*} has flat normal connection $\nabla^{*\perp}$.

Along this M^{m^*} one has $de_u = -\sum_{i^*=1}^{m^*} \lambda_{i^*u}\omega^{i^*} e_{i^*} + \omega_u^v e_v$, and therefore the field ν^q of q-dimension subspaces normal to M^{m^*} subspaces spanned by e_u is parallel with respect to $\nabla^{*\perp}$, because the normal component of de_u outside of ν^q is zero (see [LCh 81], Proposition 1).

The argument in the reverse direction works, verifying the converse assertion; for details see [Lu 87b].

Now consider a submanifold M^m in E^n with flat $\bar{\nabla}$ of Cartan type. Its m principal curvature vectors are all nonzero and mutually orthogonal, as follows from Theorem 5.1.2. Suppose that among the latter there exist m^* vectors, $m^* < m$, so that the field of m^*-dimensional tangent subspaces spanned by the corresponding principal directions is parallel with respect to the Riemannian connection ∇ induced on this submanifold M^m. This parallel property means, as is known, that for an arbitrary vector field belonging to the above-considered field of m^*-dimensional tangent subspaces, its differential has components only in this latter field and normal to M^m.

Analoguously, a field of p-dimensional normal subspaces on M^m is called parallel with respect to the normal connection $\nabla^\perp$ of M^m if for an arbitrary vector field belonging to it, the differential of the latter has components only in it and in the tangent subspace of M^m (see [LCh 81]).

Theorem 8.4.7. *Let M^m in E^n be a submanifold with flat van der Waerden–Bortolotti connection $\bar{\nabla}$ of Cartan type. The following four assertions are equivalent to each other:*

(i) *The field of m^*-dimensional tangent subspaces spanned by unit vectors e_{i^*} having the principal directions of M^m, $1 \le i^* \le m^* < m$, is parallel with respect to ∇.*
(ii) *The field of $(m - m^*)$-dimensional tangent subspaces totally orthogonal to the previous field (thus spanned by unit vectors $e_{i'}$ in the remaining principal directions) is parallel with respect to ∇.*
(iii) *The field of m^*-dimensional normal subspaces spanned by principal curvature vectors $k_1, \dots, k_{m^*}$ is parallel with respect to the connection $\nabla_0^\perp$ induced in the bundle of first normal subspaces of M^m.*
(iv) *The field of $(m - m^*)$-dimensional normal subspaces spanned by the remaining principal curvature vectors $k_{m^*+1}, \dots, k_m$ is parallel with respect to $\nabla_0^\perp$.*

Proof. The orthonormal frame can be adapted to this submanifold M^m so that e_i are in the principal directions and e_{m+i} are in the directions of the corresponding principal curvature vectors; let e_ρ be the remaining basis vectors of the frame normal to M^m. Then $k_i = \kappa_i e_{m+i}$, thus $h_{ij}^{m+k} = \kappa_i \delta_{ij} \delta_j^k$, and $h_{ij}^\rho = 0$. Substituting this into (2.2.3), one obtains $k_i \delta_{ij} \omega_{m+i}^\rho = h_{ijk}^\rho \omega^k$. Since the coefficients on the right side are symmetric with respect to lower indices, only h_{iii}^ρ can be nonzero among them. It follows that

$$\omega^{\rho}_{m+i} = l^{\rho}_{i}\omega^{i}.$$

Substitution into (4.3.3) and (4.3.4) implies, after some straightforward calculations, that

$$\omega^{j}_{i} = \kappa_i c^{j}_{i}\omega^{i} - \kappa_j c^{i}_{j}\omega^{j}, \tag{8.4.7}$$

$$\omega^{m+j}_{m+i} = \kappa_i c^{i}_{j}\omega^{i} - \kappa_j c^{j}_{i}\omega^{j}, \tag{8.4.8}$$

$$d\kappa_i = \gamma_i\omega^{i} + \sum_{j\neq i}\kappa_i^2 c^{j}_{i}\omega^{j}, \tag{8.4.9}$$

where $c^{j}_{i} = \kappa_i^{-2}L^{m+i}_{ij}$ and $\gamma_i = K^{m+i}_{i}$.

From these formulas (8.4.7) and (8.4.8) it follows that each of the four assertions of the theorem is equivalent to the system of equations

$$c^{j'}_{i*} = 0, \quad c^{i*}_{j'} = 0.$$

Remark 8.4.8. An easy computation shows that (8.4.7) and (8.4.9) yield $d(\kappa_i\omega^{i}) = 0$; therefore, at least locally, there exist functions u^{i} on M^m with flat $\bar{\nabla}$ of Cartan type, so that $\kappa_i\omega^{i} = du^{i}$. In these parameters u^{i}, the system (8.4.7)–(8.4.9) takes a simpler form, e.g., (8.4.7) is

$$\omega^{j}_{i} = c^{j}_{i}du^{i} - c^{i}_{j}du^{j}.$$

9

Umbilic-Likeness of Main Symmetric Orbits

As mentioned in the introduction, the concept of umbilic-likeness (introduced in [Lu 96c, d]) is defined as follows: A symmetric orbit is called *umbilic-like* if the second-order envelope of a family of submanifolds similar to this orbit is a single symmetric orbit or its open subset. (Note that in general such an envelope is a semiparallel submanifold.) The guiding example is the sphere, being the only submanifold all of whose points are umbilic, i.e., at each of its points the submanifold is second-order tangent to a sphere.

In this chapter the phenomenon of umbilic-likeness is investigated in detail.

9.1 Two Kinds of Symmetric Orbits

According to Theorem 8.1.2, a semiparallel Riemannian submanifold M^m in Euclidean space E^n carries a conjugate system (in the sense of [AG 93], Chapter 3) of eigenfoliations of the mean shape operator A_H. By Theorem 4.5.5, such an M^m is a second-order envelope of the corresponding parallel submanifolds, which are, if complete, product submanifolds

$$M^{m_1} \times \cdots \times M^{m_s} \times S^1(c_{s+1}) \times \cdots \times S^1(c_{s+q}) \times E^{m_0},$$

where $M^{m_1}, \ldots, M^{m_s}$ are main symmetric orbits (see Theorem 8.2.5 by Ferus). Each of the latter is minimal in a hypersphere and is a standard imbedding of a symmetric R-space.

If M^m is normally flat in E^n, then these main orbits are spheres, as is shown in [Wa 73] (see above Section 5.2). According to Theorem 5.4.1, the eigenleaves of a general normally flat semisymmetric submanifold M^m enveloped by these spheres are then these spheres themselves. At the same time, the circle components $S^1(c_{s+1}), \ldots, S^1(c_{s+q})$ envelop some curves of M^m that are orthogonal trajectories of these spheres.

This circumstance has a simple explanation, already noted in the introduction to this chapter, which is more thoroughly explained as follows.

Ü. Lumiste, *Semiparallel Submanifolds in Space Forms*,
DOI 10.1007/978-0-387-49913-0_10, © Springer Science+Business Media, LLC 2009

A point of a submanifold M^m in E^n is umbilic if and only if M^m has second-order tangency at this point with some sphere $S^m(c)$ (see Sections 2.4 and 4.5). It follows that M^m is totally umbilic if and only if M^m is a second-order envelope of m-dimensional spheres. But it is known that if $m > 1$, then a totally umbilic M^m in E^n is a sphere (see Propositions 3.1.2 and 5.2.2), thus the enveloping is trivial: the family of enveloped spheres reduces to a single sphere, the envelope itself.

For $m = 1$ the situation is different; here every curve M^1 in E^n is the second-order envelope of its circles of curvature.

Situations similar to these two occur also for other symmetric orbits. For instance, Theorem 4.6.1 above states that the second-order envelope of Segre submanifolds $S_{(p,q)}(k)$ with variable k is, in the case $p > 1$ and $q > 1$, a single Segre submanifold, and then k is a constant; and in the case $p = 1$ and $q > 1$ this envelope is a logarithmic tube. From Proposition 6.3.1 and Corollary 6.3.6 it follows that in $N^5(c)$, so also in E^5, a second-order envelope of Veronese surfaces is a single Veronese surface, and, due to Theorem 6.4.1, there exists in E^7 an envelope of Veronese surfaces that do not reduce to a single Veronese surface.

The concept of umbilic-likeness introduced above is useful in order to distinguish these two situations.

If for a main symmetric orbit, every second-order envelope of submanifolds congruent or similar to this orbit is a single such orbit or its open subset, then the orbit is umbilic-like.

So a Segre orbit without circular generators is umbilic-like, independently of the dimension of the ambient space, whereas a Segre orbit with circular generators is not umbilic-like. A Veronese surface (orbit) is umbilic-like in E^5, but in E^n with $n > 5$ it is not umbilic-like.

The problem in this chapter is to decide which main symmetric orbits are umbilic-like and which are not. Note that this is related to the extrinsic analogue of the Nomizu problem for semisymmetric Riemannian manifolds. Namely, in 1968 K. Nomizu [No 68] conjectured that if such a manifold is complete and irreducible, then it reduces to a locally symmetric manifold. This conjecture was refuted in general in [Sek 72] and [Ta 72], and so the question arose as to whether or not there are some particular cases where it is still true.

Now instead of semisymmetric (resp. locally symmetric) Riemannian manifolds, one can consider semiparallel (resp. parallel) submanifolds. Therefore, the following problem, called the *modified Nomizu problem*, arises (see [Lu 92a], [Lu 95b]): When does a semiparallel submanifold reduce to a parallel one?

The question of deciding the umbilic-likeness of main symmetric orbits is a particular case of this general problem.

Let us consider from this point of view some main symmetric orbits and their second-order envelopes in more detail.

From (4.6.1) it is seen that for a *Segre submanifold* $S_{(p,q)}(k)$ in E^{pq+m+1} the vector-valued second fundamental form has the components

$$h_{i_1 j_1} = (e_{m+1}k - cx)\delta_{i_1 j_1}, \quad h_{i_1 j_2} = e_{i_1 j_2}k, \quad h_{i_2 j_2} = (e_{m+1}k - cx)\delta_{i_2 j_2},$$

where $m = p+q, c = k^2$, and $x, e_{m+1}, e_{i_1 j_2}$ are mutually orthogonal vectors normal to $S_{(p,q)}(k)$ (see Section 3.2). Hence the mean curvature vector is $H = e_{m+1}k - cx$ and thus $h^H_{i_1 j_1} = 2c\delta_{i_1 j_1}, h^H_{i_1 j_2} = 0, h^H_{i_2 j_2} = 2c\delta_{i_2 j_2}$. This, together with the results in Section 3.2, shows that $S_{(p,q)}(k)$ is a main symmetric orbit.

As mentioned above, guided by Theorem 4.6.1, it is umbilic-like if $p > 1, q > 1$, and not umbilic-like if one of p, q is 1.

Next, let the *Plücker submanifold* $G^{2,p}(r)$ in $\wedge^2(\mathbb{R}^p)$ be considered. Here $\wedge^2(\mathbb{R}^p)$ can be made into a Euclidean vector space $\mathbb{R}^{\frac{1}{2}p(p-1)}$ (see Example 1.5.4 and Section 3.2; also [Lu 92a], [Lu 96b], [Ste 64], Chapter I, Section 4). The action of $SO(p, \mathbb{R})$ by isometries on Euclidean $\mathbb{R}^p$ also induces an action on $\wedge^2(\mathbb{R}^p)$, called the *Plücker action*, and $G^{2,p}(r)$ is its orbit, which is by Theorem 3.2.3 a parallel submanifold.

Moreover, this Plücker orbit is a main symmetric orbit. Indeed, from (3.2.8) it follows that the mean curvature vector of this orbit is $H = -\frac{1}{r}x$ and thus

$$h^H_{uv} = \frac{1}{r^2}x^2\delta_{uv}, \quad h^H_{u\bar{v}} = 0, \quad h^H_{\bar{u}\bar{v}} = \frac{1}{r^2}x^2\delta_{\bar{u}\bar{v}}.$$

Therefore, the mean shape operator A_H has only one nonzero eigenvalue $\frac{1}{r^2}x^2$, and this characterizes main symmetric orbits.

The fact that this orbit is umbilic-like will be proved in the next section.

Also the *Veronese submanifold* (see Sections 3.3 and 4.7) is a main symmetric orbit. Let us consider such a submanifold of dimension m in Euclidean space, i.e., let $s = c = 0$. Indeed, by Theorem 3.3.1 it is a parallel orbit in $E^{\frac{1}{2}m(m+3)}$, for which due to (3.3.4) one has

$$\langle h_{ij}, h_{kl}\rangle = (2\delta_{ij}\delta_{kl} + \delta_{ik}\delta_{jl} + \delta_{il}\delta_{jk})\frac{1}{r^2},$$

because now $h^*_{ij} = h_{ij}$ (cf. (4.4.13)); here the notation $\kappa = \frac{1}{r^2}$ is used. Such a Veronese orbit will be denoted by $V^m(r)$ below.

From the last formula it is seen that the vectors h_{jk} $(j \neq k)$ are mutually orthogonal vectors normal to $V^m(r)$, and $h_{11}, \ldots, h_{mm}$ are orthogonal to the vectors h_{jk} $(j \neq k)$ and form a regular simplex part in the space normal to $V^m(r)$ at an arbitrary point; the side length of this simplex is $2r^{-1}$ (cf. (4.4.8)–(4.4.11)). Here $H = \frac{1}{m}\sum_{i=1}^m h_{ii}$; therefore,

$$h^H_{11} = \cdots = h^H_{mm} = 2\frac{m+1}{m}r^{-2}, \quad h^H_{ij} = 0 \quad (i \neq j),$$

and thus $V^m(r)$ is indeed a main symmetric orbit.

The problem of its umbilic-likeness is more complicated than for the preceding examples, and will be studied separately in Section 9.7 below.

9.2 Umbilic-Likeness of Plücker Orbits

As stated above, the Plücker submanifold $G^{2,p}(r)$ is an orbit of the Plücker action of $SO(p,\mathbb{R})$ in $\mathbb{R}^{\frac{1}{2}p(p-1)}$. More precisely, it is the standardly imbedded symmetric R-space $SO(p,\mathbb{R})/SO(2,\mathbb{R})\times SO(p-2,\mathbb{R})$.

Theorem 9.2.1. *The Plücker orbit $G^{2,p}(r)$ is umbilic-like; i.e., a second-order envelope M^m of Plücker orbits $G^{2,p}(r)$ with variable r and with $m = 2(p-2)$ in an n-dimensional Euclidean space, $n \geq \frac{1}{2}p(p-1)$, is a single Plücker orbit or its open subset.*

Proof. If $p \leq 4$, the assertion follows from the known results (see, e.g., [Wo 72], 9.2) that for $p = 3$ the Plücker orbit $G^{2,3}(r)$ is equivalent to the sphere $S^2(r) \subset \mathbb{R}^3$, which is umbilic, and for $p = 4$ the Plücker orbit $G^{2,4}(r)$ is equivalent to the product $S^2(\sqrt{r}) \times S^2(\sqrt{r}) \subset \mathbb{R}^3 \times \mathbb{R}^3$, which is umbilic-like, as shown in [Ri 88].

So let $p > 4$ and let M^m be the postulated envelope. The components of its second fundamental form with respect to the adapted frame are the same as for $G^{2,p}(r)$ and thus are given in (3.2.8), where one has $x = e_1 \wedge e_2 = E_{[12]}$. Using ω in place of the notation θ of Section 3.2, it follows that this envelope M^m is defined by the system

$$\omega^{[12]} = \omega^{[uv]} = \omega^{\xi} = 0, \tag{9.2.1}$$

$$\omega_u^{[12]} = -\frac{1}{r}\omega^u, \quad \omega_{\bar{u}}^{[12]} = -\frac{1}{r}\omega^{\bar{u}}, \tag{9.2.2}$$

$$\omega_u^{[vw]} = \frac{1}{2r}(\delta_u^v \omega^{\bar{w}} - \delta_u^w \omega^{\bar{v}}), \quad \omega_{\bar{u}}^{[vw]} = \frac{1}{2r}(\delta_u^w \omega^v - \delta_u^v \omega^w), \tag{9.2.3}$$

$$\omega_u^{\xi} = \omega_{\bar{u}}^{\xi} = 0, \tag{9.2.4}$$

where ξ runs $\{\frac{1}{2}p(p-1)+1, \dots, n\}$.

Exterior differentiation of equations (9.2.1) leads to identities, but those of (9.2.2) lead to

$$d\ln r \wedge \omega^u + \omega_{[uv]}^{[12]} \wedge \omega^{\bar{v}} = 0, \quad d\ln r \wedge \omega^{\bar{u}} - \omega_{[uv]}^{[12]} \wedge \omega^v = 0.$$

Since $p > 4$, u takes more than two values here, and the same holds for $\bar{u}$; hence these exterior equations imply that

$$r = \text{const}, \quad \omega_{[uv]}^{[12]} = 0. \tag{9.2.5}$$

Equations (9.2.4) give

$$\omega_{[uv]}^{\xi} \wedge \omega^{\bar{v}} = 0, \quad \omega_{[uv]}^{\xi} \wedge \omega^v = 0,$$

thus

$$\omega_{[uv]}^{\xi} = 0. \tag{9.2.6}$$

The more complicated equations (9.2.3) lead to

$$\phi^{vw}_{ut} \wedge \omega^t + \varphi^{vw}_{ut} \wedge \omega^{\bar{t}} = 0, \quad \varphi^{vw}_{tu} \wedge \omega^t + \psi^{vw}_{ut} \wedge \omega^{\bar{t}} = 0,$$

where

$$\phi^{vw}_{ut} = \omega^{\bar{v}}_u \delta^w_t - \omega^{\bar{w}}_u \delta^v_t + \omega^{\bar{v}}_t \delta^w_u - \omega^{\bar{w}}_t \delta^v_u, \tag{9.2.7}$$

$$\varphi^{vw}_{ut} = \omega^{[vw]}_{[ut]} - \omega^v_u \delta^w_t + \omega^w_u \delta^v_t - \omega^{\bar{w}}_{\bar{t}} \delta^v_u + \omega^{\bar{v}}_{\bar{t}} \delta^w_u, \tag{9.2.8}$$

$$\psi^{vw}_{ut} = \omega^{\bar{u}}_v \delta^t_w - \omega^{\bar{u}}_w \delta^t_v + \omega^{\bar{t}}_v \delta^u_w - \omega^{\bar{t}}_w \delta^u_v \tag{9.2.9}$$

are all antisymmetric in v, w, and ϕ^{vw}_{ut}, ψ^{vw}_{ut} are also symmetric in u, t.

By Cartan's lemma,

$$\phi^{vw}_{ut} = A^{vw}_{uts} \omega^s + B^{vw}_{uts} \omega^{\bar{s}}, \tag{9.2.10}$$

$$\varphi^{vw}_{ut} = B^{vw}_{ust} \omega^s + C^{vw}_{uts} \omega^{\bar{s}}, \tag{9.2.11}$$

$$\psi^{vw}_{ut} = C^{vw}_{ust} \omega^s + D^{vw}_{uts} \omega^{\bar{s}}. \tag{9.2.12}$$

Here A^{vw}_{uts} and D^{vw}_{uts} are symmetric in u, t, s, B^{vw}_{uts} and C^{vw}_{ust} are symmetric in u, t, and all of them are antisymmetric in v, w.

If $p \geq 6$ then t, u, v, w can take four different values, for which (9.2.7) and (9.2.9) imply that $\phi^{vw}_{ut} = \psi^{vw}_{ut} = 0$, and so in (9.2.10) and (9.2.12) $A^{vw}_{uts} = B^{vw}_{uts} = C^{vw}_{uts} = D^{vw}_{uts} = 0$. The result is that also $\varphi^{vw}_{ut} = 0$, and hence

$$\omega^{[vw]}_{[ut]} = 0 \quad (t, u, v, w \text{ distinct}). \tag{9.2.13}$$

If $p = 5$, then u, v, w can take three different values, for which (9.2.7) and (9.2.10) give $\phi^{vw}_{uw} = \omega^{\bar{v}}_u = A^{vw}_{uws} \omega^s + B^{vw}_{uws} \omega^{\bar{s}}$. On the other hand, if $u = w = t \neq v$ then $\phi^{vu}_{uu} = 2\omega^{\bar{v}}_u = A^{vu}_{uus} \omega^s + B^{vu}_{uus} \omega^{\bar{s}}$. Hence

$$2A^{vw}_{uws} = A^{vu}_{uus}. \tag{9.2.14}$$

In particular, $2A^{vw}_{uww} = A^{vu}_{uuw}$. Here the roles of u and w can be interchanged; therefore, $A^{vw}_{wwu} = 2A^{vu}_{wuu}$. Due to symmetry in the lower indices, $A^{vw}_{wwu} = 2A^{vu}_{uuw} = 4A^{vw}_{uww} = 4A^{vw}_{wwu}$, thus $A^{vw}_{uww} = 0$ if u, v, w are distinct.

It follows that for every four distinct values u, v, w, s, the right side of (9.2.14) vanishes, and thus also $A^{vw}_{uws} = 0$. The result is that

$$\omega^{\bar{v}}_u = \Gamma^{\bar{v}}_{uu} \omega^u + \Gamma^{\bar{v}}_{uv} \omega^v + B^{vw}_{uws} \omega^{\bar{s}},$$

where u, v, w are distinct and $\Gamma^{\bar{v}}_{uu} = A^{vw}_{uwu}$, $\Gamma^{\bar{v}}_{uv} = A^{vw}_{uwv}$.

The same argument applied to ψ^{vw}_{ut} yields the result

$$\omega^{\bar{v}}_u = C^{vw}_{usw} \omega^s + \Gamma^{\bar{u}}_{v\bar{v}} \omega^{\bar{v}} + \Gamma^{\bar{u}}_{v\bar{u}} \omega^{\bar{u}}.$$

Hence

$$\omega^{\bar{v}}_u = \Gamma^{\bar{v}}_{uu} \omega^u + \Gamma^{\bar{v}}_{uv} \omega^v + \Gamma^{\bar{u}}_{v\bar{v}} \omega^{\bar{v}} + \Gamma^{\bar{u}}_{v\bar{u}} \omega^{\bar{u}},$$

where $u \neq v$. Thus for distinct u, v, w

$$B^{vw}_{uwv} = \Gamma^{\bar{u}}_{v\bar{v}}, \quad B^{vw}_{uwu} = \Gamma^{\bar{u}}_{v\bar{u}}, \quad C^{vw}_{uuw} = \Gamma^{\bar{v}}_{uu}, \quad C^{vw}_{uvw} = \Gamma^{\bar{v}}_{uv}, \tag{9.2.15}$$

and if u, v, s are also distinct, then

$$B^{vw}_{uws} = C^{vw}_{usw} = 0.$$

Furthermore, $\varphi^{uw}_{uw} = 0$, which implies $B^{vw}_{usw} = C^{uw}_{uws} = 0$.

One can see that for three distinct $u = v, w, t$

$$\phi^{uw}_{ut} = -\omega^{\bar{w}}_{t}, \quad \psi^{uw}_{ut} = -\omega^{\bar{t}}_{w}.$$

Therefore, $\Gamma^{\bar{t}}_{w\bar{w}} = -B^{uw}_{utw} = 0$, $\Gamma^{\bar{t}}_{ww} = -C^{uw}_{uwt} = 0$,

$$A^{uw}_{utw} = -\Gamma^{\bar{w}}_{tw}, \quad D^{uw}_{utw} = -\Gamma^{\bar{w}}_{t\bar{w}}. \tag{9.2.16}$$

Now

$$\phi^{uw}_{uw} = -\Gamma^{\bar{w}}_{sw}\omega^{s} + B^{uw}_{uws}\omega^{\bar{s}}, \quad \psi^{uw}_{uw} = C^{uw}_{usw}\omega^{s} - \Gamma^{\bar{w}}_{s\bar{w}}\omega^{\bar{s}},$$

but for the other side

$$\phi^{uw}_{uw} = \psi^{uw}_{uw} = \omega^{\bar{u}}_{u} - \omega^{\bar{w}}_{w}.$$

Thus

$$\omega^{\bar{u}}_{u} - \omega^{\bar{w}}_{w} = -\Gamma^{\bar{w}}_{sw}\omega^{s} - \Gamma^{\bar{w}}_{s\bar{w}}\omega^{\bar{s}}.$$

Similarly,

$$\omega^{\bar{w}}_{w} - \omega^{\bar{u}}_{u} = -\Gamma^{\bar{u}}_{su}\omega^{s} - \Gamma^{\bar{u}}_{s\bar{u}}\omega^{\bar{s}};$$

hence $\Gamma^{\bar{w}}_{sw} = -\Gamma^{\bar{u}}_{su}$, $\Gamma^{\bar{w}}_{s\bar{w}} = -\Gamma^{\bar{u}}_{s\bar{u}}$ for every two different u and w. In general, if for some quantities γ_i one has $\gamma_u = -\gamma_w$ for every pair of distinct u, w, then for three distinct u, v, w it follows that $\gamma_u = -\gamma_v = -(-\gamma_w) = \gamma_w = -\gamma_u$, and thus all $\gamma_u = 0$. Therefore,

$$\Gamma^{\bar{u}}_{su} = \Gamma^{\bar{u}}_{s\bar{u}} = 0,$$

and consequently all quantities in (9.2.15) and (9.2.16) are zero. All this leads to

$$\omega^{\bar{v}}_{u} = \omega^{\bar{u}}_{u} - \omega^{\bar{v}}_{v} = \omega^{\bar{v}}_{\bar{u}} - \omega^{v}_{u} = 0 \quad (u \neq v),$$

$$\omega^{[uw]}_{[uv]} - \omega^{w}_{v} = 0 \quad (u, v, w \text{ distinct}),$$

but this together with (9.2.13) verifies the assertion.

Remark 9.2.2. Theorem 9.2.1 was first proved in [Lu 92a] and then announced in [Lu 96b].

9.3 Unitary Orbits of the Plücker Action

The Plücker action in $\wedge^2(\mathbb{R}^p) = \mathbb{R}^{\frac{1}{2}p(p-1)}$ is defined by the differential equations

$$dE_{[ij]} = E_{[kj]}\theta_i^k + E_{[ik]}\theta_j^k,$$

where $E_{[ij]} = \frac{1}{r} e_i \wedge e_j$ are elements of the moving orthonormal basis in $\wedge^2(\mathbb{R}^p)$ corresponding to a moving orthonormal basis $\{e_1, \dots, e_p\}$ in $\mathbb{R}^p$.

An arbitrary element of $\wedge^2(\mathbb{R}^p)$ is then given by $c^{ij} E_{[ij]}$. If the c^{ij} are some constants satisfying $c^{ij} = -c^{ji}$, then this element describes an orbit of the Plücker action. It is known that $\{e_1, \dots, e_p\}$ can be chosen so that for this element, in general,

$$c^{ij} = \sum_{\rho=1}^{\nu} c_\rho \delta_{2\rho-1}^{[i} \delta_{2\rho}^{j]}, \quad c_\rho > 0, \quad 2\nu \le p$$

(see [Sch 24], Chapter I, Section 16; a more explicit presentation is given in [Shi 61], Sections 10 and 23). With this choice of basis, the element $c^{ij} E_{[ij]}$ assumes the canonical form

$$\sum_{\rho=1}^{\nu} c_\rho E_{[2\rho-1,2\rho]}. \tag{9.3.1}$$

Theorem 9.3.1. *If* $n = 2\nu$, *then the orbit of the Plücker action described by* (9.3.1) with $c_1 = \dots = c_\nu \neq 0$ is a symmetric orbit.

Proof. Let the common value of all c_ϱ be denoted by c. Then by (9.3.1),

$$x = c \sum_{\varrho=1}^{\nu} E_{[2\varrho-1,2\varrho]}$$

and the differential equation defining the Plücker action implies

$$dx = \sum (I_{[\varrho\sigma]_1} \omega^{[\varrho\sigma]_1} + I_{[\varrho\sigma]_2} \omega^{[\varrho\sigma]_2}), \tag{9.3.2}$$

summed over ϱ, σ for $1 \le \varrho < \sigma \le \nu$, and

$$I_{[\varrho\sigma]_1} = \frac{1}{\sqrt{2}}(E_{[2\varrho-1,2\sigma-1]} - E_{[2\varrho,2\sigma]}) = -I_{[\sigma\varrho]_1}, \tag{9.3.3}$$

$$I_{[\varrho\sigma]_2} = \frac{1}{\sqrt{2}}(E_{[2\varrho-1,2\sigma]} + E_{[2\varrho,2\sigma-1]}) = -I_{[\sigma\varrho]_2} \tag{9.3.4}$$

are mutually orthogonal unit vectors of $\mathbb{R}^{\nu(2\nu-1)}$ (i.e., elements of $\wedge^2(\mathbb{R}^p)$) tangent to the orbit, and

$$\omega^{[\varrho\sigma]_1} = c\sqrt{2}(\theta_{2\varrho-1}^{2\sigma} + \theta_{2\varrho}^{2\sigma-1}) = -\omega^{[\sigma\varrho]_1}, \tag{9.3.5}$$

$$\omega^{[\varrho\sigma]_2} = -c\sqrt{2}(\theta^{2\sigma-1}_{2\varrho-1} - \theta^{2\sigma}_{2\varrho}) = -\omega^{[\sigma\varrho]_2}. \tag{9.3.6}$$

The same differential equation yields

$$\begin{aligned}
dI_{[\varrho\sigma]_1} = {} & I_{[\varrho\sigma]_2}\omega^{[\varrho\sigma]_2}_{[\varrho\sigma]_1} \\
& + \sum_{\tau\neq\varrho,\sigma} (I_{[\tau\sigma]_1}\omega^{[\tau\sigma]_1}_{[\varrho\sigma]_1} + I_{[\varrho\tau]_1}\omega^{[\varrho\tau]_1}_{[\varrho\sigma]_1} + I_{[\tau\sigma]_2}\omega^{[\tau\sigma]_2}_{[\varrho\sigma]_1} + I_{[\varrho\tau]_2}\omega^{[\varrho\tau]_2}_{[\varrho\sigma]_1}) \\
& + \frac{1}{2c\sqrt{2}} \sum_{\tau\neq\varrho,\sigma} (J_{\langle\tau\sigma\rangle_1}\omega^{[\tau\varrho]_2} - J_{\langle\varrho\tau\rangle_1}\omega^{[\sigma\tau]_2} + J_{\langle\tau\sigma\rangle_2}\omega^{[\tau\varrho]_1} + J_{\langle\varrho\tau\rangle_2}\omega^{[\sigma\tau]_1}) \\
& - \frac{1}{2c}(J_{\langle\varrho\rangle} + J_{\langle\sigma\rangle})\omega^{[\varrho\sigma]_1}, \\
dI_{[\varrho\sigma]_2} = {} & -I_{[\varrho\sigma]_1}\omega^{[\varrho\sigma]_2}_{[\varrho\sigma]_1} \\
& + \sum_{\tau\neq\varrho,\sigma} (I_{[\tau\sigma]_1}\omega^{[\tau\sigma]_1}_{[\varrho\sigma]_2} + I_{[\varrho\tau]_1}\omega^{[\varrho\tau]_1}_{[\varrho\sigma]_2} + I_{[\tau\sigma]_2}\omega^{[\tau\sigma]_2}_{[\varrho\sigma]_2} + I_{[\varrho\tau]_2}\omega^{[\varrho\tau]_2}_{[\varrho\sigma]_2}) \\
& + \frac{1}{2c\sqrt{2}} \sum_{\tau\neq\varrho,\sigma} (J_{\langle\tau\sigma\rangle_1}\omega^{[\tau\varrho]_1} + J_{\langle\varrho\tau\rangle_1}\omega^{[\sigma\tau]_1} + J_{\langle\tau\sigma\rangle_2}\omega^{[\tau\varrho]_2} + J_{\langle\varrho\tau\rangle_2}\omega^{[\sigma\tau]_2}) \\
& - \frac{1}{2c}(J_{\langle\varrho\rangle} + J_{\langle\sigma\rangle})\omega^{[\varrho\sigma]_2},
\end{aligned}$$

where $\varrho \neq \sigma$,

$$J_{\langle\varrho\rangle} = E_{[2\varrho-1,2\varrho]},$$

$$J_{\langle\varrho\sigma\rangle_1} = \frac{1}{\sqrt{2}}(E_{[2\varrho-1,2\sigma-1]} + E_{[2\varrho,2\sigma]}) = -J_{\langle\sigma\varrho\rangle_1}, \tag{9.3.7}$$

$$J_{\langle\varrho\sigma\rangle_2} = \frac{1}{\sqrt{2}}(E_{[2\varrho-1,2\sigma]} - E_{[2\varrho,2\sigma-1]}) = J_{\langle\sigma\varrho\rangle_2} \tag{9.3.8}$$

are mutually orthogonal unit vectors of $\mathbb{R}^{\nu(\nu-1)}$ normal to the orbit, and

$$\omega^{[\varrho\tau]_1}_{[\varrho\sigma]_1} = \frac{1}{2}(\theta^{2\tau-1}_{2\sigma-1} + \theta^{2\tau}_{2\sigma}) = \omega^{[\varrho\tau]_2}_{[\varrho\sigma]_2}, \tag{9.3.9}$$

$$\omega^{[\varrho\sigma]_2}_{[\varrho\sigma]_1} = \frac{1}{2}(\theta^{2\varrho}_{2\varrho-1} + \theta^{2\sigma}_{2\sigma-1}), \quad \omega^{[\varrho\tau]_2}_{[\varrho\sigma]_1} = \frac{1}{2}(\theta^{2\tau}_{2\sigma-1} + \theta^{2\sigma}_{2\tau-1}), \ \tau \neq \sigma; \tag{9.3.10}$$

the other components of the connection form ω of ∇ vanish.

The nonzero essential components of the vector-valued second fundamental form h of the orbit are

$$h_{[\varrho\sigma]_1[\varrho\sigma]_1} = -\frac{1}{2c}(J_{\langle\varrho\rangle} + J_{\langle\sigma\rangle}) = h_{[\varrho\sigma]_2[\varrho\sigma]_2}, \tag{9.3.11}$$

$$h_{[\varrho\sigma]_1[\varrho\tau]_1} = -\frac{1}{2c\sqrt{2}}J_{\langle\sigma\tau\rangle_2} = h_{[\varrho\sigma]_2[\varrho\tau]_2}, \quad \tau \neq \sigma, \tag{9.3.12}$$

$$h_{[\varrho\sigma]_1[\varrho\tau]_2} = \frac{1}{2c\sqrt{2}}J_{\langle\sigma\tau\rangle_1}, \quad \tau \neq \sigma; \tag{9.3.13}$$

the other components of h vanish.

A direct calculation can now be used to verify that condition (3) of Proposition 3.1.1 is satisfied for the orbit being considered, and hence the orbit is symmetric.

For example, in the expression of $\overline{\nabla} h_{ab}$ for $a = [\varrho\sigma]_1, b = [\varrho\tau]_1, \tau \neq \sigma$,

$$\begin{aligned} dh_{ab} = dh_{[\varrho\sigma]_1[\varrho\tau]_1} &= -\frac{1}{4c}(dE_{[2\sigma-1,2\tau]} - dE_{[2\sigma,2\tau-1]}) \\ &= -\frac{1}{4c}\sum_{\varphi}\{[E_{[2\varphi-1,2\tau]}\theta^{2\varphi-1}_{2\sigma-1} + E_{[2\varphi,2\tau]}\theta^{2\varphi}_{2\sigma-1} \\ &\quad + E_{[2\sigma-1,2\varphi-1]}\theta^{2\varphi-1}_{2\tau} + E_{[2\sigma-1,2\varphi]}\theta^{2\varphi}_{2\tau}] \\ &\quad - [E_{[2\varphi-1,2\tau-1]}\theta^{2\varphi-1}_{2\sigma} + E_{[2\varphi,2\tau-1]}\theta^{2\varphi}_{2\sigma} \\ &\quad + E_{[2\sigma,2\varphi-1]}\theta^{2\varphi-1}_{2\tau-1} + E_{[2\sigma,2\varphi]}\theta^{2\varphi}_{2\tau-1}]\}. \end{aligned}$$

Thus the normal component of this dh_{ab} is

$$\begin{aligned} -\frac{1}{4c\sqrt{2}}\sum_{\varphi}\{&J_{\langle\varphi\tau\rangle_2}\theta^{2\varphi-1}_{2\sigma-1} + J_{\langle\varphi\tau\rangle_1}\theta^{2\varphi}_{2\sigma-1} + J_{\langle\sigma\varphi\rangle_1}\theta^{2\varphi-1}_{2\tau} + J_{\langle\sigma\varphi\rangle_2}\theta^{2\varphi}_{2\tau} \\ &- J_{\langle\varphi\tau\rangle_1}\theta^{2\varphi-1}_{2\sigma} + J_{\langle\varphi\tau\rangle_2}\theta^{2\varphi}_{2\sigma} + J_{\langle\sigma\varphi\rangle_2}\theta^{2\varphi-1}_{2\tau-1} - J_{\langle\sigma\varphi\rangle_1}\theta^{2\varphi}_{2\tau-1}\}, \end{aligned}$$

which coincides with

$$\begin{aligned} \sum_{\varphi}\{&h_{[\varrho\varphi]_1[\varrho\tau]_1}\omega^{[\varrho\varphi]_1}_{[\varrho\sigma]_1} + h_{[\varrho\varphi]_2[\varrho\tau]_1}\omega^{[\varrho\varphi]_2}_{[\varrho\sigma]_1} \\ &+ h_{[\varrho\sigma]_1[\varrho\varphi]_1}\omega^{[\varrho\varphi]_1}_{[\varrho\tau]_1} + +h_{[\varrho\sigma]_1[\varrho\varphi]_2}\omega^{[\varrho\varphi]_2}_{[\varrho\tau]_1}\}, \end{aligned}$$

as is easy to see. Hence $\overline{\nabla} h_{ab}$ is tangent to this orbit and thus condition (3) of Proposition 3.1.1 is satisfied for this choice of a and b.

For other choices of a and b, the verification is quite similar. This concludes the proof.

A group theoretic characterization of the orbit of Theorem 9.3.1 can be given as follows.

Equations (9.3.2), (9.3.5), and (9.3.6) imply that for a point x of this orbit, the isotropy subgroup K in $G = SO(n, \mathbb{R})$ is given by the completely integrable Pfaffian system

$$\theta^{2\sigma-1}_{2\varrho} = -\theta^{2\sigma}_{2\varrho-1}, \quad \theta^{2\sigma}_{2\varrho} = \theta^{2\sigma-1}_{2\varrho-1}, \quad 1 \leq \varrho < \sigma \leq \nu.$$

It follows that the skew–symmetric matrix of the Maurer–Cartan 1-forms θ_i^j of $G = SO(2\nu, \mathbb{R})$ restricted to K consists of blocks

$$\begin{pmatrix} \theta_{2\varrho-1}^{2\sigma-1} & \theta_{2\varrho-1}^{2\sigma} \\ -\theta_{2\varrho-1}^{2\sigma} & \theta_{2\varrho-1}^{2\sigma-1} \end{pmatrix} = \theta_{2\varrho-1}^{2\sigma-1} \cdot 1\!\!1 + \theta_{2\varrho-1}^{2\sigma} \cdot \overset{\circ}{u}, \quad 1 \leq \varrho < \sigma \leq \nu,$$

where

$$1\!\!1 = \begin{pmatrix} 1 & 0 \\ 0 & 1 \end{pmatrix}, \quad \overset{\circ}{u} = \begin{pmatrix} 0 & 1 \\ -1 & 0 \end{pmatrix}$$

can be identified with the complex units 1 and i, $i^2 = -1$. Thus the Lie algebra of K is isomorphic to $u(\nu, \mathbb{C}) = \{\theta : {}^t\overline{\theta} = -\theta\}$ and therefore K is isomorphic to the intersection of the unitary group $U(\nu, \mathbb{C}) = \{A : {}^t\overline{A} \cdot A = \mathrm{Id}\}$ with $SO(2\nu, \mathbb{R})$, i.e., to $SU(\nu, \mathbb{C})$. This shows that this orbit can be considered as a standardly imbedded $SO(2\nu, \mathbb{R})/SU(\nu, \mathbb{C})$. Therefore, it will be called a *unitary orbit*.

Remark 9.3.2. In [Lu 96b], where the unitary orbit is considered in this way, it is shown that the other orbits of the Plücker action, other than the Plücker orbits and unitary orbits, are nonparallel submanifolds.

9.4 Umbilic-Likeness of Unitary Orbits

It can be shown that a unitary orbit is umbilic-like, i.e., the second-order envelope of unitary orbits is a single unitary orbit. The proof is rather complicated. The first step is to prove the codimension reduction theorem.

Theorem 9.4.1. *If a submanifold $M^{\nu(\nu-1)}$ in E^n, $n > \nu(2\nu-1)$, is a second-order envelope of unitary orbits, then there exists an $E^{2\nu^2} \subset E^n$, which contains this $M^{\nu(\nu-1)}$ and all the unitary orbits enveloping it.*

Proof. Suppose the submanifold $M^{\nu(\nu-1)}$ in a Euclidean space E^n, $n > \nu(2\nu-1)$, is the second-order envelope of unitary orbits of Plücker actions. Then it has an adapted orthonormal frame bundle, whose frames have tangent vectors $I_{[\varrho\sigma]_1}$ and $I_{[\varrho\sigma]_2}$, and normal vectors $J_{\langle\varrho\rangle}$, $J_{\langle\varrho\sigma\rangle_1}$, $J_{\langle\varrho\sigma\rangle_2}$ and J_ζ, where

$$I_{[\varrho\sigma]_1} = -I_{[\sigma\varrho]_1}, \quad I_{[\varrho\sigma]_2} = -I_{[\sigma\varrho]_2}, \quad J_{\langle\varrho\sigma\rangle_1} = -J_{\langle\sigma\varrho\rangle_1}, \quad J_{\langle\varrho\sigma\rangle_2} = J_{\langle\sigma\varrho\rangle_2} \tag{9.4.1}$$

with $\varrho \neq \sigma$ (cf. (9.3.3)–(9.3.8)), $1 \leq \varrho, \sigma \leq \nu$, and $\nu(2\nu-1)+1 \leq \zeta \leq n$.

By (9.2.9), (9.3.10), the role of the Pfaffian equations $\omega_i^\alpha = h_{ij}^\alpha \omega^j$ (see (2.1.4)) for $M^{\nu(\nu-1)}$ is now played by

$$\omega_{[\varrho\sigma]_1}^{\langle\varrho\rangle} = \kappa\sqrt{2}\omega^{[\varrho\sigma]_1}, \quad \omega_{[\varrho\sigma]_2}^{\langle\varrho\rangle} = \kappa\sqrt{2}\omega^{[\varrho\sigma]_2}, \tag{9.4.2}$$

$$\omega_{[\varrho\sigma]_1}^{\langle\varrho\tau\rangle_1} = -\omega_{[\varrho\sigma]_2}^{\langle\varrho\tau\rangle_2} = \kappa\omega^{[\sigma\tau]_2}, \tag{9.4.3}$$

$$\omega^{\langle\varrho\tau\rangle_2}_{[\varrho\sigma]_1} = \omega^{\langle\varrho\tau\rangle_1}_{[\varrho\sigma]_2} = -\kappa\omega^{[\sigma\tau]_1}, \tag{9.4.4}$$

$$\text{all other} \quad \omega^{\alpha}_{[\varrho\sigma]_1} \quad \text{and} \quad \omega^{\alpha}_{[\varrho\sigma]_2} \quad \text{are zero,} \tag{9.4.5}$$

where $\kappa = -\frac{1}{2c\sqrt{2}}$; also recall that $[\varrho\sigma]_1$, $[\varrho\sigma]_2$, $\langle\varrho\sigma\rangle_1$ are skew-symmetric index pairs, and $\langle\varrho\sigma\rangle_2$ is a symmetric index pair with $\varrho \neq \sigma$, as follows from (9.4.1).

In particular, (9.4.5) implies that for the upper index ζ,

$$\omega^{\zeta}_{[\varrho\sigma]_1} = \omega^{\zeta}_{[\varrho\sigma]_2} = 0.$$

Exterior differentiation according to (1.4.3), used for the last two equations, now gives

$$\sqrt{2}(\omega^{\zeta}_{\langle\varrho\rangle} + \omega^{\zeta}_{\langle\sigma\rangle}) \wedge \omega^{[\varrho\sigma]_1} + \sum_{\tau}(\omega^{\zeta}_{\langle\varrho\tau\rangle_1} \wedge \omega^{[\sigma\tau]_2} - \omega^{\zeta}_{\langle\sigma\tau\rangle_1} \wedge \omega^{[\varrho\tau]_2}$$

$$- \omega^{\zeta}_{\langle\varrho\tau\rangle_2} \wedge \omega^{[\sigma\tau]_1} + \omega^{\zeta}_{\langle\sigma\tau\rangle_2} \wedge \omega^{[\varrho\tau]_1}) = 0, \tag{9.4.6}$$

$$\sqrt{2}(\omega^{\zeta}_{\langle\varrho\rangle} + \omega^{\zeta}_{\langle\sigma\rangle}) \wedge \omega^{[\varrho\sigma]_2} - \sum_{\tau}(\omega^{\zeta}_{\langle\varrho\tau\rangle_1} \wedge \omega^{[\sigma\tau]_1} - \omega^{\zeta}_{\langle\sigma\tau\rangle_1} \wedge \omega^{[\varrho\tau]_1}$$

$$+ \omega^{\zeta}_{\langle\varrho\tau\rangle_2} \wedge \omega^{[\sigma\tau]_2} - \omega^{\zeta}_{\langle\sigma\tau\rangle_2} \wedge \omega^{[\varrho\tau]_2}) = 0. \tag{9.4.7}$$

From (9.4.6) together with Cartan's lemma, it follows that $\theta^{\zeta}_{\varrho\sigma} = \sqrt{2}(\omega^{\zeta}_{\langle\varrho\rangle} + \omega^{\zeta}_{\langle\sigma\rangle})$ $(\varrho \neq \sigma)$ and the other forms in (9.4.6) having upper index ζ must be linear combinations of the basis 1-forms $\omega^{[\varrho\sigma]_1}, \omega^{[\varrho\tau]_1}, \omega^{[\sigma\tau]_1}, \omega^{[\varrho\tau]_2}, \omega^{[\sigma\tau]_2}$ of the coframe bundle on $M^{\nu(\nu-1)}$. Substituting this into (9.4.7), one sees that the terms obtained from $\theta^{\zeta}_{\varrho\sigma} \wedge \omega^{[\varrho\sigma]_2}$, and thus containing $\omega^{[\varrho\sigma]_2}$ as a multiplier, are all different from the terms of the other part. No cancellations can occur between these two groups of terms, and hence $\theta^{\zeta}_{\varrho\sigma} = 0$ or

$$\omega^{\zeta}_{\langle\varrho\rangle} + \omega^{\zeta}_{\langle\sigma\rangle} = 0. \tag{9.4.8}$$

So (9.4.6) yields

$$\omega^{\zeta}_{\langle\varrho\tau\rangle_1} = \sum_{\varphi\neq\varrho,\sigma} (A^{\zeta}_{\varrho\tau\varphi}\omega^{[\sigma\varphi]_2} + B^{\zeta}_{\varrho\tau\varphi}\omega^{[\varrho\varphi]_2} + C^{\zeta}_{\varrho\tau\varphi}\omega^{[\sigma\varphi]_1} + D^{\zeta}_{\varrho\tau\varphi}\omega^{[\varrho\varphi]_1}),$$

$$-\omega^{\zeta}_{\langle\sigma\tau\rangle_1} = \sum_{\varphi\neq\varrho,\sigma} (B^{\zeta}_{\varrho\varphi\tau}\omega^{[\sigma\varphi]_2} + E^{\zeta}_{\sigma\tau\varphi}\omega^{[\varrho\varphi]_2} + F^{\zeta}_{\varrho\tau\varphi}\omega^{[\sigma\varphi]_1} + G^{\zeta}_{\sigma\tau\varphi}\omega^{[\varrho\varphi]_1}),$$

$$-\omega^{\zeta}_{\langle\varrho\tau\rangle_2} = \sum_{\varphi\neq\varrho,\sigma} (C^{\zeta}_{\varrho\varphi\tau}\omega^{[\sigma\varphi]_2} + F^{\zeta}_{\sigma\varphi\tau}\omega^{[\varrho\varphi]_2} + H^{\zeta}_{\varrho\tau\varphi}\omega^{[\sigma\varphi]_1} + I^{\zeta}_{\varrho\tau\varphi}\omega^{[\varrho\varphi]_1}),$$

$$\omega^{\zeta}_{\langle\sigma\tau\rangle_2} = \sum_{\varphi\neq\varrho,\sigma} (D^{\zeta}_{\varrho\varphi\tau}\omega^{[\sigma\varphi]_2} + G^{\zeta}_{\sigma\varphi\tau}\omega^{[\varrho\varphi]_2} + I^{\zeta}_{\varrho\varphi\tau}\omega^{[\sigma\varphi]_1} + J^{\zeta}_{\sigma\tau\varphi}\omega^{[\varrho\varphi]_1}),$$

where ϱ, σ, τ are three distinct values and the coefficients in diagonal blocks are symmetric in τ and φ; recall that the index pairs are skew-symmetric, except that

$\langle\varrho\tau\rangle_2$ is symmetric. Thus from the first row, $A^\zeta_{\varrho\tau\varphi} = -A^\zeta_{\tau\varrho\varphi} = -A^\zeta_{\tau\varphi\varrho}$ for three distinct ϱ, σ, φ, i.e., a cyclic permutation of these indices changes the sign; after the third, one obtains $A^\zeta_{\varrho\tau\varphi} = 0$ for $\varphi \neq \tau$. The first row also gives $C^\zeta_{\varrho\tau\varphi} = -C^\zeta_{\tau\varrho\varphi}$, and from the third row one obtains $C^\zeta_{\varrho\varphi\tau} = C^\zeta_{\tau\varphi\varrho}$, thus $C^\zeta_{\varrho\tau\varphi} = 0$ for $\tau \neq \varphi$. The same argument shows that $E^\zeta_{\sigma\tau\varphi} = 0$, $G^\zeta_{\sigma\tau\varphi} = 0$ for $\tau \neq \varphi$. Moreover,

$$H^\zeta_{\varrho\tau\varphi} = H^\zeta_{\tau\varrho\varphi}, \quad J^\zeta_{\sigma\tau\varphi} = J^\zeta_{\tau\sigma\varphi}.$$

Interchanging ϱ and σ in the second and third rows, one obtains

$$E^\zeta_{\sigma\tau\tau} + A^\zeta_{\sigma\tau\tau} = 0, \quad B^\zeta_{\varrho\tau\varphi} + B^\zeta_{\sigma\varphi\tau} = 0, \quad D^\zeta_{\varrho\tau\varphi} + F^\zeta_{\varrho\tau\varphi} = 0,$$

$$G^\zeta_{\sigma\tau\tau} + C^\zeta_{\sigma\tau\tau} = 0, \quad I^\zeta_{\sigma\tau\varphi} + I^\zeta_{\varrho\varphi\tau} = 0, \quad H^\zeta_{\sigma\tau\varphi} + J^\zeta_{\sigma\tau\varphi} = 0.$$

All this shows that

$$\omega^\zeta_{\langle\varrho\tau\rangle_1} = A^\zeta_{\varrho\tau\tau}\omega^{[\sigma\tau]_2} + C^\zeta_{\varrho\tau\tau}\omega^{[\sigma\tau]_1} + \sum_{\varphi\neq\varrho,\sigma} (B^\zeta_{\varrho\tau\varphi}\omega^{[\varrho\varphi]_2} + D^\zeta_{\varrho\tau\varphi}\omega^{[\varrho\varphi]_1}),$$

$$\omega^\zeta_{\langle\sigma\tau\rangle_2} = -C^\zeta_{\sigma\tau\tau}\omega^{[\varrho\tau]_2} + \sum_{\varphi\neq\varrho,\sigma} (D^\zeta_{\varrho\varphi\tau}\omega^{[\sigma\varphi]_2} + I^\zeta_{\varrho\varphi\tau}\omega^{[\sigma\varphi]_1} - H^\zeta_{\sigma\tau\varphi}\omega^{[\varrho\varphi]_1}).$$

Now (9.4.7) gives

$$\sum_{\tau\neq\varrho,\sigma} \Bigg\{ -\Bigg[A^\zeta_{\varrho\tau\tau}\omega^{[\sigma\tau]_2} + C^\zeta_{\varrho\tau\tau}\omega^{[\sigma\tau]_1} + \sum_{\varphi\neq\varrho,\sigma} (B^\zeta_{\varrho\tau\varphi}\omega^{[\varrho\varphi]_2} + D^\zeta_{\varrho\tau\varphi}\omega^{[\varrho\varphi]_1}) \Bigg] \wedge \omega^{[\sigma\tau]_1}$$

$$+ \Bigg[A^\zeta_{\sigma\tau\tau}\omega^{[\varrho\tau]_2} + C^\zeta_{\sigma\tau\tau}\omega^{[\varrho\tau]_1} + \sum_{\varphi\neq\varrho,\sigma} (B^\zeta_{\sigma\tau\varphi}\omega^{[\sigma\varphi]_2} + D^\zeta_{\sigma\tau\varphi}\omega^{[\sigma\varphi]_1}) \Bigg] \wedge \omega^{[\varrho\tau]_1}$$

$$- \Bigg[G^\zeta_{\varrho\tau\tau}\omega^{[\sigma\tau]_2} + \sum_{\varphi\neq\varrho,\sigma} (D^\zeta_{\sigma\varphi\tau}\omega^{[\varrho\varphi]_2} + I^\zeta_{\sigma\varphi\tau}\omega^{[\varrho\varphi]_1} - H^\zeta_{\varrho\tau\varphi}\omega^{[\sigma\varphi]_1}) \Bigg] \wedge \omega^{[\sigma\tau]_2}$$

$$+ \Bigg[G^\zeta_{\sigma\tau\tau}\omega^{[\varrho\tau]_2} + \sum_{\varphi\neq\varrho,\sigma} (D^\zeta_{\varrho\varphi\tau}\omega^{[\sigma\varphi]_2} + I^\zeta_{\varrho\varphi\tau}\omega^{[\sigma\varphi]_1} - H^\zeta_{\sigma\tau\varphi}\omega^{[\varrho\varphi]_1}) \Bigg] \wedge \omega^{[\varrho\tau]_2} \Bigg\}$$

$$= 0.$$

This yields $A^\zeta_{\varrho\tau\tau} + H^\zeta_{\varrho\tau\tau} = 0$ and $H^\zeta_{\varrho\tau\varphi} = 0$ if $\tau \neq \varphi$, but

$$B^\zeta_{\varrho\tau\varphi} + I^\zeta_{\varrho\tau\varphi} = 0, \quad D^\zeta_{\varrho\varphi\tau} + D^\zeta_{\sigma\tau\varphi} = 0.$$

So

$$\omega^\zeta_{\langle\varrho\tau\rangle_2} = A^\zeta_{\varrho\tau\tau}\omega^{[\sigma\tau]_1} - C^\zeta_{\varrho\tau\tau}\omega^{[\sigma\tau]_2} - \sum_{\varphi\neq\varrho,\sigma} (D^\zeta_{\sigma\tau\varphi}\omega^{[\varrho\varphi]_2} + B^\zeta_{\varrho\tau\varphi}\omega^{[\varrho\varphi]_1}).$$

Interchanging ϱ and τ, one obtains

$$\omega^{\zeta}_{\langle\varrho\sigma\rangle_1} = A^{\zeta}_{\tau\varrho\varrho}\omega^{[\sigma\varrho]_2} + C^{\zeta}_{\tau\varrho\varrho}\omega^{[\sigma\varrho]_1} + \sum_{\psi\neq\tau,\sigma}(B^{\zeta}_{\tau\varrho\psi}\omega^{[\tau\psi]_2} + D^{\zeta}_{\tau\varphi\psi}\omega^{[\tau\psi]_1}),$$

$$\omega^{\zeta}_{\langle\tau\varrho\rangle_2} = A^{\zeta}_{\tau\varrho\varrho}\omega^{[\sigma\varrho]_1} - C^{\zeta}_{\tau\varrho\varrho}\omega^{[\sigma\varrho]_2} - \sum_{\psi\neq\tau,\sigma}(D^{\zeta}_{\tau\varrho\psi}\omega^{[\tau\psi]_2} + B^{\zeta}_{\tau\varphi\psi}\omega^{[\tau\psi]_1}).$$

Here

$$\omega^{\zeta}_{\langle\varrho\tau\rangle_1} + \omega^{\zeta}_{\langle\tau\varrho\rangle_1} = \omega^{\zeta}_{\langle\varrho\tau\rangle_2} - \omega^{\zeta}_{\langle\tau\varrho\rangle_2} = 0,$$

thus $A^{\zeta}_{\varrho\tau\tau} = C^{\zeta}_{\varrho\tau\tau} = 0$,

$$B^{\zeta}_{\varrho\tau\tau} - B^{\zeta}_{\tau\varrho\varrho} = 0, \quad B^{\zeta}_{\varrho\tau\varphi} = 0 \quad (\tau \neq \varphi),$$

$$D^{\zeta}_{\varrho\tau\tau} - D^{\zeta}_{\tau\varrho\varrho} = 0, \quad D^{\zeta}_{\varrho\tau\varphi} = 0 \quad (\tau \neq \varphi),$$

but at the same time

$$B^{\zeta}_{\varrho\tau\tau} + B^{\zeta}_{\tau\varrho\varrho} = 0, \quad D^{\zeta}_{\varrho\tau\tau} + D^{\zeta}_{\tau\varrho\varrho} = 0$$

and hence $B^{\zeta}_{\varrho\tau\varphi} = D^{\zeta}_{\varrho\tau\varphi} = 0$. It follows that

$$\omega^{\zeta}_{\langle\varrho\tau\rangle_1} = \omega^{\zeta}_{\langle\varrho\tau\rangle_2} = 0.$$

Together with (9.4.8), equations $\omega^{\zeta}_{\langle\sigma\rangle} + \omega^{\zeta}_{\langle\tau\rangle} = 0$, $\omega^{\zeta}_{\langle\varrho\rangle} + \omega^{\zeta}_{\langle\tau\rangle} = 0$ also hold for every three distinct values ϱ, σ, τ. This shows that

$$\omega^{\zeta}_{\langle\varrho\rangle} = 0.$$

Now $E^{2\nu^2}$ in E^n, spanned at x by all $I_{[\varrho\sigma]_1}$, $I_{[\varrho\sigma]_2}$ plus all $J_{\langle\varrho\rangle}$, $J_{\langle\varrho\sigma\rangle_1}$, $J_{\langle\varrho\sigma\rangle_2}$, is invariant, because dx and the differentials of these vectors belong to this $E^{2\nu^2}$.

This finishes the proof of Theorem 9.4.1.

It remains to give the final step in proving the assertion of this section.

The same process as above must be used, applied now to the remaining equations of the system (9.4.2)–(9.4.5). In fact, this system is a particular case of $\omega^{\alpha}_i = h^{\alpha}_{ij}\omega^j$ for the bundle of adapted orthonormal frames of $M^{\nu(\nu-1)}$ in $E^{2\nu^2}$. Now the deduction

$$\omega^{\alpha}_i = h^{\alpha}_{ij}\omega^j \Rightarrow \overline{\nabla} h^{\alpha}_{ij} \wedge \omega^j = 0 \Rightarrow \overline{\nabla} h^{\alpha}_{ij} = h^{\alpha}_{ijk}\omega^k$$

has to be performed in this particular case. Lengthy calculations will show that the result is $h^{\alpha}_{ijk} = 0$; this result proves the desired assertion.

This process will not be presented in full generality below, but rather demonstrated for a model case, namely for M^6 in E^{18}, i.e., for the case $\nu = 3$. Here the presentation can be simplified by denoting for every even permutation ϱ, σ, τ of 1, 2, 3

$$[\varrho\sigma]_1 = \tau, \quad [\varrho\sigma]_2 = \tau' \quad (= \tau + 3),$$

$$\langle\varrho\sigma\rangle_1 = \tau_1 \quad (= \tau + 9), \quad \langle\varrho\sigma\rangle_2 = \tau_2 \quad (= \tau + 12).$$

The system (9.4.2)–(9.4.5) then takes the form

$$\omega_\varrho^{\langle\varrho\rangle} = 0, \quad \omega_\varrho^{\langle\sigma\rangle} = \omega_\varrho^{\langle\tau\rangle} = \kappa\sqrt{2}\omega^\varrho, \tag{9.4.9}$$

$$\omega_\varrho^{\varrho_1} = 0, \quad \omega_\varrho^{\sigma_1} = \kappa\omega^{\tau'}, \quad \omega_\varrho^{\tau_1} = -\kappa\omega^{\sigma'}, \tag{9.4.10}$$

$$\omega_\varrho^{\varrho_2} = 0, \quad \omega_\varrho^{\sigma_2} = -\kappa\omega^{\tau}, \quad \omega_\varrho^{\tau_2} = -\kappa\omega^{\sigma}, \tag{9.4.11}$$

$$\omega_{\varrho'}^{\langle\varrho\rangle} = 0, \quad \omega_{\varrho'}^{\langle\sigma\rangle} = \omega_{\varrho'}^{\langle\tau\rangle} = \kappa\sqrt{2}\omega^{\varrho'}, \tag{9.4.12}$$

$$\omega_{\varrho'}^{\varrho_1} = 0, \quad \omega_{\varrho'}^{\sigma_1} = -\kappa\omega^{\tau}, \quad \omega_{\varrho'}^{\tau_1} = \kappa\omega^{\sigma}, \tag{9.4.13}$$

$$\omega_{\varrho'}^{\varrho_2} = 0, \quad \omega_{\varrho'}^{\sigma_2} = -\kappa\omega^{\tau'}, \quad \omega_{\varrho'}^{\tau_2} = -\kappa\omega^{\sigma'}. \tag{9.4.14}$$

Exterior differentiation of equations (9.4.9) and (9.4.12) gives

$$\sqrt{2}(\omega_{\langle\varrho\rangle}^{\langle\sigma\rangle} + \omega_{\langle\varrho\rangle}^{\langle\tau\rangle}) \wedge \omega^\varrho + (\sqrt{2}\omega_\varrho^\sigma - \omega_{\langle\varrho\rangle}^{\tau_2}) \wedge \omega^\sigma + (\sqrt{2}\omega_\varrho^\tau - \omega_{\langle\varrho\rangle}^{\sigma_2}) \wedge \omega^\tau$$
$$+ (\sqrt{2}\omega_\varrho^{\sigma'} - \omega_{\langle\varrho\rangle}^{\tau_1}) \wedge \omega^{\sigma'} + (\sqrt{2}\omega_\varrho^{\tau'} + \omega_{\langle\varrho\rangle}^{\sigma_1}) \wedge \omega^{\tau'} = 0, \tag{9.4.15}$$

$$\sqrt{2}(d\ln\kappa - \omega_{\langle\sigma\rangle}^{\langle\tau\rangle}) \wedge \omega^\varrho - (\sqrt{2}\omega_\sigma^\varrho - \omega_{\langle\sigma\rangle}^{\tau_2}) \wedge \omega^\sigma + \omega_{\langle\sigma\rangle}^{\sigma_2} \wedge \omega^\tau$$
$$- (\sqrt{2}\omega_{\sigma'}^\varrho - \omega_{\langle\sigma\rangle}^{\tau_1}) \wedge \omega^{\sigma'} - \omega_{\langle\sigma\rangle}^{\sigma_1} \wedge \omega^{\tau'} = 0, \tag{9.4.16}$$

$$\sqrt{2}(d\ln\kappa - \omega_{\langle\tau\rangle}^{\langle\sigma\rangle}) \wedge \omega^\varrho + \omega_{\langle\tau\rangle}^{\tau_2} \wedge \omega^\sigma - (\sqrt{2}\omega_\tau^\varrho - \omega_{\langle\tau\rangle}^{\sigma_2}) \wedge \omega^\tau$$
$$+ \omega_{\langle\tau\rangle}^{\tau_1} \wedge \omega^{\sigma'} - (\sqrt{2}\omega_{\tau'}^\varrho + \omega_{\langle\tau\rangle}^{\sigma_1}) \wedge \omega^{\tau'} = 0, \tag{9.4.17}$$

$$(\sqrt{2}\omega_{\varrho'}^\sigma + \omega_{\langle\varrho\rangle}^{\tau_1}) \wedge \omega^\sigma + (\sqrt{2}\omega_{\varrho'}^\tau - \omega_{\langle\varrho\rangle}^{\sigma_1}) \wedge \omega^\tau + (\omega_{\langle\varrho\rangle}^{\langle\sigma\rangle} + \omega_{\langle\varrho\rangle}^{\langle\tau\rangle}) \wedge \omega^{\varrho'}$$
$$+ (\sqrt{2}\omega_{\varrho'}^{\sigma'} - \omega_{\langle\varrho\rangle}^{\tau_2}) \wedge \omega^{\sigma'} + (\sqrt{2}\omega_{\varrho'}^{\tau'} - \omega_{\langle\varrho\rangle}^{\sigma_2}) \wedge \omega^{\tau'} = 0, \tag{9.4.18}$$

$$- (\sqrt{2}\omega_\sigma^{\varrho'} + \omega_{\langle\sigma\rangle}^{\tau_1}) \wedge \omega^\sigma + \omega_{\langle\sigma\rangle}^{\sigma_1} \wedge \omega^\tau + \sqrt{2}(d\ln\kappa - \omega_{\langle\sigma\rangle}^{\langle\tau\rangle}) \wedge \omega^{\varrho'}$$
$$- (\sqrt{2}\omega_{\sigma'}^{\varrho'} - \omega_{\langle\sigma\rangle}^{\tau_2}) \wedge \omega^{\sigma'} + \omega^{\sigma'} + \omega_{\langle\sigma\rangle}^{\sigma_2} \wedge \omega^{\tau'} = 0, \tag{9.4.19}$$

$$- \omega_{\langle\tau\rangle}^{\tau_1} \wedge \omega^\sigma - (\sqrt{2}\omega_\tau^{\varrho'} - \omega_{\langle\tau\rangle}^{\sigma_1}) \wedge \omega^\tau + \sqrt{2}(d\ln\kappa - \omega_{\langle\tau\rangle}^{\langle\sigma\rangle}) \wedge \omega^{\varrho'}$$
$$+ \omega_{\langle\tau\rangle}^{\tau_2} \wedge \omega^{\sigma'} - (\sqrt{2}\omega_{\tau'}^{\varrho'} - \omega_{\langle\tau\rangle}^{\sigma_2}) \wedge \omega^{\tau'} = 0. \tag{9.4.20}$$

The following result about the congruence of enveloping unitary orbits summarizes the intermediate stage of the proof for the model case.

Proposition 9.4.2. *If M^6 in E^{18} is a second-order envelope of unitary orbits in E^{18}, then these orbits are congruent, i.e.,* $r = \text{const}$.

Proof. From (9.4.16), (9.4.19) it follows by Cartan's lemma that $\sqrt{2}(d\ln\kappa - \omega_{\langle\sigma\rangle}^{\langle\tau\rangle})$, $\omega_{\langle\sigma\rangle}^{\sigma_1}$ and $\omega_{\langle\sigma\rangle}^{\sigma_2}$ are linear combinations of ω^ϱ, ω^σ, ω^τ, $\omega^{\sigma'}$, $\omega^{\tau'}$ and also of ω^σ, ω^τ,

$\omega^{\varrho'}, \omega^{\sigma'}, \omega^{\tau'}$. This can be repeated with (9.4.17), (9.4.20), and after permuting $(\varrho, \sigma, \tau) \mapsto (\tau, \varrho, \sigma)$, it shows that the same $\omega^{\sigma 1}_{\langle\sigma\rangle}, \omega^{\sigma 2}_{\langle\sigma\rangle}$ must be linear combinations of $\omega^{\varrho}, \omega^{\sigma}, \omega^{\tau}, \omega^{\varrho'}, \omega^{\sigma'}$ and also of $\omega^{\varrho}, \omega^{\sigma}, \omega^{\varrho'}, \omega^{\sigma'}, \omega^{\tau'}$. So

$$\sqrt{2}(d \ln \kappa - \omega^{\langle\tau\rangle}_{\langle\sigma\rangle}) = a_{\sigma\sigma}\omega^{\sigma} + a_{\sigma\tau}\omega^{\tau} + a_{\sigma\sigma'}\omega^{\sigma'} + a_{\sigma\tau'}\omega^{\tau'},$$

$$\omega^{\sigma 1}_{\langle\sigma\rangle} = b_{\sigma\sigma}\omega^{\sigma} + b_{\sigma\sigma'}\omega^{\sigma'}, \tag{9.4.21}$$

$$\omega^{\sigma 2}_{\langle\sigma\rangle} = c_{\sigma\sigma}\omega^{\sigma} + c_{\sigma\sigma'}\omega^{\sigma'}, \tag{9.4.22}$$

$$\sqrt{2}(d \ln \kappa + \omega^{\langle\tau\rangle}_{\langle\sigma\rangle}) = d_{\sigma\sigma}\omega^{\sigma} + d_{\sigma\tau}\omega^{\tau} + d_{\sigma\sigma'}\omega^{\sigma'} + d^{\tau'}_{\sigma\tau'}.$$

Further, it follows from (9.4.15), permuted by $(\varrho, \sigma, \tau) \mapsto (\sigma, \tau, \varrho)$, and from (9.4.16) that $\sqrt{2}\omega^{\varrho}_{\sigma} - \omega^{\tau 2}_{\langle\sigma\rangle}$ is a linear combination of $\omega^{\varrho}, \omega^{\sigma}, \omega^{\tau}, \omega^{\varrho'}, \omega^{\tau'}$ and also of $\omega^{\varrho}, \omega^{\sigma}, \omega^{\tau}, \omega^{\sigma'}, \omega^{\tau'}$; thus

$$\sqrt{2}\omega^{\varrho}_{\sigma} - \omega^{\tau 2}_{\langle\sigma\rangle} = e_{\sigma\varrho}\omega^{\varrho} + e_{\sigma\sigma}\omega^{\sigma} + e_{\sigma\tau}\omega^{\tau} + e_{\sigma\tau'}\omega^{\tau'}. \tag{9.4.23}$$

By the same argument, (9.4.15), permuted by $(\varrho, \sigma, \tau) \mapsto (\tau, \varrho, \sigma)$, and (9.4.17) imply

$$\sqrt{2}\omega^{\varrho}_{\tau} - \omega^{\sigma 2}_{\langle\tau\rangle} = f_{\tau\varrho}\omega^{\varrho} + f_{\tau\sigma}\omega^{\sigma} + f_{\tau\tau}\omega^{\tau} + f_{\tau\sigma'}\omega^{\sigma'}. \tag{9.4.24}$$

Similarly, (9.4.18), (9.4.19), and (9.4.20) give

$$\sqrt{2}\omega^{\varrho'}_{\sigma'} - \omega^{\tau 2}_{\langle\sigma\rangle} = e_{\sigma'\tau}\omega^{\tau} + e_{\sigma'\varrho'}\omega^{\varrho'} + e_{\sigma'\sigma'}\omega^{\sigma'} + e_{\sigma'\tau'}\omega^{\tau'}, \tag{9.4.25}$$

$$\sqrt{2}\omega^{\varrho'}_{\tau'} - \omega^{\sigma 2}_{\langle\tau\rangle} = f_{\tau'\sigma}\omega^{\sigma} + f_{\tau'\varrho'}\omega^{\varrho'} + f_{\tau'\sigma'}\omega^{\sigma'} + f_{\tau'\tau'}\omega^{\tau'}. \tag{9.4.26}$$

Substituting this into (9.4.16), one obtains

$$\begin{aligned}
&(a_{\sigma\sigma}\omega^{\sigma} + a_{\sigma\tau}\omega^{\tau} + a_{\sigma\sigma'}\omega^{\sigma'} + a_{\sigma\tau'}\omega^{\tau'}) \wedge \omega^{\varrho} \\
&\quad - (e_{\sigma\varrho}\omega^{\varrho} + e_{\sigma\tau}\omega^{\tau} + e_{\sigma\tau'}\omega^{\tau'}) \wedge \omega^{\sigma} + (c_{\sigma\sigma}\omega^{\sigma} + c_{\sigma\sigma'}\omega^{\sigma'}) \wedge \omega^{\tau} \\
&\quad - (\sqrt{2}\omega^{\varrho}_{\sigma'} - \omega^{\tau 1}_{\langle\sigma\rangle}) \wedge \omega^{\sigma'} - (b_{\sigma\sigma}\omega^{\sigma} + b_{\sigma\sigma'}\omega^{\sigma'}) \wedge \omega^{\tau'} = 0,
\end{aligned}$$

thus $a_{\sigma\sigma} + e_{\sigma\varrho} = a_{\sigma\tau} = a_{\sigma\tau'} = c_{\sigma\sigma} + e_{\sigma\tau} = b_{\sigma\sigma} - e_{\sigma\tau'} = 0$ and

$$-(\sqrt{2}\omega^{\varrho}_{\sigma'} - \omega^{\tau 1}_{\langle\sigma\rangle}) = a_{\sigma\sigma'}\omega^{\varrho} + c_{\sigma\sigma'}\omega^{\tau} + g_{\sigma\sigma'}\omega^{\sigma'} - b_{\sigma\sigma'}\omega^{\tau'}; \tag{9.4.27}$$

substitution into (9.4.17) gives

$$\begin{aligned}
&(d_{\sigma\sigma}\omega^{\sigma} + d_{\sigma\tau}\omega^{\tau} + d_{\sigma\sigma'}\omega^{\sigma'} + d_{\sigma\tau'}\omega^{\tau'}) \wedge \omega^{\varrho} + (c_{\tau\tau}\omega^{\tau} + c_{\tau\tau'}\omega^{\tau'}) \wedge \omega^{\sigma} \\
&\quad - (f_{\tau\varrho}\omega^{\varrho} + f_{\tau\sigma}\omega^{\sigma} + f_{\tau\sigma'}\omega^{\sigma'}) \wedge \omega^{\tau} + (b_{\tau\tau}\omega^{\tau} + b_{\tau\tau'}\omega^{\tau'}) \wedge \omega^{\sigma'} \\
&\quad - (\sqrt{2}\omega^{\varrho}_{\tau'} + \omega^{\sigma 1}_{\langle\tau\rangle}) \wedge \omega^{\tau'} = 0,
\end{aligned}$$

thus $d_{\sigma\sigma} = d_{\sigma\tau} + f_{\tau\varrho} = d_{\sigma\sigma'} = c_{\tau\tau} + f_{\tau\sigma} = b_{\tau\tau} + f_{\tau\sigma'} = 0$ and

$$-(\sqrt{2}\omega^{\varrho}_{\tau'} + \omega^{\sigma_1}_{\langle\tau\rangle}) = d_{\sigma\tau'}\omega^{\varrho} + c_{\tau\tau'}\omega^{\sigma} + b_{\tau\tau'}\omega^{\sigma'} + h_{\tau\tau'}\omega^{\tau'}. \tag{9.4.28}$$

From (9.4.19) and (9.4.20), one obtains $a_{\sigma\tau} = b_{\sigma\sigma'} + e_{\sigma'\tau} = a_{\sigma\sigma'} + e_{\sigma'\varrho'} = a_{\sigma\tau'} = c_{\sigma\sigma'} + e_{\sigma'\tau'} = 0$ and

$$-(\sqrt{2}\omega^{\varrho'}_{\sigma} + \omega^{\tau_1}_{\langle\sigma\rangle}) = i_{\sigma\sigma}\omega^{\sigma} + b_{\sigma\sigma}\omega^{\tau} + a_{\sigma\sigma}\omega^{\varrho'} + c_{\sigma\sigma}\omega^{\tau'}, \tag{9.4.29}$$

$d_{\sigma\sigma} = b_{\tau\tau'} - f_{\tau'\sigma} = d_{\sigma\sigma'} = d_{\sigma\tau'} + f_{\tau'\varrho'} = c_{\tau\tau'} + f_{\tau'\sigma'} = 0$ and

$$-(\sqrt{2}\omega^{\varrho'}_{\tau} - \omega^{\sigma_1}_{\langle\tau\rangle}) = -b_{\tau\tau}\omega^{\sigma} + j_{\tau\tau}\omega^{\tau} + d_{\sigma\tau}\omega^{\varrho'} + c_{\tau\tau}\omega^{\sigma'}. \tag{9.4.30}$$

Hence

$$\sqrt{2}(d\ln\kappa - \omega^{\langle\tau\rangle}_{\langle\sigma\rangle}) = a_{\sigma\sigma}\omega^{\sigma} + a_{\sigma\sigma'}\omega^{\sigma'},$$

$$\sqrt{2}(d\ln\kappa + \omega^{\langle\tau\rangle}_{\langle\sigma\rangle}) = d_{\sigma\tau}\omega^{\tau} + d_{\sigma\tau'}\omega^{\tau'},$$

and after summation one can see that the expression of $d\ln\kappa$ does not contain ω^{ϱ} and $\omega^{\varrho'}$. Similarly, after cyclic permutations of ϱ, σ, τ, it also does not contain $\omega^{\sigma}, \omega^{\sigma'}$ and $\omega^{\tau}, \omega^{\tau'}$. Hence

$$a_{\sigma\sigma} = a_{\sigma\sigma'} = d_{\sigma\tau} = d_{\sigma\tau'} = 0,$$

and thus

$$\kappa = \text{const}, \quad \omega^{\langle\tau\rangle}_{\langle\sigma\rangle} = 0. \tag{9.4.31}$$

This verifies Proposition 9.4.2.

Note that also $e_{\sigma\varrho} = f_{\tau\varrho} = e_{\sigma'\varrho'} = f_{\tau'\varrho'} = 0$.

The final result for the model case is given by the following.

Theorem 9.4.3. *If the submanifold M^6 in E^n, $n \geq 18$, is a second-order envelope of unitary orbits, then this M^6 reduces to a single unitary orbit or its open subset.*

Proof. The exterior equations (9.4.15) and (9.4.18) must be used, substituting first (9.4.23), (9.4.25), (9.4.27), (9.4.29) after permuting $(\varrho, \sigma, \tau) \mapsto (\tau, \varrho, \sigma)$, and then substituting (9.4.24), (9.4.26), (9.4.28), (9.4.30), after permuting $(\varrho, \sigma, \tau) \mapsto (\sigma, \tau, \varrho)$, and also considering (9.4.31). The result

$$\begin{aligned}
&+ (f_{\varrho\tau}\omega^{\tau} + f_{\varrho\varrho}\omega^{\varrho} + f_{\varrho\tau'}\omega^{\tau'}) \wedge \omega^{\sigma} + (e_{\varrho\varrho}\omega^{\varrho} + e_{\varrho\sigma}\omega^{\sigma} + e_{\varrho\sigma'}\omega^{\sigma'}) \wedge \omega^{\tau}\\
&\quad - (-b_{\varrho\varrho}\omega^{\tau} + j_{\varrho\varrho}\omega^{\varrho} + c_{\varrho\varrho}\omega^{\tau'}) \wedge \omega^{\sigma'} - (i_{\varrho\varrho}\omega^{\varrho} + b_{\varrho\varrho}\omega^{\sigma} + c_{\varrho\varrho}\omega^{\sigma'}) \wedge \omega^{\tau'}\\
&\quad = 0,\\
&- (c_{\varrho\varrho'}\omega^{\tau} + b_{\varrho\varrho'}\omega^{\tau'} + h_{\varrho\varrho'}\omega^{\varrho'}) \wedge \omega^{\sigma} - (c_{\varrho\varrho'}\omega^{\sigma} + g_{\varrho\varrho'}\omega^{\varrho'} - b_{\varrho\varrho'}\omega^{\sigma'}) \wedge \omega^{\tau}\\
&\quad + (f_{\varrho'\tau}\omega^{\tau} + f_{\varrho'\tau'}\omega^{\tau'} + f_{\varrho'\varrho'}\omega^{\varrho'}) \wedge \omega^{\sigma'} + (e_{\varrho'\sigma}\omega^{\sigma} + e_{\varrho'\varrho'}\omega^{\varrho'} + e_{\varrho'\sigma'}\omega^{\sigma'}) \wedge \omega^{\tau'}\\
&\quad = 0
\end{aligned}$$

yields

$$f_{\varrho\tau} - e_{\varrho\sigma} = f_{\varrho\varrho} = f_{\varrho\tau'} + b_{\varrho\varrho} = e_{\varrho\varrho} = e_{\varrho\sigma'} - b_{\varrho\varrho} = j_{\varrho\varrho} = i_{\varrho\varrho} = 0,$$
$$b_{\varrho\varrho'} + e_{\varrho'\sigma} = h_{\varrho\varrho'} = g_{\varrho\varrho'} = b_{\varrho\varrho'} - f_{\varrho'\tau} = f_{\varrho'\tau'} - e_{\varrho'\sigma'} = f_{\varrho'\varrho'} = e_{\varrho'\varrho'} = 0.$$

Hence (9.4.23)–(9.4.26) reduce to

$$\sqrt{2}\omega_\sigma^\varrho - \omega_{\langle\sigma\rangle}^{\tau_2} = -c_{\sigma\sigma}\omega^\tau + b_{\sigma\sigma}\omega^{\tau'}, \tag{9.4.32}$$

$$\sqrt{2}\omega_\tau^\varrho - \omega_{\langle\tau\rangle}^{\sigma_2} = -c_{\tau\tau}\omega^\sigma - b_{\tau\tau}\omega^{\sigma'}, \tag{9.4.33}$$

$$\sqrt{2}\omega_{\sigma'}^{\varrho'} - \omega_{\langle\sigma\rangle}^{\tau_2} = -b_{\sigma\sigma'}\omega^\tau - c_{\sigma\sigma'}\omega^{\tau'}, \tag{9.4.34}$$

$$\sqrt{2}\omega_{\tau'}^{\varrho'} - \omega_{\langle\tau\rangle}^{\sigma_2} = b_{\tau\tau'}\omega^\sigma - c_{\tau\tau'}\omega^{\sigma'}, \tag{9.4.35}$$

but (9.4.27)–(9.4.30) reduce to

$$\sqrt{2}\omega_{\sigma'}^\varrho - \omega_{\langle\sigma\rangle}^{\tau_1} = -c_{\sigma\sigma'}\omega^\tau + b_{\sigma\sigma'}\omega^{\tau'}, \tag{9.4.36}$$

$$\sqrt{2}\omega_{\tau'}^\varrho + \omega_{\langle\tau\rangle}^{\sigma_1} = -c_{\tau\tau'}\omega^\sigma - b_{\tau\tau'}\omega^{\sigma'}, \tag{9.4.37}$$

$$\sqrt{2}\omega_\sigma^{\varrho'} + \omega_{\langle\sigma\rangle}^{\tau_1} = -b_{\sigma\sigma}\omega^\tau - c_{\sigma\sigma}\omega^{\tau'}, \tag{9.4.38}$$

$$\sqrt{2}\omega_\tau^{\varrho'} - \omega_{\langle\tau\rangle}^{\sigma_1} = b_{\tau\tau}\omega^\sigma - c_{\tau\tau}\omega^{\sigma'}. \tag{9.4.39}$$

Now one has to turn to (9.4.10), (9.4.11), (9.4.13) and (9.4.14). By exterior differentiation,

$$\sqrt{2}(\omega_{\langle\sigma\rangle}^{\varrho_1} + \omega_{\langle\tau\rangle}^{\varrho_1}) \wedge \omega^\varrho + (\omega_\varrho^{\tau'} + \omega_{\varrho_1}^{\tau_2}) \wedge \omega^\sigma + (\omega_\varrho^{\sigma'} - \omega_{\varrho_1}^{\sigma_2}) \wedge \omega^\tau$$
$$+ (\omega_\varrho^\tau - \omega_{\varrho_1}^{\tau_1}) \wedge \omega^{\sigma'} - (\omega_\varrho^\sigma - \omega_{\varrho_1}^{\sigma_1}) \wedge \omega^{\tau'} = 0, \tag{9.4.40}$$

$$\sqrt{2}(\sqrt{2}\omega_{\tau'}^\varrho + \omega_{\langle\tau\rangle}^{\sigma_1}) \wedge \omega^\varrho + (\omega_\sigma^{\tau'} - \omega_{\sigma_1}^{\tau_2}) \wedge \omega^\sigma + (\omega_\tau^{\tau'} - \omega_\varrho^{\varrho'} - \omega_{\sigma_1}^{\sigma_2}) \wedge \omega^\tau$$
$$- (\omega_\varrho^\tau - \omega_{\varrho'}^{\tau'}) \wedge \omega^{\varrho'} + (\omega_{\sigma'}^{\tau'} - \omega_{\sigma_1}^{\tau_1}) \wedge \omega^{\sigma'} = 0, \tag{9.4.41}$$

$$-\sqrt{2}(\sqrt{2}\omega_{\sigma'}^\varrho - \omega_{\langle\sigma\rangle}^{\tau_1}) \wedge \omega^\varrho + (\omega_\sigma^{\sigma'} - \omega_\varrho^{\varrho'} + \omega_{\tau_1}^{\tau_2}) \wedge \omega^\sigma + (\omega_\tau^{\sigma'} + \omega_{\tau_1}^{\sigma_2}) \wedge \omega^\tau$$
$$- (\omega_\varrho^\sigma - \omega_{\varrho'}^{\sigma'}) \wedge \omega^{\varrho'} + (\omega_{\sigma'}^{\tau'} - \omega_{\sigma_1}^{\tau_1}) \wedge \omega^{\tau'} = 0, \tag{9.4.42}$$

$$\sqrt{2}(\omega_{\langle\sigma\rangle}^{\varrho_2} + \omega_{\langle\tau\rangle}^{\varrho_2}) \wedge \omega^\varrho + (\omega_\varrho^\tau + \omega_{\varrho_2}^{\tau_2}) \wedge \omega^\sigma + (\omega_\varrho^\sigma + \omega_{\varrho_2}^{\sigma_2}) \wedge \omega^\tau$$
$$+ (\omega_\varrho^{\tau'} - \omega_{\tau_1}^{\varrho_2}) \wedge \omega^{\sigma'} + (\omega_\varrho^{\sigma'} + \omega_{\sigma_1}^{\varrho_2}) \wedge \omega^{\tau'} = 0, \tag{9.4.43}$$

$$-\sqrt{2}(\sqrt{2}\omega_\tau^\varrho - \omega_{\langle\tau\rangle}^{\sigma_2} - \omega_{\langle\sigma\rangle}^{\sigma_2}) \wedge \omega^\varrho + (\omega_\sigma^\tau + \omega_{\sigma_2}^{\tau_2}) \wedge \omega^\sigma + (\omega_\varrho^{\tau'} - \omega_\tau^{\varrho'}) \wedge \omega^{\varrho'}$$
$$- (\omega_\tau^{\sigma'} + \omega_{\tau_1}^{\sigma_2}) \wedge \omega^{\sigma'} + (\omega_\varrho^{\varrho'} - \omega_\tau^{\tau'} + \omega_{\sigma_1}^{\sigma_2}) \wedge \omega^{\tau'} = 0, \tag{9.4.44}$$

$$
\begin{aligned}
&-\sqrt{2}(\sqrt{2}\omega_\sigma^\varrho - \omega_{\langle\sigma\rangle}^{\tau_2} - \omega_{\langle\tau\rangle}^{\tau_2}) \wedge \omega^\varrho - (\omega_\sigma^\tau + \omega_{\sigma_2}^{\tau_2}) \wedge \omega^\tau + (\omega_\varrho^{\sigma'} - \omega_\sigma^{\varrho'}) \wedge \omega^{\varrho'} \\
&\quad + (\omega_\varrho^{\varrho'} - \omega_\sigma^{\sigma'} - \omega_{\tau_1}^{\tau_2}) \wedge \omega^{\sigma'} - (\omega_\sigma^{\tau'} - \omega_{\sigma_1}^{\tau_2}) \wedge \omega^{\tau'} = 0, \qquad (9.4.45)
\end{aligned}
$$

$$
\begin{aligned}
&-(\omega_{\varrho'}^{\tau'} - \omega_{\varrho_1}^{\tau_1}) \wedge \omega^\sigma + (\omega_{\varrho'}^{\sigma'} - \omega_{\varrho_1}^{\sigma_1}) \wedge \omega^\tau \\
&\quad - (\omega_\tau^{\varrho'} + \omega_{\varrho_1}^{\tau_2}) \wedge \omega^{\sigma'} + (\omega_\sigma^{\varrho'} - \omega_{\varrho_1}^{\sigma_2}) \wedge \omega^{\tau'}, \qquad (9.4.46)
\end{aligned}
$$

$$
\begin{aligned}
&-(\omega_\varrho^\tau - \omega_{\varrho'}^{\tau'}) \wedge \omega^\varrho - (\omega_\sigma^\tau - \omega_{\sigma_1}^{\tau_1}) \wedge \omega^\sigma + \sqrt{2}(\sqrt{2}\omega_\tau^{\varrho'} - \omega_{\langle\tau\rangle}^{\sigma_1}) \wedge \omega^{\varrho'} \\
&\quad + (\omega_\tau^{\sigma'} - \omega_{\sigma_1}^{\tau_2}) \wedge \omega^{\sigma'} + (\omega_\tau^{\tau'} - \omega_\varrho^{\varrho'} - \omega_{\sigma_1}^{\sigma_2}) \wedge \omega^{\tau'} = 0, \qquad (9.4.47)
\end{aligned}
$$

$$
\begin{aligned}
&(\omega_\varrho^\sigma - \omega_{\varrho'}^{\sigma'}) \wedge \omega^\varrho - (\omega_\sigma^\tau - \omega_{\sigma_1}^{\tau_1}) \wedge \omega^\tau - \sqrt{2}(\sqrt{2}\omega_\sigma^{\varrho'} + \omega_{\langle\sigma\rangle}^{\tau_1}) \wedge \omega^{\varrho'} \\
&\quad + (\omega_\varrho^{\varrho'} - \omega_\sigma^{\sigma'} - \omega_{\tau_1}^{\tau_2}) \wedge \omega^{\sigma'} - (\omega_\sigma^{\tau'} + \omega_{\tau_1}^{\sigma_2}) \wedge \omega^{\tau'} = 0, \qquad (9.4.48)
\end{aligned}
$$

$$
\begin{aligned}
&(\omega_\tau^{\varrho'} - \omega_{\tau_1}^{\varrho_2}) \wedge \omega^\sigma + (\omega_\sigma^{\varrho'} + \omega_{\sigma_1}^{\varrho_2}) \wedge \omega^\tau - \sqrt{2}(\omega_{\langle\sigma\rangle}^{\varrho_2} + \omega_{\langle\tau\rangle}^{\varrho_2}) \wedge \omega^{\varrho'} \\
&\quad - (\omega_{\varrho'}^{\tau'} + \omega_{\varrho_2}^{\tau_2}) \wedge \omega^{\sigma'} - (\omega_{\varrho'}^{\sigma'} + \omega_{\varrho_2}^{\sigma_2}) \wedge \omega^{\tau'} = 0, \qquad (9.4.49)
\end{aligned}
$$

$$
\begin{aligned}
&(\omega_\varrho^{\tau'} - \omega_\tau^{\varrho'}) \wedge \omega^\varrho + (\omega_\sigma^{\tau'} + \omega_{\tau_1}^{\sigma_2}) \wedge \omega^\sigma + (\omega_\tau^{\tau'} - \omega_\varrho^{\varrho'} - \omega_{\sigma_1}^{\sigma_2}) \wedge \omega^\tau \\
&\quad - \sqrt{2}(\sqrt{2}\omega_{\tau'}^{\varrho'} - \omega_{\langle\tau\rangle}^{\sigma_2} - \omega_{\langle\sigma\rangle}^{\sigma_2}) \wedge \omega^{\varrho'} + (\omega_{\sigma'}^{\tau'} + \omega_{\sigma_2}^{\tau_2}) \wedge \omega^{\sigma'} = 0, \qquad (9.4.50)
\end{aligned}
$$

$$
\begin{aligned}
&(\omega_\varrho^{\sigma'} - \omega_\sigma^{\varrho'}) \wedge \omega^\varrho + (\omega_\sigma^{\sigma'} - \omega_\varrho^{\varrho'} + \omega_{\tau_1}^{\tau_2}) \wedge \omega^\sigma + (\omega_\tau^{\sigma'} - \omega_{\sigma_1}^{\tau_2}) \wedge \omega^\tau \\
&\quad - \sqrt{2}(\sqrt{2}\omega_{\sigma'}^{\varrho'} - \omega_{\langle\sigma\rangle}^{\tau_2} - \omega_{\langle\tau\rangle}^{\tau_2}) \wedge \omega^{\varrho'} - (\omega_{\sigma'}^{\tau'} + \omega_{\sigma_2}^{\tau_2}) \wedge \omega^{\tau'} = 0. \qquad (9.4.51)
\end{aligned}
$$

Now considering (9.4.33), (9.4.35), (9.4.37) and (9.4.39), after permuting $(\varrho, \sigma, \tau) \mapsto (\sigma, \tau, \varrho)$, together with (9.4.32), (9.4.34), (9.4.36) and (9.4.38), one sees that they yield

$$
\begin{aligned}
\omega_{\langle\varrho\rangle}^{\tau_2} + \omega_{\langle\sigma\rangle}^{\tau_2} &= (c_{\varrho\varrho} + c_{\sigma\sigma})\omega^\tau + (b_{\varrho\varrho} - b_{\sigma\sigma})\omega^{\tau'} \\
&= (b_{\sigma\sigma'} - b_{\varrho\varrho'})\omega^\tau + (c_{\varrho\varrho'} + c_{\sigma\sigma'})\omega^{\tau'}, \\
\omega_{\langle\varrho\rangle}^{\tau_1} + \omega_{\langle\sigma\rangle}^{\tau_1} &= (c_{\sigma\sigma'} - b_{\varrho\varrho})\omega^\tau + (c_{\varrho\varrho} - b_{\sigma\sigma'})\omega^{\tau'} \\
&= -(c_{\varrho\varrho'} + b_{\sigma\sigma})\omega^\tau - (b_{\varrho\varrho'} + c_{\sigma\sigma})\omega^{\tau'},
\end{aligned}
$$

thus

$$
c_{\varrho\varrho} + c_{\sigma\sigma} = b_{\sigma\sigma'} - b_{\varrho\varrho'}, \quad c_{\varrho\varrho'} + c_{\sigma\sigma'} = b_{\varrho\varrho} - b_{\sigma\sigma}.
$$

Applying the permutation $(\varrho, \sigma, \tau) \mapsto (\sigma, \tau, \varrho)$ to these results, and substituting into (9.4.43) and (9.4.22), one obtains $c_{\sigma\sigma'} + c_{\tau\tau'} = c_{\sigma\sigma} + c_{\tau\tau} = 0$ and thus

$$
\omega_{\langle\sigma\rangle}^{\varrho_2} + \omega_{\langle\tau\rangle}^{\varrho_2} = 0. \qquad (9.4.52)
$$

Of course, one also has $c_{\tau\tau'} + c_{\varrho\varrho'} = c_{\tau\tau} + c_{\varrho\varrho} = 0$, $c_{\varrho\varrho'} + c_{\sigma\sigma'} = c_{\varrho\varrho} + c_{\sigma\sigma} = 0$, and hence $c_{\varrho\varrho'} = c_{\varrho\varrho} = c_{\sigma\sigma'} = c_{\sigma\sigma} = c_{\tau\tau'} = c_{\tau\tau} = 0$. Consequently, $b_{\varrho\varrho} = b_{\sigma\sigma} =$

$b_{\tau\tau} = \beta, b_{\varrho\varrho'} = b_{\sigma\sigma'} = b_{\tau\tau'} = \beta'$ and

$$\omega^{\tau_1}_{\langle\varrho\rangle} + \omega^{\tau_1}_{\langle\sigma\rangle} = -\beta\omega^{\tau} - \beta'\omega^{\tau'}. \tag{9.4.53}$$

By (9.4.40) $\beta' = 0$; thus

$$\omega^{\sigma_1}_{\langle\sigma\rangle} = \beta\omega^{\sigma}, \quad \omega^{\sigma_2}_{\langle\sigma\rangle} = 0, \tag{9.4.54}$$

$$\sqrt{2}\omega^{\varrho}_{\sigma} - \omega^{\tau_2}_{\langle\sigma\rangle} = \beta\omega^{\tau'}, \quad \sqrt{2}\omega^{\varrho'}_{\sigma'} - \omega^{\tau_2}_{\langle\sigma\rangle} = 0, \tag{9.4.55}$$

$$\sqrt{2}\omega^{\varrho}_{\sigma'} - \omega^{\tau_1}_{\langle\sigma\rangle} = 0, \quad \sqrt{2}\omega^{\varrho'}_{\sigma} + \omega^{\tau_1}_{\langle\sigma\rangle} = -\beta\omega^{\tau} \tag{9.4.56}$$

and consequently

$$\sqrt{2}(\omega^{\varrho}_{\sigma} - \omega^{\varrho'}_{\sigma'}) = \beta\omega^{\tau'}, \quad \sqrt{2}(\omega^{\sigma'}_{\varrho} - \omega^{\varrho'}_{\sigma}) = \beta\omega^{\tau}. \tag{9.4.57}$$

From (9.4.48) and (9.4.51), it follows that $\omega^{\varrho'}_{\varrho} - \omega^{\sigma'}_{\sigma} - \omega^{\tau_2}_{\tau_1}$ is a linear combination of $\omega^{\varrho}, \omega^{\tau}, \omega^{\varrho'}, \omega^{\sigma'}, \omega^{\tau'}$ and also of $\omega^{\varrho}, \omega^{\sigma}, \omega^{\tau}, \omega^{\varrho'}, \omega^{\tau'}$; and (9.4.41) and (9.4.44), after permuting $(\varrho, \sigma, \tau) \mapsto (\sigma, \tau, \varrho)$, imply that the same form is a linear combination of $\omega^{\sigma}, \omega^{\tau}, \omega^{\varrho}, \omega^{\sigma'}, \omega^{\tau'}$, and also of $\omega^{\sigma}, \omega^{\tau}, \omega^{\sigma'}, \omega^{\tau'}, \omega^{\varrho'}$. Hence

$$\omega^{\varrho'}_{\varrho} - \omega^{\sigma'}_{\sigma} - \omega^{\tau_2}_{\tau_1} = A_{\varrho}\omega^{\varrho} + A_{\tau'}\omega^{\tau'}. \tag{9.4.58}$$

Now (9.4.47), (9.4.48), and (9.4.40), the last after cyclically permuting once and twice, give

$$\omega^{\tau}_{\sigma} - \omega^{\tau_1}_{\sigma_1} = B_{\varrho}\omega^{\varrho} + B_{\varrho'}\omega^{\varrho'}. \tag{9.4.59}$$

The same procedure applied to (9.4.44), (9.4.45), and (9.4.43) gives

$$\omega^{\tau}_{\sigma} - \omega^{\tau_2}_{\sigma_2} = C_{\varrho}\omega^{\varrho} + C_{\varrho'}\omega^{\varrho'}; \tag{9.4.60}$$

similarly, from (9.4.41), (9.4.42), and (9.4.46),

$$\omega^{\tau'}_{\sigma'} - \omega^{\tau_1}_{\sigma_1} = D_{\varrho}\omega^{\varrho} + D_{\varrho'}\omega^{\varrho'}, \tag{9.4.61}$$

and from (9.4.50), (9.4.51), and (9.4.49)

$$\omega^{\tau'}_{\sigma'} - \omega^{\tau_2}_{\sigma_2} = E_{\varrho}\omega^{\varrho} + E_{\varrho'}\omega^{\varrho'}. \tag{9.4.62}$$

Taking (9.4.42) and (9.4.44), and then (9.4.40) after $(\varrho, \sigma, \tau) \mapsto (\tau, \varrho, \sigma)$, and (9.4.49) after $(\varrho, \sigma, \tau) \mapsto (\sigma, \tau, \varrho)$, one obtains

$$\omega^{\sigma'}_{\tau} + \omega^{\sigma_2}_{\tau_1} = F_{\varrho}\omega^{\varrho} + F_{\varrho'}\omega^{\varrho'}; \tag{9.4.63}$$

similarly, from (9.4.41), (9.4.45), (9.4.40), and (9.4.49)

$$\omega^{\tau'}_{\sigma} - \omega^{\tau_2}_{\sigma_1} = G_{\varrho}\omega^{\varrho} + G_{\varrho'}\omega^{\varrho'}. \tag{9.4.64}$$

Finally, (9.4.43), (9.4.47) and (9.4.48), (9.4.50), together with (9.4.46), give

$$\omega_{\varrho}^{\tau'} - \omega_{\tau_1}^{\sigma_2} = H_{\sigma}\omega^{\sigma} + H_{\sigma'}\omega^{\sigma'}, \tag{9.4.65}$$

$$\omega_{\sigma}^{\tau'} + \omega_{\tau_1}^{\sigma_2} = K_{\varrho}\omega^{\varrho} + K_{\varrho'}\omega^{\varrho'} + L_{\sigma'}\omega^{\sigma'}. \tag{9.4.66}$$

All these expressions (9.4.52)–(9.4.66) are to be substituted into (9.4.40)–(9.4.51). The latter, except (9.4.41), (9.4.44), (9.4.47) and (9.4.50), give some relations between the coefficients of these expressions.

Namely, from (9.4.42) $A_{\varrho} = A_{\tau'} = D_{\varrho} = F_{\varrho} = F_{\varrho'} = 0$, $\beta = \sqrt{2}D_{\varrho'}$. Now (9.4.40) and (9.4.49) yield $B_{\sigma} = E_{\sigma} = G_{\sigma'} = 0$.

Further, from (9.4.45) $C_{\varrho} = 0$, $G_{\varrho} = \beta\sqrt{2}$, $\beta = -C_{\varrho'}\sqrt{2}$; from (9.4.48) and (9.4.51), respectively $K_{\varrho'} = 0$, $B_{\varrho'} = -\beta\sqrt{2}$, $\beta = K_{\varrho}\sqrt{2}$ and $E_{\varrho'} = H_{\varrho'} = 0$, $\beta = H_{\varrho}\sqrt{2}$.

Now (9.4.43) and (9.4.46) give $\beta = 0$, $L_{\varrho'} = 0$. Thus all coefficients of these expressions are zero.

The exceptional (9.4.41), (9.4.44), (9.4.47), and (9.4.50) reduce to

$$(\sqrt{2}\omega_{\tau'}^{\varrho} + \omega_{\langle\tau\rangle}^{\sigma_1}) \wedge \omega^{\varrho} = 0, \quad (\sqrt{2}\omega_{\tau}^{\varrho} - \omega_{\langle\tau\rangle}^{\sigma_2}) \wedge \omega^{\varrho} = 0,$$

$$(\sqrt{2}\omega_{\tau}^{\varrho'} - \omega_{\langle\tau\rangle}^{\sigma_1}) \wedge \omega^{\varrho'} = 0, \quad (\sqrt{2}\omega_{\tau'}^{\varrho'} - \omega_{\langle\tau\rangle}^{\sigma_2}) \wedge \omega^{\varrho'} = 0,$$

thus

$$\sqrt{2}\omega_{\tau'}^{\varrho} + \omega_{\langle\tau\rangle}^{\sigma_1} = P_{\varrho}\omega^{\varrho}, \quad \sqrt{2}\omega_{\tau}^{\varrho} - \omega_{\langle\tau\rangle}^{\sigma_2} = Q_{\varrho}\omega^{\varrho},$$

$$\sqrt{2}\omega_{\tau}^{\varrho'} - \omega_{\langle\tau\rangle}^{\sigma_1} = P_{\varrho'}\omega^{\varrho'}, \quad \sqrt{2}\omega_{\tau'}^{\varrho'} - \omega_{\langle\tau\rangle}^{\sigma_2} = Q_{\varrho'}\omega^{\varrho'}.$$

It follows that

$$\sqrt{2}(\omega_{\tau}^{\varrho'} - \omega_{\varrho}^{\tau'}) = P_{\varrho}\omega^{\varrho} + P_{\varrho'}\omega^{\varrho'},$$

$$\sqrt{2}(\omega_{\tau}^{\varrho} - \omega_{\tau'}^{\varrho'}) = Q_{\varrho}\omega^{\varrho} - Q_{\varrho'}\omega^{\varrho'};$$

comparing with (9.4.57), where $\beta = 0$, one concludes that $P_{\varrho} = P_{\varrho'} = Q_{\varrho} = Q_{\varrho'} = 0$.

This concludes the proof of Theorem 9.4.3.

It is clear that the above proof of the theorem, already complicated for the model case $\nu = 3$, will be much more complicated for the general case of arbitrary ν.

Nevertheless, it can be claimed that the general result about the triviality of the second-order envelope of unitary orbits is indeed valid for any dimension, i.e., that a unitary orbit is umbilic-like independently of the dimension.

Remark 9.4.4. The results of the last two sections were published in [Lu 96b] (with some misprints, now corrected), where the above claim was also presented. All this, together with the statement of Remark 9.3.2, can be summarized as follows.

The only symmetric orbits of the Plücker action are the Plücker orbits and the unitary orbits. All these are umbilic-like.

9.5 The Segre Action and Its Symmetric Orbits

Now it is of interest to consider the situation for other actions having symmetric orbits, such as the Segre and Veronese orbits.

First consider the Segre action. The Segre submanifolds $S_{(p,q)}(k)$ in E^{pq+m+1}, $m = p + q$, were introduced above in Section 3.2 and then investigated in Section 4.6. The orthonormal frame bundle can be adapted so that $S_{(p,q)}(k)$ is an integral submanifold of the totally integrable system

$$\omega^{m+1} = \omega^{(i_1 j_2)} = 0, \tag{9.5.1}$$

$$\omega_i^{m+1} = k\omega^i, \quad \omega_{i_1}^{(j_1 k_2)} = \delta_{i_1}^{j_1} k\omega^{k_2}, \quad \omega_{i_2}^{(j_1 k_2)} = \delta_{i_2}^{k_2} k\omega^{j_1}, \tag{9.5.2}$$

$$\omega_{i_1}^{j_2} = 0, \quad \omega_{(i_1 j_2)}^{m+1} = 0, \quad \omega_{(i_1 j_2)}^{(k_1 l_2)} = \delta_{i_1}^{k_1} \omega_{j_2}^{l_2} + \omega_{i_1}^{k_1} \delta_{j_2}^{l_2}, \tag{9.5.3}$$

where $k = \text{const}$, $1 \le i \le m$, $1 \le i_1, k_1 \le p$ and $p + 1 \le j_2, l_2 \le m$ (see (4.6.1), (4.6.6), (4.6.3), (4.6.7)–(4.6.9)). Since the coefficients here are constants, this system determines a Lie subgroup in the Lie group $O((p+1)(q+1), \mathbb{R})$ of motions in $S^{pq+p+q}(k^2)$ which is isomorphic to $O(p+1, \mathbb{R}) \times O(q+1, \mathbb{R})$, and for which $\omega^{i_1}, \omega^{j_2}, \omega_{i_1}^{k_1}, \omega_{j_2}^{l_2}$ are the Maurer–Cartan forms. This subgroup acts in E^{pq+m+1} so that $S_{(p,q)}(k)$ is its orbit. Here e_{m+1} and $e_{i_1 j_2}$ are mutually orthogonal unit vectors normal to this orbit. This action is called the *Segre action* (see [Lu 91e]).

For the frame bundle in E^{pq+m+1} adapted to $S_{(p,q)}(k)$, (9.5.1)–(9.5.3) imply that

$$dx = e_{i_1}\omega^{i_1} + e_{j_2}\omega^{j_2}, \tag{9.5.4}$$

$$de_{i_1} = e_{l_1}\omega_{i_1}^{l_1} + ke_{m+1}\omega^{i_1} + ke_{(i_1 j_2)}\omega^{j_2}, \tag{9.5.5}$$

$$de_{j_2} = e_{l_2}\omega_{j_2}^{l_2} + ke_{m+1}\omega^{j_2} + ke_{(i_1 j_2)}\omega^{i_1}, \tag{9.5.6}$$

$$de_{m+1} = -k(e_{i_1}\omega^{i_1} + e_{j_2}\omega^{j_2}), \tag{9.5.7}$$

$$de_{(i_1 j_2)} = -k(e_{i_1}\omega^{j_2} + e_{j_2}\omega^{i_1}) + e_{(k_1 j_2)}\omega_{i_1}^{k_1} + e_{(i_1 l_2)}\omega_{j_2}^{l_2}. \tag{9.5.8}$$

Every other orbit of the Segre action is described by a point x^* in the normal space of the Segre orbit $S_{(p,q)}(k)$, which has in E^{pq+m+1} the radius vector

$$x^* = x + \mu e_{m+1} + \nu^{i_1 j_2} e_{(i_1 j_2)},$$

where μ and $\nu^{i_1 j_2}$ are some real constants. Under independent orthogonal transformations of $\{e_1, \dots, e_p\}$ and $\{e_{p+1}, \dots, e_m\}$, the system of $\nu^{i_1 j_2}$ behaves as an element of the tensor product of two vector spaces of dimensions p and q. Hence for a given orbit these transformations can be chosen so as to make all $\nu^{i_1 j_2}$ zero, except possibly $\nu^{1\bar{1}} = \nu$, where $\bar{1} = p + 1$, and thus

$$x^* = x + \mu e_{m+1} + \nu e_{(1\bar{1})}.$$

Now

$$dx^* = e_1\theta^1 + e_{\bar{1}}\theta^{\bar{1}} + (1-\mu k)(e_{a_1}\omega^{a_1} + e_{b_2}\omega^{b_2}) + e_{a_1\bar{1}}\omega_1^{a_1} + e_{1b_2}\omega_{\bar{1}}^{b_2},$$

where $2 \le a_1, \ldots \le p$, $p+2 \le b_2, \ldots \le m$, and

$$\theta^1 = (1-\mu k)\omega^1 - \nu k\omega^{\bar{1}}, \quad \theta^{\bar{1}} = -\nu k\omega^1 + (1-\mu k)\omega^{\bar{1}}.$$

Denoting $\Delta = (1-\mu k)^2 - (\nu k)^2$, one gets

$$\Delta\omega^1 = (1-\mu k)\theta^1 + \nu k\theta^{\bar{1}}, \quad \Delta\omega^{\bar{1}} = \nu k\theta^1 + (1-\mu k)\theta^{\bar{1}}.$$

It is convenient here to write $\theta^{a_1} = (1-\mu k)\omega^{a_1}$, $\theta^{b_2} = (1-\mu k)\omega^{b_2}$, $\theta^{a_1\bar{1}} = \omega_1^{a_1}$, $\theta^{1b_2} = \omega_{\bar{1}}^{b_2}$.

Suppose that $\Delta \neq 0$. From (9.5.5)–(9.5.8) it follows that for the orbit described by x^*,

$$h_{11}^* = k\Delta^{-1}[(1-\mu k)e_{m+1} + \nu k e_{1\bar{1}}] = h_{\bar{1}\bar{1}}^*, \quad h_{1\bar{1}}^* = k\Delta^{-1}[\nu k e_{m+1} + (1-\mu k)e_{1\bar{1}}].$$

If in addition $1-\mu k \neq 0$, then

$$h_{a_1a_1}^* = k(1-\mu k)^{-1}e_{m+1} = h_{b_2b_2}^*, \quad h_{a_1b_2}^* = k(1-\mu k)^{-1}e_{a_1b_2},$$

and $h_{1a_1}^* = h_{\bar{1}a_1}^* = h_{1b_2}^* = h_{\bar{1}b_2}^* = 0$.

This orbit is parallel due to Proposition 3.1.1 if and only if

$$dh_{jk}^* - h_{ik}^*\theta_j^k - h_{ji}^*\theta_k^i \tag{9.5.9}$$

has zero normal part for all values of j, k. Consider this for $j = 1$ and $k = a_1$, where the normal part is

$$k\{[-(1-\mu k)^{-1} + \Delta^{-1}(1-\mu k)]e_{m+1} + \Delta^{-1}\nu k e_{(1\bar{1})}\}\omega_1^{a_1}.$$

It is zero only if $\nu = 0$. Then $\Delta = (1-\mu k)^2$ and thus the coefficient before e_{m+1} is also zero. The orbit is a Segre submanifold $S_{(p,q)}(k^*)$, which has the same center as the original $S_{(p,q)}(k)$ and is homothetic to the latter.

If $1-\mu k = 0$, then

$$dx^* = e_1\theta^1 + e_{\bar{1}}\theta^{\bar{1}} + e_{(a_1\bar{1})}\omega_1^{a_1} + e_{(1b_2)}\omega_{\bar{1}}^{b_2},$$

where $\theta^1 = -\nu k\omega^{\bar{1}}$, $\theta^{\bar{1}} = -\nu k\omega^1$. Again from (9.5.5)–(9.5.8) it follows that for the orbit under consideration,

$$h_{11}^* = -\nu^{-1}e_{(1\bar{1})} = h_{\bar{1}\bar{1}}^*, \quad h_{1\bar{1}}^* = -\nu^{-1}e_{m+1}, \quad h_{1(a_1\bar{1})}^* = e_{a_1},$$

$$h_{\bar{1}(1b_2)}^* = e_{b_2}, \quad h_{(1b_2)(1b_2)}^* = -h_{(a_1\bar{1})(a_1\bar{1})}^* = e_{(1\bar{1})}, \quad h_{(1b_2)(a_1\bar{1})}^* = e_{(a_1b_2)},$$

and the remaining vector components of h^* are zero.

The normal part of (9.5.9) for $i = 1$ and $j = \bar{1}$ is $k(\nu^{-1} - 1)(e_{a_1}\omega^{a_1} + e_{b_2}\omega^{b_2})$, thus for a parallel orbit one must have $\nu = 1$. The normal part of (9.5.9) for $i = 1$

and $j = (1b_2)$ is $-k(1+\nu^{-1})e_{(1\bar{1})}\omega^{b_2}$, which cannot be zero if $\nu = 1$. Therefore, if $1 - \nu k = 0$, then there are no parallel orbits.

Finally, suppose $\Delta = 0$. Then $1 - \mu k = \varepsilon \nu k$, where $\varepsilon = \pm 1$, thus $\theta^1 = \nu k(\varepsilon\omega^1 - \omega^{\bar{1}}) = -\varepsilon\theta^{\bar{1}}$, and

$$dx^* = e_1^*\theta^1 + \varepsilon\nu k(e_{a_1}\omega^{a_1} + e_{b_2}\omega^{b_2}) + e_{(a_1\bar{1})}\omega_1^{a_1} + e_{(1b_2)}\omega_{\bar{1}}^{b_2},$$

where $e_1^* = e_1 - \varepsilon e_{\bar{1}}$ is tangent to the orbit.

If $\nu \neq 0$, then

$$de_1^* = e_{a_1}\omega_1^{a_1} - \varepsilon e_{b_2}\omega_{\bar{1}}^{b_2} + k(e_{(1b_2)}\omega^{b_2} - \varepsilon e_{(a_1\bar{1})}\omega^{a_1}) + \varepsilon\nu^{-1}(e_{m+1} - \varepsilon e_{(1\bar{1})})\theta^1.$$

It follows that $h_{11}^* = \varepsilon\nu^{-1}(e_{m+1} - \varepsilon e_{(1\bar{1})})$ and the other components of h^* having subscript 1 are zero.

By (9.5.5), $de_{a_1} = \frac{1}{2}(e_1^* + e_{\bar{1}}^*)\omega_{a_1}^1 + e_{c_1}\omega_{a_1}^{c_1} + k(e_{m+1}\omega^{a_1} + e_{(a_1\bar{1})}\omega^{\bar{1}} + e_{(a_1b_2)}\omega^{b_2})$, where $e_{\bar{1}}^* = e_1 + \varepsilon e_{\bar{1}}$ is normal to the orbit. Therefore, $h_{a_1c_1}^* = \varepsilon\nu^{-1}e_{m+1}\delta_{a_1c_1}$, $h_{a_1b_2}^* = \varepsilon\nu^{-1}e_{(a_1b_2)}$, $h_{a_1(a_1\bar{1})}^* = \frac{1}{2}e_{\bar{1}}^*$, and the other components of h^* having subscript a_1 are zero.

Now the normal part of (9.5.9) for $i = 1$ and $j = a_1$ is

$$-\varepsilon\nu^{-1}\left(\frac{1}{2}e_{m+1} + \varepsilon e_{(1\bar{1})}\right)\omega_1^{a_1} + \nu^{-1}e_{(a_1b_2)}\omega_{\bar{1}}^{b_2} + \frac{1}{2}\varepsilon k e_{\bar{1}}^*\omega^{a_1}$$

and hence nonzero. Thus the orbit is nonparallel.

Finally, if $\nu = 0$, then $\Delta = 0$ implies $\mu = k^{-1}$, thus $x^* = x + k^{-1}e_{m+1}$ and hence $dx^* = 0$. Therefore, in this case the orbit degenerates to the center of the original $S_{(p,q)}(k)$.

All this can be summarized as follows.

Theorem 9.5.1. *Among the orbits of a Segre action in $E^{(p+1)(q+1)}$, the only parallel ones (i.e., symmetric orbits) are the Segre orbits $S_{(p,q)}(k)$ with different constants k, which constitute a cone of mutually homothetic Segre orbits, having vertex at their common center.*

Remark 9.5.2. This is the result of [Lu 91a], where the essential part of the proof was also given.

Remark 9.5.3. Theorem 9.5.1 shows that for the symmetric orbits of the Segre action, the problem of their umbilic-likeness is completely solved by Theorem 4.6.1: only the $S_{(p,q)}(k)$ with $p > 1$ and $q > 1$ are umbilic-like, and for other cases of p, q they are not.

9.6 The Veronese Action and Its Symmetric Orbits

Now consider an m-dimensional Veronese submanifold in ${}_sN^{\frac{1}{2}m(m+3)}(c)$ (cf. Sections 3.3 and 4.7). It is assumed in this section that $s = c = 0$; then the ambient space is the Euclidean space $E^{\frac{1}{2}m(m+3)}$. For a Veronese submanifold, in addition to the usual formulas (cf. Section 2.1)

$$dx = e_i\omega^i, \quad de_i = e_j\omega_i^j + h_{ij}\omega^j, \quad \omega_i^j + \omega_j^i = 0, \quad h_{ij} = h_{ji}, \quad \langle e_i, h_{jk}\rangle = 0, \tag{9.6.1}$$

there hold (3.3.4) and (3.3.2), which can be written here as

$$\langle h_{ij}, h_{kl}\rangle = r^{-2}(2\delta_{ij}\delta_{kl} + \delta_{ik}\delta_{jl} + \delta_{il}\delta_{jk}), \tag{9.6.2}$$

$$dh_{ij} = -r^{-2}(e_i\omega^j + e_j\omega^i + 2\delta_{ij}e_k\omega^k) + h_{kj}\omega_i^k + h_{ik}\omega_j^k, \tag{9.6.3}$$

since here $h^*_{ij} = h_{ij}$, $\beta = 2\alpha$, and $\alpha = r^{-2}$. A Veronese submanifold, defined here in $E^{\frac{1}{2}m(m+3)}$ by the differential system (9.6.1)–(9.6.3) together with some initial conditions, will be denoted below by $V^m(r)$.

Equations (9.6.1) and (9.6.2) show that the vectors e_i and $e_{jk} = rh_{jk}$ $(j \neq k)$ form an orthonormal part of the moving frame adapted to $V^m(r)$, and the vectors $h_{11}, \dots, h_{mm}$ are orthogonal to them and form a regular simplex part of the frame having side length $2r^{-1}$. Thus the frame made up of all these vectors moves as a rigid system in $E^{\frac{1}{2}m(m+3)}$, so that the e_i are tangent and all others are normal to $V^m(r)$.

It follows immediately from (9.6.1) and (9.6.3) that $dz_0 = 0$ for

$$z_0 = x + \frac{mr^2}{2(m+1)}H, \tag{9.6.4}$$

where $H = \frac{1}{m}\sum_{i=1}^m h_{ii}$ is the mean curvature vector of $V^m(r)$, and (9.6.3) implies $dH = -\frac{2(m+1)}{mr^2}dx$. Also (9.6.2) implies that $\|H\| = r^{-1}\sqrt{\frac{2(m+1)}{m}}$; hence $V^m(r)$ lies in a hypersphere of $E^{\frac{1}{2}m(m+3)}$ with center z_0 and radius $r\sqrt{\frac{m}{2(m+1)}}$. Moreover, $V^m(r)$ is a minimal submanifold of this hypersphere, since by (9.6.4) its mean curvature vector H is directed to the center of this hypersphere. Below, this center z_0 will be called the *center of* $V^m(r)$.

Since now $g_{ij} = \delta_{ij}$ and $c = 0$, it follows from (2.1.10) and (2.1.11) that $\Omega_i^j = -\langle h_{ik}, h_{jl}\rangle\omega^k \wedge \omega^l$. Then from (9.6.2) it follows that $\Omega_i^j = -r^{-2}\omega^i \wedge \omega^j$, and hence $V^m(r)$ is intrinsically of constant sectional curvature r^{-2}. Globally, $V^m(r)$ is an elliptic space isometrically imbedded into $E^{\frac{1}{2}m(m+3)}$, because at a point $x \in V^m(r)$, two geodesics of $V^m(r)$ going in two different directions have no other common point, as is easily seen.

Equations (9.6.1)–(9.6.3) for the moving rigid frame $\{x; e_i, h_{ij}\}$ in $E^{\frac{1}{2}m(m+3)}$ can be considered as giving infinitesimally an action of the Lie group $SO(m+1, \mathbb{R})$ in $E^{\frac{1}{2}m(m+3)}$, by rotations around the center z_0 of the Veronese submanifold $V^m(r)$, which is an orbit of the action. Therefore, this action is called the *Veronese action*.

By means of totally geodesic l-dimensional submanifolds of $V^m(r)$, it is possible to introduce an imbedding of the Grassmann manifold $G^{l,m}$ into $E^{\frac{1}{2}m(m+3)}$ as another orbit of the Veronese action.

A totally geodesic l-dimensional submanifold of $V^m(r)$ through a given point $x_0 \in V^m(r)$ is defined by adding to equations (9.6.1)–(9.6.3) the totally integrable system $\omega^p = 0$, $\omega_a^p = 0$, with $1 \le a, b, \ldots \le l$ and $l + 1 \le p, q, \ldots \le m$.

This system and these equations together yield the same equations with $a, b, \ldots$ instead of $i, j, \ldots$. Hence the totally geodesic submanifold defined is $V^l(r)$. If one considers $V^m(r)$ as an isometrically imbedded elliptic space, as above, then the submanifold $V^l(r)$ is the imbedded image of an l-dimensional plane into this space. Therefore, the manifold consisting of all such totally geodesic $V^l(r)$ realizes an imbedding of the Grassmann manifold $G^{l,m}$.

According to (9.6.4), a $V^l(r)$ can be represented by its center, whose radius vector is

$$x^* = x + \frac{r^2}{2(l+1)} \sum_{a=1}^{l} h_{aa}. \tag{9.6.5}$$

The centers of all these $V^l(r)$ of a given $V^m(r)$ form an orbit of the Veronese action, called a *Veronese–Grassmann orbit* $VGr^{l,m}(r)$ of the Veronese action.

For this orbit, (9.6.1) and (9.6.3) imply that

$$dx^* = (l+1)^{-1} e_p \omega^p + r^2 (l+1)^{-1} h_{ap} \omega_a^p.$$

The $(l+1)(m-l)$ mutually orthogonal unit vectors e_p and $e_{ap} = r h_{ap}$ are the tangent vectors of $VGr^{l,m}(r)$; its normal vectors are e_a, h_{ab} and h_{pq}; here a, b and p, q both run over their full range of values, and $h_{ab} = h_{ba}$, $h_{pq} = h_{qp}$. Denoting $(l+1)^{-1}\omega^p = \theta^p$, $r(l+1)^{-1}\omega_a^p = \theta^{ap}$, one obtains

$$dx^* = e_p \theta^p + e_{ap} \theta^{ap}. \tag{9.6.6}$$

Futhermore, from (9.6.1) and (9.6.3) one gets

$$de_p = e_q \omega_p^q + e_{ap} r^{-1} \omega^a + e_a \omega_p^a + h_{pq} \omega^q, \tag{9.6.7}$$

$$\begin{aligned} de_{ap} = & -r^{-1} e_p \omega^a + e_{bq} (\omega_a^b \delta_p^q + \delta_a^b \omega_p^q) - r^{-1} e_a \omega^p \\ & + r h_{ab} \omega_p^b + r h_{qp} \omega_a^q. \end{aligned} \tag{9.6.8}$$

The normal parts of the right-hand sides are, respectively

$$h^*_{pq} \theta^q + h^*_{p(aq)} \theta^{aq}, \quad h^*_{(ap)q} \theta^q + h^*_{(ap)(bq)} \theta^{bq},$$

where

$$\begin{aligned} h^*_{pq} &= (l+1) h_{pq}, \quad h^*_{p(aq)} = -r^{-1}(l+1)\delta_{pq} e_a, \\ h^*_{(ap)(bq)} &= (l+1)(\delta_{ab} h_{pq} - h_{ab} \delta_{pq}). \end{aligned}$$

In the tangent parts it is natural to denote

$$\theta_p^q = \omega_p^q, \quad \theta_p^{aq} = r^{-1} \delta_p^q \omega^a = -\theta_{aq}^p, \quad \theta_{ap}^{bq} = \omega_a^b \delta_p^q + \delta_a^b \omega_p^q. \tag{9.6.9}$$

Now it can be proved that $VGr^{l,m}(r)$ is a symmetric orbit, like $V^m(r)$. Indeed, (9.6.3) implies that dh^*_{pq} has normal component

$$-2r^{-2}(l+1)\delta_{pq}e_a\omega^a + h^*_{sq}\omega^s_p + h^*_{ps}\omega^s_q$$

which coincides with

$$h^*_{sq}\theta^s_p + h^*_{(as)q}\theta^{as}_p + h^*_{ps}\theta^s_q + h^*_{p(as)}\theta^{as}_q,$$

and this, according to Proposition 3.1.1, is necessary for the submanifold to be parallel. Easy calculations show that the same holds for $dh^*_{p(aq)}$ and $dh^*_{(ap)(bq)}$. By the same proposition, these results are also sufficient for the submanifold to be parallel.

Theorem 9.6.1. *A Veronese–Grassmann orbit $VGr^{l,m}(r)$ is an $(l+1)(m-l)$-dimensional parallel submanifold, thus a symmetric orbit, which is also a main symmetric orbit and lies in a hypersphere of $E^{\frac{1}{2}m(m+3)}$ as a minimal submanifold.*

Proof. The first assertion has just been verified above. It is seen that the mean curvature vector of $VGr^{l,m}(r)$ is

$$H^* = \frac{1}{m-l}\sum_p h^*_{pp} + \frac{1}{l(m-l)}\sum_{a,p} h^*_{(ap)(ap)}$$

$$= \frac{l+1}{l(m-l)}\left[(l+1)\sum_p h_{pp} - \sum_a h_{aa}\right].$$

Now one can show by using (9.6.2) that $\langle H^*, h^*_{ij}\rangle$ is proportional to δ_{ij}, as required for a main symmetric orbit.

According to a result of Ferus (see Theorem 8.2.4 above), $VGr^{l,m}(r)$ is a minimal $(l+1)(m-l)$-dimensional submanifold of a hypersphere in $E^{\frac{1}{2}m(m+3)}$ around the center of the Veronese action.

Now this can also be directly verified by means of the formulas above. Indeed, from the expression for H^* one can deduce that H^* has constant length, and (9.6.4) and (9.6.5) imply that $z_0 - x^* = \frac{r^2(m-l)}{2(l+1)(m+1)}H^*$. This finishes the proof.

A given Veronese orbit $V^m(r)$ and the corresponding Veronese–Grassmann orbit $VGr^{l,m}(r)$ are not the only symmetric orbits of the Veronese action in $E^{\frac{1}{2}m(m+3)}$. For example, under this action, every point on the straight line through a point x of $V^m(r)$ and its center z_0 describes a Veronese orbit, which is homothetic to the initial one. The same holds, of course, for Veronese–Grassmann orbits. But even these orbits do not exhaust all the symmetric orbits of this action.

Proposition 9.6.2. *The Veronese action in $E^{\frac{1}{2}m(m+3)}$ has in every hypersphere around its center z_0 two different Veronese orbits; all Veronese orbits of the action lie on two cones with a common vertex at z_0.*

Proof. In the case $l = m - 1$, due to the polarity in elliptic space, the corresponding Veronese–Grassmann orbit $VGr^{m-1,m}(r)$ is actually a $\widetilde{V}^m(\widetilde{r})$, where $\widetilde{r} = m^{-1}r$. Analytically, the unit tangent vectors in (9.6.6) are now $\widetilde{e}_m = e_m$ and $\widetilde{e}_a = e_{am}$, where $1 \leq a, \ldots \leq m - 1$; moreover, $\widetilde{\omega}^m = \theta^m = m^{-1}\omega^p$ and $\widetilde{\omega}^a = \theta^{am} = rm^{-1}\omega_a^m$. Equations (9.6.7) and (9.6.8) can be written in the form (9.6.1), where

$$\widetilde{h}_{am} = -r^{-1}me_a, \quad \widetilde{h}_{ab} = m(\delta_{ab}h_{mm} - h_{ab}), \quad \widetilde{h}_{mm} = mh_{mm}.$$

An easy calculation shows that for these vectors, equations (9.6.2) and (9.6.3) hold, with $\widetilde{\ }$. Moreover, this $\widetilde{V}^m(\widetilde{r})$ is not homothetic to $V^m(r)$ with respect to z_0. But by repeating the construction, one can see that $\widetilde{\widetilde{V}}^m(\widetilde{\widetilde{r}})$ is homothetic to $V^m(r)$.

Remark 9.6.3. In the special case $m = 2$, this result can be found in [Bre 72], Chapter IV, Exercise 8, as the following statement: An $SO(3, \mathbb{R})$-action on S^4 with a three-dimensional principal orbit has two singular orbits, which are projective planes.

In its general form, this result was established in [Lu 95a] (see Corollary 3.4 there). The following assertion was also proved in [Lu 95a] (as Theorem 1): *All other orbits of the Veronese action of $SO(m + 1, \mathbb{R})$ in $E^{\frac{1}{2}m(m+3)}$, except the Veronese and Veronese–Grassmann orbits, are nonsymmetric.*

9.7 The Problem of Umbilic-Likeness of Veronese Orbits

It was shown above that the Segre orbits $S_{(p,q)}(r)$ without circular generators (i.e., with $p > 1$ and $q > 1$), the Plücker orbits $G_{2,p}(r)$ and the unitary orbits in Euclidean space E^n are all umbilic-like. But the situation with Veronese orbits $V^m(r)$ is more complicated.

Theorem 4.7.2 and Proposition 6.3.1 together show that for $m \geq 2$, the second-order envelope of Veronese orbits $V^m(\widetilde{r})$ in $E^{\frac{1}{2}m(m+3)}$, where $\widetilde{r}$ varies, is a parallel submanifold, and hence a single Veronese orbit $V^m(r)$ or its open subset. Here $\frac{1}{2}m(m+3)$ is the minimal dimension of a Euclidean space containing $V^m(r)$. Hence in an ambient space of minimal possible dimension, a Veronese orbit $V^m(r)$ with $m \geq 2$ is umbilic-like, under certain conditions.

For dimension $m = 2$, where $\frac{1}{2}m(m+3) = 5$, Theorem 6.4.1 shows that there exist semiparallel surfaces in E^7 which are second-order envelopes of Veronese surfaces $V^2(\widetilde{r})$ and do not reduce to a single Veronese $V^2(r)$. Therefore, unless there are restrictions on the dimension n of the ambient space, a Veronese orbit is not umbilic-like, at least for $m = 2$ and $n > 5$.

The following theorem shows that this assertion generalizes to dimensions $m > 2$.

Theorem 9.7.1 (see [Lu 91b]). *In Euclidean space $E^{\frac{1}{2}m(m+3)+1}$, $m \geq 2$, there exists a second-order envelope M^m of a one-parameter family of congruent Veronese orbits $V^m(r)$ (i.e., with $r =$* const) *not reducing to a single $V^m(r)$, and it has intrinsically the same Riemannian metric as $V^m(r)$, i.e., this M^m is a constant curvature manifold.*

Proof. Let us consider such an envelope M^m in E^n, $n > \frac{1}{2}m(m+3)$. As is seen from the proof of Theorem 4.5.5, for this M^m and for the moving frame adapted to it, the expressions for dx and de_i must be the same as for a $V^m(r)$, i.e., must coincide with (9.6.1), and also (9.6.2) must hold. But now there are $n - \frac{1}{2}m(m+3)$ additional mutually orthogonal unit frame vectors e_ξ, orthogonal to all e_i and h_{ij}. So the frame vectors normal to M^m are $e_{(ij)} = h_{ij}$ and e_ξ. The first of these are not mutually orthogonal unit vectors, but satisfy the conditions

$$e_{(ij)} = e_{(ji)} \quad \langle e_{(ij)}, e_{(kl)}\rangle = r^{-2}(2\delta_{ij}\delta_{kl} + \delta_{ik}\delta_{jl} + \delta_{il}\delta_{jk}), \tag{9.7.1}$$

which follow from (9.6.1), (9.6.2); moreover

$$\langle e_i, e_j\rangle = \delta_{ij}, \quad \langle e_{(ij)}, e_k\rangle = \langle e_{(ij)}, e_\xi\rangle = 0, \quad \langle e_\xi, e_\eta\rangle = \delta_{\xi\eta}. \tag{9.7.2}$$

Equations (2.1.3), (2.1.5), and (2.1.7), the last two with $c = 0$, now give

$$\omega^{(ij)} = 0, \quad \omega^\xi = 0, \tag{9.7.3}$$

$$\omega_i^{(ij)} = \omega^j, \quad \omega_i^{(jk)} = 0 \quad (i, j, k \text{ distinct}), \quad \omega_i^\xi = 0. \tag{9.7.4}$$

For M^m one has

$$de_{(ii)} = e_i\omega^i_{(ii)} + \sum_{j\neq i} e_j\omega^j_{(ii)} + \sum_{j,k} e_{(jk)}\omega^{(jk)}_{(ii)} + \sum_\xi e_\xi\omega^\xi_{(ii)},$$

$$de_{(ij)} = e_i\omega^i_{(ij)} + e_j\omega^j_{(ij)} + \sum_{k\neq i,j} e_k\omega^k_{(ij)} + \sum_{k,l} e_{(kl)}\omega^{(kl)}_{(ij)} + \sum_\xi e_\xi\omega^\xi_{(ij)}.$$

This can be compared with (3.1.2) for a $V^m(r)$, where $h^\alpha_{ijk} = 0$, and due to the second-order tangency of M^m and $V^m(r)$ the vectors $e_{(ij)}$ are the same as the h_{ij} of $V^m(r)$ at the common point. Here equations (9.6.2) have also to be considered. This coincidence leads to

$$\omega^{(ij)}_{(ii)} = 2\omega^j_i, \quad \omega^{(ik)}_{(ij)} = \omega^k_j \quad (i \neq j), \quad \text{all other} \quad \omega^{(kl)}_{(ij)} = 0. \tag{9.7.5}$$

Differentiation of (9.7.1) and (9.7.2) leads to

$$\omega^j_i + \omega^i_j = 0, \quad \omega^\eta_\xi + \omega^\xi_\eta = 0, \tag{9.7.6}$$

$$\omega^i_{(ii)} = -4r^{-2}\omega^i, \quad 2(m+1)r^{-2}\omega^{(ii)}_\xi + m\omega^\xi_{(ii)} - \sum_{j\neq i}\omega^\xi_{(jj)} = 0, \tag{9.7.7}$$

$$\omega^j_{(ii)} = -2r^{-2}\omega^j, \quad \omega^i_{(ij)} = -r^{-2}\omega^j,$$

$$\omega^\xi_{(ij)} = -r^{-2}\omega^{(ij)}_\xi \quad (i \neq j), \tag{9.7.8}$$

$$\omega^i_{(jk)} = 0 \quad (i, j, k \text{ distinct}). \tag{9.7.9}$$

In the situation of Theorem 9.7.1, the index ξ takes only one value $\xi = \frac{1}{2}m(m+3)+1$. The envelope M^m is defined by the differential system (9.7.3)–(9.7.5). For investigation of this M^m, exterior differentiation must be used together with equations (9.7.6)–(9.7.9).

Equations (9.7.3) give identities, due to (9.7.4). The exterior equations obtained from (9.7.4) are satisfied due to (9.7.5), except $\omega_i^{(kl)} \wedge \omega_{(kl)}^{\xi} = 0$, which give

$$\omega_{(ii)}^{\xi} \wedge \omega^i + \sum_{j \neq i} \omega_{(ij)}^{\xi} \wedge \omega^j = 0.$$

From this it follows by Cartan's lemma that

$$\omega_{(ii)}^{\xi} = \rho_i \omega^i + \sum_{j \neq i} \rho_{ij} \omega^j, \tag{9.7.10}$$

$$\omega_{(ij)}^{\xi} = \rho_{ij} \omega^i + \rho_{ji} \omega^j + \sum_{k \neq i, j} \rho_{ijk} \omega^k \quad (i \neq j), \tag{9.7.11}$$

where ρ_{ijk} is symmetric in all its mutually distinct indices.

The first equation in (9.7.5) yields $\omega_{(ii)}^{\xi} \wedge \omega_{\xi}^{(ij)} = 0$. If $j = i$, then one gets from (9.7.7) that

$$\omega_{(ii)}^{\xi} \wedge \sum_{j \neq i} \omega_{(jj)}^{\xi} = 0; \tag{9.7.12}$$

if $j = k \neq i$, then by (9.7.8)

$$\omega_{(ii)}^{\xi} \wedge \omega_{(ik)}^{\xi} = 0. \tag{9.7.13}$$

From the second equation in (9.7.5) it follows that $\omega_{(ij)}^{\xi} \wedge \omega_{\xi}^{(ik)} = 0$. If $k = i$, then by (9.7.7) and (9.7.13)

$$\omega_{(ij)}^{\xi} \wedge \sum_{l \neq i, j} \omega_{(ll)}^{\xi} = 0; \tag{9.7.14}$$

if $k = j$, then (9.7.8) leads to an identity; if $k \neq i, j$, then

$$\omega_{(ij)}^{\xi} \wedge \omega_{(ik)}^{\xi} = 0. \tag{9.7.15}$$

Among the last equations in (9.7.5), $\omega_{(ii)}^{(jj)} = 0$ $(i \neq j)$ give, similarly, $\omega_{(ii)}^{\xi} \wedge \omega_{\xi}^{(jj)} = 0$. Using (9.7.7) one obtains

$$\omega_{(ii)}^{\xi} \wedge \left(m\omega_{(jj)}^{\xi} - \sum_{k \neq i, j} \omega_{(kk)}^{\xi} \right) = 0,$$

which together with (9.7.14), written in the form

$$\omega_{(ii)}^{\xi} \wedge \left(\omega_{(jj)}^{\xi} + \sum_{k \neq i, j} \omega_{(kk)}^{\xi} \right) = 0,$$

gives

$$\omega^{\xi}_{(ii)} \wedge \omega^{\xi}_{(jj)} = 0.$$

This and (9.7.13)–(9.7.15) imply the mutual proportionality of all 1-forms on the left sides of (9.7.10) and (9.7.11); recall that ξ takes only one value. Hence all rows of the matrix of coefficients of these last expressions are mutually proportional, and therefore the matrix has rank 1. It follows that the columns of this matrix are also mutually proportional.

It can be assumed that at least one of the coefficients $\rho_1, \ldots, \rho_m$ is nonzero; otherwise the proportionality of rows and columns implies $\omega^{\xi}_{(ii)} = \omega^{\xi}_{(jj)} = 0$, and thus M^m would lie in $E^{\frac{1}{2}m(m+3)}$ and be a Veronese orbit or its open subset. By renumbering the frame vectors $e_1, \ldots, e_m$ if needed, one can make $\rho_1 = \rho \neq 0$ and write $\rho_{1u} = \lambda_u \rho$, with $2 \leq u, v, \ldots \leq m$. Then

$$\omega^{\xi}_{(11)} = \rho \left(\omega^1 + \sum_v \lambda_v \omega^v \right), \tag{9.7.16}$$

and due to the proportionality,

$$\omega^{\xi}_{(1u)} = \lambda_u \omega^{\xi}_{(11)}, \tag{9.7.17}$$

$$\omega^{\xi}_{(uv)} = \lambda_u \lambda_v \omega^{\xi}_{(11)}. \tag{9.7.18}$$

It is easy to verify that the conditions obtained from the other equations (9.7.5) are satisfied because of (9.7.16)–(9.7.18). It remains to take exterior derivatives of these last equations to get

$$\theta \wedge \omega^{\xi}_{(11)} + \rho \sum_u \psi_u \wedge \omega^u = 0, \quad \psi_u \wedge \omega^{\xi}_{(11)} = 0, \tag{9.7.19}$$

$$(\lambda_u \psi_v + \lambda_v \psi_u) \wedge \omega^{\xi}_{(11)} = 0, \tag{9.7.20}$$

where

$$\theta = d \ln \rho - 3 \sum_u \lambda_u \omega^u_1, \quad \psi_u = d\lambda_u - \sum_v (\lambda^v_u - \lambda_u \lambda_v \omega^v_1) + \omega^u_1.$$

Since $\rho \neq 0$, the m 1-forms $\omega^{\xi}_{(11)}$ and ω^u are linearly independent and therefore can be taken as basis forms. Then θ and ψ_u are m linearly independent secondary forms. It is seen that equations (9.7.20) are consequences of (9.7.19), so the first character (the rank of the polar system for (9.7.19)) is $s_1 = m$ and hence the Cartan number is $Q = s_1 = m$. Then Cartan's lemma gives

$$\theta = p\omega^{\xi}_{(11)} + \rho \sum_u p_u \omega^u, \quad \psi_u = p_u \omega^{\xi}_{(11)}.$$

Here p, p_u are $N = m$ new coefficients. For the differential system under consideration, Cartan's test criterion $N = Q$ for involutivity is satisfied. Hence the second-order envelope M^m of Theorem 9.7.1, not reducing to a $V^m(r)$, does exist and is defined, up to m real holomorphic functions of one real variable (see [Ca 45], [Fin 48], [BCGGG 91], [AG 93]).

It remains to compare (9.7.1) with (9.6.2) in order to establish that the curvature 2-forms Ω_i^j of M^m are the same as for $V^m(r)$. Hence this M^m is intrinsically also of constant curvature r^{-2}.

This concludes the proof.

The assumption about congruence of enveloping Veronese orbits in Theorem 9.7.1 is superfluous for dimension $m = 2$, since Theorem 6.4.1 shows that for this dimension the conclusion of Theorem 9.7.1 is true without that assumption. On the other hand, the following result shows that for dimension $m > 2$, the congruence assumption is satisfied automatically, and thus not needed.

Proposition 9.7.2 (see [Lu 91b]). *If a submanifold M^m with $m > 2$ in E^n is a second-order envelope of Veronese orbits $V^m(\widetilde{r})$, then these orbits are congruent, i.e.,* $\widetilde{r} = r = \text{const}$.

Proof. As noted in the proof of Theorem 9.7.1, the envelope M^m and every $V^m(\widetilde{r})$ must have the same vector-valued second fundamental form h at their common point. Therefore, by (2.1.9) and (2.1.10), they also have the same curvature 2-forms Ω_i^j. For $V^m(\widetilde{r})$ these 2-forms are $\Omega_i^j = -\widetilde{r}^{-2}\omega^i \wedge \omega^j$ (see (3.3.3), where now $\varepsilon = 1$, $\beta = 2\alpha$, $\alpha = \widetilde{r}^{-2}$; cf. Section 9.6).

For M^m, the curvature forms are the same, and now exterior differentiation (leading to the Bianchi identity (1.3.5)) shows that for $d\widetilde{r} = \widetilde{r}_k\omega^k$ one has $\sum_k \widetilde{r}_k\omega^k \wedge \omega^i \wedge \omega^j = 0$. If $m > 2$ and thus i, j, k take more than two values, this implies $\widetilde{r}_k = 0$, and hence $\widetilde{r} = r = \text{const}$. This concludes the proof.

Remark 9.7.3. Theorem 9.7.1 has been generalized to pseudo-Euclidean spaces only for $m = 2$ in [Lu 99a] (Proposition 4), where it was also shown that the lines of tangency between the envelope and Veronese surfaces are geodesics of constant curvature.

For the situation of Theorem 9.7.1, it was proved in [Lu 91b] that the envelope M^m and Veronese orbits $V^m(r)$ have second-order tangency along the congruent Veronese orbits $V^{m-1}(r)$. Also some properties of the curve described by the centers of $V^m(r)$ (which are a one-parameter family) were investigated in [Lu 91b].

9.8 Umbilic-Likeness of Veronese–Grassmann Orbits

It is remarkable that the Veronese–Grassmann orbits $VGr^{l,m}(r)$ with $0 < l < m-1$, which are the other symmetric orbits of Veronese action, are umbilic-like, while the Veronese orbits are not, when the ambient space does not have minimal possible dimension, as shown above. (Recall, that $VGr^{m-1,m}(r)$ is actually a Veronese orbit $\widetilde{V}^m(\widetilde{r})$, as shown in the proof of Proposition 9.6.2, so that the assumption $0 < l < m-1$ is essential here.)

The umbilic-likeness of $VGr^{l,m}(r)$ with $0 < l < m-1$ is proved in two steps. First, the following codimension reduction theorem will be proved.

Theorem 9.8.1 (see [Lu 95a]). *If a submanifold $M^{(l+1)(m-l)}$ in E^n is a second-order envelope of Veronese–Grassmann orbits $VGr^{l,m}(r)$, $0 < l < m-1$, $n \geq \frac{1}{2}m(m+3)$, then there exists an $E^{\frac{1}{2}m(m+3)} \subset E^n$ containing this $M^{(l+1)(m-l)}$ and all these orbits.*

Proof. Any given $VGr^{l,m}(r)$ lies in an $E^{\frac{1}{2}m(m+3)} \subset E^n$ and satisfies equations (9.6.6)–(9.6.8), where e_p, $e_{ap} = rh_{ap}$ are mutually orthogonal unit tangent vectors and e_a, h_{ab}, h_{pq} are normal vectors. Moreover, e_a are also mutually orthogonal unit vectors, but h_{ab}, h_{pq} are not, as shown by (9.6.2). Indeed,

- $\langle h_{ii}, h_{ii}\rangle = 4r^{-2}$,
- $\langle h_{ii}, h_{jj}\rangle = 2r^{-2}$, $\langle h_{ij}, h_{ij}\rangle = r^{-2}$, if $i \neq j$,
- $\langle h_{ii}, h_{jk}\rangle = \langle h_{ij}, h_{ik}\rangle = 0$ for i, j, k distinct,
- $\langle h_{ij}, h_{kl}\rangle = 0$ for i, j, k, l distinct.

Here $e_{ii} = 2rh_{ii}$ are unit vectors having angle $\frac{\pi}{3}$ between each pair of vectors (they form the "unit regular simplex part" of the frame). The vectors $e_{ij} = rh_{ij}$ with $i \neq j$ are also unit vectors, mutually orthogonal, and orthogonal also to the preceding vectors.

For the moving frame in $E^{\frac{1}{2}m(m+3)}$ consisting of a point of $V^m(r)$ and of the vectors $e_1, \ldots, e_m; e_{11}, \ldots, e_{mm}; e_{12}, e_{13}, \ldots, e_{(m-1)m}$ at this point, the formulas (9.6.1) are

$$dx = \sum_i e_i\omega^i, \quad de_i = \sum_j e_j\omega_i^j + 2e_{ii}\theta^i + \sum_{j\neq i} e_{ij}\theta^j, \tag{9.8.1}$$

where $\omega_i^j + \omega_j^i = 0$, terms without $\sum$ have no summation, and $\theta^i = r^{-1}\omega^i$. For $i = j$, (9.6.3) appears as

$$de_{ii} = -\left(2e_i\theta^i + \sum_{j\neq i} e_j\theta^j\right) + 2\sum_{j\neq i} e_{ij}\omega_i^j, \tag{9.8.2}$$

and for $i \neq j$ as

$$de_{ij} = -(e_i\theta^j + e_j\theta^i) + 2(e_{jj} - e_{ii})\omega_i^j + \sum_{k\neq i,j}(e_{kj}\omega_i^k + e_{ik}\omega_j^k). \tag{9.8.3}$$

The formulas (9.6.7) and (9.6.8) now are

$$\begin{aligned} de_p = &\sum_q e_q\theta_p^q + \sum_{a,q} e_{aq}\theta_p^{aq} - \sum_a e_a(\rho\theta^{ap}) \\ &+ e_{pp}(2\rho\theta^p) + \sum_{q\neq p} e_{pq}(\rho\theta^q), \end{aligned} \tag{9.8.4}$$

$$de_{ap} = \sum_q e_q\theta_{ap}^q + \sum_b e_{bq}\theta_{ap}^{bq} - e_a(\rho\theta^p)$$
$$+ (e_{pp} - e_{aa})(2\rho\theta^{ap}) - \sum_{b\neq a} e_{ab}(\rho\theta^{bp}) + \sum_{q\neq p} e_{pq}(\rho\theta^{aq}), \tag{9.8.5}$$

where $\rho = r^{-1}(l+1)$, and any summing is indicated by $\sum$, as specified. Due to the orthonormality of the tangent frame vectors, one has

$$\theta_p^q + \theta_q^p = \theta_p^{aq} + \theta_{aq}^p = \theta_{ap}^{bq} + \theta_{bq}^{ap} = 0.$$

This follows directly from (9.6.9); moreover, it is seen that

$$\theta_p^{ap} - \theta_q^{aq} = \theta_p^{aq} = 0, \quad \theta_{ap}^{aq} - \theta_p^q = 0 \text{ for } p \neq q, \quad \theta_{ap}^{bq} = 0 \text{ for } a \neq b,\ p \neq q. \tag{9.8.6}$$

For the envelope $M^{(l+1)(m-l)}$, some complementary mutually orthogonal normal frame vectors e_ξ can occur, which are orthogonal to the previous normal vectors; their index range is $\frac{1}{2}m(m+3)+1 \leq \xi, \eta, \ldots \leq n$. From (9.6.6), (9.8.1), and (9.8.2) for $M^{(l+1)(m-l)}$,

$$\theta^a = \theta^{aa} = \theta^{ab} = \theta^{pp} = \theta^{pq} = \theta^\xi = 0, \tag{9.8.7}$$
$$\theta_p^a = -\rho\theta^{ap}, \quad \theta_p^{aa} = \theta_p^{ab} = 0, \tag{9.8.8}$$
$$\theta_p^{pp} = 2\rho\theta^p, \quad \theta_p^{qq} = 0, \quad \theta_p^{pq} = \rho\theta^q, \quad \theta_p^{qr} = 0, \quad \theta_p^\xi = 0, \tag{9.8.9}$$
$$\theta_{ap}^a = -\rho\theta^p, \quad \theta_{bp}^a = 0, \quad \theta_{ap}^{aa} = -2\rho\theta^{ap},$$
$$\theta_{ap}^{ab} = -\rho\theta^{bp}, \quad \theta_{ap}^{bb} = \theta_{ap}^{bc} = 0, \tag{9.8.10}$$
$$\theta_{ap}^{pp} = 2\rho\theta^{ap}, \quad \theta_{ap}^{pq} = \rho\theta^{aq}, \quad \theta_{ap}^{qq} = \theta_{ap}^{qr} = \theta_{ap}^\xi = 0, \tag{9.8.11}$$

where p, q, r take distinct values, as do a, b, c.

For the other frame vectors, one has

$$de_a = e_p\theta_a^p + e_{bp}\theta_a^{bp} + e_b\theta_a^b + e_{bc}\theta_a^{bc} + e_{pq}\theta_a^{pq} + e_\xi\theta_a^\xi, \tag{9.8.12}$$
$$de_{ab} = e_p\theta_{ab}^p + e_{cp}\theta_{ab}^{cp} + e_c\theta_{ab}^c + e_{cd}\theta_{ab}^{cd} + e_{pq}\theta_{ab}^{pq} + e_\xi\theta_{ab}^\xi, \tag{9.8.13}$$
$$de_{pq} = e_s\theta_{pq}^s + e_{as}\theta_{pq}^{as} + e_a\theta_{pq}^a + e_{ab}\theta_{pq}^{ab} + e_{st}\theta_{pq}^{st} + e_\xi\theta_{pq}^\xi, \tag{9.8.14}$$
$$de_\xi = e_p\theta_\xi^p + e_{ap}\theta_\xi^{ap} + e_a\theta_\xi^a + e_{ab}\theta_\xi^{ab} + e_{pq}\theta_\xi^{pq} + e_\eta\theta_\xi^\eta, \tag{9.8.15}$$

where the usual summation convention is now used for all paired upper and lower indices.

For the displacement 1-forms, the usual structure equations hold (see Section 1.2). By means of these equations, it follows from $\theta_p^\xi = 0$ that

$$-\theta^{ap} \wedge \theta_a^\xi + \theta^q \wedge \theta_{pq}^\xi = 0.$$

Cartan's lemma then gives

$$-\theta_a^\xi = A_{ab}^\xi \theta^{bp} + B_{aq}^\xi \theta^q, \quad A_{ab}^\xi = A_{ba}^\xi,$$
$$\theta_{pq}^\xi = B_{aq}^\xi \theta^{ap} + C_{pqs}^\xi \theta^s, \quad C_{pqs}^\xi = C_{psq}^\xi.$$

Since $\theta_{pFq}^\xi = \theta_{qp}^\xi$, and since $p, q, \ldots$ take more than one value, one has $A_{ab}^\xi = 0$, $B_{aq}^\xi = 0$, thus $\theta_a^\xi = 0$, $\theta_{pq}^\xi = C_{pqs}^\xi \theta^s$; the coefficients C_{pqs}^ξ are symmetric in the three subscripts.

Similarly, the equations $\theta_{ap}^\xi = 0$ yield

$$\theta^{aq} \wedge C_{pqs}^\xi \theta^s - \theta^{bp} \wedge \theta_{ab}^\xi = 0.$$

Thus $C_{pqs}^\xi = 0$ if $p \neq q$, so

$$(\theta_{ab}^\xi - \delta_{ab} C_{ppp}^\xi \theta^p) \wedge \theta^{bp} = 0$$

and hence

$$\theta_{ab}^\xi = \delta_{ab} C_{ppp}^\xi \theta^p + D_{abc}^\xi \theta^{cp}.$$

Taking this for two different values of p, one can see that $C_{ppp}^\xi = D_{abc}^\xi = 0$. Consequently,

$$\theta_p^\xi = \theta_{ap}^\xi = \theta_a^\xi = \theta_{pq}^\xi = \theta_{ab}^\xi = 0. \tag{9.8.16}$$

Hence the subspace $E^{\frac{1}{2}m(m+3)}$, spanned at a point of $M^{(l+1)(m-l)}$ by the vectors $e_p, e_{ap}, e_a, e_{ab}, e_{pq}$, is invariant, and contains all enveloping orbits $VGr^{l,m}(r)$ and the envelope $M^{(l+1)(m-l)}$. This concludes the proof.

Note that due to (9.8.16), the last terms in (9.8.12)–(9.8.15) vanish.

Now considering these formulas (9.8.12)–(9.8.15) for $VGr^{l,m}(r)$, then comparison with (9.6.1) and (9.6.3) shows that for $VGr^{l,m}(r)$,

$$\theta_p^{ap} = -\theta_{pp}^a = \theta^a, \quad \theta_p^{aq} = \theta_{pq}^a = \theta_{pp}^{aa} = \theta_{pp}^{ab} = \theta_{pq}^{aa} = \theta_{pq}^{ab} = 0, \tag{9.8.17}$$

$$\theta_a^{ap} = \theta^p, \quad \theta_a^{aa} = 2\theta^a, \quad \theta_a^{ab} = \theta^b,$$
$$\theta_a^{bp} = \theta_a^{bb} = \theta_a^{bc} = \theta_a^{pp} = \theta_a^{pq} = 0, \tag{9.8.18}$$

$$\theta_{aa}^a = -2\theta^a, \quad \theta_{aa}^b = -\theta^b, \quad \theta_{aa}^p = -\theta^p,$$
$$\theta_{aa}^{ap} = 2\theta_a^p, \quad \theta_{aa}^{ab} = 2\theta_a^b, \tag{9.8.19}$$

$$\theta_{aa}^{aa} = \theta_{aa}^{pp} = \theta_{aa}^{pq} = 0, \quad \theta_{ab}^a = -\theta^b, \tag{9.8.20}$$

$$\theta_{ab}^{aa} = 2\theta_b^a, \quad \theta_{ab}^{ap} = \theta_b^p, \quad \theta_{ab}^{ac} = \theta_b^c, \quad \theta_{ab}^p = \theta_{ab}^{pp} = \theta_{ab}^{pq} = 0, \tag{9.8.21}$$

$$\theta_{pp}^p = -2\theta^p, \quad \theta_{pp}^q = -\theta^q, \quad \theta_{pp}^a = -\theta^a, \tag{9.8.22}$$

$$\theta_{pp}^{ap} = 2\theta_p^a, \quad \theta_{pp}^{pq} = 2\theta_p^q, \quad \theta_{pp}^{aa} = \theta_{pp}^{ab} = 0, \tag{9.8.23}$$

$$\theta_{pq}^{q} = -\theta^{p}, \quad \theta_{pq}^{qq} = 2\theta_{p}^{q}, \quad \theta_{pq}^{aq} = \theta_{p}^{a}, \quad \theta_{pq}^{rq} = \theta_{p}^{r},$$

$$\theta_{pq}^{a} = \theta_{pq}^{aa} = \theta_{pq}^{ab} = 0, \tag{9.8.24}$$

where p, q, r have three distinct values, as do a, b, c.

Proposition 9.8.2. *In the situation of Theorem* 9.8.1 *one has* $\rho = \frac{l+1}{r} = \text{const}$, *i.e., all Veronese–Grassmann orbits* $VGr^{l,m}(r)$ *in* $E^{\frac{1}{2}m(m+3)}$ *are congruent.*

Proof. The proof will be obtained by differential prolongation of the system (9.8.7)–(9.8.11) which defines the envelope $M^{(l+1)(m-l)}$. Recall that this procedure consists in taking exterior derivatives of the equations of this system, and then applying Cartan's lemma to the resulting exterior equations (cf. Chapter 2 above).

The following identities for the frame vectors:

$$\langle e_a, e_b\rangle = \delta_{ab}, \quad \langle e_a, e_{pq}\rangle = \langle e_a, e_{bc}\rangle = 0,$$

$$\langle e_{ab}, e_{cd}\rangle = 2\delta_{ab}\delta_{cd} + \delta_{ac}\delta_{bd} + \delta_{ad}\delta_{bc}, \quad \langle e_{ab}, e_{pq}\rangle = 2\delta_{ab}\delta_{pq},$$

$$\langle e_{pq}, e_{rs}\rangle = 2\delta_{pq}\delta_{rs} + \delta_{pr}\delta_{qs} + \delta_{ps}\delta_{qr}$$

give, after differentiation, the following relations:

$$\theta_{a}^{b} + \theta_{b}^{a} = \theta_{pq}^{a} + 2\left(\delta_{pq}\sum_{i}\theta_{a}^{ii} + \theta_{a}^{pq}\right)$$

$$= \theta_{bc}^{a} + 2\left(\delta_{bc}\sum_{i}\theta_{a}^{ii} + \theta_{a}^{bc}\right) = 0, \tag{9.8.25}$$

$$\theta_{ab}^{cd} + \theta_{cd}^{ab} + \delta_{ab}\sum_{i}\theta_{cd}^{ii} + \delta_{cd}\sum_{i}\theta_{ab}^{ii}$$

$$= \theta_{ab}^{pq} + \theta_{pq}^{ab} + \delta_{ab}\sum_{i}\theta_{pq}^{ii} + \delta_{pq}\sum_{i}\theta_{ab}^{ii} = 0, \tag{9.8.26}$$

$$\theta_{pq}^{rs} + \theta_{rs}^{pq} + \delta_{pq}\sum_{i}\theta_{rs}^{ii} + \delta_{rs}\sum_{i}\theta_{pq}^{ii} = 0. \tag{9.8.27}$$

Exterior differentiation of equations (9.8.7) yields identities, due to the other equations of the system. For instance,

$$d\theta^{a} = \theta^{p}\wedge\theta_{p}^{a} + \theta^{ap}\wedge\theta_{ap}^{a} + \sum_{b\neq a}\theta^{bp}\wedge\theta_{bp}^{a},$$

where the right side is zero by (9.8.8) and (9.8.10). Thus $\theta^{a} = 0$ leads to an identity. The situation is similar for other equations (9.8.7).

For the first equations in (9.8.8), the structure equations

$$d\theta^{ap} = \theta^{q}\wedge\theta_{q}^{ap} + \theta^{bq}\wedge(\delta_{q}^{p}\theta_{b}^{a} + \theta_{q}^{p}\delta_{b}^{a}),$$

must be used. Since here

$$d\theta_p^a = \sum_{b,q} (\delta_b^a \theta_p^q - \theta_b^a \delta_p^q) \wedge \theta_q^b + \theta_p^{ap} \wedge \theta_{ap}^a$$

$$+ \sum_{q \neq p} \theta_p^{aq} \wedge \theta_{aq}^a + \theta_p^{pp} \wedge \theta_{pp}^a + \sum_{q \neq p} \theta_p^{pq} \wedge \theta_{pq}^a,$$

taking exterior derivatives and using (9.8.6)–(9.8.11) leads to

$$2(\theta_p^{ap} + \theta_{pp}^a) \wedge \theta^p + \sum_{q \neq p} \theta_{pq}^a \wedge \theta^q - d\ln\rho \wedge \theta^{ap} = 0.$$

Then Cartan's lemma implies

$$2(\theta_p^{ap} + \theta_{pp}^a) = P^{ap}\theta^p + \sum_{q \neq p} Q_q^{ap}\theta^q + R^{ap}\theta^{ap},$$

$$\theta_{pq}^a = Q_q^{ap}\theta^p + \sum_{r \neq p} S_{qr}^{ap}\theta^r + T_q^{ap}\theta^{ap}, \quad (p \neq q),$$

$$-d\ln\rho = R^{ap}\theta^p + \sum_{q \neq p} T_q^{ap}\theta^q + U^{ap}\theta^{ap}.$$

Since the index p takes more than one value, all coefficients in the last equality turn out to be zero. Hence $\rho = \text{const}$, thus $r = \text{const}$, and in the previous equalities the last terms disappear. This verifies the proposition.

Theorem 9.8.3. *A second-order envelope $M^{(l+1)(m-l)}$ of Veronese–Grassmann orbits $VGr^{l,m}(\widetilde{r})$ in E^n, $0 < l < m-1$, $n \geq \frac{1}{2}m(m+3)$, is a single $VGr^{l,m}(r)$ or its subset.*

Proof. By Proposition 9.8.2, all these orbits are congruent here, and hence each $\widetilde{r}$ equals r.

Now the preceding analysis can be continued. As was noted, the last terms in the equations obtained by Cartan's lemma vanish, i.e., $T_q^{ap} = 0$. Then $\theta_{pq}^a = \theta_{qp}^a$ implies $Q_q^{ap} = S_{pp}^{aq}$ and $S_{qr}^{ap} = S_{pr}^{aq}$ for three distinct p, q, r, so that

$$2(\theta_p^{ap} + \theta_{pp}^a) = P^{ap}\theta^p + \sum_{q \neq p} Q_q^{ap}\theta^q, \tag{9.8.28}$$

$$\theta_{pq}^a = Q_q^{ap}\theta^p + Q_p^{aq}\theta^q + \sum_{r \neq p,q} S_{qr}^{ap}\theta^p. \tag{9.8.29}$$

Exterior differentiation of the remaining equations (9.8.8) gives

$$2\theta_{pp}^{aa} \wedge \theta^p + \sum_{q \neq p} \theta_{pq}^{aa} \wedge \theta^q + (2\theta_p^{ap} - \theta_a^{aa}) \wedge \theta^{ap} + \sum_{b \neq a} \theta_b^{aa} \wedge \theta^{bp} = 0,$$

$$2\theta^{ab}_{pp} \wedge \theta^p + \sum_{q\neq p} \theta^{ab}_{pq} \wedge \theta^q + (\theta^{bp}_p - \theta^{ab}_a) \wedge \theta^{ap}$$
$$+ (\theta^{ap}_p - \theta^{ab}_b) \wedge \theta^{bp} + \sum_{c\neq a,b} \theta^{ab}_c \wedge \theta^{cp} = 0. \tag{9.8.30}$$

In (9.8.9), the first equations lead to

$$2\theta^{pp}_{pp} \wedge \theta^p + \sum_{q\neq p} (2\theta^p_q - \theta^{pp}_{pq}) \wedge \theta^q + \sum_a \theta^{pp}_a \wedge \theta^{ap} = 0,$$

the second equations to

$$-2\theta^{qq}_{pp} \wedge \theta^p + (2\theta^q_p - \theta^{qq}_{pq}) \wedge \theta^q - \sum_{s\neq p,q} \theta^{qq}_{ps} \wedge \theta^s + \sum_a \theta^{qq}_a \wedge \theta^{ap} = 0, \tag{9.8.31}$$

the third equation to

$$2(\theta^q_p - \theta^{pq}_{pp}) \wedge \theta^p + \sum_{r\neq p,q} (\theta^q_r - \theta^{pq}_{pr}) \wedge \theta^r + \sum_a \theta^{pq}_a \wedge \theta^{ap} - \sum_a^{r\neq p,q} \theta^{ar}_q \wedge \theta^{ar} = 0,$$

and the fourth equations to

$$-2\theta^{qr}_{pp} \wedge \theta^p + (\theta^r_p - \theta^{qr}_{pq}) \wedge \theta^q + (\theta^q_p - \theta^{qr}_{pr}) \wedge \theta^r + \sum_a \theta^{qr}_a \wedge \theta^{ap}) = 0. \tag{9.8.32}$$

In (9.8.10), the first equations lead to

$$2(\theta^a_{aa} + \theta^{ap}_p - \theta^a_{pp}) \wedge \theta^{ap} - \sum_{q\neq p} \theta^a_{pq} \wedge \theta^{aq} + \sum_{b\neq a} \theta^{bp}_p \wedge \theta^{bp} = 0,$$

the second equations to

$$(\theta^a_{ab} + \theta^{bp}_p) \wedge \theta^{ap} + 2(\theta^a_{bb} - \theta^a_{pp}) \wedge \theta^{bp} -$$
$$- \sum_{q\neq p} (\theta^{aq}_{bp} \wedge \theta^q + \theta^a_{pq} \wedge \theta^{bq}) + \sum_{c\neq a,b} \theta^a_{bc} \wedge \theta^{cp} = 0, \tag{9.8.33}$$

the third equation to

$$(\theta^{aa}_a - 2\theta^{ap}_p) \wedge \theta^p + 2(\theta^{aa}_{aa} - \theta^{aa}_{pp}) \wedge \theta^{ap} -$$
$$- \sum_{q\neq p} \theta^{aa}_{pq} \wedge \theta^{aq} + \sum_{b\neq a} (\theta^{aa}_{ab} - \theta^{ap}_{bp}) \wedge \theta^{bp} = 0,$$

the fourth equation to

$$(\theta_a^{ab} - \theta_p^{bp}) \wedge \theta^p + 2(\theta_{aa}^{ab} - \theta_{pp}^{ab} - \theta_{ap}^{bp}) \wedge \theta^{ap}$$
$$- \sum_{q \neq p} \theta_{pq}^{ab} \wedge \theta^{aq} + \sum_{c \neq a,b} (\theta_{ac}^{ab} - \theta_{cp}^{bp}) \wedge \theta^{cp} = 0,$$

the fifth equations to

$$\theta_a^{bb} \wedge \theta^p + 2(\theta_{aa}^{bb} - \theta_{pp}^{bb}) \wedge \theta^{ap} + (\theta_{ab}^{bb} - 2\theta_{ap}^{bp}) \wedge \theta^{bp} -$$
$$- \sum_{q \neq p} \theta_{pq}^{bb} \wedge \theta^{aq} + \sum_{c \neq a,b} \theta_{ac}^{bb} \wedge \theta^{cp} = 0, \tag{9.8.34}$$

and the sixth equations to

$$\theta_a^{bc} \wedge \theta^p + 2(\theta_{aa}^{bc} - \theta_{pp}^{bc}) \wedge \theta^{ap} + (\theta_{ab}^{bc} - \theta_{ap}^{cp}) \wedge \theta^{bp} +$$
$$+ (\theta_{ac}^{bc} - \theta_{ap}^{bp}) \wedge \theta^{cp} - \sum_{q \neq p} \theta_{pq}^{bc} \wedge \theta^{aq} + \sum_{d \neq a,b,c} \theta_{ad}^{bc} \wedge \theta^{dp} = 0.$$

In (9.8.11), the first equations lead to

$$\theta_a^{pp} \wedge \theta^p + 2\theta_{aa}^{pp} \wedge \theta^{ap} - 2\theta_{pp}^{pp} \wedge \theta^{ap} +$$
$$+ \sum_{q \neq p} (2\theta_{aq}^{ap} - \theta_{pq}^{pp}) \wedge \theta^{aq} + \sum_{b \neq a} \theta_{ab}^{pp} \wedge \theta^{bp} = 0, \tag{9.8.35}$$

the second equations lead to

$$\theta_a^{pq} \wedge \theta^p + 2(\theta_{ap}^{aq} + \theta_{aa}^{pq} - \theta_{pp}^{pq}) \wedge \theta^{ap} - \sum_{r \neq p,q} \theta_{pr}^{pq} \wedge \theta^{ar} + \sum_{b \neq a} \theta_{ab}^{pq} \wedge \theta^{bp} = 0,$$

the third equations lead to

$$\theta_a^{qq} \wedge \theta^p + 2(\theta_{aa}^{qq} - \theta_{pp}^{qq}) \wedge \theta^{ap} +$$
$$= (2\theta_{ap}^{aq} - \theta_{pq}^{qq}) \wedge \theta^{aq} + \sum_{b \neq a} \theta_{ab}^{qq} \wedge \theta^{bp} - \sum_{r \neq p,q} \theta_{pr}^{qq} \wedge \theta^{ar} = 0, \tag{9.8.36}$$

and the fourth equations to

$$\theta_a^{qr} \wedge \theta^p + 2(\theta_{aa}^{qr} - \theta_{pp}^{qr}) \wedge \theta^{ap} + (\theta_{ap}^{ar} - \theta_{pq}^{qr}) \wedge \theta^{aq} +$$
$$+ (\theta_{ap}^{aq} - \theta_{pr}^{qr}) \wedge \theta^{ar} + \sum_{b \neq a} \theta_{ab}^{qr} \wedge \theta^{bp} - \sum_{s \neq p,q,r} \theta_{ps}^{qr} \wedge \theta^{as} = 0. \tag{9.8.37}$$

Here the relations (9.8.6) are taken into consideration, like $\rho = \text{const}$, but (9.8.25)–(9.8.27) not yet.

The analysis of these exterior equations by means of Cartan's lemma leads to several consequences. Here all θ^p and θ^{ap}, the primary 1-forms, are linearly independent.

Let us consider (9.8.34). Here every θ_a^{bb} must be a linear combination of these primary 1-forms for some values of the subscripts p. As a result $\theta_a^{bb} = 0$. For the same reason, in the next exterior equation $\theta_a^{bc} = 0$. Now it follows from the third equation of (9.8.25) that $\theta_{bc}^a = 0$ for every three distinct a, b, c.

Substituting (9.8.29) into the equation preceding (9.8.33), one sees that the only term with $\theta^p \wedge \theta^{aq}$ has the coefficient Q_p^{aq} and the only term with $\theta^r \wedge \theta^{aq}$ has the coefficient S_{qr}^{ap}. Hence these coefficients must be zero and thus $\theta_{pq}^a = 0$, $2(\theta_p^{ap} + \theta_{pp}^a) = P^{ap}\theta^p$. Now it follows from the second equation of (9.8.25) that $\theta_a^{pq} = 0$ for every two distinct p, q.

This procedure can be continued. After a rather complicated analysis, several relations will be obtained for the secondary 1-forms in these exterior equations, which turn out to coincide with (9.8.17)–(9.8.24). This will then finally prove that the assertion of Theorem 9.8.3 is valid.

Remark 9.8.4. The final part of the proof is only given in outline. The details are left to the reader, since they are technically rather complicated, and would take up too much space here.

There is in fact an alternative way to prove Theorem 9.8.3. Namely, one could replace the "unit regular simplex part" of the frame by the orthonormal part, and then work further in the context of the orthonormal frame bundle.

One way to do that is via the formulas

$$
\begin{aligned}
e_{11} &= 2\tilde{e}_{11},\\
e_{22} &= \tilde{e}_{11} + \sqrt{3}\tilde{e}_{22},\\
e_{33} &= \tilde{e}_{11} + \frac{1}{\sqrt{3}}\tilde{e}_{22} + 2\sqrt{\frac{2}{3}}\tilde{e}_{33},\\
e_{44} &= \tilde{e}_{11} + \frac{1}{\sqrt{3}}\tilde{e}_{22} + \frac{1}{\sqrt{6}}\tilde{e}_{33} + \sqrt{\frac{5}{2}}\tilde{e}_{44},\\
&\cdots\cdots\cdots\cdots\\
e_{kk} &= \sum_{\kappa=1}^{k-1} A_\kappa \tilde{e}_{\kappa\kappa} + B_k\tilde{e}_{kk},\\
&\cdots\cdots\cdots\cdots\\
e_{mm} &= \sum_{\kappa=1}^{m-1} A_\kappa \tilde{e}_{\kappa\kappa} + B_m\tilde{e}_{mm},
\end{aligned}
$$

where $A_\kappa = \sqrt{\frac{2}{\kappa(\kappa+1)}}$ and $B_k = \sqrt{\frac{2(k+1)}{k}}$. Now from (9.6.7) and (9.6.8) it follows that the normal parts of de_p and de_{ap} are, respectively,

$$
\rho\left[-e_a\theta^{ap} + \sum_{q\neq p} e_{pq}\theta^q + \left(\sum_{\kappa=1}^{p-1} A_\kappa\tilde{e}_{\kappa\kappa} + B_p\tilde{e}_{pp}\right)\theta^p\right],
$$

$$\rho \left[-e_a\theta^p - \sum_{b\neq a} e_{ab}\theta^{bp} - \left(\sum_{\kappa=1}^{a-1} A_\kappa \widetilde{e}_{\kappa\kappa} + B_a \widetilde{e}_{aa} \right) \theta^{ap} \right.$$
$$\left. + \sum_{q\neq p} e_{pq}\theta^{aq} + \left(\sum_{\kappa=1}^{p-1} A_\kappa \widetilde{e}_{\kappa\kappa} + B_p \widetilde{e}_{pp} \right) \theta^{ap} \right],$$

where, $\rho = (l+1)r^{-1}$, as defined earlier.

Now for this new completely orthonormal frame bundle, replacing the previous "θ" by the symbol "ϑ," the equations become $\omega^\alpha = 0$, $\omega_i^\alpha = h_{ij}^\alpha \omega^j$ and we now have

$$\vartheta^a = \vartheta^{ab} = \vartheta^{pq} = 0,$$
$$\vartheta_p^a = -\rho\vartheta^{ap}, \quad \vartheta_p^{pq} = \rho\vartheta^q \quad (q \neq p), \quad \vartheta_p^{\kappa\kappa} = \rho A_\kappa \vartheta^p \quad (1 \le \kappa \le p-1)$$
$$\vartheta_p^{pp} = B_p\vartheta^p, \quad \vartheta_p^{(p+1)(p+1)} = \cdots = \vartheta_p^{mm} = 0,$$
$$\vartheta_{ap}^b = -\rho\delta_a^b\vartheta^p, \quad \vartheta_{ap}^{ab} = -\rho\vartheta^{bp} \quad (b \neq a), \quad \vartheta_{ap}^{bc} = 0 \quad (a, b, c \text{ dinstinct}),$$
$$\vartheta_{ap}^{pq} = \rho\vartheta^{aq} \quad (q \neq p), \qquad \vartheta_{ap}^{qs} = 0 \quad (p, q, s \text{ distinct}),$$
$$\vartheta_{ap}^{\kappa\kappa} = 0 \quad (1 \le \kappa \le a-1), \quad \vartheta_{ap}^{aa} = \rho(A_a - B_a)\vartheta^{ap},$$
$$\vartheta_{ap}^{\kappa\kappa} = A_\kappa\vartheta^{ap} \quad (a+1 \le \kappa \le p-1), \quad \vartheta_{ap}^{pp} = B_p\vartheta^{ap},$$
$$\vartheta_{ap}^{\kappa\kappa} = 0 \quad (p+1 \le \kappa \le m).$$

The prolongation of this system by exterior differentiation and Cartan's lemma is the alternative way to prove Theorem 9.8.3. Since this is also technically very onerous, details are also omitted here; but see below.

9.9 Detailed Analysis of a Model Case

It is instructive to give the details of the proof of Theorem 9.8.3 in a particular model case. The simplest case is $VGr^{1,3}(r)$, which can be called a *Veronese–Plücker orbit*.

For this $VGr^{1,3}(r)$ the subscript a takes the single value 1 and the subscripts p, q only the two values 2 or 3; one also assumes below that $p \neq q$, which implies that (p, q) is either $(2, 3)$ or $(3, 2)$.

For the sake of symmetry, let the "unit regular simplex part" $\{e_{11}, e_{22}, e_{33}\}$ be replaced by the orthonormal part $\{\widetilde{e}_{11}, \widetilde{e}_{22}, \widetilde{e}_{33}\}$, so that

$$e_{pp} = \widetilde{e}_{11} + c_{p2}\widetilde{e}_{22} + c_{p3}\widetilde{e}_{33},$$

where c_{p2} and c_{p3} are some suitable constants.

Then the system (9.8.7)–(9.8.11) is replaced by

$$\vartheta^1 = \vartheta^{23} = \vartheta^{11} = \vartheta^{pp} = 0, \tag{9.9.1}$$

$$\vartheta_p^1 = -\varrho\vartheta^{1p}, \quad \vartheta_p^{23} = \varrho\vartheta^q, \quad \vartheta_p^{11} = \varrho\vartheta^p,$$

$$\widetilde{\vartheta}_p^{22} = c_{p2}\varrho\vartheta^p, \quad \widetilde{\vartheta}_p^{33} = c_{p3}\varrho\vartheta^p, \tag{9.9.2}$$

$$\vartheta_{1p}^1 = -\varrho\vartheta^p, \quad \vartheta_{1p}^{23} = \varrho\vartheta^{1q}, \quad \vartheta_{1p}^{11} = -\varrho\vartheta^{1p}, \quad \widetilde{\vartheta}_{1p}^{22} = c_{p2}\varrho\vartheta^{1p}, \tag{9.9.3}$$

$$\widetilde{\vartheta}_{1p}^{33} = c_{p3}\varrho\vartheta^{1p}, \tag{9.9.4}$$

where $\varrho = 2r^{-1}$ replaces ρ. Exterior differentiation of the equations of the last column yields

$$\begin{aligned}
&(c_{p3}d\ln\varrho + \widetilde{\vartheta}_{11}^{33} + c_{p2}\widetilde{\vartheta}_{22}^{33}) \wedge \vartheta^p + [(c_{p3} - c_{q3})\vartheta_p^q + \widetilde{\vartheta}_{23}^{33}] \wedge \vartheta^q \\
&\quad - \widetilde{\vartheta}_1^{33} \wedge \vartheta^{1p} + (c_{p3} - c_{q3})\vartheta_p^{1q} \wedge \vartheta^{1q} = 0, \\
&\quad - \widetilde{\vartheta}_1^{33} \wedge \vartheta^p + (c_{q3} - c_{p3})\vartheta_q^{1p} \wedge \vartheta^q \\
&\quad + (c_{p3}d\ln\varrho - \widetilde{\vartheta}_{11}^{33} - c_{p2}\widetilde{\vartheta}_{22}^{33}) \wedge \vartheta^{1p} + [(c_{p3} - c_{q3})\vartheta_{1p}^{1q} + \widetilde{\vartheta}_{23}^{33}] \wedge \vartheta^{1q} = 0.
\end{aligned}$$

Here the simplest situation refers to $c_{23} = c_{33}$. This implies $c_{22} = -c_{32} = 1$, i.e., $c_{p2} = (-1)^p$, and $c_{23} = c_{33} = \sqrt{2}$. Now the first relation implies via Cartan's lemma that

$$\begin{aligned}
\sqrt{2}d\ln\varrho + \widetilde{\vartheta}_{11}^{33} + (-1)^p\widetilde{\vartheta}_{22}^{33} &= A_p\vartheta^p + B_p\vartheta^q + C_p\vartheta^{1p}, \\
\widetilde{\vartheta}_{23}^{33} &= B_p\vartheta^p + D_p\vartheta^q + E_p\vartheta^{1p}, \\
-\widetilde{\vartheta}_1^{33} &= C_p\vartheta^p + E_p\vartheta^q + F_p\vartheta^{1p}.
\end{aligned}$$

Putting first $p = 2$ and then $p = 3$ in the last two equations, one obtains

$$\begin{aligned}
B_2\vartheta^2 + D_2\vartheta^3 + E_2\vartheta^{12} &= B_3\vartheta^3 + D_3\vartheta^2 + E_3\vartheta^{13}, \\
C_2\vartheta^2 + E_2\vartheta^3 + F_2\vartheta^{12} &= C_3\vartheta^3 + E_3\vartheta^2 + F_3\vartheta^{13},
\end{aligned}$$

thus $D_2 = B_3$, $D_3 = B_2$, $E_2 = E_3 = 0$, $C_2 = C_3 = 0$, $F_2 = F_3 = 0$ and so $\widetilde{\vartheta}_1^{33} = 0$. The second relation yields

$$(\sqrt{2}d\ln\varrho - \widetilde{\vartheta}_{11}^{33} + (-1)^p\widetilde{\vartheta}_{22}^{33}) \wedge \vartheta^{1p} + (B_2\vartheta^2 + B_3\vartheta^3) \wedge \vartheta^{1q} = 0,$$

thus $B_2 = B_3 = 0$, so $\widetilde{\vartheta}_{23}^{33} = 0$, and

$$(\sqrt{2}d\ln\varrho - \widetilde{\vartheta}_{11}^{33} + (-1)^p\widetilde{\vartheta}_{22}^{33}) = G_p\vartheta^{1p}.$$

Now

$$\begin{aligned}
\sqrt{2}d\ln\varrho + \widetilde{\vartheta}_{11}^{33} + \widetilde{\vartheta}_{22}^{33} &= A_2\vartheta^2, \\
\sqrt{2}d\ln\varrho + \widetilde{\vartheta}_{11}^{33} - \widetilde{\vartheta}_{22}^{33} &= A_3\vartheta^3, \\
\sqrt{2}d\ln\varrho - \widetilde{\vartheta}_{11}^{33} + \widetilde{\vartheta}_{22}^{33} &= G_2\vartheta^{12},
\end{aligned}$$

$$\sqrt{2}d\ln\varrho - \widetilde{\vartheta}_{11}^{33} - \widetilde{\vartheta}_{22}^{33} = G_3\vartheta^{13}.$$

Consequently,

$$2\sqrt{2}d\ln\varrho = A_2\vartheta^2 + G_3\vartheta^{13} = A_3\vartheta^3 + G_2\vartheta^{12};$$

hence $A_2 = A_3 = G_2 = G_3 = 0$, and so

$$\varrho = \text{const}, \quad \widetilde{\vartheta}_{11}^{33} = \widetilde{\vartheta}_{22}^{33} = 0.$$

Exterior differentiation of the equations of the penultimate column gives

$$-\widetilde{\vartheta}_{11}^{22} \wedge \vartheta^p + [2(-1)^q\vartheta_p^q - \widetilde{\vartheta}_{23}^{22}] \wedge \vartheta^q + \widetilde{\vartheta}_1^{22} \wedge \vartheta^{1p} + 2(-1)^q\vartheta_p^{1q} \wedge \vartheta^{1q} = 0,$$
$$2(-1)^q\vartheta_p^{1q} \wedge \vartheta^p + \widetilde{\vartheta}_1^{22} \wedge \vartheta^q - [2(-1)^p\vartheta_{1p}^{1q} + \widetilde{\vartheta}_{23}^{22}] \wedge \vartheta^{1p} + \widetilde{\vartheta}_{11}^{22} \wedge \vartheta^{1q} = 0.$$

From the first relation

$$\begin{aligned}
-\widetilde{\vartheta}_{11}^{22} &= a_p\vartheta^p + b_p\vartheta^q + c_p\vartheta^{1p} + d_p\vartheta^{1q},\\
2(-1)^q\vartheta_p^q - \widetilde{\vartheta}_{23}^{22} &= b_p\vartheta^p + e_p\vartheta^q + f_p\vartheta^{1p} + g_p\vartheta^{1q},\\
\widetilde{\vartheta}_1^{22} &= c_p\vartheta^p + f_p\vartheta^q + h_p\vartheta^{1p} + i_p\vartheta^{1q},\\
2(-1)^q\vartheta_p^{1q} &= d_p\vartheta^p + g_p\vartheta^q + i_p\vartheta^{1p} + j_p\vartheta^{1q}.
\end{aligned}$$

Since the left sides do not change when p and q are interchanged, one has

$$a_p = b_q = e_p, \quad c_p = d_q = g_p = f_q, \quad h_p = i_q = j_p.$$

Now the second relation gives

$$\begin{aligned}
&(c_p\vartheta^q + h_q\vartheta^{1p} + h_p\vartheta^{1q}) \wedge \vartheta^p + (c_p\vartheta^p + h_p\vartheta^{1p} + h_q\vartheta^{1q}) \wedge \vartheta^q\\
&\quad - [2(-1)^p\vartheta_{1p}^{1q} + \widetilde{\vartheta}_{23}^{22}] \wedge \vartheta^{1p} - (a_p\vartheta^p + a_q\vartheta^q + c_p\vartheta^{1p}) \wedge \vartheta^{1q} = 0,
\end{aligned}$$

and yields $h_p + a_p = 0$. Hence

$$\begin{aligned}
-\widetilde{\vartheta}_{11}^{22} &= a_p\vartheta^p + a_q\vartheta^q + c_p\vartheta^{1p} + c_q\vartheta^{1q},\\
2(-1)^q\vartheta_p^q - \widetilde{\vartheta}_{23}^{22} &= a_q\vartheta^p + a_p\vartheta^q + c_q\vartheta^{1p} + c_p\vartheta^{1q},\\
\widetilde{\vartheta}_1^{22} &= c_p\vartheta^p + c_q\vartheta^q - a_p\vartheta^{1p} - a_q\vartheta^{1q},\\
2(-1)^q\vartheta_p^{1q} &= c_q\vartheta^p + c_p\vartheta^q - a_q\vartheta^{1p} - a_p\vartheta^{1q},\\
2(-1)^p\vartheta_{1p}^{1q} + \widetilde{\vartheta}_{23}^{22} &= a_q\vartheta^p + a_p\vartheta^q + c_q\vartheta^{1p} + c_p\vartheta^{1q},
\end{aligned}$$

and therefore

$$(-1)^p(\vartheta_{1p}^{1q} - \vartheta_p^q) = a_q\vartheta^p + a_p\vartheta^q + c_q\vartheta^{1p} + c_p\vartheta^{1q}.$$

Exterior differentiation of the equations of the middle column yields

$$(-1)^p \widetilde{\vartheta}_{11}^{22} \wedge \vartheta^p - \vartheta_{23}^{11} \wedge \vartheta^q - (2\vartheta_p^{1p} - \vartheta_1^{11}) \wedge \vartheta^{1p} = 0,$$

$$(2\vartheta_p^{1p} - \vartheta_1^{11}) \wedge \vartheta^p + 2\vartheta_q^{1p} \wedge \vartheta^q + (-1)^q \widetilde{\vartheta}_{11}^{22} \wedge \vartheta^{1p} + \vartheta_{23}^{11} \wedge \vartheta^{1q} = 0.$$

Substituting the expression of $\widetilde{\vartheta}_{11}^{22}$ and noting that ϑ_{23}^{11} do not change when p and q are interchanged, the first relation implies that $c_p = 0$ and

$$\vartheta_{23}^{11} = (-1)^p a_q \vartheta^p + (-1)^q a_p \vartheta^q,$$

$$2\vartheta_p^{1p} - \vartheta_1^{11} = v_p \vartheta^{1p}.$$

Now the second relation yields $v_p = a_p = 0$, thus

$$\widetilde{\vartheta}_{11}^{22} = \widetilde{\vartheta}_{23}^{22} + 2\vartheta_2^3 = \widetilde{\vartheta}_1^{22} = \vartheta_2^{13} = \vartheta_3^{12} = \vartheta_{23}^{11} = 2\vartheta_2^{12} - \vartheta_1^{11} = 2\vartheta_3^{13} - \vartheta_1^{11} = 0$$

and hence $\vartheta_2^{12} = \vartheta_3^{13}$.

The equations of the second column give

$$\vartheta_1^{23} \wedge \vartheta^{1p} = 0,$$

$$\vartheta_1^{23} \wedge \vartheta^p + (2\vartheta_{1p}^{1q} + (-1)^p \widetilde{\vartheta}_{23}^{22}) \wedge \vartheta^{1p} = 0,$$

thus $\vartheta_1^{23} = K_p \vartheta^{1p}$, $2\vartheta_{1p}^{1q} + (-1)^p \widetilde{\vartheta}_{23}^{22} = K_p \vartheta^p + L_p \vartheta^{1p}$. For $p = 2$ and $p = 3$, this is

$$\vartheta_1^{23} = K_2 \vartheta^{12} = K_3 \vartheta^{13}, \quad 2\vartheta_{12}^{13} + \widetilde{\vartheta}_{23}^{22} = K_2 \vartheta^2 + L_2 \vartheta^{12} = -(K_3 \vartheta^3 + L_3 \vartheta^{13});$$

hence $K_2 = K_3 = 0$, $L_2 = L_3 = 0$ and $\vartheta_1^{23} = 0$, $\widetilde{\vartheta}_{23}^{22} + 2\vartheta_{12}^{13} = 0$.

Now exterior differentiation of the equations of the first column gives the identity $0 = 0$.

Therefore, the system defining the given second-order envelope M^4 (where now $(l+1)(m-l) = 2 \cdot 2 = 4$), consists of (9.9.1), (9.9.2), where $\varrho = \text{const}$, and $(p, q) = (2, 3)$ or $(3, 2)$,

$$\vartheta_2^{12} = \vartheta_3^{13}, \quad \vartheta_2^{13} = \vartheta_3^{12} = 0, \quad \vartheta_{12}^{13} = \vartheta_2^3, \tag{9.9.5}$$

$$\vartheta_1^{23} = 0, \quad \vartheta_1^{11} = 2\vartheta_2^{12} (= 2\vartheta_3^{13}), \quad \widetilde{\vartheta}_1^{22} = \widetilde{\vartheta}_1^{33} = 0, \tag{9.9.6}$$

$$\vartheta_{23}^{11} = 0, \quad \widetilde{\vartheta}_{23}^{22} = -2\vartheta_2^3 (= -2\vartheta_{12}^{13}), \quad \widetilde{\vartheta}_{23}^{33} = 0, \tag{9.9.7}$$

$$\widetilde{\vartheta}_{11}^{22} = \widetilde{\vartheta}_{11}^{33} = \widetilde{\vartheta}_{22}^{33} = 0; \tag{9.9.8}$$

recall that the matrix $\|\vartheta_I^J\|$, $I, J \in \{2, 3, 12, 13, 1, 23, 11, 22, 33\}$, is skew-symmetric. This system is completely integrable, because exterior differentiation of each of its equations gives the identity $0 = 0$. This conclusion of the analysis is summarized by the following proposition verifying Theorem 9.8.3 for this particular model case.

Proposition 9.9.1. *An envelope M^4 of Veronese–Plücker orbits $VGr^{1,3}(r)$ reduces to a single such orbit $VGr^{1,3}(r)$.*

Remark 9.9.2. It is interesting that Proposition 9.9.1 can be considered as a consequence of Theorem 4.6.1. This is true because $VGr^{1,3}(r)$ can also be considered as a Segre orbit $S_{(2,2)}(\frac{r}{2\sqrt{2}})$.

To establish this, let us show first that $VG_{1,3}(r)$ carries two two-dimensional foliations. One of them is spanned at an arbitrary point x by the orthogonal unit vectors $f_1 = \frac{1}{\sqrt{2}}(e_2 - e_{13})$ and $f_2 = \frac{1}{\sqrt{2}}(e_3 + e_{12})$, the other by $\widetilde{f_1} = \frac{1}{\sqrt{2}}(e_2 + e_{13})$ and $\widetilde{f_2} = \frac{1}{\sqrt{2}}(e_3 - e_{12})$. Indeed, now

$$dx = e_2\vartheta^2 + e_3\vartheta^3 + e_{12}\vartheta^{12} + e_{13}\vartheta^{13} = f_1\phi^1 + f_2\phi^2 + \widetilde{f_1}\widetilde{\phi}^1 + \widetilde{f_2}\widetilde{\phi}^2,$$

where

$$\phi^1 = \frac{1}{\sqrt{2}}(\vartheta^2 - \vartheta^{13}), \quad \phi^2 = \frac{1}{\sqrt{2}}(\vartheta^3 + \vartheta^{12}),$$

$$\widetilde{\phi}^1 = \frac{1}{\sqrt{2}}(\vartheta^2 + \vartheta^{13}), \quad \widetilde{\phi}^2 = \frac{1}{\sqrt{2}}(\theta^3 - \theta^{12}).$$

From (9.9.6),

$$d\phi^1 = \phi^2 \wedge (-\phi_1^2), \quad d\phi^2 = \phi^1 \wedge \phi_1^2, \quad d\widetilde{\phi}^1 = \widetilde{\phi}^2 \wedge (-\widetilde{\phi}_1^2), \quad d\widetilde{\phi}^2 = \widetilde{\phi}^1 \wedge \widetilde{\phi}_1^2,$$

where $\phi_1^2 = \theta_2^3 + \theta_2^{12}$ and $\widetilde{\phi}_1^2 = \theta_2^3 - \theta_2^{12}$. Hence the differential systems $\phi^1 = \phi^2 = 0$ and $\widetilde{\phi}^1 = \widetilde{\phi}^2 = 0$ are both totally integrable and define the foliations above.

A straightforward computation shows that

$$df_1 = f_2\phi_1^2 + \rho[(f_5 + f_6)\widetilde{\phi}^2 + (f_7 - f_8)\widetilde{\phi}^1 + \sqrt{2}f_9\phi^1], \tag{9.9.9}$$
$$df_2 = -f_1\phi_1^2 + \rho[(-f_5 + f_6)\widetilde{\phi}^1 + (f_7 + f_8)\widetilde{\phi}^2 + \sqrt{2}f_9\phi^2], \tag{9.9.10}$$
$$d\widetilde{f_1} = \widetilde{f_2}\widetilde{\phi}_1^2 + \rho[(-f_5 + f_6)\phi^2 + (f_7 - f_8)\phi^1 + \sqrt{2}f_9\widetilde{\phi}^1], \tag{9.9.11}$$
$$d\widetilde{f_2} = -\widetilde{f_1}\widetilde{\phi}_1^2 + \rho[(f_5 + f_6)\phi^1 + (f_7 + f_8)\phi^2 + \sqrt{2}f_9\widetilde{\phi}^2], \tag{9.9.12}$$

where $f_5 = e_1$, $f_6 = e_{23}$, $f_7 = e_{11}$, $f_8 = e_{22}$, $f_9 = e_{33}$. Due to (9.9.2)–(9.9.8),

$$df_9 = -\sqrt{2}\rho(e_2\vartheta^2 + e_3\vartheta^3 + e_{12}\vartheta^{12} + e_{13}\vartheta^{13}) = -\sqrt{2}\rho dx.$$

Hence $d[x + (\sqrt{2}\rho)^{-1}f_9] = 0$, where $\rho = 2r^{-1}$ as above. Thus $c = x + \frac{r}{2\sqrt{2}}f_9$ is the radius vector of a fixed point in E^9. So x is a point of a hypersphere $S^8(r^*)$, $r^* = \frac{r}{2\sqrt{2}}$, and $VG_{1,3}(r)$ is a submanifold of this hypersphere.

From (9.9.9)–(9.9.12) it follows that the leaves of the two foliations above are great 2-spheres in $S^8(r^*)$, totally orthogonal at the arbitrary point $x \in VG_{1,3}(r)$.

All this shows that $VGr_{1,3}(r)$ is actually the Segre submanifold $S_{(2,2)}(r^*)$.

Therefore, Proposition 9.9.1, which is Theorem 9.8.3 for the model case, can indeed be considered as a consequence of Theorem 4.6.1.

10

Geometric Descriptions in General

Normally flat semiparallel submanifolds were geometrically described in Section 5.4 as warped products which are second-order envelopes of products of several spheres, circles, and a plane. In this chapter this kind of descriptions is investigated in some more general situations.

10.1 Products of Umbilic-Like Orbits

According to Theorem 4.5.5, every semiparallel submanifold M^m in E^n is a second-order envelope of parallel submanifolds. By Proposition 5.2.1, every parallel normally flat submanifold in E^n is a product of several spheres and, possibly, some circles and a plane. Therefore, every normally flat semiparallel M^m in E^n is a second-order envelope of such products. In Theorem 5.4.1 these envelopes were described as certain warped products, which in general are not parallel submanifolds. The simplest situation occurs when the enveloping parallel submanifolds are products of multidimensional spheres only, i.e., there are no circles, nor a plane. The spheres are umbilic submanifolds, and now the problem arises: are their products umbilic-like? In other words: is a second-order envelope of products of spheres a single such product?

A similar problem also arises in a more general setting. According to a result of Ferus (see Theorem 8.2.5 above), a complete parallel submanifold in E^n is a product of several main symmetric orbits which are standardly imbedded symmetric R-spaces, and possibly a plane and some circles. Therefore, every semiparallel submanifold in E^n is a second-order envelope of such parallel products. The simplest situation occurs when the latter are the products only of main symmetric orbits, and even simpler is the case where all these are umbilic-like orbits. So the following generalization of our problem arises: Are products of umbilic-like main symmetric orbits also umbilic-like?

The following result gives a positive answer to the first problem.

Proposition 10.1.1. *Every product of multidimensional spheres is umbilic-like in E^n.*

Proof. A second-order envelope of products of spheres $S^{m_1}(c_1) \times \cdots \times S^{m_r}(c_r)$ with $m_\rho > 1$ and variable c_ρ, $\rho \in \{1, \cdots, r\}$, in E^n is a semiparallel normally flat

Ü. Lumiste, *Semiparallel Submanifolds in Space Forms*,
DOI 10.1007/978-0-387-49913-0_11,

submanifold, and therefore the results of Chapter 5 can be used, where all $k_{(\rho)}$ are now nonsimple, and $\langle k_{(\rho)}, k_{(\sigma)}\rangle = 0$ for $\rho \neq \sigma$ (see (5.1.5), where now $c = 0$ in E^n and therefore $k_i^* = k_i$). In particular, formulas (5.1.1)–(5.1.3) hold, together with Lemma 5.1.2.

By differentiation, one obtains $\langle dk_{(\rho)}, k_{(\sigma)}\rangle + \langle k_{(\rho)}, dk_{(\sigma)}\rangle = 0$. Now substituting from (5.1.2) and applying Lemma 5.1.2 gives

$$\left\langle \sum_{\tau\neq\rho} \lambda_{(\rho)j_\tau}(k_{(\rho)} - k_{(\tau)})\omega^{j_\tau}, k_{(\sigma)} \right\rangle + \left\langle k_{(\rho)}, \sum_{\tau\neq\sigma} \lambda_{(\sigma)j_\tau}(k_{(\sigma)} - k_{(\tau)})\omega^{j_\tau} \right\rangle = 0.$$

This leads to

$$\sum_{j_\sigma} \lambda_{(\rho)j_\sigma} k_{(\sigma)}^2 \omega^{j_\sigma} + \sum_{j_\rho} \lambda_{(\sigma)j_\rho} k_{(\rho)}^2 \omega^{j_\rho} = 0.$$

Here the superscripts j_ρ and j_σ run over disjoint index ranges.

Therefore, ω^{j_σ} and ω^{j_ρ} are linearly independent, and since $k_{(\sigma)}^2$ and $k_{(\rho)}^2$ are nonzero, $\lambda_{(\rho)j_\sigma} = \lambda_{(\sigma)j_\rho} = 0$. Thus, due to (5.1.6), $L_{i_\rho j_\sigma} = 0$ for every pair $\rho \neq \sigma$; also $Li_\rho j_\rho = K_{i_\rho} = 0$ for every ρ. This implies that all $h_{ijk} = 0$; hence the above envelope is a parallel submanifold and so is a single product of spheres, as claimed.

Now consider the problem in a more general setting, where it turns out that the second problem also has a positive answer.

Theorem 10.1.2. *Every product of umbilic-like main symmetric orbits in E^n is umbilic-like, i.e., if a submanifold M^m in E^n is a second-order envelope of products $M^{m_1} \times \cdots \times M^{m_r}$, with every M^{m_ρ} an umbilic-like main symmetric orbit for $1 \leq \rho \leq r$, $m = m_1 + \cdots + m_r$, then this M^m is a single such product.*

Proof. The vector-valued second fundamental form of this envelope M^m is at every point the same as for such a product. Consider the orthonormal frame bundle adapted to M^m so that the tangent vectors e_{i_ρ} are also tangent to M^{m_ρ} at their common point x, the normal vector $e_{m+\rho}$ is collinear with the mean curvature vector H^ρ of M^{m_ρ}, and e_{α_ρ} are the remaining frame vectors in the principal normal space of M^{m_ρ} at x. If there are more frame vectors at x normal to M^m, they will be denoted by e_ξ. Then M^m is defined in E^n by the Pfaffian system

$$\omega^{m+\rho} = \omega^{\alpha_\rho} = \omega^\xi = 0,$$

$$\omega_{i_\sigma}^{m+\rho} = \delta_\sigma^\rho \kappa_\sigma \omega^{i_\sigma}, \quad \omega_{i_\sigma}^{\alpha_\rho} = \delta_\sigma^\rho h_{i_\rho j_\rho}^{\alpha_\rho} \omega^{j_\rho}, \quad \omega_{i_\rho}^\xi = 0, \tag{10.1.1}$$

so that $h_{i_\rho j_\rho} = \kappa_\rho \delta_{i_\rho j_\rho} e_{m+\rho} + h_{i_\rho j_\rho}^{\alpha_\rho} e_{\alpha_\rho}$ and $h_{i_\rho j_\sigma} = 0$, if $\rho \neq \sigma$. This implies $H^\rho = \kappa_\rho e_{m+\rho} + m_\rho^{-1} e_{\alpha_\rho} \sum_{i_\rho} h_{i_\rho i_\rho}^{\alpha_\rho}$, and since $e_{m+\rho}$ is taken collinear with H^ρ, one concludes that

$$\sum_{i_\rho} h_{i_\rho i_\rho}^{\alpha_\rho} = 0. \tag{10.1.2}$$

Exterior differentiation of equations (10.1.1) gives the covariant exterior equations. The first ones with $\rho \neq \sigma$ lead to

$$\omega_{j_\rho}^{i_\sigma} \wedge \omega^{j_\rho} + \kappa_\rho^{-1} \sum_{j_\sigma}^{\rho \neq \sigma} (\delta_{i_\sigma j_\sigma} \kappa_\sigma \omega_{m+\sigma}^{m+\rho} + h_{i_\sigma j_\sigma}^{\alpha_\sigma} \omega_{\alpha_\sigma}^{m+\rho}) \wedge \omega^{j_\sigma} = 0.$$

By Cartan's lemma

$$\omega_{j_\rho}^{i_\sigma} = \lambda_{j_\rho k_\rho}^{i_\sigma} \omega^{k_\rho} + \mu_{j_\rho j_\sigma}^{i_\sigma} \omega^{j_\sigma},$$

$$\kappa_\rho^{-1} (\delta_{i_\sigma j_\sigma} \kappa_\sigma \omega_{m+\sigma}^{m+\rho} + h_{i_\sigma j_\sigma}^{\alpha_\sigma} \omega_{\alpha_\sigma}^{m+\rho}) = \mu_{j_\rho j_\sigma}^{i_\sigma} \omega^{j_\rho} + \nu_{j_\sigma k_\sigma}^{i_\sigma} \omega^{k_\sigma},$$

where $\lambda_{j_\rho k_\rho}^{i_\sigma} = \lambda_{k_\rho j_\rho}^{i_\sigma}$, $\nu_{j_\sigma k_\sigma}^{i_\sigma} = \nu_{k_\sigma j_\sigma}^{i_\sigma}$. Then $\omega_{j_\rho}^{i_\sigma} + \omega_{i_\sigma}^{j_\rho} = 0$ implies $\mu_{j_\rho j_\sigma}^{i_\sigma} = -\lambda_{i_\sigma j_\sigma}^{j_\rho}$, and thus

$$\omega_{j_\rho}^{i_\sigma} = \lambda_{j_\rho k_\rho}^{i_\sigma} \omega^{k_\rho} - \lambda_{i_\sigma k_\sigma}^{j_\rho} \omega^{k_\sigma}, \quad \rho \neq \sigma. \tag{10.1.3}$$

The same first equations (10.1.1) with $\rho = \sigma$ lead to

$$\left(\delta_{i_\rho j_\rho} d \ln \kappa_\rho + \kappa_\rho^{-1} h_{i_\rho j_\rho}^{\alpha_\rho} \omega_{\alpha_\rho}^{m+\rho} - \sum_{k_\tau, \tau \neq \rho} \lambda_{i_\rho j_\rho}^{k_\tau} \omega^{k_\tau} \right) \wedge \omega^{j_\rho} = 0,$$

and so

$$\delta_{i_\rho j_\rho} d \ln \kappa_\rho + \kappa_\rho^{-1} h_{i_\rho j_\rho}^{\alpha_\rho} \omega_{\alpha_\rho}^{m+\rho} - \sum_{k_\sigma, \sigma \neq \rho} \lambda_{i_\rho j_\rho}^{k_\sigma} \omega^{k_\sigma} = \pi_{i_\rho j_\rho k_\rho} \omega^{k_\rho}, \tag{10.1.4}$$

where the last coefficients are symmetric in all the indices.

Similarly, the second equations of (10.1.1) with $\rho \neq \sigma$ imply

$$\delta_{i_\sigma j_\sigma} \kappa_\sigma \omega_{m+\sigma}^{\alpha_\rho} + h_{i_\sigma j_\sigma}^{\beta_\sigma} \omega_{\beta_\sigma}^{\alpha_\rho} + \lambda_{i_\sigma j_\sigma}^{j_\rho} h_{j_\rho k_\rho}^{\alpha_\rho} \omega^{k_\rho} = \chi_{i_\sigma j_\sigma k_\sigma}^{\alpha_\rho} \omega^{k_\sigma}, \tag{10.1.5}$$

$$\sum_{j_\rho} (\lambda_{j_\rho k_\rho}^{i_\sigma} h_{j_\rho l_\rho}^{\alpha_\rho} - \lambda_{j_\rho l_\rho}^{i_\sigma} h_{j_\rho k_\rho}^{\alpha_\rho}) = 0, \tag{10.1.6}$$

and the same equations with $\rho = \sigma$ imply

$$\bar{\nabla}_\rho h_{i_\rho j_\rho}^{\alpha_\rho} + \delta_{i_\rho j_\rho} \kappa_\rho \omega_{m+\rho}^{\alpha_\rho} - \sum_{l_\rho, k_\tau}^{\tau \neq \rho} h_{i_\rho l_\rho}^{\alpha_\rho} \lambda_{l_\rho j_\rho}^{k_\tau} \omega^{k_\tau} = \chi_{i_\rho j_\rho k_\rho}^{\alpha_\rho} \omega^{k_\rho}, \tag{10.1.7}$$

where

$$\bar{\nabla}_\rho h_{i_\rho j_\rho}^{\alpha_\rho} = d h_{i_\rho j_\rho}^{\alpha_\rho} - h_{k_\rho j_\rho}^{\alpha_\rho} \omega_{i_\rho}^{k_\rho} - h_{i_\rho k_\rho}^{\alpha_\rho} \omega_{j_\rho}^{k_\rho} + h_{i_\rho j_\rho}^{\beta_\rho} \omega_{\beta_\rho}^{\alpha_\rho};$$

moreover, the coefficients on the right sides are symmetric in all three subscripts.

From the last equations (10.1.1) it follows that

$$\delta_{i_\rho j_\rho} \kappa_\rho \omega_{m+\rho}^{\xi} + h_{i_\rho j_\rho}^{\alpha_\rho} \omega_{\alpha_\rho}^{\xi} = h_{i_\rho j_\rho k_\rho}^{\xi} \omega^{k_\rho}. \tag{10.1.8}$$

Now consider on M^m the distribution of all tangent subspaces M^{m_ρ} for a fixed value of ρ. It is defined by the Pfaffian system $\omega^{i_1} = \cdots = \omega^{i_{\rho-1}} = \omega^{i_{\rho+1}} = \cdots = \omega^{i_r} = 0$, and it is a foliation, since by (10.1.3) for $\sigma \neq \rho$ the differentials

$$d\omega^{i_\sigma} = \omega^{j_\sigma} \wedge \left(\omega^{i_\sigma}_{j_\sigma} + \sum_{j_\rho}^{\rho \neq \sigma} \lambda^{j_\rho}_{i_\sigma j_\sigma} \omega^{j_\rho} \right)$$

vanish due to the equations of the same system. The leaves of this foliation are second-order envelopes of M^{m_ρ}, for every fixed value of ρ. Since all M^{m_ρ} were assumed to be umbilic-like, these leaves are the exemplars of the single M^{m_ρ}, hence parallel ones. This implies that

$$h_{i_\rho j_\rho k_\rho} = 0. \tag{10.1.9}$$

In particular, $h^{m+\rho}_{i_\rho j_\rho k_\rho} = 0$; but for a leaf with a fixed value of ρ, this together with (2.2.2), (2.2.3), and $h^{m+\rho}_{i_\rho j_\rho} = \kappa_\rho \delta_{i_\rho j_\rho}$ implies

$$d\kappa_\rho \delta_{i_\rho j_\rho} + h^{\alpha_\rho}_{i_\rho j_\rho} \omega^{m+\rho}_{\alpha_\rho} = 0, \tag{10.1.10}$$

which by (10.1.2) leads to $d\kappa_\rho = 0$, and also to $h^{\alpha_\rho}_{i_\rho j_\rho} \omega^{m+\rho}_{\alpha_\rho} = 0$. In the last equation, the matrix of coefficients with row pair-index $(i_\rho j_\rho)$ and column index α_ρ has maximal rank; hence it follows that

$$d\kappa_\rho = \omega^{m+\rho}_{\alpha_\rho} = 0. \tag{10.1.11}$$

Substitution into (10.1.4) gives $-\sum_{k_\sigma}^{\sigma \neq \rho} \lambda^{k_\sigma}_{i_\rho j_\rho} \omega^{k_\sigma} = \pi_{i_\rho j_\rho k_\rho} \omega^{k_\rho}$. Since all ω^{k_σ} and ω^{k_ρ} with $\sigma \neq \rho$ are linearly independent, it follows that

$$\lambda^{k_\sigma}_{i_\rho j_\rho} = \pi_{i_\rho j_\rho k_\rho} = 0 \tag{10.1.12}$$

for every pair of distinct values of ρ, σ. Now (10.1.3) implies $\omega^{i_\sigma}_{j_\rho} = 0$, and the equations preceding (10.1.2) give

$$\kappa_\rho^{-1}(\delta_{i_\sigma j_\sigma} \kappa_\sigma \omega^{m+\rho}_{m+\sigma} + h^{\alpha_\sigma}_{i_\sigma j_\sigma} \omega^{m+\rho}_{\alpha_\sigma}) = \nu^{i_\sigma}_{j_\sigma k_\sigma} \omega^{k_\sigma}.$$

Summing over $i_\sigma = j_\sigma$, one gets $\omega^{m+\rho}_{m+\sigma} = \kappa_\rho (m_\sigma \kappa_\sigma)^{-1} \nu_{k_\sigma} \omega^{k_\sigma}$; and since $\omega^{m+\rho}_{m+\sigma} + \omega^{m+\sigma}_{m+\rho} = 0$, one has $\nu_{k_\rho} = \nu_{k_\sigma} = 0$, whence

$$\omega^{m+\rho}_{m+\sigma} = 0, \quad \omega^{m+\rho}_{\alpha_\sigma} = \nu^\rho_{k_\sigma} \omega^{k_\sigma}, \tag{10.1.13}$$

where $\rho \neq \sigma$.

Now for all fixed values of $\rho \in \{1, \ldots, r\}$, the foliations on M^m tangent to M^{m_ρ} are parallel in the Riemannian connection ∇ induced on M^m by immersion. The argument used in the proof of Theorem 8.1.5 then shows that M^m is the product of the leaves of these foliations. For these leaves, equation (10.1.9) holds; hence they are parallel submanifolds, which must therefore coincide with M^{m_ρ}. This concludes the proof.

In addition, note that (10.1.6) is now satisfied due to (10.1.11), and since $h^{\alpha_\rho}_{i_\rho j_\rho k_\rho} = h^{\xi}_{i_\rho j_\rho k_\rho} = 0$, the relations (10.1.7) and (10.1.8) become trivial identities $0 = 0$. Moreover, since $h_{i_\rho j_\rho} = \delta_{i_\rho j_\rho} \kappa_\rho e_{m+\rho} + h^{\alpha_\rho}_{i_\rho j_\rho} r_{\alpha_\rho}$ span the principal normal subspace of M^{m_ρ}, the matrix of coefficients on the right-hand side of (10.1.8) has maximal rank, as noted above, and therefore $\omega^{\xi}_{m+\rho} = \omega^{\xi}_{\alpha_\rho} = 0$.

10.2 General Semiparallel Submanifolds and Their Adapted Frame Bundles

A general semiparallel submanifold M^m in E^n is, by Theorems 4.5.5 and 8.2.5, the second-order envelope of products

$$M^{m_1} \times \cdots \times M^{m_s} \times S^1(c_{s+1}) \times \cdots \times S^1(c_{s+q}) \times E^{m_0}$$

of main symmetric orbits M^{m_ρ}, $\rho = 1, \ldots, s$, each imbedded as a minimal submanifold in a sphere $S^{n_\rho}(c_\rho)$, and of circles $S^1(c_{s+a})$, $1 \leq a \leq q$, and an m_0-dimensional plane E^{m_0}; here $m = m_0 + m^* + q$, $m^* = \sum_\rho m_\rho$, and c_ρ, c_{s+a} can vary on M^m.

The orthonormal frame bundle can be adapted to such an M^m so that, as in Section 10.1, the tangent vectors e_{i_ρ} are also tangent to M^{m_ρ} at their common point x, the normal vector $e_{m+\rho}$ is collinear with the mean curvature vector H^ρ of M^{m_ρ}, and e_{α_ρ} are the remaining frame vectors in the principal normal space of M^{m_ρ} at x. Moreover, let e_{m^*+a} and e_{m+s+a} be the tangent and the normal, respectively, of the circle $S^1(c_{s+a})$ in its plane at x, and let e_{i_0} belong to E^{m_0}, $m^* + q + 1 \leq i_0 \leq m^* + q + m_0$. If there are additional frame vectors normal to M^m at x, they will be denoted by e_ξ.

Then M^m is defined in E^n by the Pfaffian system consisting of equations (10.1.1) plus the equations

$$\omega^{m+s+a} = \omega_{i_0}^{m+\rho} = \omega_{i_0}^{\alpha_\rho} = \omega_{i_0}^{m+s+a} = \omega_{i_0}^{\xi} = 0, \tag{10.2.1}$$

$$\omega_{i_\rho}^{m+s+a} = \omega_{m^*+a}^{m+\rho} = \omega_{m^*+a}^{\alpha_\rho} = \omega_{m^*+a}^{\xi} = 0, \tag{10.2.2}$$

$$\omega_{m^*+a}^{m+s+b} = \delta_a^b \kappa_{s+a} \omega^{m^*+a}. \tag{10.2.3}$$

These additional equations do not alter the conclusion obtained by exterior differentiation of the first equations (10.1.1) with $\rho \neq \sigma$; namely, as above, (10.1.2) and (10.1.3) still hold.

Equations (10.2.1) give

$$\sum_{j_\rho} \omega_{j_\rho}^{i_0} \wedge \omega^{j_\rho} = 0, \quad \sum_{i_\rho, j_\rho} h_{i_\rho j_\rho}^{\alpha_\rho} \omega_{i_0}^{i_\rho} \wedge \omega^{j_\rho} = 0, \quad \omega_{m^*+a}^{i_0} \wedge \omega^{m^*+a} = 0,$$

and therefore

$$\omega_{j_\rho}^{i_0} = \sum_{k_\rho} \lambda_{j_\rho k_\rho}^{i_0} \omega^{k_\rho}, \quad \sum_{i_\rho} h_{i_\rho j_\rho}^{\alpha_\rho} \omega_{i_0}^{i_\rho} = \sum_{k_\rho} \lambda_{j_\rho k_\rho}^{\alpha_\rho, i_0} \omega^{k_\rho}, \quad \omega_{m^*+a}^{i_0} = \lambda_a^{i_0} \omega^{m^*+a}, \tag{10.2.4}$$

where those coefficients on the right sides that have two subscripts are symmetric in those subscripts, so that, in particular,

$$\sum_{i_\rho} (h_{i_\rho j_\rho}^{\alpha_\rho} \lambda_{i_\rho k_\rho}^{i_0} - h_{i_\rho k_\rho}^{\alpha_\rho} \lambda_{i_\rho j_\rho}^{i_0}) = 0. \tag{10.2.5}$$

From (10.2.2) it follows that

$$\kappa_\rho \omega^{i_\rho} \wedge \omega_{m+\rho}^{m+s+a} + \sum_{j_\rho, \alpha_\rho} h_{i_\rho j_\rho}^{\alpha_\rho} \omega^{j_\rho} \wedge \omega_{\alpha_\rho}^{m+s+a} + \kappa_{s+a} \omega_{i_\rho}^{m^*+a} \wedge \omega^{m^*+a} = 0,$$

$$-\kappa_\rho \sum_{i_\rho} \omega_{i_\rho}^{m^*+a} \wedge \omega^{i_\rho} + \kappa_{s+a} \omega^{m^*+a} \wedge \omega_{m+s+a}^{m+\rho} = 0, \tag{10.2.6}$$

$$\kappa_{s+a} \omega^{m^*+a} \wedge \omega_{m+s+a}^{\alpha_\rho} + \sum_{i_\rho, j_\rho} \omega_{m^*+a}^{i_\rho} \wedge h_{i_\rho j_\rho}^{\alpha_\rho} \omega^{j_\rho} = 0, \tag{10.2.7}$$

$$\omega^{m^*+a} \wedge \omega_{m+s+a}^{\xi} = 0. \tag{10.2.8}$$

By Cartan's lemma, one gets from (10.2.6) that

$$\omega_{i_\rho}^{m^*+a} = \sum_{j_\rho} \lambda_{i_\rho j_\rho}^a \omega^{j_\rho} + \mu_{i_\rho}^a \omega^{m^*+a}, \tag{10.2.9}$$

$$\kappa_\rho^{-1} \kappa_{s+a} \omega_{m+s+a}^{m+\rho} = \sum_{i_\rho} \mu_{i_\rho}^a \omega^{i_\rho} + \nu_a^\rho \omega^{m^*+a}, \tag{10.2.10}$$

and from (10.2.7) and (10.2.8) similarly

$$\sum_{i_\rho} h_{i_\rho j_\rho}^{\alpha_\rho} \omega_{m^*+a}^{i_\rho} = \sum_{k_\rho} \lambda_{j_\rho k_\rho}^{\alpha_\rho, a} \omega^{k_\rho} + \mu_a^{\alpha_\rho} \omega^{m^*+a},$$

$$\kappa_{s+a} \omega_{m+s+a}^{\alpha_\rho} = \mu_a^{\alpha_\rho} \omega^{j_\rho} + \varphi_a^{\alpha_\rho} \omega^{m^*+a}, \quad \omega_{m+s+a}^{\xi} = \nu_a^{\xi} \omega^{m^*+a};$$

here the coefficients with two subscripts on the right sides are symmetric in those subscripts.

In the penultimate equation, the index j_ρ takes more than one value. This implies that $\mu_a^{\alpha_\rho} = 0$, and so

$$\sum_{i_\rho} h_{i_\rho j_\rho}^{\alpha_\rho} \omega_{m^*+a}^{i_\rho} = \sum_{k_\rho} \lambda_{j_\rho k_\rho}^{\alpha_\rho, a} \omega^{k_\rho}, \quad \omega_{m+s+a}^{\alpha_\rho} = \nu_a^{\alpha_\rho} \omega^{m^*+a}, \quad \omega_{m+s+a}^{\xi} = \nu_a^{\xi} \omega^{m^*+a}. \tag{10.2.11}$$

Substituting (10.2.9) into the first equation (10.2.11), one obtains $\sum_{i_\rho} h_{i_\rho j_\rho}^{\alpha_\rho} \mu_{i_\rho}^a = 0$. Here the matrix of coefficients with the row pair index $(i_\rho j_\rho)$ and the column index α_ρ has maximal rank; therefore, $\mu_{i_\rho}^a = 0$, and thus

$$\omega_{i_\rho}^{m^*+a} = \sum_{k_\rho} \lambda_{i_\rho k_\rho}^a \omega^{k_\rho}, \quad \sum_{i_\rho} (h_{i_\rho j_\rho}^{\alpha_\rho} \lambda_{i_\rho k_\rho}^a - h_{i_\rho k_\rho}^{\alpha_\rho} \lambda_{i_\rho j_\rho}^a) = 0. \tag{10.2.12}$$

It remains to apply the same procedure to (10.2.3). If $a \neq b$, this gives

$$\omega_{m^*+a}^{m^*+b} \wedge \kappa_{s+b} \omega^{m^*+b} + \kappa_{s+a} \omega^{m^*+a} \wedge \omega_{m+s+a}^{m+s+b} = 0$$

and further, due to $\omega^{m^*+b}_{m^*+a} + \omega^{m^*+a}_{m^*+b} = 0$ and $\omega^{m+s+b}_{m+s+a} + \omega^{m+s+a}_{m+s+b} = 0$ one gets

$$\omega^{m^*+b}_{m^*+a} = \varphi^b_a \kappa_{s+b} \omega^{m^*+b} - \varphi^a_b \kappa_{s+a} \omega^{m^*+a},$$

$$\omega^{m+s+b}_{m+s+a} = \varphi^a_b \kappa_{s+b} \omega^{m^*+b} - \varphi^b_a \kappa_{s+a} \omega^{m^*+a}.$$

For $a = b$, exterior differentiation of (10.2.3) leads to

$$d \ln \kappa_{s+a} - \sum_{i_\rho} \mu^a_{i_\rho} \omega^{i_\rho} - \kappa_{s+a} \sum_{b \neq a} \varphi^a_b \omega^{m^*+b} + \sum_{i_0} \lambda^{i_0}_a \omega^{i_0} = \psi_a \omega^{m^*+a}.$$

Now the first equations of (10.1.1) with $\rho = \sigma$ imply that on the left side of (10.1.4) some new terms must be added, due to (10.1.2), (10.2.4), and (10.2.12), so as to obtain

$$\delta_{i_\rho j_\rho} d \ln \kappa_\rho + \kappa_\rho^{-1} h^{\alpha_\rho}_{i_\rho j_\rho} \omega^{m+\rho}_{\alpha_\rho} - \sum_{k_\sigma, \sigma \neq \rho} \lambda^{k_\sigma}_{i_\rho j_\rho} \omega^{k_\sigma} - \sum_a \lambda^a_{i_\rho j_\rho} \omega^{m^*+a} - \sum_{j_0} \lambda^{j_0}_{i_\rho j_\rho} \omega^{j_0}$$
$$= \pi_{i_\rho j_\rho k_\rho} \omega^{k_\rho}. \qquad (10.2.13)$$

The same argument applied to (10.1.7) yields

$$\bar{\nabla}_\rho h^{\alpha_\rho}_{i_\rho j_\rho} + \delta_{i_\rho j_\rho} \kappa_\rho \omega^{\alpha_\rho}_{m+\rho}$$
$$- \sum_{l_\rho} h^{\alpha_\rho}_{i_\rho l_\rho} \left(\sum_{k_\tau}^{\tau \neq \rho} \lambda^{k_\tau}_{l_\rho j_\rho} \omega^{k_\tau} + \sum_a \lambda^a_{i_\rho j_\rho} \omega^{m^*+a} + \sum_{i_0} \lambda^{i_0}_{l_\rho j_\rho} \omega^{i_0} \right)$$
$$= \chi^{\alpha_\rho}_{i_\rho j_\rho k_\rho} \omega^{k_\rho}. \qquad (10.2.14)$$

The second equations of (10.1.1) with $\rho \neq \sigma$ now give the same equations (10.1.5) and (10.1.6); and finally, the last equations of (10.1.1) give the same equation (10.1.8).

The above argument allows some geometric conclusions to be drawn about the foliations on the given M^m.

Equations (10.1.2), (10.2.4), and (10.2.12) imply the expressions

$$d\omega^{i_\sigma} = \sum_{j_\sigma} \omega^{j_\sigma} \wedge \left(\omega^{i_\sigma}_{j_\sigma} + \sum_{j_\rho}^{\rho \neq \sigma} \lambda^{j_\rho}_{i_\sigma j_\sigma} \omega^{j_\rho} + \sum_a \lambda^a_{i_\sigma j_\sigma} \omega^{m^*+a} + \sum_{i_0} \lambda^{i_0}_{i_\sigma j_\sigma} \omega^{i_0} \right), \qquad (10.2.15)$$

$$d\omega^{m^*+a} = \sum_b \omega^{m^*+b} \wedge \left(\omega^{m^*+a}_{m^*+b} + \delta^a_b \sum_{i_0} \lambda^{i_0}_a \omega^{i_0} \right), \qquad (10.2.16)$$

$$d\omega^{i_0} = \omega^{j_0} \wedge \omega^{i_0}_{j_0}, \qquad (10.2.17)$$

and so the following geometric consequences can be formulated.

Theorem 10.2.1. *A semiparallel submanifold M^m in E^n, as a second-order envelope of products*

$$M^{m_1} \times \cdots \times M^{m_s} \times S^1(c_{s+1}) \times \cdots \times S^1(c_{s+q}) \times E^{m_0},$$

carries two totally orthogonal foliations, whose leaves are second-order envelopes of

(1) *the products $M^{m_1} \times \cdots \times M^{m_s}$ of main symmetric orbits, and*
(2) *the products $S^1(c_{s+1}) \times \cdots \times S^1(c_{s+q}) \times E^{m_0}$.*

- (i) *The last ones are generated by m_0-dimensional planes, have flat van der Waerden–Bortolotti connection $\bar{\nabla}$, and carry q families of mutually orthogonal lines of curvature, orthogonal also to the generating planes.*
- (ii) *The tangent subspaces of these leaves are invariant along every plane generator.*
- (iii) *If all main symmetric orbits $M^{m_1}, \ldots, M^{m_s}$ are umbilic-like, then the leaves of the first set are also umbilic-like and hence parallel submanifolds.*
- (iv) *Then for every fixed value $\rho \in \{1, \ldots, s\}$, the tangent subspaces of M^{m_ρ} are parallel along the leaves* (2).

Proof. Consider the Pfaffian system $\omega^{i_\rho} = 0$, where $1 \leq \rho \leq s$ and every i_ρ takes all its values. This system is totally integrable due to to (10.2.15), and so defines a foliation on M^m. Its leaves are $(m_0 + q)$-dimensional submanifolds that are second-order envelopes of the products (2).

For one of these leaves, consider the Pfaffian system $\omega^{m^*+a} = 0$, where $1 \leq a \leq q$. Then (10.2.16) implies that this system defines a foliation. For each of its leaves one has

$$dx = e_{i_0}\omega^{i_0}, \quad de_{i_0} = e_{j_0}\omega^{j_0}_{i_0},$$

due to (10.2.1) and (10.2.4). Therefore, these leaves are m_0-dimensional planes, as asserted in (i). Also the other statements of (i) hold, since they are valid for the products (2), and so can be transferred also to their second-order envelopes.

The distribution defined by

$$\omega^{m^*+a} = \omega^{i_0} = 0, \ 1 \leq a \leq q, \ m^* + q + 1 \leq i_0 \leq m,$$

is also a foliation, due to (10.2.16) and (10.2.17). Its leaves are second-order envelopes of the products (1). These leaves are orthogonal to the previous normally flat locally Euclidean submanifolds.

If all main symmetric orbits $M^{m_1}, \ldots, M^{m_s}$ are umbilic-like, then by Theorem 10.1.2, their product is also umbilic-like. Hence, due to umbilic-likeness, the above leaves reduce to these products, and therefore are parallel submanifolds, as asserted in (iii).

This implies that modulo ω^{m^*+a} and ω^{i_0}, equation (10.2.13) reduces to (10.1.10). It follows that $\lambda^{k_\sigma}_{i_\rho j_\rho} = 0$, if $\rho \neq \sigma$, and $\pi_{i_\rho j_\rho k_\rho} = 0$ (cf. (10.1.12)). This together with (10.1.2) implies $\omega^{i_\sigma}_{j_\rho} = 0$ for $\rho \neq \sigma$, and shows, together with (10.2.12), (10.2.4), and (10.1.1), that

$$de_{i_\rho} = e_{j_\rho}\omega^{j_\rho}_{i_\rho} + \sum_{k_\rho}\left(\sum_a e_{m^*+a}\lambda^a_{i_\rho k_\rho} + \sum_{i_0} e_{i_0}\lambda^{i_0}_{i_\rho k_\rho}\right)\omega^{k_\rho}$$
$$+ e_{m+\rho}\kappa_\rho\omega^{i_\rho} + e_{\alpha_\rho}h^{\alpha_\rho}_{i_\rho k_\rho}\omega^{k_\rho}.$$

Since along the leaves (2) one has $\omega^{k_\rho} = 0$ for $\rho = 1, \dots, s$ and all values of k_ρ, one now gets $de_{i_\rho} = e_{j_\rho}\omega^{j_\rho}_{i_\rho}$, as asserted in (iv).

That concludes the proof.

Remark 10.2.2. Most parts of Theorem 10.2.1 were first announced in [Lu 96d], many of them without detailed proofs. Some consequences on the special geometric structure of semiparallel submanifolds and their intrinsic semisymmetric Riemannian manifold structure were also stated in [Lu 96d].

10.3 Warped Products and Immersed Fibre Bundles

The foliations introduced on M^m in the previous section provide a special geometric structure for the Riemannian manifold as well as its immersion, which will be considered now.

A Riemannian manifold M is said to be reducible to a product if $M = M_1 \times \cdots \times M_k$, and the component submanifolds $M_1, \dots, M_k$ are mutually orthogonal in the Riemannian metric of M.

A generalization of this is the concept of a semireducible Riemannian manifold (see [Kr 57]) or warped product (see [BiO'N 69], [DN 93], [Nö 96], [Ch 2000], 3.5).

Let $M_0, \dots, M_k$ be Riemannian manifolds, $M = M_0 \times \cdots \times M_k$ their product, and $\pi_i : M \to M_i$ the canonical projections, $i = 0, 1, \dots, k$. If $\varphi_1, \dots, \varphi_k : M_0 \to \mathbb{R}_+$ are positive real-valued functions, then

$$\langle X, Y\rangle := \langle \pi_{0*}X, \pi_{0*}Y\rangle + \sum_{i=1}^{k}(\varphi_i \circ \pi_0)^2\langle \pi_{i*}X, \pi_{i*}Y\rangle$$

defines a Riemannian metric on M called a warped product metric; M with this metric is called a *warped product*, denoted $M_0 \times_{\varphi_1} M_1 \times_{\varphi_2} \cdots \times_{\varphi_k} M_k$.

Theorem 10.3.1. *If a semiparallel submanifold M^m in E^n is a second-order envelope of parallel $M^{m_1} \times \cdots \times M^{m_s} \times S^1(c_{s+1}) \times \cdots \times S^1(c_{s+q}) \times E^{m_0}$ whose main symmetric orbits M^{m_ρ} are umbilic-like, then M^m is intrinsically a warped product $M^{n_0} \times_{\varphi_1} M^{m_1} \times_{\varphi_2} \cdots \times_{\varphi_s} M^{m_s}$, where M^{n_0}, $n_0 = m_0 + q$, is locally Euclidean, and M^{m_ρ}, $\rho = 1, \dots, s$, are symmetric R-spaces.*

Proof. M^{n_0} is defined as a leaf with flat $\bar{\nabla}$ of the foliation (2) of Theorem 10.2.1. Orthogonal to these leaves are the leaves of the foliation (1), for which, due to umbilic-likeness, equations (10.2.13) and (10.2.14) reduce to

$$\delta_{i_\rho j_\rho} d\ln\kappa_\rho + \kappa_\rho^{-1} h^{\alpha_\rho}_{i_\rho j_\rho}\omega^{m+\rho}_{\alpha_\rho} - \sum_a \lambda^a_{i_\rho j_\rho}\omega^{m^*+a} - \sum_{j_0} \lambda^{j_0}_{i_\rho j_\rho}\omega^{j_0} = 0, \tag{10.3.1}$$

$$\bar{\nabla}_\rho h^{\alpha_\rho}_{i_\rho j_\rho} + \delta_{i_\rho j_\rho}\kappa_\rho \omega^{\alpha_\rho}_{m+\rho} - \sum_{l_\rho} h^{\alpha_\rho}_{i_\rho l_\rho}\left(\sum_a \lambda^a_{i_\rho j_\rho}\omega^{m^*+a} + \sum_{i_0} \lambda^{i_0}_{l_\rho j_\rho}\omega^{i_0}\right) = 0. \tag{10.3.2}$$

In (10.3.1), setting $i_\rho = j_\rho$, summing, and using the relation (10.1.2), one obtains $d\ln\kappa_\rho = \sum_a \lambda^{m^*+a}_\rho \omega^{m^*+a} + \sum_{j_0} \lambda^{j_0}_\rho \omega^{j_0}$, where $\lambda^{m^*+a}_\rho = m_\rho^{-1}\sum_{i_\rho} \lambda^a_{i_\rho i_\rho}$ and $\lambda^{j_0}_\rho = m_\rho^{-1}\sum_{i_\rho} \lambda^{j_0}_{i_\rho i_\rho}$. This can be written more compactly as

$$d\kappa_\rho = \kappa_\rho \sum_u \lambda^u_\rho \omega^u, \tag{10.3.3}$$

where the index u runs over the ranges first of $m^* + a$ and then of i_0.

Similar summing in (10.3.2) leads to

$$\omega^{\alpha_\rho}_{m+\rho} = \sum_u \mu^u_{\alpha_\rho}\omega^u, \tag{10.3.4}$$

where $\mu^u_{\alpha_\rho} = (m_\rho\kappa_\rho)^{-1}\sum_{l_\rho, i_\rho} h^{\alpha_\rho}_{l_\rho i_\rho}\lambda^u_{l_\rho i_\rho}$.

The leaves orthogonal to all M^{n_0} are the products of main symmetric orbits. According to a result of Ferus, formulated as Theorem 3.6.1, every main symmetric orbit is a standardly imbedded symmetric R-space. Each orbit is pseudoumbilic (see Theorems 8.2.2 and 8.2.4) and hence minimal in a sphere, by Proposition 8.2.1. For one of these pseudoumbilic orbits,

$$de_{i_\rho} = e_{j_\rho}\omega^{j_\rho}_{i_\rho} + e_u \lambda^u_{i_\rho j_\rho}\omega^{j_\rho} + e_{m+\rho}\kappa_\rho\omega^{i_\rho} + e_{\alpha_\rho} h^{\alpha_\rho}_{i_\rho j_\rho}\omega^{j_\rho},$$

due to (10.2.4) and (10.2.12). Here the vector-valued second fundamental tensor is $e_u\lambda^u_{i_\rho j_\rho} + e_{m+\rho}\kappa_\rho\delta_{i_\rho j_\rho} + e_{\alpha_\rho} h^{\alpha_\rho}_{i_\rho j_\rho}$. Summing with $i_\rho = j_\rho$ and dividing by m_ρ, one obtains the mean curvature vector $H_\rho = e_{m+\rho}\kappa_\rho + e_u\lambda^u_\rho$, for which $\langle H_\rho, H_\rho\rangle = \kappa_\rho^2 + \sum_u (\lambda^u_\rho)^2$. As can be seen from the proof of Proposition 8.2.1, this pseudoumbilic orbit lies in a sphere with radius $r_\rho = [\kappa_\rho^2 + \sum_u (\lambda^u_\rho)^2]^{-\frac{1}{2}}$. The metric form of this orbit can be obtained by multiplying by $\varphi_\rho = r_\rho^2$ the metric form of a standard orbit, which corresponds to the value $r_\rho = 1$.

From (10.3.3) one gets via exterior differentiation and Cartan's lemma

$$d\lambda^u_\rho = \sum_v (\lambda^v_\rho \omega^v_u + \lambda^u_{\rho v}\omega^v).$$

Therefore, $d\sum_u (\lambda^u_\rho)^2 = 2\sum_u \lambda^u_\rho \lambda^u_{\rho v}\omega^v$. This and (10.3.3) show that φ_ρ are functions on M^{n_0}, thus concluding the proof.

Remark 10.3.2. Theorem 10.3.1 can be considered as a generalization of Theorem 5.4.1, which deals with the case where (due to Proposition 5.2.1) the umbilic-like main symmetric orbits are spherical submanifolds, hence simply spheres in Euclidean space E^n. But there is a difference between these theorems. Namely, Theorem 5.4.1 describes a normally flat semiparallel submanifold as an immersed warped product submanifold, while Theorem 10.3.1 concerns only the inner geometry of the given semiparallel submanifold, where instead of spheres one has general umbilic-like symmetric orbits.

Remark 10.3.3. The semiparallel submanifolds M^m of Theorem 10.3.1 can also be considered as immersed fibre bundles with homogeneous fibres, associated to some principal bundles. These kinds of fibre bundles have been studied in [KN 63], [Hu 66] (see also [Lu 66], [Lu 71]).

Recall that every main symmetric orbit is a standardly imbedded symmetric R-space K/K_0 and can be realized as a minimal submanifold of a sphere. Therefore, the Lie group K is a subgroup of the orthogonal group $O(n, \mathbb{R})$. If a submanifold is a second-order envelope of main umbilic-like symmetric orbits, then it reduces to such an orbit, according to the definition of umbilic-likeness (see Section 9.1). By Theorem 10.1.2, a product of main umbilic-like symmetric orbits is umbilic-like. Thus the leaves in M^m, being second-order envelopes of the products of these main umbilic-like symmetric orbits (i.e., the leaves of the foliation (2) of Theorem 10.2.1), reduce to such products, and therefore are products of standardly imbedded symmetric R-spaces. Hence each of these leaves is a symmetric space G/G_0, where $G = K_1 \times \cdots \times K_s$ and $G_0 = K_{01} \times \cdots \times K_{0s}$. Now considering the principal bundle with structure Lie group G and its associated bundle with fibres G/G_0, one can see that M^m of Theorem 10.3.1 is an immersion of this associated fibre bundle.

The leaves of the foliation (1) in Theorem 10.2.1 are orthogonal to these fibres above. This foliation can be considered as the horizontal distribution of an inner connection (in the sense of [KN 63], [Lu 66], [Lu 71]) for M^m as an immersed associated fibre bundle. Since the horizontal distribution is now a foliation, this inner connection is locally flat, i.e., has zero curvature 2-form with values in the Lie algebra of G.

This point of view for the semiparallel submanifolds of Theorem 10.3.1 was briefly described in [Lu 96d], where only three particular cases of umbilic-like main symmetric orbits were considered, namely spheres, Segre orbits, and Plücker orbits.

10.4 Semiparallel Submanifolds of Cylindrical or Toroidal Segre Type

The above general investigations will now be illustrated by considering a particular case. Namely, let us consider the semiparallel submanifold M^m of Theorem 10.2.1 for a single main symmetric orbit M^{m_1} which is a Segre orbit $S_{(p,\bar{p})}(k)$, plus a single other component which is either a straight line E^1, or a circle $S^1(c)$. Such an M^m is said to be of *cylindrical* or *toroidal* Segre type, respectively.

The Segre orbit $S_{(p,\bar{p})}(k)$ was introduced above in Section 3.2 as a $(p+\bar{p})$-dimensional submanifold of the sphere $S^{p\bar{p}+m}(k^2)$, $m = p+\bar{p}$, having two families of generating great spheres, of dimensions p and $\bar{p}$, respectively, and totally orthogonal at every point $x \in S_{(p,\bar{p})}(k)$ (see also Section 4.6). Note that in this section $\bar{p}$ is used in place of the notation q used in Sections 3.2 and 4.6; also $\pi, \sigma, \ldots$ replace $i_1, j_1, \ldots$ and $\bar{\pi}, \bar{\sigma}, \ldots$ replace $i_2, j_2, \ldots$. Recall that by Theorem 4.6.1, this $S_{(p,\bar{p})}(k)$ is always umbilic-like only if $p > 1$, $\bar{p} > 1$ (see Section 9.1); otherwise it is in general not umbilic-like.

10.4.1 The case of umbilic-like Segre orbits

Suppose the Segre orbit $S_{(p,\bar{p})}(k)$ is umbilic-like. Then $p > 1$ and $\bar{p} > 1$. In this case, the semiparallel submanifolds of cylindrical or toroidal Segre type are described geometrically by the following two theorems.

Theorem 10.4.1. *If $p > 1$, $\bar{p} > 1$, then a second-order envelope $M^{p+\bar{p}+1}$ of products $S_{(p,\bar{p})}(k) \times E^1$ with variable k is either*

(i) *a single $S_{(p,\bar{p})}(k) \times E^1$ in $E^{(p+1)(\bar{p}+1)+1}$ or its open subset, or*
(ii) *an open subset of a cone in $E^{(p+1)(\bar{p}+1)+1}$ with a point vertex and one-dimensional straight generators, intersected orthogonally by Segre orbits $S_{(p,\bar{p})}(\widetilde{k})$.*

In case (i) this $M^{p+\bar{p}+1}$ is a parallel submanifold, and, if complete, then a symmetric product–orbit. In case (ii) $M^{p+\bar{p}+1}$ is a semiparallel but not parallel submanifold, which is not complete (the vertex of the cone is a singular point). Note that (i) can be considered as the limiting case of (ii), where the vertex of the cone has moved to infinity.

Theorem 10.4.2. *If $p > 1$, $\bar{p} > 1$, then the second-order envelope $M^{p+\bar{p}+1}$ of products $S_{(p,\bar{p})}(k) \times S^1(c)$ with variable k and c is either*

(i) *a product $S_{(p,\bar{p})}(k) \times M^1$ or its subset in E^n, where $k =$ const, and M^1 is a curve in an $E^{n-(p+1)(\bar{p}+1)}$ totally orthogonal to the subspace $E^{(p+1)(\bar{p}+1)}$ of E^n containing $S_{(p,\bar{p})}(k)$, $n > (p+1)(\bar{p}+1)$, or*
(ii) *a subset of a bundle of Segre orbits over a base curve, immersed into E^n, $n > (p+1)(\bar{p}+1)$, so that*
   (1) *these orbits have their centers on the base curve and lie in parallel $(p+1)(\bar{p}+1)$-subspaces orthogonal to the subspace $E^{n-(p+1)(\bar{p}+1)}$ containing the whole base curve,*
   (2) *the radius of the orbit is a linear function on the base curve,*
   (3) *if one considers tangent lines to the orthogonal trajectories of the orbit fibres, then the set of such tangent lines taken at all points of an orbit fibre lies on a cone, and the vertices of all such cones lie on an evolvent of the base curve.*

Here in case (ii) this $M^{p+\bar{p}+1}$ is a semiparallel but not parallel submanifold, which is not complete (the singular point of the evolvent is a singular point of this "warped cone").

As preparation for the proofs, one derives a Pfaffian system describing such a second-order envelope in a suitably adapted bundle of orthonormal frames. So let M^{m+1} be a second-order envelope of products $S_{(p,\bar{p})}(k) \times M^1$ in E^n, where $m = p+\bar{p}, n > (p+1)(\bar{p}+1)+1$, and M^1 is either a straight line E^1 or a circle $S^1(c)$. According to the adaptation in the proof of Theorem 3.2.1 and equations (4.6.1), the required system is the following:

$$\omega^{m+1} = \omega^{\pi\bar{\pi}} = \omega^{2^*} = \omega^{\xi} = 0, \tag{10.4.1}$$

$$\omega_{\pi}^{m+1} = k\omega^{\pi}, \quad \omega_{\bar{\pi}}^{m+1} = k\omega^{\bar{\pi}}, \quad \omega_{1^*}^{m+1} = 0, \tag{10.4.2}$$

$$\omega_{\pi}^{\sigma\bar{\tau}} = \delta_{\pi}^{\sigma} k\omega^{\bar{\tau}}, \quad \omega_{\bar{\pi}}^{\sigma\bar{\tau}} = \delta_{\bar{\pi}}^{\bar{\tau}} k\omega^{\sigma}, \quad \omega_{1^*}^{\sigma\bar{\tau}} = 0, \tag{10.4.3}$$

$$\omega_{\pi}^{2^*} = 0, \qquad \omega_{\bar{\pi}}^{2^*} = 0, \qquad \omega_{1^*}^{2^*} = c\omega^{1^*}, \tag{10.4.4}$$

$$\omega_{\pi}^{\xi} = 0, \qquad \omega_{\bar{\pi}}^{\xi} = 0, \qquad \omega_{1^*}^{\xi} = 0. \tag{10.4.5}$$

Indices $1^* = (p+1)(\bar{p}+1)+1$ and $2^* = (p+1)(\bar{p}+1)+2$ refer to the unit tangent and normal vectors e_{1^*} and e_{2^*} of $S^1(c)$, and $\xi \in \{2^*+1, \ldots, n\}$ refers to the other normal vectors of M^{m+1}.

The case $M^1 = S^1(c)$ corresponds to $c \neq 0$, and $M^1 = E^1$ to $c = 0$, in which case ξ can be replaced by $\xi' \in \{2^*, 2^*+1, \ldots, n\}$.

This Pfaffian system will be investigated via exterior differentiation and Cartan's lemma.

Exterior differentiation of equations (10.4.1) yields identities, but differentiation of (10.4.5) leads, for fixed values π and $\bar{\pi}$, to

$$\omega^{\pi} \wedge \omega_{m+1}^{\xi} + \sum_{\bar{\sigma}} \omega^{\bar{\sigma}} \wedge \omega_{\pi\bar{\sigma}}^{\xi} = 0,$$

$$\omega^{\bar{\pi}} \wedge \omega_{m+1}^{\xi} + \sum_{\sigma} \omega^{\sigma} \wedge \omega_{\sigma\bar{\pi}}^{\xi} = 0,$$

$$c\omega^{1^*} \wedge \omega_{2^*}^{\xi} = 0.$$

Since now $p > 1, \bar{p} > 1$, the indices π and $\bar{\pi}$ can both take more than one value. Therefore, one gets by Cartan's lemma

$$\omega_{m+1}^{\xi} = \omega_{\pi\bar{\sigma}}^{\xi} = 0, \quad c\omega_{2^*}^{\xi} = B^{\xi}\omega^{1^*}. \tag{10.4.6}$$

The last equations $\omega_{1^*}^{\sigma\bar{\tau}} = 0$ in (10.4.3) imply that for every fixed pair of values of $\sigma, \bar{\tau}$

$$\omega_{1^*}^{\sigma} \wedge \omega^{\bar{\tau}} + \omega_{1^*}^{\bar{\tau}} \wedge \omega^{\sigma} + ck^{-1}\omega^{1^*} \wedge \omega_{2^*}^{\sigma\bar{\tau}} = 0. \tag{10.4.7}$$

From $p > 1, \bar{p} > 1$ it then follows that

$$\omega_{1^*}^{\sigma} = \lambda\omega^{\sigma} - ck^{-1}\mu^{\sigma}\omega^{1^*}, \tag{10.4.8}$$

$$\omega_{1^*}^{\bar{\tau}} = \lambda\omega^{\bar{\tau}} - ck^{-1}\nu^{\bar{\tau}}\omega^{1^*}. \tag{10.4.9}$$

The equations $\omega_{\sigma}^{2^*} = 0$ together with $\omega_{\bar{\tau}}^{2^*} = 0$ now give

$$\omega_{2^*}^{\sigma\bar{\tau}} = 0, \qquad k\omega_{2^*}^{m+1} = -c\lambda\omega^{1^*}. \tag{10.4.10}$$

Substituting all this into (10.4.7), one obtains (1) in case $c = 0$, an identity, (2) in case $c \neq 0$, due to $p > 1$, $\bar{p} > 1$ the relations $\mu^{\sigma} = \nu^{\bar{\tau}} = 0$; thus from (10.4.8) and (10.4.9) for both these cases,

$$\omega_{1^*}^{\sigma} = \lambda\omega^{\sigma}, \qquad \omega_{1^*}^{\bar{\tau}} = \lambda\omega^{\bar{\tau}}. \tag{10.4.11}$$

The equation $\omega_{1^*}^{2^*} = c\omega^{1^*}$ yields

$$dc = \kappa\omega^{1^*}. \tag{10.4.12}$$

Equations (10.4.2) give

$$d\ln k = -\lambda\omega^{1^*}, \qquad \omega_{\pi\bar{\sigma}}^{m+1} = 0. \tag{10.4.13}$$

The equations $\omega_{\pi}^{\sigma\bar{\tau}} = \delta_{\pi}^{\sigma} k\omega^{\bar{\tau}}$ and $\omega_{\bar{\pi}}^{\sigma\bar{\tau}} = \delta_{\bar{\pi}}^{\bar{\tau}} k\omega^{\sigma}$ imply

$$\omega_{\pi}^{\bar{\tau}} = 0, \quad \omega_{\varphi\bar{\pi}}^{\sigma\bar{\pi}} = \omega_{\varphi}^{\sigma}, \quad \omega_{\pi\bar{\psi}}^{\pi\bar{\tau}} = \omega_{\bar{\psi}}^{\bar{\tau}}, \quad \omega_{\pi\bar{\psi}}^{\sigma\bar{\tau}} = 0 \quad (\sigma \neq \pi, \bar{\tau} \neq \bar{\psi}). \tag{10.4.14}$$

This concludes the first differential prolongation.

The second differential prolongation deals with the additional equations. Taking exterior derivatives and using the equations of the extended system, almost all of them give identities. Exceptions to this are the equations

$$c\omega_{2^*}^{\xi} = B^{\xi}\omega^{1^*}, \quad dc = \kappa\omega^{1^*}, \tag{10.4.15}$$

$$d\ln k = -\lambda\omega^{1^*}, \quad \omega_{1^*}^{\pi} = \lambda\omega^{\pi}, \quad \omega_{1^*}^{\bar{\pi}} = \lambda\omega^{\bar{\pi}}, \quad k\omega_{2^*}^{m+1} = -c\lambda\omega^{1^*}. \tag{10.4.16}$$

The first equation (10.4.16) yields $d\lambda \wedge \omega^{1^*} = 0$, thus $d\lambda = \gamma\omega^{1^*}$. Now from the next two groups of equations (10.4.16), $(\gamma + \lambda^2)\omega^{1^*} \wedge \omega^{\sigma} = 0$; hence $\gamma = -\lambda^2$ and

$$d\lambda = -\lambda^2\omega^{1^*}. \tag{10.4.17}$$

This is one of the results of the second prolongation. After that, all equations (10.4.16) give, by exterior differentiation, identities, plus the new additional equation (10.4.17).

If $c \equiv 0$ on M^{m+1}, then $2^* = (p+1)(p+2) + 2$ can be considered as the first value of $\xi' \in \{(p+1)(\bar{p}+1)+2, \ldots, n\}$, equations (10.4.15) vanish, and the whole extended system is totally integrable. It follows that in this case the required M^{m+1} exists and depends on some constants.

Suppose $c \neq 0$ on M^{m+1}. If $B^{\xi} e_{\xi} = 0$, then equations (10.4.15) reduce to

$$\omega_{2^*}^{\xi} = 0, \quad dc = \kappa\omega^{1^*}. \tag{10.4.18}$$

The first one gives identities, and the second gives $d\kappa \wedge \omega^{1^*} = 0$, which is the only essential covariant equation of the whole extended system. By the Cartan theory, this shows that in this case M^{m+1} exists and depends on a real function of one real argument.

Suppose $c \neq 0$ and $B^\xi e_\xi \neq 0$ on M^{m+1}. Then (10.4.15) gives $dB^\xi \wedge \omega^{1^*} = 0$, $d\kappa \wedge \omega^{1^*} = 0$; hence M^{m+1} exists and depends on $q+1$ real functions of one real argument, where q is the number of linearly independent differentials among dB^ξ.

It remains to interpret these results geometrically. They show that

$$dx = e_\pi \omega^\pi + e_{\bar\pi}\omega^{\bar\pi} + e_{1^*}\omega^{1^*}, \tag{10.4.19}$$

$$de_\pi = e_\sigma \omega_\pi^\sigma - \lambda e_{1^*}\omega^\pi + k e_{m+1}\omega^\pi + k e_{\pi\bar\tau}\omega^{\bar\tau}, \tag{10.4.20}$$

$$de_{\bar\pi} = e_{\bar\sigma}\omega_{\bar\pi}^{\bar\sigma} - \lambda e_{1^*}\omega^{\bar\pi} + k e_{m+1}\omega^{\bar\pi} + k e_{\sigma\bar\pi}\omega^\sigma, \tag{10.4.21}$$

$$de_{1^*} = \lambda(e_\pi \omega^\pi + e_{\bar\pi}\omega^{\bar\pi}) + c e_{2^*}\omega^{1^*}, \tag{10.4.22}$$

$$de_{m+1} = -k(e_\pi\omega^\pi + e_{\bar\pi}\omega^{\bar\pi}) + ck^{-1}\lambda e_{2^*}\omega^{1^*}, \tag{10.4.23}$$

$$de_{\pi\bar\sigma} = -k(e_\pi \omega^{\bar\sigma} + e_{\bar\sigma}\omega^\pi) + e_{\pi\bar\varphi}\omega_{\bar\sigma}^{\bar\varphi} + e_{\tau\bar\sigma}\omega_\pi^\tau. \tag{10.4.24}$$

Since $d\omega^\pi = \omega^\sigma \wedge \omega_\sigma^\pi + \lambda\omega^{1^*} \wedge \omega^\pi$, $d\omega^{\bar\pi} = \omega^{\bar\sigma} \wedge \omega_{\bar\sigma}^{\bar\pi} + \lambda\omega^{1^*} \wedge \omega^{\bar\pi}$, and $d\omega^{1^*} = 0$, the Pfaffian systems $\omega^{\bar\pi} = 0$, $\omega^{1^*} = 0$ and $\omega^\pi = 0$, $\omega^{1^*} = 0$ are both totally integrable on M^{m+1}. For integral submanifolds of the first system, one has

$$dx = e_\pi\omega^\pi, \quad de_\pi = e_\sigma\omega_\pi^\sigma + (-\lambda e_{1^*} + k e_{m+1})\omega^\pi;$$

hence each of them is totally umbilic and thus is an $S^p(\widetilde{k})$ or its subset, where $\widetilde{k} = (\lambda^2 + k^2)^{\frac{1}{2}}$. Similarly, every integral submanifold of the second system is an $S^{\bar p}(\widetilde{k})$ or its subset. Both of the spheres $S^p(\widetilde{k})$ and $S^{\bar p}(\widetilde{k})$ through a given point $x \in M^{m+1}$ are totally orthogonal in M^{m+1} and have the same center y with radius vector $y = x + (\widetilde{k})^{-1}\widetilde{e}_{m+1}$, where

$$\widetilde{e}_{m+1} = \widetilde{k}(-\lambda e_{1^*} + k e_{m+1}) \tag{10.4.25}$$

is the unit vector along the radius of this $S^{\bar p}(\widetilde{k})$. It follows that every integral submanifold of the Pfaffian equation $\omega^{1^*} = 0$ on M^{m+1} is a Segre orbit $S_{(p,\bar p)}(\widetilde{k})$ or its subset.

Proof of Theorem 10.4.1. Suppose $c = 0$ on M^{m+1}, i.e., consider a second-order envelope M^{m+1} of products $S_{(p,\bar p)}(k) \times E^1$. Then $\omega_{1^*}^{2^*} = 0$, and $2^* = (p+1)(\bar p + 1) + 2$ can be included in the set $\{(p+1)(\bar p+1)+2, \dots, n\}$ of values of ξ', so that

$$\omega_\pi^{\xi'} = \omega_{\bar\pi}^{\xi'} = \omega_{1^*}^{\xi'} = 0.$$

This shows that de_π, $de_{\bar\pi}$ and de_{1^*} have zero components in the subspace spanned by the $e_{\xi'}$, and likewise for de_{m+1} and $de_{\pi\bar\sigma}$, due to (10.4.23) and (10.4.24). Thus M^{m+1} lies in an $E^{(p+1)(\bar p+1)+1} \subset E^n$.

For the integral curves of the system $\omega^\pi = \omega^{\bar\pi} = 0$ on M^{m+1}, one has

$$dx = e_{1^*}\omega^{1^*}, \quad de_{1^*} = 0,$$

therefore these curves are straight lines.

If $\lambda \neq 0$ on M^{m+1}, then all of them go through a fixed point z with radius vector $z = x - \lambda^{-1}e_{1^*}$, because $dz = 0$. Hence M^{m+1} is a subset of a cone with the vertex z and one-dimensional generators. This cone consists of Segre orbits $S_{(p,\bar{p})}(\tilde{k})$, intersecting the generators orthogonally.

If $\lambda = 0$ on M^{m+1}, then all the straight lines above are mutually parallel, all Segre orbits are congruent due to $k = \text{const}$, and M^{m+1} is a product $S_{(p,\bar{p})}(k) \times E^1$ or its subset.

This ends the proof of Theorem 10.4.1.

Proof of Theorem 10.4.2. Suppose $c \neq 0$ on M^{m+1}, i.e., consider a second-order envelope M^{m+1} of products $S_{(p,\bar{p})}(k) \times S^1(c)$. To the derivation formulas (10.4.19)–(10.4.24), one must now add

$$de_{2^*} = [-c(e_{1^*} + k^{-1}\lambda e_{m+1}) + c^{-1}B^{\xi}e_{\xi}]\omega^{1^*}.$$

If x moves freely on M^{m+1}, then for $y = x + (\widetilde{k})^{-1}\widetilde{e}_{m+1}$ one has

$$dy = k(\widetilde{k})^{-1}\widetilde{e}_{1^*}\omega^{1^*},$$

where $\widetilde{e}_{1^*} = (\widetilde{k})^{-1}(ke_{1^*} + \lambda e_{m+1})$ is a unit vector orthogonal to $\widetilde{e}_{m+1}$. This shows that the centers y of the Segre orbits $S_{(p,\bar{p})}(k)$ whose one-parameter family generates M^{m+1}, $m = p + \bar{p}$, describe a curve with unit tangent vector $\widetilde{e}_{1^*}$ at y, and with arclength parameter $\widetilde{s}$, where $d\widetilde{s} = k(\widetilde{k})^{-1}\omega^{1^*}$. This curve is called the base curve for M^{m+1}, and M^{m+1} can be considered as (a subset of) a bundle of Segre orbits on this curve.

Since $d(\widetilde{k})^{-1} = \lambda(\widetilde{k})^{-1}\omega^{1^*}$, the function $(\widetilde{k})^{-1}$ on the base curve has the derivative $\frac{d(\widetilde{k})^{-1}}{d\widetilde{s}} = \frac{\lambda}{k}$, which is a constant because $d\frac{\lambda}{k} = 0$. Thus $(\widetilde{k})^{-1}$ is a linear function on the base curve.

Moreover, the orbit fibre $S_{p,\bar{p})}(\widetilde{k})$ lies in an $E^{(p+1)(\bar{p}+1)}$ whose vector space is spanned by $e_{\pi}, e_{\bar{\pi}}, \widetilde{e}_{m+1}, e_{\pi\bar{\sigma}}$. Due to equations (10.4.20), (10.4.21), (10.4.24), and

$$d\widetilde{e}_{m+1} = -(\widetilde{k})^{-1}(e_{\pi}\omega^{\pi} + e_{\bar{\pi}}\omega^{\bar{\pi}}),$$

which follows easily from (10.4.25), this vector space is invariant for M^{m+1}. Thus all orbit fibres $S_{(p,\bar{p})}(\widetilde{k})$ lie in parallel $(p+1)(\bar{p}+1)$-dimensional subspaces of E^n, totally orthogonal to the subspace of the base curve.

The orthogonal trajectories of the orbit fibres $S_{(p,\bar{p})}(\widetilde{k})$ are defined by $\omega^{\pi} = \omega^{\bar{\pi}} = 0$ on M^{m+1}. At $x \in M^{m+1}$, such a trajectory has unit tangent vector e_{1^*}.

If $\lambda \neq 0$ on M^{m+1}, there is a point z along this tangent with radius vector $z = x - \lambda^{-1}e_{1^*}$, whose differential is

$$dz = -c\lambda^{-1}e_{2^*}\omega^{1^*}.$$

On the other hand, $y - z = \varrho\lambda^{-1}\widetilde{r}\widetilde{e}_{1^*}$, so that z belongs to a tangent line of the base curve. If x moves freely on M^{m+1}, this point describes a curve whose tangent at z is orthogonal to the tangent of the base curve at the corresponding point y. Hence this curve of z is an evolvent of the base curve. At the singular point of this evolvent, one has $\widetilde{k} = 0$, so this point is also a singular point of M^{m+1}. The latter can be considered as a warped "cone" of orbit fibres $S_{(p,\bar{p})}(\widetilde{k})$, whose "axis" is the base curve and whose "vertex" is the singular point.

If $\lambda \to 0$, then $\widetilde{k} \to k = \text{const}$, $\widetilde{e}_{m+1} \to e_{m+1}$, $\widetilde{e}_{1*} \to e_1$, and in the limiting case $\lambda = 0$, this M^{m+1} is the ordinary product of $S_{(p,\bar{p})}(k)$ and the base curve.

This concludes the proof of Theorem 10.4.2.

Remark 10.4.3. For normally flat semiparallel submanifolds M^m in $N^n(c)$, Theorem 5.4.1 holds, which for the case where $N^n(c)$ is a Euclidean space E^n states that such an M^m is in general a warped product submanifold $B^{m'} \times_{r_1} S_1^{\nu_1}(1) \times_{r_2} \cdots \times_{r_p} S_p^{\nu_p}(1)$, where the base submanifold $B^{m'}$ has flat $\bar{\nabla}$, the warping functions $r_1, \ldots, r_p$ are nonconstant linear functions (with respect to some local affine coordinates in $B^{m'}$), and the fibres are products of spheres, hence umbilic-like.

Theorems 10.4.1 and 10.4.2 show that in two particular cases, similar assertions are true, where instead of spheres one takes other umbilic-like symmetric orbits, namely, a Segre orbit. It is an open problem whether this analogy can be generalized to other umbilic-like main orbits. In [Lu 96d] it was asserted, without a detailed proof, that this can be done at least for products of umbilic-like Segre orbits and Plücker orbits.

10.4.2 The case of nonumbilic-like Segre orbits

Now consider the case of nonumbilic-like $S_{(1,\bar{p})}(k)$; here $p = 1$, $\bar{p} > 1$, so that the index π takes only one value 1, and $\bar{\pi}$ takes more than one value. Differential prolongation of equations (10.4.5) then leads to

$$\omega^1 \wedge \omega^\xi_{m+1} + \sum_{\bar{\sigma}} \omega^{\bar{\sigma}} \wedge \omega^\xi_{1\bar{\sigma}} = 0,$$

$$\omega^{\bar{\pi}} \wedge \omega^\xi_{m+1} + \omega^1 \wedge \omega^\xi_{1\bar{\pi}} = 0,$$

$$c\omega^{1^*} \wedge \omega^\xi_{2*} = 0.$$

Therefore, instead of (10.4.6) one obtains

$$\omega^\xi_{m+1} = A^\xi \omega^1, \quad \omega^\xi_{1\bar{\sigma}} = A^\xi \omega^{\bar{\sigma}}, \quad c\omega^\xi_{2*} = B^\xi \omega^{1^*}; \tag{10.4.26}$$

and (10.4.8) and (10.4.9) are replaced by

$$\omega^1_{1*} = \lambda\omega^1 - ck^{-1}\mu\omega^{1^*}, \quad \omega^{\bar{\tau}}_{1*} = \kappa^{\bar{\tau}}\omega^1 + \lambda\omega^{\bar{\tau}} - ck^{-1}\nu^{\bar{\tau}}\omega^{1^*}, \tag{10.4.27}$$

and (10.4.10) by

$$\omega^{1\bar{\tau}}_{2*} = c\phi\omega^{\bar{\tau}}, \quad \omega^{m+1}_{2*} = c(\phi\omega^1 - \lambda k^{-1}\omega^{1^*}). \tag{10.4.28}$$

Substituting into (10.4.7) then gives $c\phi = \mu$, $\nu^{\bar{\tau}} = 0$, and hence

$$\omega_{1^*}^{\bar{\tau}} = \kappa^{\bar{\tau}}\omega^1 + \lambda\omega^{\bar{\tau}}, \quad \omega_{2^*}^{1\bar{\tau}} = \mu\omega^{\bar{\tau}}, \quad \omega_{2^*}^{m+1} = \mu\omega^1 - c\lambda k^{-1}\omega^{1^*};$$

but from $\omega_{1^*}^{2^*} = c\omega^{1^*}$ it follows after prolongation that $\kappa^{\bar{\tau}} = 0$, so that, in particular,

$$\omega_{1^*}^{\bar{\tau}} = \lambda\omega^{\bar{\tau}}. \tag{10.4.29}$$

Moreover, equations (10.4.14) are replaced by

$$\omega_1^{\bar{\pi}} = -\nu\omega^{\bar{\pi}}, \quad \omega_{1\bar{\pi}}^{1\bar{\tau}} = \omega_{\bar{\pi}}^{\bar{\tau}}. \tag{10.4.30}$$

From these last equations, an interesting geometric conclusion can be drawn. Namely, on the envelope M^{m+1}, the distribution defined by the system $\omega^{\bar{\pi}} = 0$ is a foliation, because then all $d\omega^{\bar{\pi}} = \omega^1 \wedge (-\nu\omega^{\bar{\pi}}) + \omega^{1^*} \wedge \lambda\omega^{\bar{\pi}}$ become zero due to the equations of the same system. Moreover, the leaves of this foliation are intrinsically locally Euclidean, because it follows from (10.4.29), (10.4.30), and the last equations in (10.4.2) and (10.4.3) that $d\omega_1^{1^*} = 0$ holds.

The differential prolongation of the whole system above must be carried out, of course, and the results can be interpreted further geometrically. This was done in [Lu 96c], where the following two theorems were proved (with notations differing somewhat from those used above).

Theorem 10.4.4 (see [Lu 96c], Theorem 3). *A second-order envelope of products $S_{(1,\bar{p})}(k) \times E^1$ in a Euclidean space, is an open subset either*

- *of a cylinder over a $(1+\bar{p})$-dimensional logarithmic spiral tube (see Theorem 4.6.1), or*
- *of a cone $C^{\bar{p}+2}$ with a point vertex z in $E^{2(\bar{p}+2)}$, consisting of a one-parameter family of $(\bar{p}+1)$-dimensional round cones with a vertex z, whose axes belong to a plane angular domain D and vertex angles χ vary according to $\sin^2\chi = \sin^2\chi_0 - \cos^2\chi_0 \cdot \tan^2\psi$, where ψ is the angle between the axis and bisectrix of D, $0 \leq \psi \leq \chi_0 = \mathrm{const.}$*

The $(\bar{p}+1)$-dimensional submanifolds which are cut from this cone $C^{\bar{p}+2}$ by the hyperspheres $S^{2\bar{p}+3}$ around z were also described geometrically in [Lu 96c].

Theorem 10.4.5 (see [Lu 96c], Theorem 4). *A second-order envelope of products $S_{(1,\bar{p})}(k) \times S^1(c)$ in a Euclidean space is an open subset either*

- *of a product of a $(1+\bar{p})$-dimensional logarithmic spiral tube and a curve, or*
- *of a sphere bundle, whose base is a developable surface M^2 and whose $\bar{p}$-dimensional sphere fibres have their centers on M^2, and their $(\bar{p}+1)$-dimensional subspaces totally orthogonal to the osculating subspace of M^2.*

Some particular subcases were also described in [Lu 96c] (see Propositions 5 and 6 there), e.g., the subcases where the base is a plane domain with a base curve on it, or a cylinder which can be bent onto a plane angular domain.

11

Isometric Semiparallel Immersions of Riemannian Manifolds of Conullity Two

A new point of view for semiparallel submanifolds is to consider them as special isometric immersions of semisymmetric Riemannian manifolds. Each of these is locally isometric to a direct product of infinitesimally irreducible simple semisymmetric leaves and, possibly, a Euclidean space; moreover, each such leaf is, due to Szabó [Sza 82] (see Theorem 1.6.1 above) either (a) locally symmetric, (b) an elliptic, hyperbolic, Euclidean cone, (c) a Kählerian cone, or (d) of conullity two (i.e., foliated by Euclidean leaves of codimension 2).

For type (a), the problem of parallel immersions was solved by Ferus [Fe 74c, 80] (see Theorem 3.6.1): such immersions exist only for symmetric R-spaces and are their standard immersions. Hence by Theorem 4.5.5, a locally symmetric Riemannian manifold can be isometrically immersed as a semiparallel submanifold only if its metric is a second-order envelope of the family of metrics of symmetric R-spaces, in the sense of [KoN 98].

For types (b) and (c), their isometric immersions can be taken as submanifold immersions of the corresponding cones. So type (d) appears to be the most interesting case, and this will be the topic of this chapter.

11.1 Semiparallel Submanifolds with Plane Generators of Codimension 2

To start, consider the case where the immersion is into Euclidean space and Euclidean leaves are immersed as Euclidean planes. In this case the semiparallel submanifold M^m in E^n is generated by planes which are of codimension 2 in M^m, and thus M^m is intrinsically of conullity two.

Let the frame of $O(M^m, E^n)$ be adapted further so that e_u $(3 \leq u, v, \ldots \leq m)$ belong to the $(m-2)$-plane through $x \in M^m$. Then these planes are the leaves of the foliation defined by the differential system $\omega^a = 0$ $(1 \leq a \leq 2)$. Therefore,

$$de_u = e_a\omega_u^a + e_v\omega_u^v + h_{ua}\omega^a + h_{uv}\omega^v, \qquad (11.1.1)$$

Ü. Lumiste, *Semiparallel Submanifolds in Space Forms*,
DOI 10.1007/978-0-387-49913-0_12,

taken mod$\{\omega^1, \omega^2\}$, must be expressed only in terms of $e_3, \ldots, e_m$, whence, using the notation in (1.6.4), one gets

$$\omega_u^1 = A_u\omega^1 + B_u\omega^2, \quad \omega_u^2 = C_u\omega^1 + F_u\omega^2, \quad h_{uv} = 0. \tag{11.1.2}$$

Now suppose additionally that M^m is semiparallel, i.e., it satisfies condition (4.1.2), which is equivalent to

$$\sum_p (H_{i[k,l]p}h_{pj} + H_{j[k,l]p}h_{ip} - H_{ij,p[k}h_{l]p}) = 0, \tag{11.1.3}$$

where $H_{ik,lj} = \langle h_{ik}, h_{lj}\rangle$ (cf. with (7.3.1), where now $c = 0$ and so $h^*_{ij} = h_{ij}$; see Remark 4.2.3).

For such an M^m with generator $(m-2)$-planes in E^n, condition (11.1.3) reduces for $(k, l) = (a, u)$ to

$$\sum_p [(H_{ia,up} - H_{iu,ap})h_{pj} + (H_{ja,up} - H_{ju,ap})h_{ip} - H_{ij.pa}h_{up} + H_{ij.pu}h_{ap}] = 0,$$

and for $(i, j) = (v, w)$ this gives, due to (11.1.2),

$$\sum_b (H_{va,ub}h_{wb} + H_{wa,ub}h_{vb}) = 0.$$

Setting $u = v = w$ in the last equation leads to the pair of equations

$$\langle h_{u1}, h_{u1}\rangle h_{u1} + \langle h_{u1}, h_{u2}\rangle h_{u2} = 0, \tag{11.1.4}$$

$$\langle h_{u2}, h_{u1}\rangle h_{u1} + \langle h_{u2}, h_{u2}\rangle h_{u2} = 0. \tag{11.1.5}$$

Now the following lemma can be applied.

Lemma 11.1.1. *If two vectors p, q in a real Euclidean vector space satisfy the two equations $\langle p, p\rangle p + \langle p, q\rangle q = 0$ and $\langle p, q\rangle p + \langle q, q\rangle q = 0$, then $p = q = 0$.*

Proof. The two vectors p and q lie in a two-dimensional vector subspace. An orthonormal basis can be chosen such that $p = (p_1, 0), q = (q_1, q_2)$. The two equations then become

$$p_1^2(p_1, 0) + p_1q_1(q_1, q_2) = 0, \quad p_1q_1(p_1, 0) + (q_1^2 + q_2^2)(q_1, q_2) = 0.$$

For second coordinates, this means that $p_1q_1q_2 = (q_1^2 + q_2^2)q_2 = 0$ and leads to $q_2 = 0$; and then for the first coordinates, $(p_1^2+q_1^2)p_1 = (p_1^2+q_1^2)q_1 = 0$; therefore, $p_1 = q_1 = 0$.

Theorem 11.1.2. *If a submanifold M^m with generator $(m-2)$-planes in E^n is semiparallel, then its tangent m-planes along each of its $(m-2)$-plane generators coincide, so that the tangent plane of M^m depends on at most two parameters.*

Proof. Indeed, then the pair of equations (11.1.4) and (11.1.5) must be satisfied, and Lemma 11.1.1 then leads to $h_{ua} = 0$. Now

$$de_1 = -\sum_u (A_u\omega^1 + B_u\omega^2)e_u + \omega_1^2 e_2 + (h_{11}\omega^1 + h_{12}\omega^2),$$

$$de_2 = -\sum_u (C_u\omega^1 + F_u\omega^2)e_u - \omega_1^2 e_1 + (h_{12}\omega^1 + h_{22}\omega^2),$$

$$de_u = e_v\omega_u^v + (A_u\omega^1 + B_u\omega^2)e_1 + (C_u\omega^1 + F_u\omega^2)e_2;$$

the latter due to (11.1.1) and (11.1.2). This shows that the two subspaces of T_xM^m spanned by e_a $(1 \le a, b \le 2)$ and by e_u $(3 \le u, v \le m)$ are invariant along each generator $(m-2)$-plane, since the latter are defined by $\omega^b = 0$. This concludes the proof.

The main result of this section is the following statement.

Theorem 11.1.3. *Each semiparallel submanifold M^m with generator $(m-2)$-planes in E^n is intrinsically a Riemannian manifold of conullity two of the planar type.*

Proof. In this setting, equations (3.1.2) and (3.1.3) give

$$dh_{ij} = -\sum_k e_k\langle h_{ij}, h_{kl}\rangle\omega^l + h_{kj}\omega_i^k + h_{ik}\omega_j^k + h_{ijk}\omega^k,$$

where $h_{ijk} = e_\alpha h_{ijk}^\alpha$ are symmetric in their indices. Since $h_{uv} = h_{ua} = 0$ for the submanifold M^m considered here, this gives, for $(i, j) = (u, v)$ and for $(i, j) = (u, a)$, respectively, $h_{uvw} = h_{uva} = 0$ and $-h_{ac}\omega_u^c = h_{uab}\omega^b$. Hence by (11.1.2) (where now $A_u = A_{u1}^1$, $B_u = A_{u2}^1$, $C_u = A_{u1}^2$, $F_u = A_{u2}^2$), $h_{uab} = -h_{ac}A_{ub}^c$; and from here via symmetry, $h_{ac}A_{ub}^c = h_{bc}A_{ua}^c$, with $a, b, c \in \{1, 2\}$. Therefore,

$$h_{11}B_u + h_{12}(F_u - A_u) - h_{22}C_u = 0. \tag{11.1.6}$$

Suppose that $\operatorname{span}\{h_{11}, h_{12}, h_{22}\}$ has the maximal possible dimension 3 at every point $x \in M^m$. Then (11.1.6) yields $B_u = C_u = 0$, $F_u = A_u$, and this shows via (1.6.5) that M^m is indeed of the planar type (see Remark 1.6.11).

Therefore, further analysis is needed only for the cases where this span has dimension ≤ 2.

If the span has dimension 0, the submanifold M^m is totally geodesic, and thus an open subset of an m-dimensional plane, and belongs to a special case not of conullity two.

Suppose the span dimension is 1. Then each of the vectors h_{ab} has only one coordinate, and the symmetric matrix of these coordinates can be diagonalized by a suitable orthogonal transformation of $\{e_1, e_2\}$. (Note that the relations (11.1.2) are invariant with respect to this transformation; this also follows from the fact that these relations have invariant geometric meaning.) Consequently, M^m is in this case defined by the equations

$$\omega^\alpha = 0, \quad \omega_1^{m+1} = \kappa_1 \omega^1, \quad \omega_2^{m+1} = \kappa_2 \omega^2, \quad \omega_a^\xi = \omega_u^\alpha = 0,$$

where $\xi \in \{m+2, \ldots, n\}$. By exterior differentiation, one gets

$$(d\kappa_1 + \kappa_1 A^1_{u1}\omega^u) \wedge \omega^1 + [(\kappa_1 - \kappa_2)\omega_1^2 + \kappa_1 A^1_{u2}\omega^u] \wedge \omega^2 = 0,$$

$$[(\kappa_1 - \kappa_2)\omega_1^2 + \kappa_2 A^2_{u1}\omega^u] \wedge \omega^1 + (d\kappa_2 + \kappa_2 A^2_{u2}\omega^u) \wedge \omega^2 = 0.$$

The semiparallel condition (11.1.3) now reduces to $(\kappa_1 - \kappa_2)\kappa_1\kappa_2 = 0$. Here $\kappa_1\kappa_2 = 0$ implies $\Omega_{12} = 0$; moreover, $h_{uv} = h_{ua} = 0$ imply $\Omega_{uv} = \Omega_{ua} = 0$, so that $\Omega_{ij} = 0$; hence M^m is intrinsically locally Euclidean and not of conullity two. Therefore, $\kappa_1 = \kappa_2 = \kappa \neq 0$, and the exterior equations reduce to

$$(d\ln\kappa + A^1_{u1}\omega^u) \wedge \omega^1 + A^1_{u2}\omega^u \wedge \omega^2 = 0,$$

$$A^2_{u1}\omega^u \wedge \omega^1 + (d\ln\kappa + A^2_{u2}\omega^u) \wedge \omega^2 = 0.$$

From here it follows that

$$d\ln\kappa + A^1_{u1}\omega^u = P\omega^1, \quad A^1_{u2} = A^2_{u1} = 0, \quad d\ln\kappa + A^2_{u2}\omega^u = Q\omega^2,$$

and hence $A^1_{u1} - A^2_{u2} = P = Q = 0$; and comparison with (1.6.5) shows that M^m is intrinsically of conullity two of the planar type.

Suppose the dimension of span$\{h_{11}, h_{12}, h_{22}\}$ is 2. The orthonormal frame can be further adapted to M^m, taking e_{m+1} and e_{m+2} as belonging to this span. Then $h^\xi_{ij} = 0$ for $m+3 \leq \xi \leq n$, and hence among Ω^α_β only $\Omega^{m+1}_{m+2} = \sum_i h^{m+1}_{i[k} h^{m+2}_{l]i} \omega^k \wedge \omega^l$ can be nonzero.

Summing over $i = j$ in the semiparallel condition (4.1.2), and then using the symmetry of h_{ij} and antisymmetry of Ω_{ij} in i, j, it follows that $H^\beta \Omega^\alpha_\beta = 0$, where $H^\beta = \frac{1}{m}\sum_i h^\beta_{ii}$ are components of the mean curvature vector H of M^m. For the current case, this reduces, due to antisymmetry of Ω^α_β in α, β, to

$$\Omega^{m+1}_{m+2} H^{m+2} = 0, \quad \Omega^{m+1}_{m+2} H^{m+1} = 0.$$

The semiparallel submanifold in E^n is minimal (i.e., has $H = 0$) only if it is an open part of a plane (see Theorem 4.1.7) and thus is not of conullity two. Therefore, only the case where $\Omega^{m+1}_{m+2} = 0$ is possible here. This leads to the consequence that the matrices $\|h^{m+1}_{ab}\|$ and $|h^{m+2}_{ab}\|$ commute and therefore can be simultaneously diagonalized by a suitable orthogonal transformation of $\{e_1, e_2\}$. Subsequently, $h_{ab} = k_a \delta_{ab}$, and the semiparallel condition (11.1.3) reduces to $(k_1 - k_2)\langle k_1, k_2\rangle = 0$. Here $k_1 - k_2 = 0$ is impossible for the current case, since span$\{k_1, k_2\}$ has dimension 2; therefore, $\langle k_1, k_2\rangle = 0$, so $\Omega^2_1 = 0$. Moreover, $\Omega^v_u = \Omega^a_u = 0$, since $h_{uv} = h_{ua} = 0$. Hence the submanifold M^m is locally Euclidean and cannot be of conullity two. This finishes the proof.

Remark 11.1.4. Theorem 11.1.3 was stated in [Lu 2001], where the following conjecture was also first formulated: *If a semiparallel submanifold M^m in E^n is intrinsically a Riemannian manifold of conullity two, then it can only be of planar type.* Later this conjecture was repeated and confirmed in [Lu 2002a, b] and [Lu 2003] in some particular cases, which will be considered in the following section.

11.2 Some Particular Cases

Proposition 6.2 of [Lu 2002b] states in particular that if a semiparallel submanifold M^m in E^{m+2} (i.e., of codimension 2) is intrinsically of conullity two, then it must be of planar type. Here due to Proposition 5.1.3 and Remark 5.1.4, every semiparallel submanifold of codimension 2 in Euclidean space is normally flat. General normally flat semiparallel submanifolds were investigated in [Lu 2002a] from the point of view of their inner Riemannian geometry. It was proved there (see Theorem 3.2) that if such an M^m in E^n is intrinsically of conullity two, then it is of planar type.

This result will be proved below based on the material in Chapter 5.

For normally flat semiparallel M^m in E^n, the principal curvature vectors k_i can be introduced so that $h_{ij} = k_i \delta_{ij}$. If there are r distinct principal curvature vectors $k_{(1)}, \dots, k_{(r)}$, then (5.1.6) holds. In Section 5.3, these vectors are divided into three groups: (1) $k_{(\rho)}$ which are nonzero and of multiplicity > 1, (2) k_a which are nonzero and of multiplicity 1, and (3) one $k_{(0)} = 0$. Correspondingly, equations (5.1.6) then reduce to (5.3.2), (5.3.3).

Due to (5.1.4), then $\Omega_i^j = -\langle k_i, k_j \rangle \omega^i \wedge \omega^j$ and this gives

$$\Omega_{i_\rho}^{j_\rho} = -(k_\rho)^2 \omega^{i_\rho} \wedge \omega^{j_\rho} \neq 0, \qquad \Omega_{i_\rho}^{j_\tau} = 0, \quad \rho \neq \tau;$$

$$\Omega_a^b = \Omega_a^{j_0} = \Omega_{i_0}^{j_0} = 0. \tag{11.2.1}$$

It is seen that the distribution of the tangent subspaces spanned by e_a and e_{i_0} has Euclidean leaves, and if $r = 1$ with $k_{(1)}$ of multiplicity 2, then M^m is intrinsically of conullity two. The ranges of indices $a, b, \dots$ and $i_0, j_0, \dots$ can be joined by introducing indices $u, v, \dots$ running through the union of these ranges. Then (5.3.3) yields

$$\omega_u^1 = -\lambda_{(1)u}\omega^1, \quad \omega_u^2 = -\lambda_{(1)u}\omega^2.$$

Comparing this with (1.6.4), one sees that now $B_u = C_u = 0$ and $A_u = F_u = -\lambda_{(1)u}$. Hence equation (1.6.5) becomes an identity, and this shows that M^m is intrinsically of conullity two, of planar type. This leads to the following.

Theorem 11.2.1. *If a normally flat semiparallel submanifold M^m in E^n is intrinsically of conullity two, then it is of planar type.*

Remark 11.2.2. The statement of Theorem 11.2.1 can be considered as part of Theorem 6.2 in [Lu 2002b]; that article also contains Proposition 6.4, stating that among nonsemiparallel normally flat submanifolds M^m in E^{m+2}, there exist intrinsically semisymmetric M^m of conullity two, whose Euclidean leaves of codimension 2 are $(m-2)$-dimensional planes in E^{m+2}, and which are not of parabolic, but of hyperbolic type.

The other particular case where the problem is already solved is the case of three-dimensional semiparallel submanifolds of conullity two in Euclidean space E^n. The complete classification of all semiparallel three-dimensional submanifolds M^3 in E^n

was done in [LR 90], [Lu 90b] (see also [Lu 2000a], Chapter 20), and has been presented here in Chapter 7. Their inner semisymmetric Riemannian geometry was characterized in [Lu 2003], whose Main Theorem states that such an M^3 is intrinsically either locally of constant curvature, or semiparallel of conullity two of planar type.

The second part of this theorem will now be proved here using the results of Chapter 7.

In Chapter 7 the semiparallel three-dimensional submanifolds were examined first according to their principal codimension $m_1 = \dim(\text{span}\{h_{ij}\})$. If $m_1 \leq 2$, then such a submanifold is normally flat, due to Proposition 5.1.3 and Remark 5.1.4; therefore, Theorem 11.2.1 can be applied here, in the particular case of $m = 3$.

If $m_1 = 3$ and $\nabla^\perp$ is nonflat, then M^3 is a logarithmic spiral tube in E^6, due to Proposition 7.2.1, and equations (7.2.3) hold with $c = 0$:

$$\omega_1^2 = -a\omega^2, \qquad \omega_1^3 = -a\omega^3. \tag{11.2.2}$$

This shows that the differential system $\omega^2 = \omega^3 = 0$ defines a two-parameter system of curves in M^3, which are logarithmic spirals, as shown in Section 4.6 (for the case of $p = 1, q > 1$). Hence M^3 is of conullity two. It remains to exchange in (11.2.2) the roles of the indices 1 and 3, and to compare the result with (1.6.4), in order to establish that now $B_3 = F_3 = 0$ and $A_3 = C_3 = -a$. By (1.6.5), this implies that the logarithmic spiral tube M^3 is of planar type.

If $m_1 = 4$, then Theorem 7.3.1 holds, together with equations (7.3.16), which can be written as

$$\omega_3^1 = -\tau\omega^1, \qquad \omega_3^2 = -\tau\omega^2. \tag{11.2.3}$$

This shows that the differential system $\omega^1 = \omega^2 = 0$ defines a two-parameter system of curves in M^3, which is thus of conullity two. Now comparing (11.2.3) with (1.6.4), one gets $B_3 = F_3 = 0$ and $A_3 = C_3 = -\tau$. Hence by (1.6.5), M^3 is of planar type.

The analysis of the other possibilities for semisymmetric M^3 in E^n, done in Chapter 7 and formulated in Theorem 7.4.1, shows that all such M^3 must locally be manifolds of constant curvature. (See also [Lu 2003], where inner Riemannian geometry was emphasized.)

These results are summarized in the following theorem.

Theorem 11.2.3. *Consider a three-dimensional semisymmetric Riemannian manifold immersed isometrically in E^n as a semiparallel submanifold. Then it is either locally a space of constant curvature, or a manifold of conullity two and of the planar type.*

11.3 Semiparallel Manifolds of Conullity Two in General

The geometric description of a general semiparallel submanifold M^m in E^n is given by Theorem 10.2.1. According to this theorem, such an M^m is a second-order envelope of products

$$M^{m_1} \times \cdots \times M^{m_s} \times S^1(c_{s+1}) \times \cdots \times S^1(c_{s+q}) \times E^{m_0}.$$

The $(q + m_0)$-dimensional leaves enveloped by the products $S^1(c_{s+1}) \times \cdots \times S^1(c_{s+q}) \times E^{m_0}$ have flat van der Waerden–Bortolotti connection $\bar{\nabla}$, thus also flat ∇, and therefore they are intrinsically locally Euclidean.

The chief possibility for semiparallel isometric immersion of a manifold of conullity two is the case where the above leaves are of codimension 2 in M^m, in other words, when the product $M^{m_1} \times \cdots \times M^{m_s}$ of main symmetric orbits has dimension 2. Here this product is a parallel submanifold, and thus Theorem 6.2.1 can be used, since it classifies all semiparallel surfaces, including all parallel surfaces. Among them, the main symmetric orbits are the complete irreducible and not totally geodesic ones. These are the spheres $S^2(c)$ and the Veronese orbits $V^2(k)$.

Here the analysis in Section 10.2 leading to Theorem 10.2.1 is useful. The subindex ρ now takes only one value 1, while $i_1, j_1, \ldots$ take two values 1, 2, which are denoted by $a, b, \ldots,$ below.

The first equations in (10.2.4) and (10.2.12) can be summarized as

$$\omega_a^u = \sum_b \lambda_{ab}^u \omega^b, \tag{11.3.1}$$

where the index u runs over the ranges of i_0 and $m^* + a$ (here a in the old sense) and the coefficients λ_{ab}^u are symmetric in a, b. Equations (10.2.5) and the last equations in (10.2.12) can be summarized as

$$\sum_a (h_{ab}^\alpha \lambda_{ac}^u - h_{ac}^\alpha \lambda_{ab}^u) = 0, \tag{11.3.2}$$

with α replacing α_1.

Consider first the case where the only two-dimensional main symmetric orbit $M^{m_1} = M^2$ is a sphere $S^2(c)$. Then the given second-order envelope M^m has flat $\nabla^\perp$, and by Theorem 11.2.1 it is intrinsically of conullity two of the planar type.

Now suppose the main symmetric orbit $M^{m_1} = M^2$ is a Veronese orbit $V^2(k)$. Then $m^* = m_1 = 2$. By Proposition 6.3.1 and Corollary 6.3.6, the adapted orthonormal frame bundle for such a $V^2(k)$ satisfies (6.3.1) and (6.3.2) with $c = 0$. Therefore, the matrices $h^\alpha = \|h_{ab}^\alpha\|$ take the form

$$h^{3^*} = \begin{pmatrix} k\sqrt{3} & 0 \\ 0 & k\sqrt{3} \end{pmatrix}, \qquad h^{4^*} = \begin{pmatrix} k & 0 \\ 0 & -k \end{pmatrix}, \qquad h^{5^*} = \begin{pmatrix} 0 & k \\ k & 0 \end{pmatrix}, \qquad h^\xi = 0.$$

Equation (11.3.2) can be considered as the condition that the product of the two symmetric matrices h^α and $\lambda^u = \|\lambda_{ab}^u\|$ is a symmetric matrix. For $\alpha = 3^*$ and $\alpha = \xi$, this condition is trivially satisfied. Since

$$h^{4^*} \cdot \lambda^u = \begin{pmatrix} k\lambda_{11}^u & k\lambda_{12}^u \\ -k\lambda_{21}^u & -k\lambda_{22}^u \end{pmatrix}, \qquad h^{5^*} \cdot \lambda^u = \begin{pmatrix} k\lambda_{21}^u & k\lambda_{22}^u \\ k\lambda_{11}^u & k\lambda_{12}^u \end{pmatrix},$$

the same condition for $\alpha = 4^*$ and $\alpha = 5^*$ implies $\lambda_{21}^u = -\lambda_{12}^u$ and $\lambda_{11}^u = \lambda_{22}^u$. This, and the symmetricity $\lambda_{21}^u = \lambda_{12}^u$, imply that $\lambda_{12}^u = \lambda_{21}^u = 0$, so that

$$\omega_1^u = \lambda^u \omega^1, \qquad \omega_2^u = \lambda^u \omega^2, \tag{11.3.3}$$

where λ^u is the common value of λ_{11}^u and λ_{22}^u. Comparing with (1.6.4) shows that $B_u = C_u = 0$ and $A_u = F_u = -\lambda^u$; hence the given M^m is, by (1.6.5), intrinsically of conullity two of the planar type.

There is another possibility for semiparallel isometric immersion of a manifold of conullity two, namely the case where $s = 1$ as above, but M^{m_1} carries a foliation whose leaves have flat ∇, and which, together with the leaves enveloped by $S^1(c_2) \times \cdots \times S^1(c_{1+q}) \times E^{m_0}$, generate the locally Euclidean submanifolds of codimension 2 in M^m.

This occurs if $m_1 = 3$ and $M^{m_1} = M^3$ is a Segre orbit $S_{(1,2)}(k)$. So let M^m be the second-order envelope of products $S_{(1,2)}(k) \times S^1(c_2) \times \cdots \times S^1(c_{1+q}) \times E^{m_0}$, $m = 3 + q + m_0$. The orthonormal frame bundle will be adapted to this M^m so that, at a point $x \in M^m$, the vectors e_a for $1 \leq a, b, \ldots \leq 2$ are tangent to the generator sphere $S^2(k)$, and e_3 is tangent to the generator circle $S^1(k)$ of $S_{(1,2)}(k)$, going through x. Moreover, let e_f for $4 \leq f, g, \ldots \leq 3 + q$ be tangent to the circle $S^1(c_{f-2})$, and let e_{i_0} for $3 + q + 1 \leq i_0, j_0, \ldots \leq 3 + q + m_0$ be in E^{m_0}. Among the basis vectors of the orthonormal frame normal to M^m at x, let e_{m+1} be directed to the center of the sphere $S^5(k^2)$ containing $S_{(1,2)}(k)$ (see Theorem 3.2.1), and let the vectors $e_{(a3)}$ in the proof of Theorem 3.2.1 be denoted now by e_{m+1+a} with $1 \leq a \leq 2$. Finally, let e_{m+f} be normal to the circle $S^1(c_{f-2})$ at x, and e_ξ be the remaining basis vectors normal to M^m in E^n. Then the given M^m is defined by the following Pfaffian system, as one of its integral submanifolds:

$$\omega^{m+1} = \omega^{m+1+a} = \omega^{m+f} = \omega^\xi = 0,$$

$$\omega_a^{m+1} = k\omega^a, \quad \omega_3^{m+1} = k\omega^3, \quad \omega_b^{m+1+a} = \delta_b^a k\omega^3, \quad \omega_3^{m+1+a} = k\omega^a, \tag{11.3.4}$$

$$\omega_a^{m+f} = \omega_3^{m+f} = \omega_a^\xi = \omega_3^\xi = 0, \tag{11.3.5}$$

$$\omega_f^{m+1} = \omega_f^{m+1+a} = \omega_f^{m+f} - \gamma_f \omega^f = \omega_f^\xi = 0, \tag{11.3.6}$$

$$\omega_{i_0}^{m+1} = \omega_{i_0}^{m+1+a} = \omega_{i_0}^{m+f} = \omega_{i_0}^\xi = 0 \tag{11.3.7}$$

(equations (11.3.4) correspond to (4.6.1)).

The first equations (11.3.4) give by exterior differentiation

$$-d \ln k \wedge \omega^a + \omega_{m+1}^{m+1+a} \wedge \omega^3 + \omega_f^a \wedge \omega^f + \omega_{i_0}^a \wedge \omega^{i_0} = 0,$$

$$-d \ln k \wedge \omega^3 + \sum_a \omega_{m+1}^{m+1+a} \wedge \omega^a + \omega_f^3 \wedge \omega^f + \omega_{i_0} \wedge \omega^{i_0} = 0,$$

and then by Cartan's lemma

$$-d \ln k = \kappa\omega^3 + A_f \omega^f + A_{i_0}\omega^{i_0}, \tag{11.3.8}$$

$$\omega_{m+1}^{m+1+a} = \kappa\omega^a + B_f^a \omega^f + B_{i_0}^a \omega^{i_0}, \tag{11.3.9}$$

$$\omega_f^a = A_f\omega^a + B_f^a\omega^3 + C_{fg}^a\omega^g + C_{fi_0}^a\omega^{i_0}, \tag{11.3.10}$$

$$\omega_{i_0}^a = A_{i_0}\omega^a + B_{i_0}^a\omega^3 + C_{fi_0}^a\omega^f + C_{i_0j_0}^a\omega^{j_0}, \tag{11.3.11}$$

$$\omega_f^3 = \sum_a B_f^a\omega^a + A_f\omega^3 + D_{fg}\omega^g + E_{fi_0}\omega^{i_0}, \tag{11.3.12}$$

$$\omega_{i_0}^3 = \sum_a B_{i_0}^a\omega^a + A_{i_0}\omega^3 + E_{fi_0}\omega^f + F_{i_0j_0}\omega^{j_0}, \tag{11.3.13}$$

where the coefficients are symmetric in f, g and in i_0, j_0.

The remaining equations (11.3.4) give by exterior differentiation

$$\begin{aligned}&(\delta_b^a\omega_c^3 + \delta_c^a\omega_b^3 - \delta_c^b\omega_{m+1}^{m+1+a}) \wedge \omega^c + [-\delta_b^a d\ln k + (\omega_b^a - \omega_{m+1+b}^{m+1+a})] \wedge \omega^3\\ &\quad + \delta_b^a(\omega_f^3 \wedge \omega^f + \omega_{i_0}^3 \wedge \omega^{i_0}) = 0,\end{aligned}$$

$$\begin{aligned}&[-\delta_b^a d\ln k + (\omega_b^a - \omega_{m+1+b}^{m+1+a})] \wedge \omega^b + (2\omega_3^a - \omega_{m+1}^{m+1+a}) \wedge \omega^3\\ &\quad + \omega_f^a \wedge \omega^f + \omega_{i_0}^a \wedge \omega^{i_0} = 0.\end{aligned}$$

For $a = b = 1$ and $a = b = 2$, the first terms in the first exterior equations reduce respectively to

$$(2\omega_1^3 - \omega_{m+1}^{m+2}) \wedge \omega^1 + \omega_2^3 \wedge \omega^2, \quad \omega_1^3 \wedge \omega^1 + (2\omega_2^3 - \omega_{m+1}^{m+3}) \wedge \omega^2,$$

and thus these equations together with (11.3.7)–(11.3.12), give

$$\omega_a^3 = P_{(a)}\omega^a + B_f^a\omega^f + B_{i_0}^a\omega^{i_0}.$$

Substituting $a = b$ into the same equation leads to $B_f^a = B_{i_0}^a = 0$, $P_{(a)} = -\kappa$, and $\omega_1^2 = \omega_{m+2}^{m+3}$, so that

$$\omega_3^a = \kappa\omega^a, \tag{11.3.14}$$

$$\omega_f^a = A_f\omega^a + C_{fg}^a\omega^g + C_{fi_0}^a\omega^{i_0},$$

$$\omega_{i_0}^a = A_{i_0}\omega^a + C_{fi_0}^a\omega^f + C_{i_0j_0}^a\omega^{j_0}. \tag{11.3.15}$$

The first equations in (11.3.5) give

$$k^{-1}\omega_a^f \wedge \gamma_f\omega^f + \omega^a \wedge \omega_{m+1}^{m+f} + \omega^3 \wedge \omega_{m+1+a}^{m+f} = 0,$$

$$k^{-1}\omega_3^f \wedge \gamma_f\omega^f + \omega^3 \wedge \omega_{m+1}^{m+f} + \omega^a \wedge \omega_{m+1+a}^{m+f} = 0,$$

and hence $C_{fg}^a = C_{fi_0}^a = D_{fd} = E_{fi_0} = 0$, $\omega_{m+1}^{m+f} - k^{-1}A_f\gamma_f\omega^f = 0$, $\omega_{m+1+a}^{m+f} = 0$.

The first equations in (11.3.7) lead to $\sum_a \omega_{i_0}^a \wedge \omega^a + \omega_{i_0}^a \wedge \omega^3 = 0$, and thus $C_{i_0j_0}^a = F_{i_0j_0} = 0$. Hence (11.3.15), (11.3.12), (11.3.13) reduce to

$$\omega_f^a = A_f \omega^a, \quad \omega_{i_0}^a = A_{i_0} \omega^a, \quad \omega_f^3 = A_f \omega^3, \quad \omega_{i_0}^3 = A_{i_0} \omega^3. \tag{11.3.16}$$

The distribution defined by the system $\omega^a = 0$, $1 \le a \le 2$, is a foliation, since by (11.3.14) and (11.3.16)

$$d\omega^a = \omega^b \wedge \omega_b^a + \omega^3 \wedge \kappa\omega^a + \omega^f \wedge A_f \omega^a + \omega^{i_0} \wedge A_{i_0} \omega^a.$$

The leaves of this foliation are generated by the second-order envelopes of the products of circular generators of $S_{(1,2)}(k)$ and of $S^1(c_2) \times \cdots \times S^1(c_{1+q}) \times E^{m_0}$.

For these leaves, indices $u, v, \ldots$ can be introduced, whose range consists of the value 3 and the ranges of f and i_0. These leaves are intrinsically locally Euclidean, because $e_a, e_{m+1}, e_{m+1+a}, e_{m+f}, e_\xi$ are normal to them, and by (11.3.4)–(11.3.7)

$$\begin{aligned}\Omega_u^v = \omega_u^a \wedge \omega_a^v + \omega_u^{m+1} \wedge \omega_{m+1}^v + \omega_u^{m+1+a} \wedge \omega_{m+1+a}^v \\ + \omega_u^{m+f} \wedge \omega_{m+f}^v + \omega_u^\xi \wedge \omega_\xi^v = 0.\end{aligned}$$

Hence the given M^m in E^n is intrinsically of conullity two. Since (11.3.14) and the first equations in (11.3.16) can be put together as $\omega_u^a = A_u \omega^a$, where $\kappa = A_3$, this M^m is of planar type.

All these results are summarized in the following theorem.

Theorem 11.3.1. *If a submanifold M^m in E^n is the second-order envelope of products $M^{m_1} \times S^1(c_2) \times \cdots \times S^1(c_{1+q}) \times E^{m_0}$, where M^{m_1} is either $S^2(c)$, or $V^2(k)$, or $S_{(1,2)}(k)$, then it is intrinsically of conullity two of the planar type.*

In connection with this theorem, the following problem arises: is the conjecture stated in Remark 11.1.4 completely verified by Theorem 11.3.1?

For the particular case $m_1 = 2$, the answer is positive, since the spheres and Veronese orbits are the only two-dimensional main symmetric orbits.

The three-dimensional semiparallel submanifolds in E^n are classified above by Theorem 7.4.1, including the parallel submanifolds. Among the latter, the only ones which satisfy the conditions stated above for the case $m_1 = 3$ are the Segre orbits $S_{(1,2)}(k)$. Hence for this case also the answer is positive.

Only the case of $m_1 > 3$ remains open. Here the question is whether there exist main symmetric orbits satisfying the above conditions. They are the standardly imbedded symmetric R-spaces (see Section 3.6). A list of these was given in [TK 68] (see also [Ch 2000] and [Na 84]), but it is not known which ones among them satisfy the conditions.

Nevertheless, it is very plausible that the submanifolds of Theorem 11.3.1 are the only semiparallel submanifolds in E^n which are intrinsically of conullity two.

Remark 11.3.2. Theorem 11.3.1 was proved in the recent paper [Lu 2004], where the above conjecture was also stated, that this theorem probably describes all semiparallel submanifolds that are intrinsically of conullity two.

Remark 11.3.3. A special case among submanifolds of Theorem 11.3.1, namely the four-dimensional second-order envelope of $V^2(k) \times S^1(c_2) \times S^1(c_2)$ in E^n, was investigated in [Ri 99] and [Ri 2000]. A reducibility function was introduced for this case, which can be extended to each submanifold M^m of Theorem 11.3.1 with $M^{m_1} = V^2(k)$, as follows.

From equations (11.3.3), exterior differentiation and Cartan's lemma produce $d\lambda^u = -\sum_v \lambda^v(\omega_v^u - \lambda^u\omega^v)$. Introducing the function γ by $\gamma^2 = \sum_u(\lambda^u)^2$, it follows that $d\gamma = \gamma^2 \sum_v \lambda^v\omega^v$. Hence the function γ on M^m is constant along every leaf of the foliation defined by $\omega^v = 0$, with v running over its range. Each of these leaves is a second-order envelope of Veronese surfaces.

If $\gamma \equiv 0$, then all $\lambda^u \equiv 0$, thus $\omega_1^u = \omega_2^u = 0$; and this means that M^m reduces to a product of one of these leaves and of a locally Euclidean leaf (see Section 8.1 or [Lu 88a]). Therefore, following [Ri 2000], the function γ is called the *function of reducibility*. Some of its other properties were also established in [Ri 99] and [Ri 2000].

12

Some Generalizations

In this last chapter, some generalizations of semiparallel submanifolds in space forms are considered. The semiparallel condition is generalized to higher orders, and then transferred to tensor fields defined by the fundamental forms on the submanifold.

12.1 k-Semiparallel Submanifolds

Recall the concept of a k-parallel submanifold (see Section 3.1): it is a submanifold M^m in $N^n(c)$ with $\bar{\nabla}^{k-1}h \neq 0$, $\bar{\nabla}^k h = 0$ $(k \geq 1)$. Here $\bar{\nabla}^k h$ has the components $h^\alpha_{ijp_1\ldots p_k}$, and therefore 1-parallel means simply parallel. By (2.2.7), the integrability condition of the differential system $\bar{\nabla}^k h = 0$ is

$$\bar{\Omega} \circ h^\alpha_{ijp_1\ldots p_{k-1}} = 0, \tag{12.1.1}$$

which says that the curvature 2-form operator $\bar{\Omega}$ acting on the $(k+1)$st-order fundamental form $\bar{\nabla}^{k-1}h$ gives the result 0. The condition (12.1.1) can be written more compactly as $\bar{\Omega} \circ \bar{\nabla}^{k-1}h = 0$ or, using the curvature tensor $\bar{R}$ of $\bar{\nabla}$ (the tensor of coefficients in $\bar{\Omega}$), also as $\bar{R}(X, Y) \circ \bar{\nabla}^{k-1}h = 0$. For $k = 1$, this condition reduces to the semiparallel condition (4.1.2) (in (0.4) in the introduction).

A submanifold M^m in $N^n(c)$ satisfying condition (12.1.1) for $k \geq 1$ is called a *k-semiparallel submanifold*, according to [Mi 96]. Here 1-semiparallel means simply semiparallel.

Proposition 4.1.3 states that every parallel submanifold is semiparallel; and by Proposition 4.1.5 every 2-parallel submanifold is also semiparallel. At the same time, this 2-parallel submanifold is 2-semiparallel. Indeed, for $h^\alpha_{ijkl} = 0$, therefore $\bar{\nabla} h^\alpha_{ijkl} = 0$; hence from the first equation in (2.2.7) it follows that $\bar{\Omega} \circ h^\alpha_{ijk} = 0$, and this is 2-semiparallel.

Similarly, every k-parallel submanifold is k-semiparallel and also $(k-1)$-semiparallel, because $h^\alpha_{ijp_1\ldots p_k p_{k+1}} = 0$ implies $h^\alpha_{ijp_1\ldots p_k p_{k+1} p_{k+2}} = 0$, and this leads to $\bar{\Omega} \circ h^\alpha_{ijp_1\ldots p_{k-1}} = 0$ and $\bar{\Omega} \circ h^\alpha_{ijp_1\ldots p_k} = 0$.

Ü. Lumiste, *Semiparallel Submanifolds in Space Forms*,
DOI 10.1007/978-0-387-49913-0_13, © Springer Science+Business Media, LLC 2009

The main purpose in [Mi 96] was to extend Theorem 4.5.5 (in the particular case $s = 0$) to k-semiparallel submanifolds. For this purpose, higher-order tangency of two submanifolds at their common point is defined by means of higher-order fundamental forms.

A common point x of two submanifolds M^m and $\bar{M}^m$ in $N^n(c)$ is called a *tangency point of order* k, if at x

(1) the tangent subspaces $T_x M^m$ and $T_x \bar{M}^m$ coincide,
(2) all the corresponding fundamental forms of M^m and $\bar{M}^m$ of orders up to k coincide, i.e., at x

$$\bar{\nabla}^{s-2}h = \bar{\nabla}^{s-2}\bar{h} \quad (s = 2, \dots, k). \tag{12.1.2}$$

A submanifold M^m in $N^n(c)$ is said to be an *envelope of order* k of a family of m-dimensional submanifolds if each of its points x is a tangency point of order k for M^m and some submanifold of this family.

In the particular case $k = 2$, this defines a second-order envelope, introduced in Section 4.5.

Theorem 12.1.1 (see [Mi 96], Theorem 2). *A submanifold M^m in $N^n(c)$ is an envelope of order k of some family of m-dimensional k-parallel submanifolds ($k > 2$) if and only if it is $(k-1)$- and k-semiparallel simultaneously.*

Proof. If M^m is such an envelope and x one of its points, then (12.1.2) is satisfied. Since $\bar{M}^m$ is k-parallel here, it is $(k-1)$-semiparallel and k-semiparallel, i.e., one has $\bar{\Omega}^* \circ \bar{h}^\alpha_{ijp_1\dots p_{k-1}} = 0$ and $\bar{\Omega}^* \circ \bar{h}^\alpha_{ijp_1\dots p_k} = 0$. Due to (12.1.2), (2.1.10) and (2.1.11), these conditions are not differential but algebraic, thus pointwise, and therefore they hold also for M^m, i.e., this M^m is indeed $(k-1)$- and k-semiparallel.

Conversely, let M^m in $N^n(c)$ be $(k-1)$- and k-semiparallel simultaneously. The concept of the Euclidean fundamental triplet introduced in Section 4.2 can be generalized to higher orders. Considering a real Euclidean vector space V and its subspace T, let $h_s : T \times \cdots \times T \to T^\perp$ $(s = 2, \dots, k)$ be s-linear mappings, where $T^\perp$ is the orthogonal complement of T in V, such that h_2 and h_3 are symmetric. Then $(V, T, h_2, \dots, h_k)$ is called a *Euclidean fundamental k-plet* (for $k = 3$ quadruplet, for $k = 4$ quintuplet, etc.). At any point $x \in M^m$, a k-plet is defined by $T = T_x M^m$, $V = T_x N^n(c)$, second fundamental form $h_2 = h$, third fundamental form $h_3 = \bar{\nabla} h$, and so on, up to the k-order fundamental form $h_k = \bar{\nabla}^{(k-2)} h$.

Two submanifolds have tangency of order k at a common point x if their k-plets (as just defined) coincide at x.

For a point x in $N^n(c)$, the pair consisting of x and the fundamental k-plet $(V, T, h_2, \dots, h_k)$, with $V = T_x N^n(c)$ and m-dimensional T, is called a *centered fundamental k-plet* for $N^n(c)$. The manifold of all centered fundamental k-plets will be denoted by Φ^k. Taking for each of them an adapted orthonormal frame with origin x, having the first m basis vectors e_i in T, and the next $n - m$ basis vectors e_α in $T^\perp$, one obtains a *framed fundamental k-plet* for $N^n(c)$. The manifold of all such objects will be denoted by Ψ^k. The local coordinates in Ψ^k are the local coordinates $\{x^I\}$ in $N^n(c)$, the elements of the regular matrix $\|X^J_I\|$ which transforms the natural basis of $\partial/\partial x^I$ into the basis adapted to $(V, T, h_2, \dots, h_k)$ as above, and the components $h_{ijp_1\dots p_{(s-2)}}$ of h_s $(s = 2, \dots, k)$ in this basis.

Now the following differential system can be considered for Ψ^k:

$$\begin{aligned}
&\omega^\alpha = 0,\\
&\omega_i^\alpha - h_{ij}^\alpha = 0,\\
&dh_{ij}^\alpha - h_{kj}^\alpha\omega_i^k - h_{ik}^\alpha\omega_j^k + h_{ij}^\beta\omega_\beta^\alpha - h_{ijk}^\alpha\omega^k = 0,\\
&\dots\dots\dots\dots\dots\dots\dots\dots \qquad (12.1.3)\\
&dh_{ijp_1\dots p_{k-1}}^\alpha - h_{ljp_1\dots p_{k-1}}^\alpha\omega_i^l - \dots - h_{ijp_1\dots l}^\alpha\omega_{p_{k-1}}^l\\
&\quad + h_{ijp_1\dots p_{k-1}}^\beta\omega_\beta^\alpha - h_{ijp_1\dots p_{k-1}p_k}^\alpha\omega^{p_k} = 0,\\
&dh_{ijp_1\dots p_k}^\alpha - h_{ljp_1\dots p_k}^\alpha\omega_i^l - \dots - h_{ijp_1\dots l}^\alpha\omega_{p_k}^l + h_{ijp_1\dots p_k}^\beta\omega_\beta^\alpha = 0.
\end{aligned}$$

This system is based on formulas (2.1.3), (2.1.4), (2.2.3), where (2.2.2)–(2.2.7) in general are written for a k-parallel submanifold, taking $h_{ijp_1\dots p_k p_{k+1}}^\alpha = 0$. Motivated by those same formulas, in particular by (2.2.4) with (2.2.5) and (2.2.6), it is assumed that

$$\begin{aligned}
&h_{kj}^\alpha\Omega_i^k + h_{ik}^\alpha\Omega_j^k - h_{ij}^\beta\Omega_\beta^\alpha + h_{ijkl}^\alpha\omega^l \wedge \omega^k = 0,\\
&\dots\dots\dots\dots\dots\dots\dots\dots \qquad (12.1.4)\\
&h_{ljp_1\dots p_{k-1}}^\alpha\Omega_i^l + \dots + h_{ijp_1\dots l}^\alpha\Omega_{p_{k-1}}^l - h_{ijp_1\dots p_{k-1}}^\beta\Omega_\beta^\alpha + h_{ijp_1\dots p_{k-1}p_k l}^\alpha\omega^l \wedge \omega^{p_k} = 0,
\end{aligned}$$

where according to (2.1.10) and (2.1.11), Ω_i^k and Ω_β^α can be expressed just in terms of the coordinates h_{pq}^γ.

A framed (resp. centered) fundamental k-plet is said to be *k-semiparallel* if the last equation in (12.1.4) is satisfied for $h_{ijp_1\dots p_{k-1}p_k l}^\alpha = 0$ (cf. with (12.1.1)). The manifold of all framed (resp. centered) semiparallel fundamental k-plets will be denoted by Ψ_S^k (resp. Φ_S^k).

Now consider the differential system (12.1.3) with assumptions (12.1.4) on $\Phi_S^k \cap \Phi_S^{k-1}$. A direct computation shows that the exterior differentials on the left sides of the equations in (12.1.3) reduce to zero due to these equations and assumptions. Therefore, this system is totally integrable on $\Phi_s^k \cap \Phi_S^{k-1}$.

Two framed fundamental k-plets are said to be equivalent if $e_i' = e_j A_i^j$, $e_\alpha' = e_\beta A_\alpha^\beta$ with regular coefficient matrices, and

$$'h_{klq_1\dots q_{s-2}}^\beta = A_\alpha^\beta h_{ijp_1\dots p_{s-2}}^\alpha A_k^i A_l^j A_{q_1}^{p_1} \dots A_{q_{s-2}}^{p_{s-2}}.$$

This equivalence defines a map $\Psi_S^s \to \Phi_S^s$ for every $s \in \{2, \dots, k\}$ and this map projects the system (12.1.3) onto a differential system on Φ_S^s, which is totally integrable on $\Phi_S^k \cap \Phi_S^{k-1}$.

Now take a submanifold M^m in $N^n(c)$ which is $(k-1)$- and k-semiparallel simultaneously, and fix one of its points x. Then a centered fundamental k-plet $(T_xN^n(c), T_xM^m, h_x, \ldots, h_x^k)$ is defined, which is simultaneously $(k-1)$- and k-semiparallel. This can be taken as the initial condition for the totally integrable system, and so it determines an integral submanifold of the system. The special nature of this system guarantees that this submanifold is k-parallel having tangency of order k with M^m at x, due to the initial condition. Since this is true at any point of M^m, this manifold is the envelope of order k of k-parallel submanifolds, as claimed.

12.2 On 2-Semiparallel Submanifolds

The 2-semiparallel condition $\bar{\Omega} \circ \bar{\nabla} h = 0$ written out explicitly is

$$h^\alpha_{ljk}\Omega^l_i + h^\alpha_{ilk}\Omega^l_j + h^\alpha_{ijl}\Omega^l_k - h^\beta_{ijk}\Omega^\alpha_\beta = 0; \tag{12.2.1}$$

this condition was already used in Section 6.7 and coincides with (6.7.2).

It is seen immediately that there exist two trivial subclasses of 2-semiparallel submanifolds: the parallel ones, for which all $h^\alpha_{ijk} = 0$, and the submanifolds with flat $\bar{\nabla}$, for which all $\Omega^j_i = 0$ and all $\Omega^\beta_\alpha = 0$. The question arises, do there exist any nontrivial 2-semiparallel submanifolds M^m in $N^n(c)$?

The answer is positive, as shown by the following.

Theorem 12.2.1. *A normally flat semiparallel submanifold M^m in $N^n(c)$ with $c \geq 0$ is also 2-semiparallel.*

Proof. Normally flat semiparallel submanifolds were considered in Chapter 5. From (5.3.4)–(5.3.6) it follows that for them, among h^α_{ij} only

$$h^{m+\rho}_{i_\rho i_\rho} = \kappa_\rho, \qquad h^{m+\rho+a^*}_{aa} = \kappa_a \tag{12.2.2}$$

can be nonzero. Now equations (2.2.3) together with (2.2.2) show, via (5.3.7)–(5.3.11), that

$$h^{m+\rho}_{i_\rho i_\rho u} = \kappa_\rho \lambda_{(\rho)u}, \qquad h^{m+\rho}_{aaa} = \kappa^2_a \kappa^{-1}_\rho \lambda_{(\rho)a},$$

$$h^{m+\rho+a^*}_{i_\rho i_\rho a} = -\kappa_a \lambda_{(\rho)a}, \qquad h^{m+\rho+a^*}_{aaa} = \psi_a, \qquad h^{m+\rho+a^*}_{aai_0} = \kappa_a \lambda_{(a)i_0},$$

$$h^{m+\rho+b^*}_{aaa} = \kappa^2_a \gamma^a_b, \qquad h^{m+\rho+b^*}_{aab} = -\kappa_a \kappa_b \gamma^b_a \quad (a \neq b),$$

and all other h^α_{ijk} are zero.

Among the curvature 2-forms Ω^j_i and Ω^β_α, only $\Omega^{j_\rho}_{i_\rho} = -\kappa^2_\rho \omega^{i_\rho} \wedge \omega^{j_\rho}$, $i_\rho \neq j_\rho$, are nonzero, as follows from (2.1.10), (2.1.11), and (12.2.2). Now a straightforward checking shows that the 2-semiparallel condition (12.2.1) is satisfied. It is clear that a normally flat semiparallel submanifold, lika a 2-semiparallel submanifold, is nontrivial in general. Indeed, it is not locally Euclidean, as was shown above, and it is not parallel, as follows from comparison of Sections 5.2 and 5.4.

On the whole, the 2-semiparallel submanifolds are not yet classified. This has been done only for dimension 2, i.e., for the surfaces. The classification presented in the next section shows that there also exist normally nonflat 2-semiparallel surfaces.

12.3 2-Semiparallel Surfaces in Space Forms

In Section 12.1 it was stated that every k-parallel submanifold is also k-semiparallel. Now in the particular case $k = 2$ it will be shown that the converse does not hold. This will follow from the classification of 2-semiparallel surfaces.

For surfaces M^2 in $N^n(c)$, the bundle of orthonormal frames can be adapted to M^2 so that (6.1.8) hold, together with (6.2.1) and (6.2.2), where now $\varepsilon_4 = \varepsilon_5 = 1$. Then in equation (12.2.1) only $\Omega_1^2 = -\Omega_2^1$ and $\Omega_4^5 = -\Omega_5^4$ can be nonzero, and therefore this condition reduces to

$$3h^4_{112}\Omega^2_1 + h^5_{111}\Omega^5_4 = 0, \tag{12.3.1}$$

$$2h^4_{122}\Omega^2_1 - h^4_{111}\Omega^2_1 + h^5_{112}\Omega^5_4 = 0, \tag{12.3.2}$$

$$-2h^4_{112}\Omega^2_1 + h^4_{222}\Omega^2_1 + h^5_{122}\Omega^5_4 = 0, \tag{12.3.3}$$

$$-3h^4_{122}\Omega^2_1 + h^5_{222}\Omega^5_4 = 0, \tag{12.3.4}$$

$$-h^4_{111}\Omega^5_4 + 3h^5_{112}\Omega^2_1 = 0, \tag{12.3.5}$$

$$-h^4_{112}\Omega^5_4 - h^5_{111}\Omega^2_1 + 2h^5_{122}\Omega^2_1 = 0, \tag{12.3.6}$$

$$-h^4_{122}\Omega^5_4 - 2h^5_{112}\Omega^2_1 + h^5_{222}\Omega^2_1 = 0, \tag{12.3.7}$$

$$-h^4_{222}\Omega^5_4 - 3h^5_{122}\Omega^2_1 = 0, \tag{12.3.8}$$

$$\begin{aligned} 3h^\xi_{112}\Omega^2_1 &= (2h^\xi_{122} - h^\xi_{111})\Omega^2_1 \\ &= (-2h^\xi_{112} + h^\xi_{222})\Omega^2_1 = 3h^\xi_{122}\Omega^2_1 = 0, \end{aligned} \tag{12.3.9}$$

where

$$\xi \in \{3\} \cup \{6, \ldots, n\}.$$

It is seen that every surface M^2 with flat connection $\bar{\nabla}$ in $N^n(c)$, which is characterized by $\Omega^2_1 = \Omega^5_4 = 0$, is 2-semiparallel. Also every parallel surface M^2 in $N^n(c)$, characterized by $h^\alpha_{ijk} = 0$, is 2-semiparallel. These are called the trivial cases (cf. Section 12.2).

The system of equations (12.3.1)–(12.3.9) is a homogeneous linear system for h^4_{ijk} and h^5_{ijk}, having determinant

$$D = \begin{vmatrix} 0 & 3\Omega_1^2 & 0 & 0 & \Omega_4^5 & 0 & 0 & 0 \\ -\Omega_1 & 0 & 2\Omega_1^2 & 0 & 0 & \Omega_4^5 & 0 & 0 \\ 0 & -2\Omega_1^2 & 0 & \Omega_1^2 & 0 & 0 & \Omega_4^5 & 0 \\ 0 & 0 & -3\Omega_1^2 & 0 & 0 & 0 & 0 & \Omega_4^5 \\ -\Omega_4^5 & 0 & 0 & 0 & 0 & 3\Omega_1^2 & 0 & 0 \\ 0 & -\Omega_4^5 & 0 & 0 & -\Omega_1^2 & 0 & 2\Omega_1^2 & 0 \\ 0 & 0 & -\Omega_4^5 & 0 & 0 & -2\Omega_1^2 & 0 & \Omega_1^2 \\ 0 & 0 & 0 & -\Omega_4^5 & 0 & 0 & -3\Omega_1^2 & 0 \end{vmatrix},$$

equal to $D = [9(\Omega_1^2)^2 - (\Omega_4^5)^2]^2[(\Omega_1^2)^2 - (\Omega_4^5)^2]^2$.

If this determinant D is nonzero, then $h^4_{ijk} = h^5_{ijk} = 0$. If, moreover, $\Omega_1^2 \neq 0$, then (12.2.1) also implies $h^\xi_{ijk} = 0$, and this gives the trivial case of a parallel surface.

It is seen that in order to have a nontrivial case, only these three possibilities can occur:

(1) $(\Omega_4^5)^2 = (\Omega_1^2)^2 \neq 0$, or
(2) $(\Omega_4^5)^2 = (3\Omega_1^2)^2 \neq 0$, or
(3) $\Omega_1^2 = 0$, $\Omega_4^5 \neq 0$.

It turns out that nontrivial 2-semiparallel surfaces of the first two possibilities, i.e., with $D = 0$, do not exist.

Indeed, in Section 6.1 one can take $k \geq 0$ and $l \geq 0$. Then for a nontrivial surface, two principal cases are distinguished:

$$\text{(I)}\ k > l \geq 0, \qquad \text{and} \qquad \text{(II)}\ k = l > 0.$$

The case $k = l = 0$ is ruled out, because it leads to a totally umbilic (in fact geodesic) surface, which is parallel (see Section 6.1).

Consider first possibility (1) above. Then $\Omega_4^5 = \varepsilon\Omega_1^2 \neq 0$, where $\varepsilon = 1$ or $\varepsilon = -1$; thus $kl \neq 0$. Denoting $h^4_{112} = a$, $h^5_{112} = b$, it follows from (12.3.1)–(12.3.9) that

$$h^4_{111} = 3\varepsilon b, \qquad h^4_{122} = \varepsilon b, \qquad h^4_{222} = 3a,$$

$$h^5_{111} = -3\varepsilon a, \qquad h^5_{122} = -\varepsilon a, \qquad h^5_{222} = 3b.$$

Now for $\alpha = 4$, $\{ij\} = \{12\}$, and also for $\alpha = 5$, $i = j$, (2.2.3) leads to

$$2k\omega_1^2 - l\omega_4^5 = a\omega^1 + \varepsilon b\omega^2, \quad 2k\omega_1^2 - l\omega_4^5 = \varepsilon a\omega^1 + b\omega^2.$$

Therefore, $2k\omega_1^2 - l\omega_4^5 = \varepsilon(2l\omega_1^2 - k\omega_4^5)$, and so

$$2(k - \varepsilon l)\omega_1^2 = -\varepsilon(k - \varepsilon l)\omega_4^5. \qquad (12.3.10)$$

From this one obtains for case (I), and also for (II) with $\varepsilon = -1$, that $k - \varepsilon l \neq 0$, and so $2\omega_1^2 = -\varepsilon\omega_4^5$. By exterior differentiation, this gives

$$2\Omega_1^2 = -\varepsilon\left(\Omega_4^5 - \sum_\xi \omega_4^\xi \wedge \omega_5^\xi\right),$$

and from (12.3.9), $h_{ijk}^\xi = 0$. Hence for $\{ij\} = \{12\}$, $\alpha = \xi$, formula (2.2.3) leads to $-l\omega_5^\xi = 0$, and thus $\omega_5^\xi = 0$ (since here $0 \neq \Omega_4^5 = -2kl\omega^1 \wedge \omega^2$). The result is a contradiction $4(\Omega_1^2)^2 = (\Omega_4^5)^2$ to the hypothesis $(\Omega_4^5)^2 = (\Omega_1^2)^2 \neq 0$ of possibility (1).

In the same way it can be shown that in case (I), and in case (II) with $\varepsilon = -1$, the possibility (2) above also leads to a contradiction. Indeed, then $\Omega_4^5 = 3\varepsilon\Omega_1^2$, and from (12.3.1)–(12.3.9) it follows that

$$h_{111}^5 = -\varepsilon a, \quad h_{122}^5 = \varepsilon a, \quad h_{222}^5 = -b, \quad h_{122}^4 = -\varepsilon b,$$

so

$$2k\omega_1^2 - l\omega_4^5 = a\omega^1 - \varepsilon b\omega^2, \quad 2l\omega_1^2 - k\omega_4^5 = \varepsilon a\omega^1 - b\omega^2.$$

Therefore, $2k\omega_1^2 - l\omega_4^5 = \varepsilon(2l\omega_1^2 - k\omega_4^5)$, as above, which gives the same equation $4(\Omega_1^2)^2 = (\Omega_4^5)^2$, contradicting $(\Omega_4^5)^2 = 9(\Omega_1^2)^2 \neq 0$.

The analysis is more complicated in case (II) with $\varepsilon = 1$. Then for $k = l$, one gets from (2.2.3) that for $\rho \in \{6, \dots, n\}$,

$$\begin{aligned} -(\beta + k)\omega_4^\rho - \gamma\omega_5^\rho - \alpha\omega_3^\rho &= 0, \\ -k\omega_5^\rho &= 0, \\ -(\beta - k)\omega_4^\rho - \gamma\omega_5^\rho - \alpha\omega_3^\rho &= 0. \end{aligned}$$

Thus $\alpha\omega_3^\rho = \omega_4^\rho = \omega_5^\rho = 0$. Likewise, for $\alpha = 3$, (2.2.3) gives $\omega_3^5 = \omega_4^5 = d\alpha = 0$, and the other equations of (2.2.3) reduce, in the case of possibility (1), to

$$\begin{aligned} dk &= b\omega^1 - a\omega^2. \\ a(2\omega_1^2 - \omega_4^5) &= a\omega^1 + b\omega^2, \\ d\beta &= \gamma\omega_4^5 + 2(b\omega^1 + a\omega^2), \\ d\gamma &= -\beta\omega_4^5 + 2(-a\omega^1 + b\omega^2). \end{aligned}$$

By exterior differentiation, these equations lead to

$$\begin{aligned} &(db + a\omega_1^2) \wedge \omega^1 - (da - b\omega_1^2) \wedge \omega^2 = 0, \\ &[da + b(\omega_1^2 - \omega_4^5)] \wedge \omega^1 + [db - a(\omega_1^2 - \omega_4^5) + 2k^3\omega^1] \wedge \omega^2 = 0, \\ &[db - a(\omega_1^2 - \omega_4^5)] \wedge \omega^1 + [da + b(\omega_1^2 - \omega_4^5) - k^2\gamma\omega^1] \wedge \omega^2 = 0, \\ &-[da + b(\omega_1^2 - \omega_4^5)] \wedge \omega^1 + [db - a(\omega_1^2 - \omega_4^5) + k^2\beta\omega^1] \wedge \omega^2 = 0. \end{aligned} \tag{12.3.11}$$

By Cartan's lemma,

$$db + a\omega_1^2 = p\omega^1 + q\omega^2, \quad -(da - b\omega_1^2) = q\omega^1 + r\omega^2, \tag{12.3.12}$$

and now substitution into the last three exterior equations gives, due to the expression of $k(2\omega_1^2 - \omega_4^5)$, that

$$p = k^{-1}a^2 + \frac{1}{2}k^2(2k - \beta),$$

$$q = k^{-1}ab - \frac{1}{2}k^2\gamma,$$

$$r = k^{-1}b^2 + \frac{1}{2}k^2(2k + \beta).$$

By substituting into (12.3.12) and taking exterior derivatives, one obtains

$$a\gamma - b(\beta + 8k) = 0, \tag{12.3.13}$$

$$a(\beta - 8k) + b\gamma = 0. \tag{12.3.14}$$

Now if $\beta^2 + \gamma^2 \neq 64k^2$, then $a = b = 0$, but if $\beta^2 + \gamma^2 = 64k^2$, then differentiation gives

$$\beta b - \gamma a = 32kb, \quad \beta a + \gamma b = -32ka,$$

which together with (12.3.13) and (12.3.14) leads to $40kb = 40ka = 0$, thus again to $a = b = 0$. As a result, (12.3.11) reduces to the contradiction $k^3\omega^1 \wedge \omega^2 = 0$.

For possibility (2), where $\Omega_4^5 = 3\Omega_1^2 \neq 0$ and thus

$$8k^2 = 3(c + \alpha^2 + \beta^2 + \gamma^2), \tag{12.3.15}$$

the system (2.2.3) reduces to

$$dk = b\omega^1 + a\omega^2,$$

$$k(2\omega_1^2 - \omega_4^5) = a\omega^1 + b\omega^2,$$

$$d\beta = \gamma\omega_4^5, \quad d\gamma = -\beta\omega_4^5,$$

and, as before, $\omega_3^\xi = \omega_4^\xi = \gamma\omega_5^\xi = d\alpha = 0$. The equations with $d\beta, d\gamma$ lead by exterior differentiation to $2k^2\gamma = 2k^2\beta = 0$, and hence to $\beta = \gamma = 0$. From (12.3.15) it follows that $8k^2 = 3(c + \alpha^2) = \text{const}$, thus $a = b = 0$. But instead of (12.3.11) now

$$[da + b(\omega_1^2 - \omega_4^5)] \wedge \omega^1 + \left[db + a(3\omega_1^2 - \omega_4^5) - \frac{2}{3}k^3\omega^1\right] \wedge \omega^2 = 0,$$

and this again leads to a contradiction.

As a result, both possibilities (1) and (2) imply contradictions, and so this follows.

Proposition 12.3.1. *A surface M^2 in $N^n(c)$ is nontrivial and 2-semiparallel only if the possibility (3) above is realized, i.e., only if $\Omega_1^2 = 0$, $\Omega_4^5 \neq 0$; or more generally, only if M^2 is locally Euclidean and has nonflat normal connection.*

Then from (12.3.1)–(12.3.9) it follows that $h^4_{ijk} = h^5_{ijk} = 0$, which shows that span$\{h_{ijk}\}$ reduces to span$\{h^{\xi}_{ijk}e_{\xi}\}$ and therefore is orthogonal to span$\{A, B\}$ = span$\{e_4, e_5\}$ at each point $x \in M^2$. Observing that due to $\Omega^2_1 = 0$, the conditions (12.3.9) leave h^{ξ}_{ijk} free, it follows that if this orthogonality holds, then conversely $h^4_{ijk} = h^5_{ijk} = 0$, which implies that the given M^2 is 2-semiparallel. The result can be formulated as follows.

Proposition 12.3.2. *A locally Euclidean surface M^2 with nonflat normal connection in a space form $N^n(c)$ is a nontrivial 2-semiparallel surface if and only if its* span$\{h_{ijk}\}$ *is orthogonal to the plane of the normal curvature indicatrix at each point $x \in M^2$, i.e., is orthogonal to* span$\{A, B\}$.

Now (2.2.3) imply that

$$dk = dl = 2k\omega^2_1 - l\omega^5_4 = 2l\omega^2_1 - k\omega^5_4 = 0, \tag{12.3.16}$$

$$d\beta = \gamma\omega^5_4 + \alpha\omega^3_4, \quad d\gamma = -\beta\omega^5_4 + \alpha\omega^3_5, \tag{12.3.17}$$

$$d\alpha + \beta\omega^3_4 + \gamma\omega^3_5 = \frac{1}{2}(p^3 + r^3)\omega^1 + \frac{1}{2}(q^3 + s^3)\omega^2, \tag{12.3.18}$$

$$l\omega^{\xi}_5 = q^{\xi}\omega^1 + r^{\xi}\omega^2, \tag{12.3.19}$$

$$k\omega^{\xi}_4 = \frac{1}{2}(p^{\xi} - r^{\xi})\omega^1 + \frac{1}{2}(q^{\xi} - s^{\xi})\omega^2, \tag{12.3.20}$$

$$\alpha\omega^{\rho}_3 + \beta\omega^{\rho}_4 + \gamma\omega^{\rho}_5 = \frac{1}{2}(p^{\rho} + r^{\rho})\omega^1 + \frac{1}{2}(q^{\rho} + s^{\rho})\omega^2, \tag{12.3.21}$$

where $\rho \in \{6, \dots, n\}$ and the notation $p^{\xi} = h^{\xi}_{111}, q^{\xi} = h^{\xi}_{112}, r^{\xi} = h^{\xi}_{122}, s^{\xi} = h^{\xi}_{222}$ is used.

Equations (12.3.16) give by exterior differentiation the relation $d\omega^5_4 = 0$, which is equivalent to

$$4kl + \mathbf{p}\cdot\mathbf{r} + \mathbf{q}\cdot\mathbf{s} - \mathbf{q}^2 - \mathbf{r}^2 = 0, \tag{12.3.22}$$

where $\mathbf{p}\cdot\mathbf{r} = \sum_{\xi} p^{\xi}r^{\xi}$, etc.

Equations (12.3.17) lead by exterior differentiation to

$$\left[-\gamma\omega^3_5 + \frac{1}{2}(p^3+r^3)\omega^1+\frac{1}{2}(q^3+s^3)\omega^2\right] \wedge \omega^3_4 + \alpha\omega^{\rho}_4 \wedge \omega^3_{\rho} = 0, \tag{12.3.23}$$

$$\left[-\beta\omega^3_4 + \frac{1}{2}(p^3+r^3)\omega^1+\frac{1}{2}(q^3+s^3)\omega^2\right] \wedge \omega^3_5 + \alpha\omega^{\rho}_5 \wedge \omega^3_{\rho} = 0, \tag{12.3.24}$$

where $\omega^3_4, \omega^{\rho}_4, \omega^3_5, \omega^{\rho}_5$ can be expressed via (12.3.19) and (12.3.20); then (12.3.21) implies

$$-2kl\alpha\omega^{\rho}_3 = \{l[(\beta - k)p^{\rho} - (\beta + k)r^{\rho}] + 2k\gamma q^{\rho}\}\omega^1$$

$$+\{l[(\beta-k)q^\rho-(\beta+k)s^\rho]+2k\gamma r^\rho\}\omega^2.$$

Substitution into (12.3.23) and (12.3.24) produces relations between $k, l, \beta, \gamma, p^\xi$, q^ξ, r^ξ and s^ξ. Some of these are rather complicated, making it difficult to study locally Euclidean 2-semiparallel surfaces M^2 with nonflat $\nabla^\perp$ in general. But one of the properties of these surfaces can be shown easily: from (12.3.16) it follows that k and l are nonzero constants. Recalling the content of Remark 6.2.3, this property is equivalent to the statement that the normal curvature indicatrices of the surface at any two of its points are congruent ellipses.

All this can be formulated as the following.

Theorem 12.3.3. *A surface M^2 in a space form $N^n(c)$ is 2-semiparallel if and only if it belongs to one of the following three mutually exclusive classes:*

(i) *surfaces with flat van der Waerden–Bortolotti connection $\bar{\nabla}$,*
(ii) *parallel surfaces with nonflat $\bar{\nabla}$, i.e., totally umbilical surfaces and Veronese surfaces,*
(iii) *for sufficiently high dimension n, the locally Euclidean surfaces (i.e., with flat ∇, or equivalently with vanishing Gaussian curvature) whose normal connection $\nabla^\perp$ is nonflat and* span$\{h_{ijk}\}$ *at any point $x \in M$ is orthogonal to the plane of the normal curvature ellipse at x; these ellipses at any two points of such an M^2 are congruent.*

The difficulty of a general approach to nontrivial 2-semiparallel surfaces brings up the problem of their existence.

Consider this problem for case (iii), with the restrictions that $n = 6$, and that the mean curvature vector H is orthogonal to the plane of the normal curvature ellipse at each point $x \in M^2$. These restrictions mean that $\beta = \gamma = 0$, and thus $\alpha^2 = k^2 + l^2 - c = \text{const} \geq 0$, due to $\Omega_1^2 = 0$.

First, suppose $\alpha \neq 0$, so that $k^2 + l^2 > 0$. From (12.3.17) then $\omega_4^3 = \omega_5^3 = 0$; hence by (12.3.19) and (12.3.20), $p^3 = q^3 = r^3 = s^3 = 0$, and so (12.3.18) becomes an identity.

Moreover, since $n = 6$, it is convenient to denote p^6, q^6, etc., by p, q, etc. Then equations (12.3.19)–(12.3.22) reduce to

$$l\omega_5^6 = q\omega^1 + r\omega^2, \tag{12.3.19'}$$

$$k\omega_4^6 = \frac{1}{2}(p-r)\omega^1 + \frac{1}{2}(q-s)\omega^2, \tag{12.3.20'}$$

$$\alpha\omega_3^6 = \frac{1}{2}(p+r)\omega^1 + \frac{1}{2}(q+s)\omega^2, \tag{12.3.21'}$$

$$4kl + pr + qs - q^2 - r^2 = 0, \tag{12.3.22'}$$

and then (12.3.23) and (12.3.24) give

$$ps - qr = 0, \quad pr - qs - q^2 + r^2 = 0. \tag{12.3.23'}$$

Exterior differentiation applied to (12.3.19′)–(12.3.21′) yields, via (12.3.16), the following exterior equations:

$$[dq + (p - 2r)\omega_1^2] \wedge \omega^1 + [dr + (2q - s)\omega_1^2] \wedge \omega^2 = 0,$$

$$[d(p - r) + (s - 5q)\omega_1^2] \wedge \omega^1 + [d(q - s) + (p - 5r)\omega_1^2] \wedge \omega^2 = 0,$$

$$[d(p + r) - (q + s)\omega_1^2] \wedge \omega^1 + [d(q + s) + (p + r)\omega_1^2] \wedge \omega^2 = 0.$$

Therefore, by Cartan's lemma

$$dp = 3q\omega_1^2 + P\omega^1 + Q\omega^2, \tag{12.3.25}$$

$$dq = (2r - p)\omega_1^2 + Q\omega^1 + R\omega^2, \tag{12.3.26}$$

$$dr = (s - 2q)\omega_1^2 + R\omega^1 + S\omega^2, \tag{12.3.27}$$

$$ds = -3r\omega_1^2 + S\omega^1 + T\omega^2. \tag{12.3.28}$$

Differentiating (12.3.22′) and (12.3.23′), and then using (12.3.25)–(12.3.28), (12.3.16), and the same (12.3.23′), one finds that all terms containing ω_1^2 cancel, while the coefficients of ω^1 and ω^2 lead to

$$rP + (s - 2q)Q + (p - 2r)R + qS = 0, \tag{12.3.29}$$

$$rQ + (s - 2q)R + (p - 2r)S + qT = 0, \tag{12.3.30}$$

$$-sP + rQ + qR - pS = 0, \tag{12.3.31}$$

$$-sQ + rR + qS - pT = 0, \tag{12.3.32}$$

$$rP - (s + 2q)Q + (p + 2r)R - qS = 0, \tag{12.3.33}$$

$$rQ - (s + 2q)R + (p + 2r)S - qT = 0. \tag{12.3.34}$$

This is a homogeneous linear system for P, Q, R, S, and T, whose determinant turns out to be nonzero, in general. Therefore, $P = Q = R = S = T = 0$, and so (12.3.25)–(12.3.28) reduce to

$$dp = 3q\omega_1^2, \quad dq = (2r - p)\omega_1^2, \quad dr = (s - 2q)\omega_1^2, \quad ds = -3r\omega_1^2,$$

This last differential system is totally integrable, since $d\omega_1^2 = \Omega_1^2 = 0$.

This system coincides with $\bar{\nabla} h_{ijk}^6 = 0$; recall that $h_{ijk}^3 = h_{ijk}^4 = h_{ijk}^5 = 0$. However, $\bar{\nabla} h_{ijk}^3$, $\bar{\nabla} h_{ijk}^4$, $\bar{\nabla} h_{ijk}^5$ need not be zero in general.

The result can be formulated as follows.

Proposition 12.3.4. *In $N^6(c)$ there exist, and depend on some constants, 2-semiparallel locally Euclidean surfaces M^2 with nonflat $\nabla^\perp$, whose mean curvature vector H at each point $x \in M^2$ is orthogonal to the plane of the normal curvature ellipse. These ellipses are congruent at any two points of M^2, and the length of H (i.e., the distance of x from the plane of the ellipse) is constant. Such an M^2 is neither parallel, nor 2-parallel, nor semiparallel.*

Among these surfaces there are the minimal ones, for which also $\alpha = 0$. Then $c = k^2 + l^2$, so that these can exist only in elliptic spaces. Now equations (12.3.17) are satisfied, and from (12.3.18) and (12.3.21) it follows that $r^\xi = -p^\xi$, $s^\xi = -q^\xi$. So the only essential equations among (12.3.15)–(12.3.21) are

$$k\omega_4^\xi = p^\xi\omega^1 + q^\xi\omega^2, \quad l\omega_5^\xi = q^\xi\omega^1 - p^\xi\omega^2.$$

By exterior differentiation they give, by (12.3.16), that

$$(dp^\xi - 3q^\xi\omega_1^2) \wedge \omega^1 + (dq^\xi + 3p^\xi\omega_1^2) \wedge \omega^2 = 0,$$

$$(dq^\xi + 3p^\xi\omega_1^2) \wedge \omega^1 - (dp^\xi - q^\xi\omega_1^2) \wedge \omega^2 = 0;$$

from here one gets by Cartan's lemma

$$dp^\xi - 3q^\xi\omega_1^2 = P^\xi\omega^1 + Q^\xi\omega^2,$$

$$dq^\xi + 3p^\xi\omega_1^2 = Q^\xi\omega^1 - P^\xi\omega^2.$$

Now suppose $n = 5$, so that ξ takes only the one value 3; denote p^3, q^3, etc., by p, q, etc. Equation (12.3.22), which reduces to $2kl - p^2 - q^2 = 0$, gives by differentiation $pP + qQ = 0$, $pQ - qP = 0$. Here $p^2 + q^2 = 0$ leads to the contradiction $kl = 0$. Therefore, $P = Q = 0$, and the investigation ends with the completely integrable system $dp = 3q\omega_1^2$, $dq = -3p\omega_1^2$.

The result is the following.

Proposition 12.3.5. *In elliptic space $N^5(c)$, $c > 0$, there exist, and depend on some constants, 2-semiparallel minimal locally Euclidean surfaces M^2 with nonflat $\nabla^\perp$, whose normal curvature ellipses at any two points are congruent. Such an M^2 is neither parallel, nor 2-parallel, nor semiparallel.*

Remark 12.3.6. The results of this section were proved in [Lu 2000b]. Two particular cases of 2-semiparallel surfaces were considered in [ALM 2000]. These surfaces were obtained again in [ÖA 2002a, b].

Remark 12.3.7. The minimal locally Euclidean surfaces M^2 with nonflat $\nabla^\perp$ of Proposition 12.3.5 were found earlier in the paper [Lu 62], which dealt with minimal surfaces in space forms whose normal curvature ellipses are at least similar at any two points of the surface. The special case where these ellipses are circles (and thus the surface is pointwise isotropic) was considered in [Bor 28].

Remark 12.3.8. In Theorem 12.1.1 it was shown, following Mirzoyan [Mi 96], that a submanifold M^m in $N^n(c)$ is an envelope of order s for some family of m-dimensional s-parallel submanifolds if and only if it is q-semiparallel for the values $q = s - 1$ and $q = s$, if $s \geq 2$, and for the value $q = s$, if $s = 1$ (here 1-parallel and 1-semiparallel mean simply parallel and semiparallel, respectively).

Propositions 12.3.4 and 12.3.5 now show that the q-semiparallel condition for both values $q = s - 1$ and $q = s \geq 2$ in Mirzoyan's theorem is essential, at least for $q = 2$, and cannot be weakened just to $q = s$.

12.4 Recurrent and Pseudoparallel Submanifolds

J. Deprez, after introducing the concept of semiparallel submanifolds in [De 85] (see also [De 86]), gave the following generalization of parallel submanifolds in his third paper [De 89].

He called a submanifold M^m in Euclidean space E^n *recurrent* if there exists a 1-form μ on M^m such that $\bar{\nabla} h = h \otimes \mu$.

If $\mu = 0$, the recurrent submanifold is parallel. On the other hand *every recurrent submanifold is semiparallel*, as is shown in [De 89]. Indeed, it is easy to see that $d\|h\|^2 = d\langle h, h\rangle = 2\langle \bar{\nabla} h, h\rangle = \mu\|h\|^2$, thus $\mu = d(\ln \|h\|)$ is a total differential. Hence, setting $d\mu = \mu_k \omega^k$, it follows that in (2.3.1), $\bar{\nabla}_k \bar{\nabla}_l h^\alpha_{ij} = \nabla_k \mu_l h^\alpha_{ij} + \mu_l \mu_k h^\alpha_{ij}$ is symmetric in k, l, which by (4.1.1) implies the semiparallel condition.

There are rather few recurrent submanifolds M^m in E^n which are not parallel. Namely, Deprez has proved the following.

Theorem 12.4.1. *A recurrent submanifold M^m in E^n is either*

(i) *parallel, or*
(ii) *there is a dense open subset $U \subset M^m$ such that for each point x of U there is a neighborhood W of x such that*
  (a) *W is totally geodesic, or*
  (b) *W is locally congruent to a cylinder over a plane curve, and is contained in an $(m+1)$-dimensional totally geodesic subspace of E^n.*

Proof. See in [De 89].

Remark 12.4.2. Previously, the concept of recurrence was introduced in [Mat 85] for hypersurfaces in space forms. This paper also contained the following generalization of 2-parallel hypersurfaces in space forms.

Namely, a hypersurface is called *birecurrent* if there exists a covariant order-2 tensor field μ such that $\nabla^2 h = h \otimes \mu$. It was proved that, in a real space form, a recurrent hypersurface is locally symmetric, and a complete irreducible birecurrent hypersurface is recurrent.

Further, it was shown in [Mat 90] that if a birecurrent hypersurface M^m in $N^{m+1}(c)$, with $c \leq 0$ (resp. with $c > 0$) has constant (resp. nonzero) curvature, then it is parallel; and if $N^{m+1}(c)$ is simply connected, M^m is complete, and $m \geq 3$, then M^m is totally umbilic, or is a Riemannian product of two totally umbilic submanifolds of constant curvature.

Recurrent (resp. birecurrent) submanifolds generalize the parallel (resp. 2-parallel) ones in the class of semiparallel submanifolds. Recently, a generalization has been given also for semiparallel submanifolds; this is done in the same way as the generalization of semisymmetric Riemannian manifolds to pseudosymmetric manifolds by R. Deszcz in [Des 92]. In that paper, a Riemannian manifold (M, g) with curvature tensor R is called *pseudosymmetric* if

$$R \cdot R = \lambda Q(g, R), \tag{12.4.1}$$

where λ is a smooth function and $Q(g, R)(X_1, X_2, X_3, X_4; X, Y) = -R((X \wedge Y)X_1, X_2, X_3, X_4) - R(X_1, (X \wedge Y)X_2, X_3, X_4) - R(X_1, X_2, (X \wedge Y)X_3, X_4) - R(X_1, X_2, X_3, (X \wedge Y)X_4)$, where $(X \wedge Y)Z = g(Y, Z)X - g(X, Z)Y$ (see [Des 92], [Ver 94]). For $\lambda \equiv 0$, the pseudosymmetric condition (12.4.1) reduces to the semisymmetric one (see (1.6.3)).

According to [ALM 99], [ALM 2002], a submanifold M^m in $N^n(c)$ is called *pseudoparallel* if $\bar{R}(X, Y) \cdot h(Z, W) = -\phi[h((X \wedge Y)Z, W) + h((X \wedge Y)W, Z)]$, where ϕ is a smooth function on M^m.

Componentwise, this condition is

$$\begin{aligned} &R^p_{i,kl}h^\alpha_{pj} + R^p_{j,kl}h^\alpha_{ip} - R^\alpha_{\beta,kl}h^\beta_{ij} \\ &= -\phi[g_{il}h^\alpha_{kj} - g_{ik}h^\alpha_{lj} + g_{jl}h^\alpha_{ki} - g_{jk}h^\alpha_{li}]; \qquad (12.4.2) \end{aligned}$$

if one takes here the orthonormal frame bundle $O(M^m, N^n(c))$, then $g_{ij} = \delta_{ij}$. The special case $\phi \equiv 0$ gives the semiparallel condition (4.1.1).

Theorem 12.4.3. *Every pseudoparallel submanifold M^m in $N^n(c)$ is intrinsically a pseudosymmetric Riemannian manifold.*

Proof. Using the Gauss identity $R_{ij,kl} = \langle h_{ik}, h_{jl}\rangle - \langle h_{il}, h_{jk}\rangle$, one can derive from (12.4.1), after some calculation, the following equation for the inner metric:

$$\begin{aligned} &R_{sj,kl}R^s_{i,pq} + R_{is,kl}R^s_{j,pq} + R_{ij,sl}R^s_{k,pq} + R_{ij,ks}R^s_{l,pq} \\ &= \phi(g_{ip}R_{kl.qj} + g_{jp}R_{kl,iq} + g_{kp}R_{ij,ql} + g_{lp}R_{ij,kq} \\ &\quad - g_{iq}R_{kl.pj} - g_{jq}R_{kl,ip} - g_{kq}R_{ij,pl} - g_{lq}R_{ij,kp}). \qquad (12.4.3) \end{aligned}$$

After contracting with the coordinates of vector fields $X^p, Y^q, X^i_1, X^j_2, X^k_3, X^l_4$ one obtains, e.g., via $g_{ip}X^i_1X^p = \langle X, X_1\rangle$, etc., that this condition is exactly (12.4.1) with $\lambda = \phi$.

In the papers where pseudoparallel submanifolds were introduced, the first results about them were also obtained. In [ALM 99], examples were given of pseudoparallel submanifolds that are not semiparallel, namely, certain hypersurfaces of revolution. In [ALM 2002] it was shown that every pseudoparallel hypersurface is either quasi-umbilic or a cyclide of Dupin. Some classifications were also given there, e.g., for all pseudoparallel surfaces M^2 in $N^5(c)$ for which ϕ is constant, ≥ 0, and $c + \phi > 0$. A topological classification was derived for all complete simply connected Riemannian manifolds admitting a pseudoparallel immersion into a space form with $\phi \geq 0$ and $c + \phi > 0$.

Very recently, all normally flat pseudoparallel submanifolds M^m in space forms $N^n(c)$ were classified and described in [LT 2006], where the following theorem was stated and proved.

Theorem 12.4.4. *Let M^m be a normally flat pseudoparallel submanifold in $N^n(c)$. Then either $m = 2$ and $\phi = K$ on the open subset of nonumbilic points, where K is the Gaussian curvature of M^2, or else there exists an open dense subset $\bar{M}^m$ of M^m where one of the following holds locally:*

(i) $\bar{M}^m$ *is umbilical,*
(ii) $\bar{M}^m$ *is a cyclide of Dupin in* $\bar{N}^{m+1}(\bar{c}) \subset N^n(c)$*, and* $\bar{N}^{m+1}(\bar{c})$ *is either umbilical or totally geodesic,*
(iii) $\bar{M}^m$ *is a quasi-umbilic hypersurface in* $N^{m+1}(c) \subset N^n(c)$*,*
(iv) $\phi = k \in \mathbb{R}$*, and* $\bar{M}^m$ *has constant sectional curvature k,*
(v) $\phi = 0$*, and* $\bar{M}^m$ *is an extrinsic product of spherical submanifolds of* $N^n(c)$*,*
(vi) $\phi = k \in \mathbb{R}$*, and* $\bar{M}^m$ *is the restriction of a multirotational submanifold*

$$V \times_{\rho_1} N_1 \times_{\rho_2} \cdots \times_{\rho_l} N_l \to N^n(c) \approx N_0 \times_{\sigma_1} N_1 \times_{\sigma_2} \cdots \times_{\sigma_l} N_l,$$

where $V \subset N^p(k)$*,* $p = m - \sum_{i=1}^{l} \dim N_i \geq 1$*, and the profile* $V \to N_0$ *is an isometric immersion with flat normal bundle, satisfying the k-helix property with respect to the mean curvature vectors* $H_1, \ldots, H_l$ *of* $N_1, \ldots, N_l$ *in the flat space* $\mathbb{O}_0 \supset N^n(c)$*.*

This theorem generalizes the results obtained above in Section 5.4 (see Theorem 5.4.1 and Remark 5.4.4), and in part (vi) uses concepts and terminology from [DN 93].

Remark 12.4.5. Pseudoparallel submanifolds were also introduced in [ÖAM 2002a], for ambient Euclidean space and under the name *extended semiparallel.* Extended semiparallel surfaces M^2 in E^n were classified there. It was proved that such a surface is locally either (i) semiparallel, or (ii) normally flat of Gaussian curvature $K \equiv \phi$, or (iii) an isotropic surface of codimension at least 3 and $H^2 = 3K - 2\phi$, where H is the mean curvature vector.

On the whole, the theory of pseudoparallel submanifolds is still in the stage of creation.

12.5 Submanifolds with Semiparallel Tensor Fields

In the differential geometry of submanifolds M^m in $N^n(c)$ some tensor fields are known, which are derived from the second fundamental form h and the metric form g by means of tensor algebra operations. These are the curvature tensor R and the normal curvature tensor $R^\perp$, whose components $R^j_{i,pq}$ and $R^\beta_{\alpha,pq}$ are defined by (2.1.10), the Ricci tensor *Ric* with components $R_{ip} = R^s_{i,ps}$, and the mean curvature vector H with components $H^\alpha = \frac{1}{m} g^{ij} h^\alpha_{ij}$. By means of H and $R^\perp$, the tensor field $T = HR^\perp$ can be formed with components $T^\alpha_{ij} = H^\beta R^\alpha_{\beta,ij}$.

The Levi-Civita connection ∇, the normal connection $\nabla^\perp$, and their pair—the van der Waerden–Bortolotti connection $\bar{\nabla} = (\nabla, \nabla^\perp)$ make it possible to introduce covariant differentiation of the abovementioned tensor fields, and the concept of covariant constant or (*parallel*) tensor field as having zero covariant derivative.

The submanifolds M^m in $N^n(c)$ with parallel R, or $R^\perp$, or *Ric*, or H, or T will be called, respectively, *R-parallel* (or intrinsically symmetric; cf. Section 5.1), or $R^\perp$*-parallel*, or *Ric-parallel* (or, considered intrinsically, *Ric*-symmetric), or

H-parallel, or *T-parallel*. Here, for instance, H-parallel submanifolds are characterized by $dH^\alpha + H^\beta \omega_\beta^\alpha = 0$ and T-parallel ones by

$$dT_{ij}^\alpha + T_{ij}^\beta \omega_\beta^\alpha - T_{kj}^\alpha \omega_i^k - T_{ik}^\alpha \omega_j^k = 0. \tag{12.5.1}$$

Analogous differential systems characterize submanifolds with other parallel conditions.

By exterior differentiation, these differential systems lead to corresponding conditions, which are actually algebraic in the components of the second fundamental tensor h, characterizing the corresponding semiparallel submanifolds. For instance,

$$H^\beta \Omega_\beta^\alpha = 0, \qquad T_{ij}^\beta \Omega_\beta^\alpha - T_{kj}^\alpha \Omega_i^k - T_{ik} \Omega_j^k = 0 \tag{12.5.2}$$

characterize *H-semiparallel* and *T-semiparallel* submanifolds, respectively. *R-semiparallel* submanifolds (or intrinsically semisymmetric ones; cf. Section 1.6, especially (1.6.2)), *$R^\perp$-semiparallel*, T-semiparallel, and *Ric-semiparallel* submanifolds will be introduced in a similar way.

A direct computation shows that Propositions 4.1.3 and 4.1.4 can be generalized as follows.

Proposition 12.5.1. *Let Λ be one of the tensor fields R, $R^\perp$, Ric, H, T on a submanifold M^m in $N^n(c)$. The class of Λ-semiparallel submanifolds includes*

- *all Λ-parallel submanifolds (in particular, those with $\Lambda \equiv 0$),*
- *all submanifolds with flat $\bar{\nabla}$, i.e., with $\Omega_i^j = \Omega_\alpha^\beta = 0$.*

Proof. The first assertion was already proved above for H- and T-parallel submanifolds. The proof is analogous for R-, $R^\perp$-, and *Ric*-parallel submanifolds. The last two assertions are obvious; see, e.g., (12.5.2).

Mirzoyan has given in [Mi 98b] generalizations of Theorem 4.5.5 for Λ-semiparallel submanifolds.

First, the concept of *Λ-tangency* of submanifolds M^m in $N^n(c)$ for a tensor field Λ was introduced in [Mi 98b] as follows.

A common point x of two submanifolds M^m and $\bar{M}^m$ is said to be a *Λ-tangency* point if the tangent spaces $T_x M^m$ and $T_x \bar{M}^m$ coincide, along with the tensors Λ_x and $\bar{\Lambda}_x$ (cf. with Proposition 4.5.1).

A submanifold M^m in $N^n(c)$ is called the *Λ-envelope* of a family of m-dimensional submanifolds if at each point $x \in M^m$ the submanifold M^m has Λ-tangency with a submanifold of this family.

Theorem 12.5.2 (Mirzoyan [Mi 98b]). *If a submanifold M^m in $N^n(c)$ is an R-, $R^\perp$-, and Λ-envelope of a family of m-dimensional Λ-parallel submanifolds, then it is a Λ-semiparallel submanifold.*

Proof. Consider first the case where $\Lambda = T$. By Proposition 12.5.1, every T-parallel submanifold $\bar{M}^m$ of the family is also T-semiparallel, i.e., satisfies the second condition (12.5.2), which is equivalent to

$$T^{\alpha}_{kj} R^{k}_{i,pq} + T^{\alpha}_{ik} R^{k}_{j,pq} - T^{\beta}_{ij} R^{\alpha}_{\beta,pq} = 0. \tag{12.5.3}$$

For an arbitrary point $x \in M^m$, there is an $\bar{M}^m$ having R-, $R^{\perp}$- and T-tangency with M^m at x. This means that in the adapted orthonormal frame, common to M^m and $\bar{M}^m$ at x, the components $R^{j}_{i,pq}$, $R^{\alpha}_{\beta,pq}$ and T^{α}_{ij} are the same for M^m and $\bar{M}^m$ at x. Since for $\bar{M}^m$ they satisfy (12.5.3), then this condition (12.5.3) is also satisfied for M^m; and since x is an arbitrary point, it follows that M^m is T-semiparallel.

For the other cases, where Λ is R, or $R^{\perp}$, or *Ric*, or H, the proof is analogous and is left to the reader. Note that in the condition analogous to (12.5.3), if $R^{\alpha}_{\beta,pq}$ is not needed, as for R and *Ric*, or $R^{j}_{i,pq}$ is not needed, as for H, then in the statement of the theorem, the tensor fields $R^{\perp}$ or R would be omitted correspondingly.

In the converse direction, the concept of Λ-tangency must be extended to the higher orders.

A common point x of two submanifolds M^m and $\bar{M}^m$ is said to be a *Λ-tangency point* of order s ($s \geq 0$), if

(1) the tangent spaces $T_x M^m$ and $T_x \bar{M}^m$ coincide,
(2) the tensors Λ_x and $\bar{\Lambda}_x$ and their corresponding covariant derivatives of order less than $s+1$ coincide.

A submanifold M^m in $N^n(c)$ is called the *Λ-envelope* of order s ($s \geq 0$) for some family of m-dimensional submanifolds if at each point $x \in M^m$ the submanifold M^m has Λ-tangency of order s with a submanifold of this family.

Theorem 12.5.3 (Mirzoyan [Mi 98b]). *Each Λ-semiparallel submanifold M^m in $N^n(c)$ is an Λ-envelope of some order s ($s \geq 2$) for a family of m-dimensional Λ-parallel submanifolds.*

Proof. For any submanifold, one has equations (2.1.3), (2.1.4), (2.2.3) with (2.2.2), etc., in general (2.2.7), which can be written in detail as (12.1.3), where instead of the last equation (expressing the k-parallel property), the previous ones are continued for higher values of k. This process of continuation terminates at some value of k, because for some order s, the osculating subspaces of higher order (see Section 2.4) fill the tangent space of $N^n(c)$, and therefore the components $h_{ijp_1 \dots p_t}$ for $t \geq s+1$ can be expressed in terms of the components of lower-order fundamental forms.

By exterior differentiation, these equations (12.1.3), except for the last ones, lead to (12.1.4), which can also be continued for higher values of k.

The Λ-semiparallel condition is added here. Suppose, for instance, that $\Lambda = T$, where this condition is

$$T^{\alpha}_{kj} \Omega^{k}_{i} + T^{\alpha}_{ik} \Omega^{k}_{j} - T^{\beta}_{ij} \Omega^{\alpha}_{\beta} = 0,$$

which is equivalent to (12.5.3).

Now consider the differential system consisting of equations (12.1.3), except for the last ones, and of

$$dT^{\alpha}_{ij} + T^{\beta}_{ij}\omega^{\alpha}_{\beta} - T^{\alpha}_{kj}\omega^{k}_{i} - T^{\alpha}_{ik}\omega^{k}_{j} = 0,$$

which is equivalent to (12.5.1). This system is totally integrable, since its integrability conditions are satisfied due to (12.1.4) and (12.5.3).

Therefore, this system defines, for any initial conditions, a submanifold $\bar{M}^m$, which is T-parallel due to (12.5.1). Taking for initial conditions the k-tuple of the given T-semiparallel submanifold M^m at an arbitrary point $x \in M^m$, one guarantees that $\bar{M}^m$ and M^m have T-tangency of order s at this common point x. Thus M^m is a T-envelope of order s of the family of all such T-parallel $\bar{M}^m$.

The proof is similar for other choices of Λ.

Remark 12.5.4. Theorem 12.5.3 was given by Mirzoyan [Mi 98b]. He called it a generalization of Lumiste's theorem, i.e., in [Lu 90a]; see Theorem 4.5.5.

Remark 12.5.5. This Theorem 4.5.5 has inspired a purely intrinsic analogue, proved in [KoN 98]: each semisymmetric Riemannian metric on a manifold is a second-order envelope of a family of locally symmetric Riemannian metrics.

Remark 12.5.6. Mirzoyan gave in [Mi 99] and [Mi 2002] a generalization of Theorem 12.5.3 for the case where the tensor field Λ on the submanifold M^m in $N^n(c)$ is derived from the second fundamental form h and metric form not only by operations of tensor algebra but also by some arbitrary operations of tensor calculus (including covariant differentiation of arbitrary order). It was proved that if M^m in $N^n(c)$ is $\bar{\nabla}^{s-1}\Lambda$- and $\bar{\nabla}^{s}\Lambda$-semiparallel ($s \geq 0$; for $s = 0$ only the Λ-semiparallel condition is assumed), then M^m is an envelope of arbitrary order for a family of m-dimensional $\bar{\nabla}^{s}\bar{\Lambda}$-parallel submanifolds. A converse result was also proved.

Theorem 12.1.1 can now be considered as a particular case of these general results.

12.6 Examples: The Surfaces

The results of Section 12.5 for Λ-parallel and Λ-semiparallel submanifolds will now be illustrated by some examples consisting of Λ-parallel and Λ-semiparallel surfaces M^2 in a space form $N^n(c)$, where Λ is one of the tensor fields R, $R^{\perp}$, Ric, H, T.

For an M^2 in $N^n(c)$, the orthonormal frame bundle can be adapted so that equations (6.1.9) and (6.1.10) hold, where one may assume $k \geq l \geq 0$, by renumbering and redirecting the frame vectors e_4 and e_5, if needed; then $\varepsilon_4 = \varepsilon_5 = 1$. By (6.1.8), (6.2.1), and (6.2.2), then

$$H = \alpha e_3 + \beta e_4 + \gamma e_5, \quad \Omega^2_1 = (k^2 + l^2 - H^2 - c)\omega^1 \wedge \omega^2, \quad \Omega^5_4 = -2kl\omega^1 \wedge \omega^2;$$

the other curvature and normal curvature 2-forms are zero. Therefore, the only possible nonzero components of R, $R^{\perp}$, Ric, H, T are, correspondingly,

$$R^2_{1,12} = -R^1_{2,12} = -R^2_{1,21} = R^1_{2,21} = K = c + H^2 - k^2 - l^2,$$

$$R^5_{4,12} = -R^4_{5,12} = -R^5_{4,21} = R^4_{5,21} = 2kl,$$

$$R_{11} = R_{22} = K,$$
$$H^3 = \alpha, \quad H^4 = \beta, \quad H^5 = \gamma,$$
$$T^4_{12} = -T^4_{21} = -2\gamma kl, \quad T^5_{12} = -T^5_{21} = 2\beta kl.$$

Hence the Λ-parallel conditions are

$$dK = 0 \quad (\text{for } \Lambda = R \text{ and } \Lambda = Ric), \tag{12.6.1}$$
$$d(kl) = 0, \ kl\omega^\xi_4 = kl\omega^\xi_5 = 0, \ \xi \in \{3, 6, 7, \dots, n\} \quad (\text{for } \Lambda = R^\perp), \tag{12.6.2}$$
$$d\alpha - \beta\omega^4_3 - \gamma\omega^5_3 = d\beta + \alpha\omega^4_3 - \gamma\omega^5_4 = d\gamma + \alpha\omega^5_3 + \beta\omega^5_4 = 0,$$
$$\alpha\omega^\eta_3 + \beta\omega^\eta_4 + \gamma\omega^\eta_5 = 0, \ \eta \in \{6, 7, \dots, n\} \qquad (\text{for } \Lambda = H), \tag{12.6.3}$$
$$\beta kl\omega^2_1 = \gamma kl\omega^2_1 = 0, \quad kl(\gamma\omega^\xi_4 - \beta\omega^\xi_5) = 0, \ \xi \in \{3, 6, 7, \dots, n\},$$
$$d(\gamma kl) + \beta kl\omega^5_4 = d(\beta kl) - \gamma kl\omega^5_4 = 0 \quad (\text{for } \Lambda = T). \tag{12.6.4}$$

A straightforward computation shows that the Λ-semiparallel conditions are trivially satisfied for R, $R^\perp$, and Ric, i.e., every M^2 in $N^n(c)$ is R-, $R^\perp$- and Ric-semiparallel. For H, the semiparallel conditions are $H^\beta \Omega^\alpha_\beta = 0$, and reduce to

$$\beta kl = \gamma kl = 0; \tag{12.6.5}$$

and for T they are (12.5.3), and reduce to

$$\beta(kl)^2 = \gamma(kl)^2 = \beta kl K = \gamma kl K = 0. \tag{12.6.6}$$

12.6.1 *H*-semiparallel and *H*-parallel surfaces

The following proposition can be deduced from (12.6.5).

Proposition 12.6.1. *For an H-semiparallel M^2 in $N^n(c)$ (or for some of its open subsets), the following two possibilities exist: either*

(I) $kl = 0$, *thus $\Omega^5_4 = 0$ and so $\nabla^\perp$ is flat, or*
(II) $kl \neq 0$, $\beta = \gamma = 0$, *thus $H = \alpha e_3$ is orthogonal to the 2-plane of the normal curvature indicatrix spanned by $A = ke_4$ and $B = le_5$ (see Remark 6.4.3), i.e., under the orthogonal projection onto this 2-plane at $x \in M^2$, the point x maps into the center of the normal curvature ellipse or its degenerate form.*

Every H-parallel M^2 is by Proposition 12.5.1 also H-semiparallel; therefore, only (I) and (II) are possible in this case.

One should note here the special subclass characterized by $H = 0$, i.e., by $\alpha = \beta = \gamma = 0$; it therefore contains all minimal surfaces M^2 in $N^n(c)$, which are obviously H-semiparallel and H-parallel.

Hence in the further investigation below, it can be assumed that $H \neq 0$. The two possibilities above will be considered separately.

(I) If $\nabla^{\perp}$ is flat then $l = 0$, and at an arbitrary point $x \in M^2$, the frame can be adapted further, so that its vector e_3 belongs to the subspace spanned by $A = ke_4$ and $H \neq 0$. Then $\gamma = 0$, and (12.6.3) reduce to

$$d\alpha - \beta\omega_3^4 = d\beta + \alpha\omega_3^4 = 0, \quad \alpha\omega_3^\zeta + \beta\omega_4^\zeta = 0, \ \zeta \in \{5, 6, \ldots, n\}. \qquad (12.6.7)$$

These equations are added to (6.1.9) and (6.1.10), which now appear as

$$\omega_1^3 = \alpha\omega^1, \quad \omega_1^4 = (\beta + k)\omega^1, \quad \omega_1^\zeta = 0, \qquad (12.6.8)$$

$$\omega_2^3 = \alpha\omega^2, \quad \omega_2^4 = (\beta - k)\omega^2, \quad \omega_2^\zeta = 0. \qquad (12.6.9)$$

Exterior differentiation leads from these to

$$k\omega_3^4 \wedge \omega^1 = 0, \quad dk \wedge \omega^1 + 2k\omega_1^2 \wedge \omega^2 = 0, \quad k\omega_4^\zeta = 0, \qquad (12.6.10)$$

$$k\omega_3^4 \wedge \omega^2 = 0, \quad 2k\omega_1^2 \wedge \omega^1 - dk \wedge \omega^2 = 0, \quad k\omega_4^\zeta \wedge \omega^2 = 0. \qquad (12.6.11)$$

Hence due to Cartan's lemma

$$k\omega_3^4 = k\omega_4^\zeta = 0, \quad dk = \varphi\omega^1 + \psi\omega^2, \quad 2k\omega_1^2 = \psi\omega^1 - \varphi\omega^2. \qquad (12.6.12)$$

(Ia) If $k \equiv 0$, then the normal curvature ellipse at an arbitrary point $x \in M^2$ degenerates to a point, and formulas (2.1.5) reduce to

$$de_1 = e_2\omega_1^2 + H^*\omega^1, \quad de_2 = -e_1\omega_1^2 + H^*\omega^2,$$

where $H^* = H - cx$ and $H = \alpha e_3 + \beta e_4$. Hence M^2 is totally umbilic (see Section 2.4) and, due to Propositions 3.1.4 and 3.1.5, is a standard model $N^2(c_1) \subset N^n(c)$ or its open subset. From Remark 3.1.6 it follows that if $c \geq 0$, then this $N^2(c_1)$ is a sphere, and if $c < 0$ then either a sphere, or an equidistant surface, or a horosphere in a Lobachevsky space.

(Ib) If $k \neq 0$, then the normal curvature ellipse degenerates to a straight line segment, (12.6.12) gives $\omega_3^4 = \omega_4^\zeta = 0$, and thus by (12.6.7), $d\alpha = d\beta = 0$, $\alpha\omega_3^\zeta = 0$.

(Ib_1) If $\alpha \neq 0$ everywhere, then (2.1.5) reduces to

$$de_1 = e_2\omega_1^2 + [\alpha e_3 + (\beta + k)e_4 - xc]\omega^1,$$

$$de_2 = -e_1\omega_1^2 + [\alpha e_3 + (\beta - k)e_4 - xc]\omega^2,$$

and $dx = e_1\omega^1 + e_2\omega^2$,

$$de_3 = -\alpha(e_1\omega^1 + e_2\omega^2) = -\alpha dx, \quad de_4 = -(\beta + k)e_1\omega^1 - (\beta - k)e_2\omega^2.$$

It follows that M^2 is contained in a 5-dimensional plane of ${}_\sigma E^{n+1}$ through the center o and spanned by e_1, e_2, x, e_3, e_4, which intersects $N^n(c)$ along an $N^4(c)$. Since the point z with radius vector $z = x + \alpha^{-1}e_3$ is a fixed point in this 5-dimensional plane and $\|z - x\| = \alpha^{-1} = \text{const}$, the surface M^2 belongs to a three-dimensional sphere of this $N^4(c)$, which in case $c < 0$ can be, in particular, an equidistant submanifold or a horosphere. With respect to this sphere, the surface M^2 has unit normal vector e_4 and principal curvatures $\beta + k$, $\beta - k$, thus its mean curvature is β, which is a constant.

(Ib$_2$) If $\alpha \equiv 0$, then M^2 belongs to a four-dimensional plane at o spanned by e_1, e_2, x, e_4, thus to an $N^3(c)$, in which it similarly has constant mean curvature.

(II) If $kl \neq 0$, $\beta = \gamma = 0$, then for $H = \alpha e_3 \neq 0$ one gets from (12.6.3)

$$\alpha = \text{const} \neq 0, \quad \omega_3^4 = \omega_3^5 = \omega_3^\eta = 0.$$

Hence

$$de_1 = e_2\omega_1^2 + e_3\alpha\omega^1 + e_4k\omega^1 + e_5l\omega^2 - xc,$$

$$de_2 = -e_1\omega_1^2 + e_3\alpha\omega^2 - e_4k\omega^2 + e_5l\omega^1 - xc$$

and

$$de_3 = -\alpha(e_1\omega^1 + e_2\omega^2) = -\alpha dx.$$

The point z with radius vector $z = x + \alpha^{-1}e_3$ is a fixed point, as above, with constant α. Hence the surface M^2 belongs to a hypersphere of $N^n(c)$ (which in case $c < 0$ can be an equidistant submanifold or a horosphere). With respect to this hypersphere, the vector valued second fundamental form h of M^2 has the components $h_{11} = ke_4$, $h_{12} = le_5$, $h_{22} = -ke_4$, and thus its mean curvature vector is zero. Hence the surface M^2 is a minimal surface of this hypersphere.

These results can be summarized as follows.

Theorem 12.6.2. *An H-parallel surface M^2 in $N^n(c)$ consists of its open subsets or their closures, each of which is either a minimal surface, i.e., with $H = 0$, or*

(Ia) *a totally umbilic surface, or*
(Ib) *a surface of constant mean curvature in an $N^3(c) \subset N^n(c)$ or in a three-dimensional sphere of an $N^4(c) \subset N^n(c)$, or*
(II) *a minimal surface of a hypersphere in $N^n(c)$.*

Remark 12.6.3. This theorem was proved by Chen [Ch 72], [Ch 73a] and Yau [Yau 74], and was presented (with a proof and some complements) in [Ch 73b], and again (without proof) in [Ch 2000]. The proof given above, using Cartan's method of adapted moving frame bundles and exterior calculus, is simpler.

[Ch 2000] also gives some information about H-parallel submanifolds M^m. It does not, however, mention the paper [Lu 93], in which all H-parallel canal submanifolds M^m with arbitrary m in E^n are classified. General H-parallel manifolds M^m in $N^n(c)$ with $m \geq 3$ are not yet classified, according to [Ch 2000]. In [RV 70] they are characterized for the case $c = 0$ as having harmonic Gauss maps. But there exist some results under some additional conditions; see [Ch 2000]. Recently, in [She 2002] they have been studied for finite total curvature.

Remark 12.6.4. Theorem 12.5.3 can now be illustrated as follows.

If a surface M^2 is H-semiparallel of the class (I) of Proposition 12.6.1, then its normal curvature ellipse at an arbitrary point reduces either (a) to a point, or (b) to a straight line segment. This M^2 is, correspondingly, H-parallel of type (Ia), or an H-envelope of order 2 of a family of H-parallel surfaces of type (Ib).

If a surface M^2 is H-semiparallel of the class (II) of Proposition 12.6.1, then it is either a minimal surface, hence H-parallel, or an H-envelope of order 2 of a family of H-parallel surfaces of type (II) of Theorem 12.6.2.

12.6.2 $R^\perp$-parallel surfaces

For $R^\perp$-parallel surfaces, the conditions (12.6.2) must be satisfied. Thus either

(i) $l = 0$, or
(ii) $kl = \text{const} > 0$, $\omega_4^\xi = \omega_5^\xi = 0$, $\xi \in \{3, 6, 7, \dots, n\}$.

If (i) holds on an open subset of M^2, then $\nabla^\perp$ is flat on this subset.

Suppose the conditions (ii) hold on some open subset of M^2. Among equations (6.1.9) there are $\omega_1^3 = \alpha\omega^1$, $\omega_2^3 = \alpha\omega^2$. By exterior differentiation they give $d\alpha \wedge \omega^1 = d\alpha \wedge \omega^2 = 0$, thus $\alpha = \text{const}$; and the equations $\omega_1^\eta = \omega_2^\eta = 0$, $\eta \in \{6, 7, \dots, n\}$ give in the same way

$$\alpha\omega^1 \wedge \omega_3^\eta = \alpha\omega^2 \wedge \omega_3^\eta = 0. \tag{12.6.13}$$

Here the conditions $kl = \text{const}$ and $\alpha = \text{const}$ have interesting equivalent geometrical interpretations. The first says that the normal curvature ellipse, with semiaxes k and l, has constant area, and the second means that the distance of the point $x \in M^2$ from the 2-plane of this ellipse is also a constant.

The remaining conditions in (ii) together with (12.6.13) imply a reduction of the codimension.

Namely, on this open subset $dx = e_1\omega^1 + e_2\omega^2$,

$$\begin{aligned}
de_1 &= e_2\omega_1^2 + e_4\omega_1^4 + e_5\omega_1^5 + (e_3\alpha - xc)\omega^1,\\
de_2 &= -e_1\omega_1^2 + e_4\omega_2^4 + e_5\omega_2^5 + (e_3\alpha - xc)\omega^2,\\
de_4 &= -e_1\omega_1^4 - e_2\omega_2^4 + e_5\omega_4^5, \quad de_5 = -e_1\omega_1^5 - e_2\omega_2^5 - e_4\omega_4^5.
\end{aligned}$$

If $\alpha \equiv 0$ on an open subset, then this subset lies in an $N^4(c)$, which is intersected from $N^n(c)$ by an ${}_\sigma E^5$ through the origin o, spanned by x, e_1, e_2, e_4, e_5.

If $\alpha \neq 0$ on an open subset, then for this subset

$$de_3 = -e_1\omega_1^3 - e_2\omega_2^3 = -\alpha(e_1\omega^1 + e_2\omega^2) = -\alpha dx, \quad \alpha = \text{const}.$$

Hence the subset lies in an $N^5(c)$ which is intersected from $N^n(c)$ by an ${}_\sigma E^6$ through the origin o, spanned by $x, e_1, e_2, e_3, e_4, e_5$.

On the other hand, for $z = x + \alpha^{-1}e_3$ one has $dz = dx + \alpha^{-1}(-\alpha dx) = 0$. Thus the point z with this radius vector in ${}_\sigma E^6$ is fixed, and $\|z - x\| = \alpha^{-1} = \text{const}$. Hence the above subset belongs to a four-dimensional sphere, which in the case of $c < 0$ can be, in particular, an equidistant submanifold, or a horosphere.

The results can be summarized as follows.

Theorem 12.6.5. *An $R^\perp$-parallel surface M^2 in $N^n(c)$ consists of its open subsets or their closures, each of which is either*

(i) *normally flat, i.e., has zero $R^\perp$, or*
(ii) *lies in an $N^4(c)$, or in a four-dimensional sphere of an $N^5(c)$, and has the property that the normal curvature ellipses at every point x have the same area and their 2-planes have the same distance from x.*

Remark 12.6.6. Theorem 12.6.5 is proved in [Lu 95c], where the above constancy of distance was not explicitly formulated in Theorem A, but nevertheless was established in course of the proof.

Remark 12.6.7. Recall that every surface M^2 in $N^n(c)$ is $R^\perp$-semiparallel. If it is normally flat, then it is also $R^\perp$-parallel. Otherwise, it has at any arbitrarily fixed point the configuration in an ${}_\sigma E^6$ consisting of this point x, the tangent 2-plane at x, and the normal curvature ellipse. For this configuration, there exists an $R^\perp$-parallel surface of type (*ii*) having at x the same configuration, and at all other points the properties stated in (*ii*) above. Therefore, this surface is an $R^\perp$-envelope of these $R^\perp$-parallel surfaces.

12.6.3 *R*- or *Ric*-parallel surfaces.

These are surfaces of constant Gaussian curvature K, due to (12.6.1). Here the semiparallel condition is trivial: every surface M^2 in $N^n(c)$ is R- and *Ric*-semiparallel. It is obviously a K-envelope of the family of surfaces with $K = \text{const}$; namely, at each point $x \in M^2$, there exists a surface with constant Gaussian curvature having at x the same tangent plane as M^2 and the same K as M^2.

12.6.4 *T*-semiparallel surfaces

Due to (12.6.6), only the possibilities (I) and (II) of Proposition 12.6.1 can occur here. Each of them leads to $T = 0$, and thus to trivially T-parallel surfaces. So the result is also trivial.

Proposition 12.6.8. *The only T-semiparallel surfaces M^2 in $N^n(c)$ are the surfaces with zero T. All of them are T-parallel and are described by* (I) *and* (II) *of Proposition* 12.6.1.

12.7 *Ric*-Semiparallel Hypersurfaces and Ryan's Problem

The hypersurfaces M^{n-1} in $N^n(c)$ are all normally flat, i.e., have $R^\perp = 0$; consequently, $T = 0$. The mean curvature vector H reduces to the mean curvature; thus its parallel condition means that the latter is constant.

Therefore, among the above-considered tensor fields Λ, only R and *Ric* can have some interest for hypersurfaces, from the point of view of their parallel and semiparallel conditions.

Here an R-parallel or -semiparallel submanifold M^m is simply an M^m with locally symmetric or semisymmetric intrinsic Riemannian metric, respectively, i.e., with $\nabla R = 0$ or $R(X.Y) \circ R = 0$.

For a normally flat submanifold M^m in $N^n(c)$, the curvature tensor R has by (5.1.4) and (2.1.11) the components

$$R_{ij,kl} = \langle k_i^*, k_j^* \rangle (\delta_{ik}\delta_{jl} - \delta_{il}\delta_{jk}). \tag{12.7.1}$$

Correspondingly, the Ricci tensor *Ric* has the components

$$R_{jk} = \sum_i R_{ij,ki} = (\langle k_j^*, k_k^* \rangle - \langle mH^*, k_j^* \rangle)\delta_{jk}, \tag{12.7.2}$$

where H^* is the outer mean curvature vector of the immersion:

$$H^* = \frac{1}{m} g^{ij} h_{ij}^* = \frac{1}{m}(k_1^* + \cdots + k_m^*).$$

Since $k_i^* = k_i - xc$ and hence $\langle k_i^*, k_j^* \rangle = \langle k_i, k_j \rangle + c$, the semiparallel condition (5.1.5) reduces to

$$(k_i - k_j)(\langle k_i, k_j \rangle + c) = 0. \tag{12.7.3}$$

The R-semiparallel and *Ric*-semiparallel conditions reduce, correspondingly, to

$$\langle k_i - k_j, k_k \rangle (\langle k_i, k_j \rangle + c) = 0, \tag{12.7.4}$$

$$\langle k_i + k_j - mH, k_i - k_j \rangle (\langle k_i, k_j \rangle + c) = 0, \tag{12.7.5}$$

for every distinct pair i, j and for every distinct triple i, j, k; moreover, $mH = k_1 + \cdots + k_m$.

Remark 12.7.1. It is known that for a Riemannian M^m, semisymmetric implies *Ric*-semisymmetric, and that these conditions are equivalent if $m = 3$.

For normally flat submanifolds M^3 in $N^n(c)$, this follows easily. Indeed, (12.7.4) implies (12.7.5); and if $m = 3$, then $k_i + k_j - mH = k_k$ for every three distinct values of i, j, k, and the conditions (12.7.4) and (12.7.5) coincide.

An interesting question is the situation for $m \geq 4$. This problem has been actively investigated in the particular case of hypersurfaces M^m in $N^{m+1}(c)$.

For hypersurfaces, $k_i = \lambda_i e_{m+1}$, where $\lambda_1, \ldots, \lambda_m$ are the principal curvatures. The conditions (12.7.3), (12.7.4) and (12.7.5) are now, respectively,

$$(\lambda_i - \lambda_j)(\lambda_i\lambda_j + c) = 0, \quad (12.7.6)$$

$$\lambda_k(\lambda_i - \lambda_j)(\lambda_i\lambda_j + c) = 0, \quad (12.7.7)$$

$$(\lambda_i - \lambda_j)(\lambda_i + \lambda_j - mH)(\lambda_i\lambda_j + c) = 0, \quad (12.7.8)$$

for distinct pairs i, j and for distinct triples i, j, k, and where $mH = \lambda_1 + \cdots + \lambda_m$.

Lemma 12.7.2. *Among the principal curvatures $\lambda_1, \ldots, \lambda_m$ of a Ric-semiparallel hypersurface M^m in $N^{m+1}(c)$, there can be at most two distinct values, if $c \neq 0$, or at most two distinct nonzero values, if $c = 0$.*

Proof. From (12.7.8) with $c = 0$, it follows for any two λ_i and λ_j, that either they are equal, or one of them is zero, or their sum is equal to mH. If one assumes that there are three distinct nonzero λ_i, λ_j and λ_k, then by (12.7.8),

$$(\lambda_i + \lambda_j - mH)(\lambda_i\lambda_j + c) = 0,$$

$$(\lambda_i + \lambda_k - mH)(\lambda_i\lambda_k + c) = 0,$$

$$(\lambda_k + \lambda_j - mH)(\lambda_k\lambda_j + c) = 0.$$

Here at least two first multiplicands are zero, e.g., $\lambda_i + \lambda_j - mH = \lambda_i + \lambda_k - mH = 0$, or at least two other multiplicands are zero, e.g., $\lambda_i\lambda_j + c = \lambda_i\lambda_k + c = 0$. Both possibilities lead to a contradiction, which in the first case is $\lambda_j = \lambda_k$. The second possibility leads to $\lambda_i(\lambda_j - \lambda_k) = 0$, thus to the same contradiction, or else to the now impossible $\lambda_i = 0 \Rightarrow c = 0$.

All R-semiparallel hypersurfaces M^m in $N^{m+1}(c)$, i.e., satisfying (12.7.7), were classified by Ryan [Ry 69], and then for the case of $c = 0$, i.e., in E^{m+1}, for a complementary condition of completeness, by Szabó [Sza 84].

For hypersurfaces M^m with positive scalar curvature in E^{m+1}, the equivalence of (12.7.7) and (12.7.8) was proved by Tanno [Tan 69], and then generalized by Ryan [Ry 71] to the case of nonnegative scalar curvature, and also of constant scalar curvature, or of nonzero constant sectional curvature. Another important result in [Ry 71] asserts that in a space form $N^{m+1}(c)$ with $c \neq 0$, every *Ric*-semiparallel hypersurface M^m is R-semiparallel, i.e., (12.7.7) and (12.7.8) are equivalent in the case $c \neq 0$. Therefore, the following problem is essential only for $c = 0$, i.e., in Euclidean space E^{m+1}:

Do there exist Ric-semiparallel hypersurfaces that are not R-semiparallel?

This problem was stated as Problem P808 in [Ry 72] and became known as *Ryan's problem.*

The most general solution of this problem can be obtained from a classification of all *Ric*-semiparallel hypersurfaces M^m in E^{m+1}. Such a classification is given by Mirzoyan [Mi 99]. The following exposition here is based on the analysis in [Mi 99], with some additions and refinements. Some of these were presented in [Lu 2002b] along with Mirzoyan's analysis (see Remark 12.7.9 below).

Due to Lemma 12.7.2, at most two distinct nonzero principal curvatures can occur here.

The situation is simple if there is only one principal curvature λ, or one nonzero λ of multiplicity p and a zero principal curvature of multiplicity $m-p$. Then (12.7.6) with $c=0$ is satisfied. Therefore, this M^m is a semiparallel hypersurface in E^{m+1}, and thus is one of the hypersurfaces described in Theorem 5.5.1.

Suppose that there are two distinct nonzero principal curvatures λ and μ, of multiplicities p and q, respectively; the other $m-p-q$ principal curvatures are zero. Due to the *Ric*-semiparallel condition (12.7.8), with $c=0$, here $\lambda+\mu-a=0$, where $a=mH=p\lambda+q(a-\lambda)$. This implies

$$(p-1)\lambda+(q-1)(a-\lambda)=0.$$

If $p=1$ then also $q=1$, and conversely, if $q=1$ then $p=1$. Therefore, only two of the principal curvatures $\lambda_1,\dots,\lambda_m$ are nonzero, the ohers are zero. Then (12.7.5) with $c=0$ is satisfied, implying that the hypersurface M^m is R-semiparallel. Its rank is 2.

Most interesting is the general case where $p\geq 2$ and $q\geq 2$. Then

$$a=\frac{p-q}{1-q}\lambda,\quad \mu=a-\lambda=\frac{p-1}{1-q}\lambda.$$

Here the frame vectors e_i can be renumbered so that $\lambda_b=\lambda$, $\lambda_u=\mu$, and $\lambda_s=0$, where $b,\dots\in\{1,\dots,p\}$, $u,\dots\in\{p+1,\dots,p+q\}$, and $s,\dots\in\{p+q+1,\dots,m\}$.

From (12.7.2), it follows that the Ricci tensor now has diagonal form with diagonal elements $\rho_j=\lambda_j^2-a\lambda_j$, which are

$$\rho_b=\rho_u=\frac{p-1}{1-q}\lambda^2<0,\quad \rho_s=0. \tag{12.7.9}$$

This implies that the hypersurface M^m is intrinsically a semi-Einstein Riemannian manifold. From (2.2.3) it follows due to $h_{ij}^{m+1}=\lambda_i\delta_{ij}$ that

$$\delta_{ij}d\lambda_i+(\lambda_j-\lambda_i)\omega_j^i=h_{ijk}^{m+1}\omega^k,$$

where $i,j,k\in\{1,\dots,m\}$, h_{ijk}^{m+1} are symmetric, and there is no summation on the left-hand side. After renumbering e_i, as above, this gives for $b\neq c$: $h_{bck}^{m+1}=0$, $h_{bbk}^{m+1}=h_{cck}^{m+1}$, for $u\neq v$: $h_{uvk}^{m+1}=0$, $h_{uuk}^{m+1}=h_{vvk}^{m+1}$; further, $h_{stk}^{m+1}=0$, and due to $(1-q)\mu=(p-1)\lambda$ here $h_{bbu}^{m+1}=h_{buu}^{m+1}=0$, $(1-q)h_{uus}^{m+1}=(p-1)h_{ccs}$. This finally gives

$$d\lambda=\chi_s\omega^s,\quad (\mu-\lambda)\omega_u^b=\chi_{bus}\omega^s, \tag{12.7.10}$$

$$\lambda\omega_b^s=\chi_s\omega^b+\chi_{bus}\omega^u,\quad \mu\omega_u^s=\chi_{bus}\omega^b+\frac{p-1}{1-q}\chi_s\omega^u, \tag{12.7.11}$$

where $\chi_s=h_{bbs}^{m+1}$ does not depend on b, and $\chi_{bus}=h_{bus}^{m+1}$. Exterior differentiation of the first equation of (12.7.10) leads to

$$0 = (d\chi_s - \chi_t \omega_s^t) \wedge \omega^s + (\lambda^{-1} - \mu^{-1}) \sum_s \chi_s \chi_{bus} \omega^b \wedge \omega^u. \tag{12.7.12}$$

Thus

$$d\chi_s - \chi_t \omega_s^t = \psi_{st} \omega^t, \tag{12.7.13}$$

$$\sum_s \chi_s \chi_{bus} = 0, \tag{12.7.14}$$

since ω^b, ω^u, ω^s are linearly independent, and λ and μ are distinct.

From (12.7.10), (12.7.11) it follows that the three orthogonal eigendistributions of the eigenvalues λ, $\mu = a - \lambda$, and 0 are foliations, and the leaves of the first two are locally the spheres of dimension p and q, respectively (for details, see [Mi 99]).

The third foliation is defined by $\omega^b = \omega^u = 0$ for all values of b and u. Due to (12.7.13) on every of its leaves an invariant vector field is defined by $\chi = \sum_s \chi_s e_s$ with $d\chi = (\sum_s e_s \psi_{st})\omega^t$.

Here (12.7.14) shows that this χ is orthogonal to all $\chi_{bu} = \sum_s \chi_{bus} e_s$. For rank $(\chi_{bu}) = r$ there are three possibilities: (a) $r = 0$, (b) r has maximal value pq, (c) $0 < r < pq$.

Let us consider first the possibility (a)—a subclass, where (12.7.14) is satisfied by $\chi_{bus} = 0$, or, equivalently, by

$$\omega_u^b = 0. \tag{12.7.15}$$

(Below it will be proved that this is the only one which is really possible.) Exterior differentiation now leads to

$$\sum_{s=p+q+1}^{m} \chi_s^2 = \lambda^4 \frac{p-1}{q-1}. \tag{12.7.16}$$

Here the equation $p + q = m$ cannot hold, since then the left side would be zero, while the right side is not. Hence at least one zero eigenvalue λ_{p+q+1} must exist. Thus the case $m = 4$ is impossible here.

Equations (12.7.11) reduce to

$$\lambda \omega_b^s = \chi_s \omega^b, \quad \lambda_u^s = \chi_s \omega^u, \tag{12.7.17}$$

and imply by exterior differentiation that $\chi_{st} = 2\lambda^{-1} \chi_s \chi_t$.

The orthonormal frame bundle can be adapted so that $\chi = \nu e_m$. Then $\chi_m = \nu$, where $\nu^2 = \lambda^4 \frac{p-1}{q-1}$ by (12.7.16), and $\chi_{s'} = 0$ for all $s' \in \{p + q + 1, \ldots, m - 1\}$. Hence equations (12.7.10), (12.7.17) reduce to

$$d\lambda = \nu \omega^m, \quad \lambda \omega_b^m = \nu \omega^b, \quad \lambda \omega_u^m = \nu \omega^u, \quad \omega_u^b = \omega_b^{s'} = \omega_u^{s'} = 0, \tag{12.7.18}$$

where

$$d\nu = 2\lambda^{-1} \nu^2 \omega^m, \quad \omega_{s'}^m = 0.$$

Substituting e_m for $-e_m$, if needed, one can obtain $\nu = \kappa\lambda^2$, where $\kappa = \sqrt{\frac{p-1}{q-1}} =$ const. Then $\mu = -\kappa^2\lambda$, and so the given hypersurface M^m in E^{m+1}, with $m > 4$, is defined by the differential system

$$\omega^{m+1} = 0, \quad \omega_b^{m+1} = \lambda\omega^b, \quad \omega_u^{m+1} = -\kappa^2\lambda\omega^u, \quad \omega_{s'}^{m+1} = \omega_m^{m+1} = 0, \tag{12.7.19}$$

$$d\lambda = \kappa\lambda^2\omega^m, \quad \omega_b^m = \kappa\lambda\omega^b, \quad \omega_u^m = \kappa\lambda\omega^u, \quad \omega_u^b = \omega_b^{s'} = \omega_u^{s'} = \omega_{s'}^m = 0. \tag{12.7.20}$$

It is easy to check that the exterior equations obtained by exterior differentiation from the equations of this system are satisfied due to the equations of the same system. Therefore, the Frobenius theorem implies that this system is totally integrable and defines the considered hypersurface up to some constants.

Taking $i = k = 1$ and $j = p+1$, so that $\lambda_i = \lambda_k = \lambda$ and $\lambda_j = \mu = \frac{p-1}{1-q}\lambda \neq \lambda$, it can be seen that (12.7.7) with $c = 0$ is not satisfied, since $\lambda \neq 0$. Therefore, this M^m is not R-semiparallel.

It remains to investigate the geometric structure of the hypersurface M^m of this subclass. The infinitesimal displacement equations (1.2.1) for the orthonormal frame bundle adapted as above to this M^m in E^{m+1} are, by (12.7.19) and (12.7.20),

$$dx = e_b\omega^b + e_u\omega^u + e_m\omega^m + e_{s'}\omega^{s'}, \tag{12.7.21}$$

$$de_b = e_f\omega_b^f + (e_m\kappa + e_{m+1})\lambda\omega^b, \tag{12.7.22}$$

$$de_u = e_v\omega_u^v + (e_m - e_{m+1}\kappa)\kappa\lambda\omega^u, \tag{12.7.23}$$

$$de_m = -(e_b\omega^b + e_u\omega^u)\kappa\lambda, \tag{12.7.24}$$

$$de_{s'} = e_{t'}\omega_{s'}^{t'}, \tag{12.7.25}$$

$$de_{m+1} = -e_b\lambda\omega^b + e_u\kappa^2\lambda\omega^u. \tag{12.7.26}$$

The distribution on M^m defined by $\omega^b = \omega^u = \omega^m = 0$ is a foliation, since this differential system is totally integrable. For its leaves, $dx = e_{s'}\omega^{s'}$, $de_{s'} = e_{t'}\omega_{s'}^{t'}$; therefore, they are parallel $(m-p-q-1)$-dimensional planes $E^{m-p-q-1}$ in E^{m+1}.

The distribution orthogonal to these planes is defined by $\omega^{s'} = 0$. It is also a foliation, and for its $(p+q+1)$-dimensional leaves one has (12.7.22)–(12.7.26), and therefore they are the congruent hypersurfaces M^{p+q+1} in parallel planes E^{p+q+2} orthogonal to the above $E^{m-p-q-1}$. Hence M^m is a product submanifold $M^{p+q+1} \times E^{m-p-q-1}$, namely, a cylinder on M^{p+q+1}.

On M^{p+q+1}, the distribution defined by $\omega^u = \omega^m = 0$ is a foliation, whose leaves are by (12.7.22) totally umbilic, with mean curvature vector $(e_m\kappa + e_{m+1})\lambda$. Hence these leaves are p-dimensional spheres of radius $(\lambda\sqrt{\kappa^2+1})^{-1}$ in the $(p+1)$-dimensional planes spanned at x by e_b and this mean curvature vector. Similarly, the distribution defined by $\omega^b = \omega^m = 0$ is a foliation, whose leaves are by (12.7.23) also totally umbilic, with mean curvature vector $(e_m - e_{m+1}\kappa)\kappa\lambda$; so they too are

q-dimensional spheres of radius $(\kappa\lambda\sqrt{1+\kappa^2})^{-1}$ in the $(q+1)$-dimensional planes spanned at x by e_u and this mean curvature vector.

These $(p+1)$- and $(q+1)$-dimensional planes are totally orthogonal. Therefore, the leaves of the foliation defined on M^{p+q+1} by $\omega^m = 0$ are the products of pairs of the above spheres.

The curves in M^{p+q+1}, orthogonal to these products, are defined by $\omega^b = \omega^u = 0$, whence (12.7.24) implies that $dx = e_m\omega^m$ and $de_m = 0$, thus they are straight lines. Therefore, M^{p+q+1} is a ruled hypersurface in E^{p+q+2}.

For the point z with radius vector $z = x + (\kappa\lambda)^{-1}e_m$, equation (12.7.20) implies $dz = 0$. The point is hence a fixed point in E^{p+q+2}, and so M^{p+q+1} is a cone over the product of two spheres.

The results of the above analysis are formulated in the following theorem (if we suppose that possibility (a) above is the only possible one that holds).

Theorem 12.7.3. *A hypersurface M^m in E^{m+1} is Ric-semiparallel if and only if it is an open subset of one of the following hypersurfaces:*

(I) semiparallel:
- (1) *a hypersphere S^m in E^{m+1};*
- (2) *a hypercone of rotation C^m in E^{m+1};*
- (3) *a product $S^k \times E^{m-k}$, where S^k is a hypersphere in E^{k+1} and E^{m-k} is an $(m-k)$-dimensional plane, totally orthogonal to E^{k+1}, $2 \le k \le m-1$;*
- (4) *a product $C^k \times E^{m-k}$, where C^k is a hypercone of revolution in E^{k+1} and E^{m-k} is an $(m-k)$-dimensional plane, totally orthogonal to E^{k+1}, $2 \le k \le m-1$;*

(II) R-semiparallel:
- (5) *a hypersurface whose second fundamental form h has the matrix $\|h_{ij}\|$ of rank ≤ 2;*

(III) not R-semiparallel:
- (6) *a ruled hypersurface which is a cone with point-vertex over a product of two spheres in E^{m+1} $(m \ge 5)$;*
- (7) *the product of a k-dimensional cone of* (6) *in E^{k+1} and an $(m-k)$-dimensional plane E^{m-k} totally orthogonal to E^{k+1}, $5 \le k \le m-1$.*

Proof. Parts (I) and (II) summarize the results obtained by the analysis above, like the geometric descriptions in (6) and (7). It remains to show that in part (III) possibility (a) is really the only possible one, i.e., that possibilities (b) and (c) above lead to contradictions.

First let us consider (b), when due to (12.7.13) $\chi_s = 0$. Then via (12.7.10) $\lambda = \text{const}$, thus also $\mu = \frac{p-1}{1-q}\lambda = \text{const}$, but (12.7.11) reduce to

$$\lambda\omega_b^s = \chi_{bus}\omega^u, \quad \mu\omega_u^s = \chi_{bus}\omega^b. \tag{12.7.27}$$

Here exterior differentiation leads, due to (12.7.10), to

$$(d\chi_{bus} - \chi_{cus}\omega_b^c - \chi_{bvs}\omega_u^v - \chi_{but}\omega_s^t) \wedge \omega^u$$

$$+\frac{\lambda}{\mu(\mu-\lambda)}\sum_u(\chi_{but}\chi_{cus}+\chi_{bus}\chi_{cut})\omega^t\wedge\omega^c=0.$$

Hence $\sum_u(\chi_{but}\chi_{cus}+\chi_{bus}\chi_{cut})=0$. From here for $c=b$ and $t=s$ one obtains $\sum_u\chi_{bus}^2=0$, thus $\chi_{bus}=0$. Now (12.7.16) is valid and leads to contradiction: the left side is zero, but the right side is not!

It remains to consider possibility (c). Then after the adaption above of the orthonormal frame bundle $\chi_{s'}=0$ for all $s'\in\{p+q+1,\dots,m-1\}$, $\chi_m\neq 0$. From (12.7.13) now $\chi_{bum}=0$, and so (12.7.10), (12.7.11) reduce to

$$d\lambda=\chi_m\omega^m,\quad(\mu-\lambda)\omega_u^b=\chi_{bus}\omega^s, \tag{12.7.10$'$}$$

$$\lambda\omega_b^m=\chi_m\omega^b,\quad\lambda\omega_b^{s'}=\chi_{bus'}\omega^u, \tag{12.7.11$'$}$$

$$\mu\omega_u^m=\frac{p-1}{1-q}\chi_m\omega^u,\quad\mu\omega_u^{s'}=\chi_{bus'}\omega^b; \tag{12.7.11$''$}$$

moreover, here

$$d\chi_m=\psi_{ms'}\omega^{s'}+\psi_{mm}\omega^m,\quad\chi_m\omega_{s'}^m=\psi_{s't'}\omega^{t'}+\psi_{s'm}\omega^m. \tag{12.7.28}$$

Since the leaves of the foliation, defined by $\omega^b=\omega^u=0$ in M^m, are the Euclidean planes, along each of which $de_m=0$, due to (12.7.11') and (12.7.11''), the orthonormal frame bundle can be further adapted so that along each of these planes also $de_{s'}=0$ for all values s'. After that (12.7.27) reduce to

$$\omega_{s'}^m=0,\quad d\chi_m=\psi_{mm}\omega^m. \tag{12.7.29}$$

If we replace $\mu=\frac{p-1}{1-q}\lambda$ in the first equation (12.7.11'') and differentiate exteriorily after that, we obtain

$$(d\chi_m+2\lambda^{-1}\chi_m^2\omega^u)\wedge\omega^m+(\lambda\mu^{-1}\chi_{cut'}\omega_m^{t'}-\mu^{-1}\chi_m\chi_{cut'}\omega^{t'})\wedge\omega^c=0,$$

which due to (12.7.29) reduces to

$$2\lambda^{-1}\chi_m^2\omega^u\wedge\omega^m-\mu^{-1}\chi_m\chi_{cut'}\omega^{t'}\wedge\omega^c=0.$$

This is a contradiction, because χ_m is supposed to be nonzero, but $\omega^u\wedge\omega^m$ and $\omega^{t'}\wedge\omega^c$ are linearly independent. This finishes the proof.

Now one can show how some of the previously known results about Ryan's problem can be deduced from Theorem 12.7.3 (and its proof).

Corollary 12.7.4. *Every Ric-semiparallel hypersurface M^m with positive scalar curvature in E^{m+1} is R-semiparallel.*

Indeed, it can be seen from (12.7.9) that the non-R-semiparallel hypersurfaces are ruled out here. This was a result of [Tan 69].

The same argument shows that in this corollary *positive scalar curvature* can be replaced with *nonnegative scalar curvature*; this was a result of [Ry 71].

Corollary 12.7.5. *Every Ric-semisymmetric hypersurface M^m with constant scalar curvature in E^{m+1} is R-semiparallel.*

Indeed, from (12.7.10) it is seen that $\lambda = \text{const}$ implies $\chi_s = 0$, but due to (12.7.15) this is impossible for (6) and (7). So follows another result of [Ry 71].

Corollary 12.7.6. *If a Ric-semisymmetric hypersurface M^m in E^{m+1} is complete, then it is R-semiparallel.*

Indeed, from the geometrical description of hypersurfaces (6) and (7) it is seen that they, as cones, are noncomplete. This gives the result of [Mat 83].

Corollary 12.7.7. *Every Ric-semisymmetric hypersurface M^4 in E^5 is R-semiparallel.*

Indeed, above there is shown that from (12.7.15) it follows that for (6) and (7) the case $m = 4$ is impossible. This is the result of [DSVY 97].

The most important conclusion from Theorem 12.7.3 is the full solution of Ryan's problem, stated as follows.

Theorem 12.7.8. *There do exist Ric-semiparallel but not R-semiparallel hypersurfaces M^m in E^{m+1}, $m > 4$.*

Proof. The existence of hypersurfaces (6) and (7) was established above in the course of the analysis, before Theorem 12.7.3 was formulated. Moreover, it was established that they are not R-semiparallel.

Remark 12.7.9. Recall that Theorem 12.7.3 is based on the results stated by Mirzoyan in [Mi 99]; some additions and refinements were made then in [Lu 2002b]. Recently V. Mirzoyan indicated in private correspondence that there is a mistake in his paper [Mi 99] (the conclusion $d\omega^s = 0$ in the proof of Theorem 1 is not correct), and also in the presentation in [Lu 2002b] there is a defect (from the conclusion (12.7.13) only a subcase $\chi_{bus} = 0$ is considered, i.e., possibilities (b) and (c) are excluded from the proof). Now his last defect is removed by the complementary proof of Theorem 12.7.3; here it must be noted that the idea of this proof was given by Mirzoyan.

Remark 12.7.10. The assertion of Theorem 12.7.8 has been announced by F. Defever in [Def 2000] (see also [DKV 2000]). Note that the example given in [Def 2000] by Theorem 3.2 coincides with one of the hypersurfaces (6) of our Theorem 12.7.3.

12.8 Extended Ryan's Problem for Normally Flat Submanifolds

Having now the complete solution of the classical Ryan's problem, it is natural to pose the problem in a more general setting, namely to ask whether the Ric-semiparallel normally flat submanifolds M^m in E^n must be R-semiparallel or not. Recall that the answer is positive for dimension $m = 3$, as shown above (see Remark 12.7.1).

In the present section, this extended Ryan's problem will be analysed for the dimension $m = 4$. Then $4H = k_1 + k_2 + k_3 + k_4$, and so the *Ric*-semiparallel condition (12.7.5) reduces to

$$\langle k_i, k_j\rangle\langle k_i - k_j, k_k + k_l\rangle = 0 \tag{12.8.1}$$

for every four distinct values of i, j, k, l. Permuting indices in (12.8.1), first k, j, i, l and then i, l, k, j, one can see that the set of equations (12.8.1) is equivalent to the set

$$\langle k_i, k_j\rangle = \langle k_k, k_l\rangle,$$

which for $m = 4$ gives

$$\langle k_1, k_2\rangle = \langle k_3, k_4\rangle, \quad \langle k_1, k_3\rangle = \langle k_2, k_4\rangle, \quad \langle k_1, k_4\rangle = \langle k_2, k_3\rangle. \tag{12.8.2}$$

This set of three equations is invariant under interchange of indices 1,2, and of 3,4, as well as of the pairs $\{1, 2\}$ and $\{3, 4\}$.

Suppose that the R-semiparallel condition (12.7.4) is not satisfied for at least one triple of different values i, j, k. After a renumbering, if needed, one may assume that $i = 1$, $j = 2$, so that

$$\langle k_1, k_2\rangle\langle k_1 - k_2, k_k\rangle \neq 0, \tag{12.8.3}$$

where the subscript k is either 3 or 4. In particular, $\langle k_1, k_2\rangle \neq 0$.

It is convenient to call these k_1 and k_2 the *distinguished* principal curvature vectors, for a normally flat *Ric*-semiparallel but not R-semiparallel submanifold M^4 in E^n.

In the rest of this section, it will be assumed that $n = 6$ and that the distinguished principal curvature vectors are collinear. This leads to $k_2 = \kappa k_1 \neq 0$, whence by (12.8.3), $(1 - \kappa)\langle k_1, k_k\rangle \neq 0$. On the other hand, (12.8.2) implies $\langle k_1, k_3\rangle = \kappa\langle k_1, k_4\rangle$ and $\langle k_1, k_4\rangle = \kappa\langle k_1, k_3\rangle$. Hence $\langle k_1, k_k\rangle = \kappa^2\langle k_1, k_k\rangle$, which is equivalent to $(1 + \kappa)(1 - \kappa)\langle k_1, k_k\rangle = 0$, and therefore $\kappa = -1$. Thus $k_2 = -k_1$, and by (12.8.2), $\langle k_1, k_3 + k_4\rangle = 0$.

The orthonormal frame bundle can be further adapted to the given M^4 in E^6, so that at an arbitrary point $x \in M^4$, one has $k_1 = -k_2 = \lambda e_5$. Then $k_3 = \mu e_5 + \nu_3 e_6$, $k_4 = -\mu e_5 + \nu_4 e_6$. Thus M^4 is defined by the differential system

$$\omega^5 = \omega^6 = 0,$$

$$\omega_1^5 = \lambda\omega^1, \quad \omega_2^5 = -\lambda\omega^2, \quad \omega_3^5 = \mu\omega^3, \quad \omega_4^5 = -\mu\omega^4, \tag{12.8.4}$$

$$\omega_1^6 = \omega_2^6 = 0, \quad \omega_3^6 = \nu_3\omega^3, \quad \omega_4^6 = \nu_4\omega^4. \tag{12.8.5}$$

By (12.8.2),

$$-\lambda^2 = -\mu^2 + \nu_3\nu_4,$$

and in general there exists a function ν such that $\nu_3 = \nu(\mu - \lambda)$, $\nu_4 = \nu^{-1}(\mu + \lambda)$.

By exterior differentiation, equations (12.8.4) give the following exterior equations:

$$d\lambda \wedge \omega^1 + 2\lambda\omega_1^2 \wedge \omega^2 + (\lambda - \mu)\omega_1^3 \wedge \omega^3$$

$$+ (\lambda + \mu)\omega_1^4 \wedge \omega^4 = 0, \tag{12.8.6}$$

$$2\lambda\omega_1^2 \wedge \omega^1 - d\lambda \wedge \omega^2$$
$$- (\lambda + \mu)\omega_2^3 \wedge \omega^3 - (\lambda - \mu)\omega_2^4 \wedge \omega^4 = 0, \tag{12.8.7}$$

$$(\lambda - \mu)\omega_1^3 \wedge \omega^1 - (\lambda + \mu)\omega_2^3 \wedge \omega^2$$
$$+ [d\mu + \nu(\lambda - \mu)\omega_5^6] \wedge \omega^3 + 2\mu\omega_3^4 \wedge \omega^4 = 0, \tag{12.8.8}$$

$$(\lambda + \mu)\omega_1^4 \wedge \omega^1 - (\lambda - \mu)\omega_2^4 \wedge \omega^2 + 2\mu\omega_3^4 \wedge \omega^3$$
$$- [d\mu + \nu^{-1}(\lambda + \mu)\omega_5^6] \wedge \omega^4 = 0. \tag{12.8.9}$$

The same procedure for (12.7.5) leads to

$$\lambda\omega_5^6 \wedge \omega^1 - \nu(\lambda - \mu)\omega_1^3 \wedge \omega^3 + \nu^{-1}(\lambda + \mu)\omega_1^4 \wedge \omega^4 = 0, \tag{12.8.10}$$

$$\lambda\omega_5^6 \wedge \omega^2 - \nu(\lambda - \mu)\omega_2^3 \wedge \omega^3 + \nu^{-1}(\lambda + \mu)\omega_2^4 \wedge \omega^4 = 0, \tag{12.8.11}$$

$$\nu(\lambda - \mu)[\omega_1^3 \wedge \omega^1 + \omega_2^3 \wedge \omega^2] + [-(\lambda - \mu)d\nu - \nu(d\lambda - d\mu) + \mu\omega_5^6] \wedge \omega^3$$
$$+ (\nu_3 - \nu_4)\omega_3^4 \wedge \omega^4 = 0, \tag{12.8.12}$$

$$- \nu^{-1}(\lambda + \mu)[\omega_1^4 \wedge \omega^1 + \omega_2^4 \wedge \omega^2] + (\nu_3 - \nu_4)\omega_3^4 \wedge \omega^3$$
$$+ [-\nu^{-2}(\lambda + \mu)d\nu + \nu^{-1}(d\lambda + d\mu) - \mu\omega_5^6] \wedge \omega^4 = 0. \tag{12.8.13}$$

From (12.8.6) it follows by Cartan's lemma that

$$d\lambda = A\omega^1 + B\omega^2 + C\omega^3 + D\omega^4,$$
$$2\lambda\omega_1^2 = B\omega^1 + E\omega^2 + F\omega^3 + G\omega^4,$$
$$(\lambda - \mu)\omega_1^3 = C\omega^1 + F\omega^2 + H\omega^3 + I\omega^4,$$
$$(\lambda + \mu)\omega_1^4 = D\omega^1 + G\omega^2 + I\omega^3 + J\omega^4.$$

It follows similarly from (12.8.7), that $E = -A$, and

$$- (\lambda + \mu)\omega_2^3 = F\omega^1 - C\omega^2 + K\omega^3 + L\omega^4,$$
$$- (\lambda - \mu)\omega_2^4 = G\omega^1 - D\omega^2 + L\omega^3 + M\omega^4.$$

Now substitution into (12.8.10) gives $F = G = I = 0$, and

$$\lambda\omega_5^6 = Q\omega^1 - \nu C\omega^3 + \nu^{-1}D\omega^4,$$

and further substitution into (12.8.11) adds $C = D = L = Q = 0$.

The result is

$$d\lambda = A\omega^1 + B\omega^2, \quad 2\lambda\omega_1^2 = B\omega^1 - A\omega^2, \quad \omega_5^6 = 0, \tag{12.8.14}$$

$$(\lambda - \mu)\omega_1^3 = H\omega^3, \quad -(\lambda + \mu)\omega_2^3 = K\omega^3, \tag{12.8.15}$$

$$(\lambda + \mu)\omega_1^4 = J\omega^4, \quad -(\lambda - \mu)\omega_2^4 = M\omega^4. \tag{12.8.16}$$

Now (12.8.8) and (12.8.9) reduce to

$$(d\mu - H\omega^1 - K\omega^2) \wedge \omega^3 + 2\mu\omega_3^4 \wedge \omega^4 = 0$$

$$2\mu\omega_3^4 \wedge \omega^3 - (d\mu + J\omega^1 + M\omega^2) \wedge \omega^4 = 0;$$

from here, Cartan's lemma gives $J = -H$, $M = -K$,

$$d\mu = H\omega^1 + K\omega^2 + R\omega^3 + S\omega^4, \quad 2\mu\omega_3^4 = S\omega^3 + T\omega^4, \tag{12.8.17}$$

so that

$$(\lambda + \mu)\omega_1^4 = -H\omega^4, \quad (\lambda - \mu)\omega_2^4 = K\omega^4. \tag{12.8.18}$$

Finally, via some calculations, (12.8.12) and (12.8.13) lead to $A = B = 0$, and therefore from (12.8.14) to $\omega_1^2 = 0$. Taking the exterior derivative, and using equations (12.8.15)–(12.8.18), and (12.8.4), one gets a contradiction: $\lambda^2\omega^1 \wedge \omega^2 = 0$. Hence the following statement holds.

Theorem 12.8.1. *There do not exist any submanifolds M^4 in E^6 which are normally flat, Ric-semiparallel, but not R-semiparallel, and whose distinguished principal curvature vectors are collinear.*

Of course, this theorem does not solve the extended Ryan's problem in general (i.e., without assuming the collinearity of the distinguished principal curvature vectors): do there exist *Ric*-semiparallel but not *R*-semiparallel normally flat submanifolds M^4 in E^6? The problem is open all the more so for general dimensions m and n.

Remark 12.8.2. This extended Ryan's problem was posed in [Lu 2002b], where Theorem 12.8.1 was also proved.

12.9 *R*-Semiparallel but Not Semiparallel Normally Flat Submanifolds of Codimension 2

The present chapter, and with it the book, will be concluded by considering the relationship between the semiparallel condition (12.7.3) and the *R*-semiparallel condition (12.7.4) in the case of normally flat submanifolds.

It is known that every semiparallel submanifold is also *R*-semiparallel, i.e., intrinsically a semisymmetric Riemannian manifold (see Proposition 4.1.2; for normally flat submanifolds, this follows from the implication (12.7.3) $\Longrightarrow$ (12.7.4)).

Now the following problem arises. Do there exist isometrically immersed semisymmetric Riemannian manifolds which are not semiparallel as submanifolds? For hypersurfaces, a positive answer can be deduced from the classification results in [Ry 69] and [Sza 84]. In Chapter 10 of [BKV 96], a detailed analysis was given for hypersurfaces in E^4 which are intrinsically of conullity two. The existence of such hypersurfaces of parabolic and hyperbolic types was proved, and the problem of their local rigidity was studied.

This problem is investigated below in the case of normally flat submanifolds M^m in E^{m+2}. It will be proved that there exist normally flat submanifolds M^m in E^{m+2}, which are not semiparallel, but are intrinsically of conullity two of hyperbolic type.

As preparation, first consider a semiparallel M^m in E^{m+2}. By Proposition 5.1.3 and Remark 5.1.4, such an M^m is normally flat. Hence Theorems 5.5.1 and 5.6.1 can be used here; they give the extrinsic characterization of such an M^m in E^{m+2}.

The results can be summarized as follows.

Theorem 12.9.1. *Let M^m be a semiparallel submanifold in E^{m+2}.*

(1) *If M^m has only one nonzero principal curvature vector of multiplicity s, then M^m is*
- *for $s = 1$, an envelope of a one-parameter family of m-dimensional planes, thus intrinsically locally Euclidean;*
- *for $s = m$, a sphere, thus of positive constant curvature;*
- *for $2 < s < m$, a product of a round cone (or a round cylinder) and a plane;*
- *for $s = 2$, a second-order envelope of products $S^2(c) \times E^{m-2}$, thus intrinsically a manifold of conullity two of the planar type.*

(2) *If M^m has two orthogonal nonzero principal curvature vectors of multiplicities p and q, then M^m is*
- *for $p = q = 1$, an envelope with flat $\bar{\nabla}$ of a two-parameter family of m-dimensional planes, thus intrinsically locally Euclidean;*
- *for $p > 1$ and $q > 1$, a product of two round cones of dimensions $p+1$ and $q+1$ (possibly degenerated to round cylinders) and a plane;*
- *for $p > 2$ and $q = 1$, an envelope of orthogonal type of a one-parameter family of semiparallel submanifolds of part* (1) *above (where s is replaced by $p = s+1$), thus intrinsically a product of a $(p+1)$-dimensional elliptic cone and an $(m-p-1)$-dimensional locally Euclidean manifold, in general;*
- *for $p = 2$ and $q = 1$, a second-order envelope of products $S^2(c) \times S^1(c_2) \times E^{m-3}$, thus intrinsically a manifold of conullity two of the planar type.*

Corollary 12.9.2. *If a semiparallel submanifold M^m in E^{m+2} is intrinsically a Riemannian manifold of conullity two, then it is of planar type.*

Returning now to the above problem of existence of certain nonsemiparallel normally flat submanifolds, the following theorem can be proved.

Theorem 12.9.3. *Among the nonsemiparallel normally flat submanifolds M^m in E^{m+2}, there exist R-semiparallel M^m of conullity two whose Euclidean leaves of codimension 2 are $(m-2)$-dimensional planes in E^{m+2}, and which are of hyperbolic type.*

Proof. For such an M^m, equations (12.7.4) hold with $c = 0$, but not (12.8.1), i.e., $(k_i - k_j)\langle k_i, k_j\rangle \neq 0$ for at least one pair (i, j). After renumbering if needed, this gives $(k_1 - k_2)\langle k_1, k_2\rangle \neq 0$. Due to (12.7.4) (with $c = 0$), all $k_3, \dots, k_m$ must be orthogonal to this nonzero vector.

The orthonormal frame of $O(M^m, E^{m+2})$ can be adapted further so that at an arbitrary point $x \in M^m$, the unit normal vector e_{m+1} is collinear to $k_1 - k_2 \neq 0$. Then

$$k_1 = \lambda_1 e_{m+1} + \kappa e_{m+2}, \quad k_2 = \lambda_2 e_{m+1} + \kappa e_{m+2}, \quad k_u = \mu_u e_{m+2}, \tag{12.9.1}$$

where $(\lambda_1 - \lambda_2)(\lambda_1\lambda_2 + \kappa^2) \neq 0$ and the index $u \in \{3, \dots, m\}$.

Now (12.7.4) applied to the triples $(1, u, 2)$ and $(2, u, 1)$ leads to

$$\kappa\mu_u(\lambda_1\lambda_2 + \kappa^2 - \kappa\mu_u) = 0, \tag{12.9.2}$$

but, by $(u, v, 1)$ and $(u, v, 2)$, to

$$\kappa(\mu_u - \mu_v)\mu_u\mu_v = 0. \tag{12.9.3}$$

It is sufficient to take here the subcase when $\mu_u = 0$ for every value $u \in \{3, \dots, m\}$. Then the given M^m in E^{m+2} is defined by the differential system

$$\omega^{m+1} = \omega^{m+2} = 0,$$

$$\omega_1^{m+1} = \lambda_1\omega^1, \quad \omega_2^{m+1} = \lambda_2\omega^2, \quad \omega_u^{m+1} = 0, \tag{12.9.4}$$

$$\omega_1^{m+2} = \kappa\omega^1, \quad \omega_2^{m+2} = \kappa\omega^2, \quad \omega_u^{m+2} = 0. \tag{12.9.5}$$

The last equations of (12.9.4) and (12.9.5) give by exterior differentiation

$$\omega_u^1 \wedge \lambda_1\omega^1 + \omega_u^2 \wedge \lambda_2\omega^2 = 0, \quad \kappa(\omega_u^1 \wedge \omega^1 + \omega_u^2 \wedge \omega^2) = 0. \tag{12.9.6}$$

Exterior differentiation of the first two equations (12.9.4) gives

$$(d\lambda_1 - \kappa\omega_{m+1}^{m+2}) \wedge \omega^1 + (\lambda_1 - \lambda_2)\omega_1^2 \wedge \omega^2 - \lambda_1 \sum_u \omega_u^1 \wedge \omega^u = 0, \tag{12.9.7}$$

$$(\lambda_1 - \lambda_2)\omega_1^2 \wedge \omega^1 + (d\lambda_2 - \kappa\omega_{m+1}^{m+2}) \wedge \omega^2 - \lambda_2 \sum_u \omega_u^2 \wedge \omega^u = 0, \tag{12.9.8}$$

and the first two equations (12.9.5) lead to

$$(d\kappa + \lambda_1\omega_{m+1}^{m+2}) \wedge \omega^1 - \kappa \sum_u \omega_u^1 \wedge \omega^u = 0, \tag{12.9.9}$$

$$(d\kappa + \lambda_2\omega_{m+1}^{m+2}) \wedge \omega^2 - \kappa \sum_u \omega_u^2 \wedge \omega^u = 0. \tag{12.9.10}$$

Suppose the essential codimension of M^m is two. Then by (12.8.1), $\kappa \neq 0$, and now the second equation (12.9.6) and Cartan's lemma give

$$\omega_u^1 = a_u\omega^1 + b_u\omega^2, \quad \omega_u^2 = b_u\omega^1 + e_u\omega^2, \tag{12.9.11}$$

and substitution into the first equation (12.9.6) leads to $(\lambda_1 - \lambda_2)b_u = 0$, thus to $b_u = 0$.

The differential system $\omega^1 = \omega^2 = 0$ is totally integrable, since $d\omega^1$ and $d\omega^2$ vanish as algebraic consequences of the equations of this system. For the leaves of the foliation defined by this system, one has $dx = \sum_u e_u\omega^u$, $de_u = \sum_v e_v\omega_u^v$; hence these leaves are generating $(m-2)$-planes. Analysis of the system of exterior equations (12.9.6)–(12.9.10) shows that the characters here are $s_1 = 2m$ and $s_2 = 1$, and Cartan's number $Q = s_1 + 2s_2 = 2(m+1)$ is equal to the number of new coefficients after developing these exterior equations by Cartan's lemma. Hence (see [IL 2003], [BCGGG 91], [Ca 45], [Fin 48]) this M^m exists and depends on one real analytic function of two real arguments. The generating $(m-2)$-planes are its Euclidean leaves, so that M^m is intrinsically of conullity two. Now equations (12.9.11) are equations (11.1.2) with $C_u = b_u$, and since here $b_u = 0$, comparison with (11.1.2) shows that this M^m is of the hyperbolic type in general, where $e_u \neq a_u$ for at least one value of u. This concludes the proof.

Corollary 12.9.4. *There exist Riemannian manifolds M^m of conullity two, which have an isometric immersion into E^{m+2} as a normally flat, R-semiparallel, and nonsemiparallel submanifold.*

Indeed, by Corollary 12.9.2, a submanifold of Theorem 12.9.3 of hyperbolic type cannot be semiparallel.

Remark 12.9.5. The last section of this book can be considered as a slight expansion of Section 6 of [Lu 2002b]. Recall also that the case of hypersurfaces was analysed in [Ry 69], [Sza 84], and [BKV 96] (Chapter 10), as already noted above.

References

[Ab 71] V. N. Abdullin, Symmetric Riemannian spaces V_4, *Izv. Vyssh. Uchebn. Zaved. Mat.*, **1971**-2 (1971), 3–12 (in Russian).

[Ak 76] S. Akiba, Submanifolds with flat normal connection and parallel second fundamental tensor, *Sci. Repts Yokohama Nat. Univ. Sec.* I, **23** (1976), 7–14.

[AG 93] M. A. Akivis and V. V. Goldberg, *Projective Differential Geometry of Submanifolds*, North-Holland, Amsterdam, 1993.

[AG 96] M. A. Akivis and V. V. Goldberg, *Conformal Differential Geometry and Its Generalizations*, Wiley, New York, 1996.

[ALM 2000] K. Arslan, Ü. Lumiste, C. Murathan, and C. Özgür, 2-semiparallel surfaces in space forms 1: Two particular cases, *Proc. Estonian Acad. Sci. Phys. Math.*, **49** (2000), 139–148.

[As 93] A. C. Asperti, Semi-parallel surfaces in space forms, in VIII *School on Differential Geometry* (*Campinas*, 1992), Soc. Brasil. Mat., Rio de Janeiro, 1993, 21–25.

[ALM 99] A. C. Asperti, G. A. Lobos, and F. Mercuri, Pseudo-parallel immersions in space forms, *Mat. Contemp.*, **17** (1999), 59–70.

[ALM 2002] A. C. Asperti, G. A. Lobos, and F. Mercuri, Pseudo-parallel submanifolds of a space form, *Adv. Geom.*, **2** (2002), 57–71.

[AM 94] A. C. Asperti and F. Mercuri, Semi-parallel immersions into space forms, *Boll. Unione Mat. Ital.* (7), **8-B** (1994), 833–895.

[Ast 73] V. V. Astrakhantsev, Pseudo-Riemannian symmetric spaces with commutative holonomy group, *Mat. Sb.*, **90** (1973), 288–305 (in Russian).

[Ba 83] E. Backes, Geometric applications of euclidean Jordan triple systems, *Manuscripta Math.*, **42** (1983), 265–272.

[BR 83] E. Backes and H. Reckziegel, On symmetric submanifolds of spaces of constant curvature, *Math. Ann.*, **263** (1983), 419–433.

[BG 95] M. Barros and O. J. Garay, On submanifolds with harmonic mean curvature, *Proc. Amer. Math. Soc.*, **129** (1995), 2545–2549.

[Be 99] M. Belkhelfa, Parallel and minimal surfaces in Heisenberg space, in *Summer School on Differential Geometry* (*Coimbra*, 1999), Universidade de Coimbra, Coimbra, Portugal, 1999, 67–76.

[BeD 2002] M. Belkhelfa and F. Dillen, Parallel surfaces in the real special linear group $SL(2, R)$, *Bull. Austral. Math. Soc.*, **65** (2002), 183–189.

[Be 57] M. Berger, Les espaces symétriques non compacts, *Ann. Sc. École Norm. Sup.*, **64** (1957), 85–177.

[Bern 2003] J. Berndt, Symmetric submanifolds of symmetric spaces, in *Proceedings of the 7th International Workshop on Differential Geometry* (*KMS Special Session on Geometry*), Kyungpook National University, Taegu, South Korea, 2003, 1–15.

[BCO 2003] J. Berndt, S. Console, and C. Olmos, *Submanifolds and Holonomy*, Chapman and Hall/CRC, Boca Raton, FL, London, New York, Washington, DC, 2003.

[BENT 2005] J. Berndt, J.-H. Eschenburg, H. Naitoh, and K. Tsukada, Symmetric submanifolds associated with irreducible symmetric R-spaces, *Math. Ann.*, **332** (2005), 721–737.

[BiO'N 69] R. L. Bishop and B. O'Neill, Manifolds of negative curvature, *Trans. Amer. Math. Soc.*, **145** (1969), 1–49.

[Bla 53] D. Blanusha, Les espaces elliptiques plongés isométriquement dans des espaces euclidiennes, *Glasnik Mat.-Fiz. Astron.*, **8**-2 (1953), 3–23, 81–114.

[Blo 85] C. Blomstrom, Symmetric immersions in pseudo-Riemannian space forms, in D. Ferus, R. B. Gardner, S. Helgason, and U. Simon, eds., *Global Differential Geometry and Global Analysis* (*Berlin*, 1984), Lecture Notes in Mathematics, Vol. 1156, Springer-Verlag, Berlin, New York, 1985, 30–45.

[Blo 86] C. Blomstrom, Planar geodesic immersions in pseudo-Euclidean spaces, *Math. Ann.*, **274** (1986), 585–598.

[Bo 95] E. Boeckx, *Foliated Semi-Symmetric Spaces*, Ph.D. thesis, Katholieke Universiteit Leuven, Leuven, Belgium, 1995.

[BKV 96] E. Boeckx, O. Kowalski, and L. Vanhecke, *Riemannian Manifolds of Conullity Two*, World Scientific, London, 1996.

[Bo 27] E. Bortolotti, Spazi subordinati: equazioni di Gauss e Codazzi, *Boll. Unione Mat. Ital.*, **6** (1927), 134–137.

[Bor 28] O. Boruvka, Sur une classe de surfaces minima plongées dans un espace à quatre dimensions à courbure constante, *C. R. Acad. Sci.*, **187** (1928), 334–336.

[Bre 72] G. E. Bredon, *Introduction to Compact Transformation Groups*, Academic Press, New York, London, 1972.

[Br 85] R. Bryant, Minimal surfaces of constant curvature in S^n, *Trans. Amer. Math. Soc.*, **290** (1985), 259–271.

[BCGGG 91] R. L. Bryant, S. S. Chern, R. B. Gardner, H. L. Goldsmith, and P. A. Griffiths, *Exterior Differential Systems*, Springer-Verlag, New York, 1991.

[CML 68] M. Cahen and R. McLenaghan, Métriques des espaces lorentziens symetriques á quatre dimensions, *C. R. Acad. Sci.*, **266** (1968), A1125–A1128.

[CP 70] M. Cahen and M. Parker, Sur des classes d'espaces pseudo-riemanniens symétriques, *Bull. Soc. Math. Belg.*, **22** (1970), 339–354.

[CMR 94] A. Carfagna D'Andrea, R. Mazzocco, and G. Romani, Some characterizations of 2-symmetric submanifolds in spaces of constant curvature, *Czech. Math. J.*, **44** (1994), 691–711.

[CW 71] M. do Carmo and N. Wallach, Minimal immersions of spheres, *Ann. Math.*, **93** (1971), 43–62.

[Ca 19] É. Cartan, Sur les variétés de courbure constante d'un espace euclidien ou noneuclidien, *Bull. Soc. Math. France*, **47** (1919), 125–160, **48** (1920), 132–208.

[Ca 26] É. Cartan, Sur une classe remarquable d'espaces de Riemann, *Bull. Soc. Math. France*, **54** (1926), 214–264, **55** (1927), 114–134.

[Ca 45] É. Cartan, *Les systèmes différentiels extérieurs et leurs applications géométriques*, Hermann, Paris, 1945; 2nd ed., 1971.

[Ca 46] É. Cartan, *Leçons sur la géométrie des espaces de Riemann*, 2nd ed., Gauthier-Villars, Paris, 1946.

[Ca 60] É. Cartan, *Riemannian Geometry in Orthogonal Frame: After the Lectures by E. Cartan Delivered in Sorbonne in* 1926–27, translated, revised, and edited by S. P. Finikov, Mockovskii Universitet, Moscow, 1960 (in Russian).

[CGR 90] I. Cattaneo Gasparini and G. Romani, Normal and osculating maps for submanifolds of R^N, *Proc. Roy. Soc. Edinburgh Ser.* A, **114** (1990), 39–55.

[Ch 72] B.-Y. Chen, Surfaces with parallel mean curvature vector, *Bull. Amer. Math. Soc.*, **78** (1972), 709–710.

[Ch 73a] B.-Y. Chen, On the surfaces with parallel mean curvature vector, *Indiana Univ. Math. J.*, **22** (1973), 655–666.

[Ch 73b] B.-Y. Chen, *Geometry of Submanifolds*, Marcel Dekker, New York, 1973.

[Ch 2000] B.-Y. Chen, Riemannian submanifolds, in F. J. E. Dillen and L. C. A. Verstraelen, eds., *Handbook of Differential Geometry*, Vol. I, Elsevier Science, Amsterdam, Lausanne, New York, Oxford, 2000, 187–418.

[CY 83] B.-Y. Chen and S. Yamaguchi, Classification of surfaces with totally geodesic Gauss image, *Indiana Univ. Math. J.*, **32** (1983), 143–154.

[CY 84] B.-Y. Chen and S. Yamaguchi, Submanifolds with totally geodesic Gauss image, *Geom. Dedic.*, **15** (1984), 313–322.

[Che 47] S. S. Chern, Sur une classe remarquable de variétés dans l'espace projectif à *n* dimensions, *Sci. Rep. Tsing Hua Univ.*, **4** (1947), 328–336.

[ChdCK 70] S. S. Chern, M. P. do Carmo, and S. Kobayashi, Minimal submanifolds of a sphere with second fundamental form of constant length, in F. E. Browder, ed., *Functional Analysis and Related Fields* (*Chicago* 1968), Springer-Verlag, Berlin, Heidelberg, New York, 1970, 59–75.

[ChK 52] S. S. Chern and N. H. Kuiper, Some theorems on the isometric imbedding of compact Riemann manifolds in Euclidean space, *Ann. Math.*, **56** (1952), 422–430.

[CGa 89] P. Coulton and H. Gauchman, Submanifolds of quaternion projective space with bounded second fundamental form, *Kodai Math. J.*, **12** (1989), 296–307.

[CGl 90] P. Coulton and J. Glazebrook, Submanifolds of the Cayley projective plane with bounded second fundamental form, *Geom. Dedic.*, **33** (1990), 265–275.

[Co 57] R. Couty, Sur les transformations définies par le groupe d'holonomie infinitesimale, *C. R. Acad. Sci.*, **244** (1957), 553–555.

[DaN 81] M. Dajczer and K. Nomizu, On flat surfaces in S_1^3 and H_1^3, in *Manifolds and Lie Groups: Papers in Honor of Y. Matsushima*, Birkhäuser, Basel, 1981, 71–108.

[Def 2000] F. Defever, Ricci-semisymmetric hypersurfaces, *Balkan J. Geom. Appl.*, **5** (2000), 81–91.

[DSVY 97] F. Defever, R. Deszcz, Z. Sentürk, L. Verstraelen, and S. Yaprak, On a problem of P. J. Ryan, *Kyungpook Math. J.*, **37** (1997), 371–376.

[DKV 2000] F. Defever, R. Deszcz, D. Kowalczyk, and L. Verstraelen, Semisymmetry and Ricci-semisymmetry for hypersurfaces of semi-Riemannian space forms, *Arab J. Math. Sci.*, **6** (2000), 1–16.

[DelP 1886] P. Del-Pezzo, Sugli spazii tangenti ad una superficie o ad una varieta immersa in uno spazio di piu dimensioni, *Rend. Accad. Napoli*, **25** (1886), 176–180.

[De 85] J. Deprez, Semi-parallel surfaces in Euclidean space, *J. Geom.*, **25** (1985), 192–200.

[De 86] J. Deprez, Semi-parallel hypersurfaces, *Rend. Semin. Mat. Univ. Politec. Torino*, **44** (1986), 303–316.

[De 89] J. Deprez, Semi-parallel immersions, in *Geometric Topology of Submanifolds (Proceedings of the Meeting at Luminy, Marseille,* 18–23 *May*, 1987), World Scientific, Singapore, 1989, 73–88.

[Des 92] R. Deszcz, On pseudosymmetric spaces, *Bull. Soc. Belg. Math.*, **A44** (1992), 1–34.

[Di 90a] F. Dillen, Sur les hypersurfaces paralleles d'ordre supérieur, *C. R. Acad. Sci. Ser.* 1, **311** (1990), 185–187.

[Di 90b] F. Dillen, The classification of hypersurfaces of a Euclidean space with parallel higher order fundamental form, *Math. Z.*, **203** (1990), 635–643.

[Di 91a] F. Dillen, The classification of hypersurfaces of a real space form with parallel higher order fundamental form, in *Differential Geometry: In Honor of Radu Rosca*, Katholieke Universiteit Leuven, Leuven, Belgium, 1991, 83–100.

[Di 91b] F. Dillen, Semi-parallel hypersurfaces of a real space form, *Israel J. Math.*, **75** (1991), 193–202.

[Di 91c] F. Dillen, Higher order parallel submanifolds, in *Geometric Topology of Submanifolds* III, World Scientific, Singapore, 1991, 148–152.

[Di 92] F. Dillen, Hypersurfaces of a real space form with parallel higher order fundamental form, *Soochow J. Math.*, **18** (1992), 321–338.

[DN 93] F. Dillen and S. Nölker, Semi-parallelity, multi-rotation surfaces and the helix-property, *J. Reine Angew. Math.*, **435** (1993), 33–63.

[DPV 97] F. Dillen, M. Petrović, and L. Verstraelen, Einstein, conformally flat and semi-symmetric submanifolds satisfying Chen's equality, *Israel J. Math.*, **100** (1997), 163–169.

[DV 90] F. Dillen and L. Vrancken, Higher order parallel submanifolds of a complex space form, *Results Math.*, **18** (1990), 202–208.

[DV 91] F. Dillen and L. Vrancken, Generalized Cayley surfaces, in B. Wegner, D. Ferus, U. Pinkall, and U. Simon, eds., *Global Differential Geometry and Global Analysis*, Lecture Notes in Mathematics, Vol. 1481, Springer-Verlag, Berlin, New York, 1991, 36–47.

[DB 59] J. Dubnov and L. Beskin, Solution of a problem, *Mat. Prosvesch.*, **4** (1959), 267–269 (in Russian).

[EH 95] J. H. Eschenburg and E. Heintze, Extrinsic symmetric spaces and orbits of s-representations, *Manuscripta Math.*, **88** (1995), 517–524.; erratum, **92** (1997), 408.

[Er 71] J. Erbacher, Isometric immersions of constant mean curvature and triviality of the normal connection, *Nagoya Math. J.*, **45** (1971), 139–165.

[Fa 36] F. Fabricius-Bierre, Sur variétés a torsion nulle, *Acta Math.*, **66** (1936), 49–77.

[Fav 57] J. Favard, *Course de géométrie différentielle locale*, Gauthier-Villars, Paris, 1957.

[Fed 56] A. S. Fedenko, Symmetric spaces with simple non-compact fundamental groups, *Dokl. Akad. Nauk SSSR*, **108** (1956), 1026–1028 (in Russian).

[Fed 59] A. S. Fedenko, Symmetric spaces with simple fundamental groups, *Uchen. Zap. Byelorussk. Univ.*, **3** (1959), 3–25 (in Russian).

[Fed 77] A. S. Fedenko, *Spaces with Symmetries*, Belarusian State University, Minsk, 1977 (in Russian).

[Fe 74a] D. Ferus, Immersionen mit paralleler zweiter Fundamentalform: Beispiele und Nicht-Beispiele, *Manuscripta Math.*, **12** (1974), 153–162.

[Fe 74b] D. Ferus, Produkt-Zerlegung von Immersionen mit paralleler zweiter Fundamentalform, *Math. Ann.*, **211** (1974), 1–5.

[Fe 74c] D. Ferus, Immersions with parallel second fundamental form, *Math. Z.*, **140** (1974), 87–93.

[Fe 80] D. Ferus, Symmetric submanifolds of Euclidean space, *Math. Ann.*, **247** (1980), 81–93.

[Fil 95] E. Filonenko, *Semiparallel Space-Like Surfaces in Pseudo-Euclidean Space*, Master's thesis, University of Tartu, Tartu, Estonia, 1995.

[Fin 48] S. P. Finikov, *Cartan's Method of Exterior Forms in Differential Geometry: The Theory of Compatibility of Total and Partial Differential Equations*, OGIZ, Moscow, 1948 (in Russian)

[Fu 72] S. Fujimura, On Riemannian manifolds satisfying the condition $R(X, Y)\cdot R = 0$, *J. Fac. Sci. Hokkaido Univ. Ser.* 1 **22** (1972), 1–8.

[GLOP 70] R. Galchenkova, Ü. Lumiste, J. Ozhigova, and I. Pogrebysskii, *Ferdinand Minding* 1806–1885, Nauka, Leningrad, 1970 (in Russian).

[Gri 83] P. A. Griffiths, *Exterior Differential Systems and the Calculus of Variations*, Birkhäuser, Basel, Stuttgart, Cambridge, MA, 1983.

[GJ 87] P. A. Griffiths and G. R. Jensen, *Differential Systems and Isometric Embeddings*, Priceton University Press, Princeton, NJ, 1987.

[Ha 65] T. Hangan, Structures pseudoriemanniennes sur l'ensemble des p-plans d'un espace pseudoeuclidien, *Bull. Math. Soc. Sci. Math. RSR*, **9** (1965), 265–278.

[HE 73] S. W. Hawking and G. F. R Ellis, *The Large Scale Structure of Space-Time*, Cambridge University Press, Cambridge, UK, 1973.

[He 62] S. Helgason, *Differential Geometry and Symmetric Spaces*, Academic Press, New York, 1962.

[He 78] S. Helgason, *Differential Geometry, Lie Groups, and Symmetric Spaces*, Academic Press, New York, 1978.

[Hie 79] S. Hiepko, Eine innere Kennzeichnung der verzerrten Produkte, *Math. Ann.*, **241** (1979), 209–215.

[Hou 72] C.-S. Houh, Pseudo-umbilical surfaces with parallel second fundamental form, *Tensor*, **26** (1972), 262–266.

[Hu 66] D. Husemoller, *Fibre Bundles*, McGraw–Hill, New York, St. Louis, San Francisco, Toronto, London, Sydney, 1966.

[HT 97] J.-T. Hyun and R. Takagi, Hypersurfaces of a real space form with parallel higher order fundamental form, *Yokahama Math. J.*, **44** (1997), 5–20.

[It 75] T. Itoh, On Veronese manifolds, *J. Math. Soc. Japan*, **27** (1975), 497–506.

[IL 2003] T. A. Ivey and J. M. Landsberg, *Cartan for Beginners: Differential Geometry via Moving Frames and Exterior Differential Aystems*, Graduate Studies in Mathematics, Vol. 61, American Mathematical Society, Providence, RI, 2003.

[Je 77] G. R. Jensen, *Higher Order Contact of Submanifolds of Homogeneous Spaces*, Lecture Notes in Mathematics, Vol. 610, Springer-Verlag, Berlin, New York, 1977.

[JR 2006] T. Jentsch and H. Reckziegel, Submanifolds with parallel second fundamental form studied via the Gauss map, *Ann. Global Anal. Geom.*, **29** (2006), 51–93.

[Ka 48] V. F. Kagan, *Foundations of Theory of Surfaces in Tensor Representation*, Part 2, *Gos. Izd. Tekhn.-Teor. Lit.*, Moscow, Leningrad, 1948 (in Russian).

[Kai 78] V. R. Kaigorodov, Semisymmetric Lorentzian spaces with perfect holonomy group, *Gravit. i Teoriya Otnos.*, Vol. 14–15, Kazan University Press, Kazan, Russia, 1978, 113–120 (in Russian).

[Kai 83] V. R. Kaigorodov, Curvature structure of space-time, in *Problems in Geometry*, Vol. 14, Vsesoyuzn. Inst. Nauchn. Tekhn. Inform., Akad. Nauk SSSR, Moscow, 1983, 177–204 (in Russian).

[KP 87] U.-H. Ki and J. S. Pak, Submanifolds of a Euclidean m-space with totally umbilical Gauss image, *Tensor*, **44** (1987), 233–239.

[KALM 03] B. Kilic, K. Arslan, Ü. Lumiste, and C. Murathan, On weak biharmonic submanifolds and 2-parallelity, *Differential Geom. Dynam. Systems*, **5** (2003), 39–48.

[Ko 68] S. Kobayashi, Isometric imbeddings of compact symmetric spaces, *Tohoku Math. J.*, **20** (1968), 21–25.

[KNa 64, 65] S. Kobayashi and T. Nagano, On filtered Lie algebras and geometric structures I, II, *J. Math. Mech.*, **13** (1964), 875–908, **14** (1965), 513–522.

[KN 63, 69] S. Kobayashi and K. Nomizu, *Foundations of Differential Geometry*, Vols. I and II, Interscience, New York, London, Sydney, 1963, 1969.

[Kon 74] M. Kon, On some complex submanifolds in Kaehler manifolds, *Canad. J. Math.*, **26** (1974), 1442–1449.

[Kon 75] M. Kon, Totally real minimal submanifolds with parallel second fundamental form, *Atti Accad. Naz. Lincei. Rend. Cl. Sci. Fis. Mat. Natur.*, **57** (1974–1975), 70–74.

[Kow 80] O. Kowalski, *Generalized Symmetric Spaces*, Lecture Notes in Mathematics, Vol. 805, Springer-Verlag, Berlin, New York, 1980.

[Kow 96] O. Kowalski, An explicit classification of 3-dimensional Riemannian spaces satisfying $R(X, Y) \cdot R = 0$, *Czech. Math. J.* **46**-121 (1996), 427–474.

[KoK 87] O. Kowalski and A. Kulich, Generalized symmetric submanifolds of Euclidean spaces, *Math. Ann.*, **277** (1987), 67–78.

[KoN 98] O. Kowalski and S. Ž. Nikćević, Contact homogeneity and envelopes of Riemannian metrics, *Beitr. Algebra Geom.*, **39** (1998), 155–167.

[KoTV 90] O. Kowalski, F. Tricceri, and L. Vanhecke, Examples nouveaux de varietes riemanniennes non homogenes dont le tenseur de courbure est celui d'un espace symetrique riemnnien, *C. R. Acad. Sci. Ser.* I, **311** (1990), 355–360.

[KoTV 92] O. Kowalski, F. Tricceri, and L. Vanhecke, Curvature homogeneous Riemannian manifolds, *J. Math. Pures Appl.*, **71** (1992), 471–501.

[Kr 57] G. I. Krućković, On semi-reducible Riemannian spaces, *Dokl. Akad. Nauk SSSR*, **115** (1957), 862–865 (in Russian).

[Le 25] H. Levy, Symmetric tensors of the second order whose covariant derivatives vanish, *Ann. Math.* (2), **27** (1925), 91–98.

[Le 61] K. Leichtweiss, Zur Riemannschen Geometrie in Grassmannschen Mannigfaltigkeiten, *Math. Z.*, **76** (1961), 334–336.

[L-C 17] T. Levi-Civita, Nozione di parallelismo in una varieta qualunque e conseguente specificazione geometrica della curvatura Riemanniana, *Rend. Palermo*, **42** (1917), 173–205.

[L-C 25] T. Levi-Civita, *Lezioni di calcolo differenziale assoluto*, Stock, Rome, 1925 (in Italian); *Der absolute Differentialkalkül*, A. Duschek, transl., Springer-Verlag, Berlin, 1928 (in German); *The Absolute Differential Calculus*, M. Long, transl., Blackie, London, 1929 (in English).

[Li 2001] G. Li, Semi-parallel, semi-symmetric immersions and Chen's equality, *Results Math.*, **40** (2001), 257–264.

[Li 52] A. Lichnerowicz, Courbure nombres de Betti et espaces symetriques, in *Proceedings of the International Congress of Mathematicians* (*Cambridge*, 1950), Vol. 2, American Mathematical Society, Providence, 1952, 216–223.

[Li 55] A. Lichnerowicz, *Theorie globale des connexions et des groupes d'holonomie*, Edizioni Cremonese, Rome, 1955 (in French); *Global Theory of Connections and Holonomy Groups*, Noordhoff, Groningen, the Netherlands, 1976 (in English).

[Li 58] A. Lichnerowicz, *Geometrie des groupes des transformations*, Dunod, Paris, 1958.

[LMSS 96] H. Liu, M. Magid, Ch. Scharlach, and U. Simon, Recent developments in affine differential geometry, in Geometric Topology of Submanifolds VIII, World Scientific, Singapore, 1996, 1–15, 393–408.

[LT 2006] G. A. Lobos and R. Tojeiro, Pseudo-parallel submanifolds with flat normal bundle of space forms, *Glasgow Math. J.*, **48** (2006), 171–177.

[Lu 62] Ü. Lumiste, Zur Theorie der zweidimensionalen Minimalflächen II: Flächen fester Krümmung, *Tartu Ülik. Toim. Acta Comm. Univ. Tartuensis*, **102** (1962), 16–28 (in Russian; summary in German).

[Lu 66] Ü. Lumiste, Connections in homogeneous fibre bundles, *Mat. Sb.*, **69** (1966), 419–454 (in Russian); *Amer. Math. Soc. Transl.* (2), **92** (1970), 231–274 (in English).

[Lu 71] Ü. Lumiste, Theory of connections in fibre bundles, in *Itogi Nauki i Tekhniki, Algebra, Topologiya, Geometriya* 1969, Akad Nauk SSSR Inst. Nauchn. Informatsii, Moscow, 1971, 123–168 (in Russian); *J. Soviet Math.*, **1** (1973), 363–390 (in English).

[Lu 75] Ü. Lumiste, Differential geometry of submanifolds, in *Itogi Nauki i Tekhniki, Algebra, Topologiya, Geometriya*, Vol. 13, Akad Nauk SSSR Inst. Nauchn. Informatsii, Moscow, 1975, 273–340 (in Russian); *J. Soviet Math.*, **7**-4 (1977), 654–677 (in English).

[Lu 86] Ü. Lumiste, Small-dimensional irreducible submanifolds with parallel third fundamental form, *Tartu Ülik. Toim. Acta Comm. Univ. Tartuensis*, **734** (1986), 50–62 (in Russian; summary in English).

[Lu 87a] Ü. Lumiste, Decomposition and classification theorems for semi-symmetric immersions, *Eesti TA Toim. Füüs. Mat. Proc. Acad. Sci. Estonia Phys. Math.*, **36** (1987), 414–417.

[Lu 87b] Ü. Lumiste, Submanifolds with a van der Waerden–Bortolotti plane connection and parallelism of the third fundamental form, *Izv. Vyssh. Uchebn. Zaved. Mat.*, **31**-1 (1987), 18–27 (in Russian); *Soviet Math.* (*Iz. VUZ*), **31**-1 (1987), 25–35 (in English).

[Lu 87c] Ü. Lumiste, Reducibility of submanifolds with parallel third fundamental form, *Izv. Vyssh. Uchebn. Zaved. Mat.*, **31**-11 (1987), 32–41 (in Russian); *Soviet Math.* (*Iz. VUZ*), **31**-11 (1987), 40–52 (in English).

[Lu 88a] Ü. Lumiste, Decomposition of semi-symmetric submanifolds, *Tartu Ülik. Toim. Acta Comm. Univ. Tartuensis*, **803** (1988), 69–78.

[Lu 88b] Ü. Lumiste, Classification of two-codimensional semi-symmetric submanifolds, *Tartu Ülik. Toim. Acta Comm. Univ. Tartuensis*, **803** (1988), 79–94.

[Lu 89a] Ü. Lumiste, Normally flat submanifolds with parallel third fundamental form, *Eesti TA Toim. Füüs. Mat. Proc. Acad. Sci. Estonia Phys. Math.*, **38** (1989), 129–138.

[Lu 89b] Ü. Lumiste, Normally flat semi-symmetric submanifolds, in *Proceedings of the Conference on Differential Geometry and Its Applications* (*Dubrovnik, June 26–July* 3, 1988), University of Belgrade/University of Novi Sad, Novi Sad, Yugoslavia, 1989, 159–171.

[Lu 89c] Ü. Lumiste, Semi-symmetric submanifolds with maximal first normal space, *Eesti TA Toim. Füüs. Mat. Proc. Acad. Sci. Estonia Phys. Math.*, **38** (1989), 453–457.

[Lu 90a] Ü. Lumiste, Semi-symmetric submanifold as the second order envelope of symmetric submanifolds, *Eesti TA Toim. Füüs. Mat. Proc. Acad. Sci. Estonia Phys. Math.*, **39** (1990), 1–8.

[Lu 90b] Ü. Lumiste, Classification of three-dimensional semi-symmetric submanifolds in Euclidean spaces, *Tartu Ülik. Toim. Acta Comm. Univ. Tartuensis*, **899** (1990), 29–44.

[Lu 90c] Ü. Lumiste, Three-dimensional submanifolds with parallel third fundamental form in Euclidean spaces, *Tartu Ülik. Toim. Acta Comm. Univ. Tartuensis*, **899** (1990), 45–56.

[Lu 90d] Ü. Lumiste, Irreducible normally flat semi-symmetric submanifolds I, *Izv. Vyssh. Uchebn. Zaved. Mat.*, **34**-8 (1990), 45–53 (in Russian); *Soviet Math. (Iz. VUZ)*, **34**-8 (1990), 50–59 (in English).

[Lu 90e] Ü. Lumiste, Irreducible normally flat semi-symmetric submanifolds II, *Izv. Vyssh. Uchebn. Zaved. Mat.*, **34**-9 (1990), 32–40 (in Russian); *Soviet Math. (Iz. VUZ)*, **34**-9 (1990), 35–47 (in English).

[Lu 91a] Ü. Lumiste, Second order envelopes of symmetric Segre submanifolds, *Tartu Ülik. Toim. Acta Comm. Univ. Tartuensis*, **930** (1991), 15–26.

[Lu 91b] Ü. Lumiste, Second order envelopes of m-dimensional Veronese submanifolds, *Tartu Ülik. Toim. Acta Comm. Univ. Tartuensis*, **930** (1991), 35–46.

[Lu 91c] Ü. Lumiste, On submanifolds with parallel higher order fundamental form in Euclidean spaces, in B. Wegner, D. Ferus, U. Pinkall, and U. Simon, eds., *Global Differential Geometry and Global Analysis*, Lecture Notes in Mathematics, Vol. 1481, Springer-Verlag, Berlin, New York, 1991, 126–137.

[Lu 91d] Ü. Lumiste, Semi-symmetric envelopes of some symmetric cylindrical submanifolds, *Eesti TA Toim. Füüs. Mat. Proc. Acad. Sci. Estonia Phys. Math.*, **40** (1991), 245–257.

[Lu 91e] Ü. Lumiste, Symmetric orbits of the orthogonal Segre action and their second order envelopes, *Rend. Semin. Mat. Messina Ser.* II, **1** (1991), 142–150.

[Lu 91f] Ü. Lumiste, Semi-symmetric submanifolds, in *Problems in Geometry*, Vol. 23, Vsesoyuzn. Inst. Nauchn. Tekhn. Inform., Akad. Nauk SSSR, Moscow, 1991, 3–28 (in Russian); *J. Math. Sci. New York* **70**-2 (1994), 1609–1623.

[Lu 92a] Ü. Lumiste, Semi-symmetric submanifolds and modified Nomizu problem, in *Proceedings of the 3rd Congress of Geometry (Thessaloniki* 1991), Aristotle University of Thessaloniki, Thessaloniki, Greece, 1992, 263–274.

[Lu 92b] Ü. Lumiste, Semi-symmetric fundamental triplets, *Tartu Ülik. Toim. Acta Comm. Univ. Tartuensis*, **953** (1992), 7–18.

[Lu 93] Ü. Lumiste, Canal submanifolds with parallel mean curvature vector, *Eesti TA Toim. Füüs. Mat. Proc. Acad. Sci. Estonia Phys. Math.*, **42** (1993), 222–228.

[Lu 95a] Ü. Lumiste, Symmetric orbits of orthogonal Veronese actions and their second order envelopes, *Results Math.*, **27** (1995), 284–301.

[Lu 95b] Ü. Lumiste, Modified Nomizu problem for semi-parallel submanifolds, in *Geometric Topology of Submanifolds* VII: *Differential Geometry: In Honor of Professor Katsumi Nomizu*, World Scientific, Singapore, 1995, 176–181.

[Lu 95c] Ü. Lumiste, Surfaces with a parallel normal curvature tensor, *Eesti TA Toim. Füüs. Mat. Proc. Acad. Sci. Estonia Phys. Math.*, **44** (1995), 411–419

[Lu 96a] Ü. Lumiste, Semi-parallel pseudo-Riemannian submanifolds with non-null principal normals of extremal dimension, *Prepr. Ser. Inst. Math. Univ. Oslo*, **1** (1996), 1–34.

[Lu 96b] Ü. Lumiste, Symmetric orbits of orthogonal Plücker action and triviality of their second order envelopes, *Ann. Global Anal. Geom.*, **14** (1996), 237–256.

[Lu 96c] Ü. Lumiste, Semi-parallel submanifolds of cylindrical or toroidal Segre type, *Eesti TA Toim. Füüs. Mat. Proc. Acad. Sci. Estonia Phys. Math.*, **45** (1996), 161–177.

[Lu 96d] Ü. Lumiste, Semi-parallel submanifolds as some immersed fibre bundles with flat connections, in *Geometric Topology of Submanifolds* VIII, World Scientific, Singapore, 1996, 236–244.

[Lu 96e] Ü. Lumiste, A classification of real semi-symmetric curvature operators in dimension four, in *Geometric Topology of Submanifolds* VIII, World Scientific, Singapore, 1996, 245–256.

[Lu 96f] Ü. Lumiste, Semisymmetric curvature operators and Riemannian 4-spaces elementarily classified, *Algebras Groups Geom.*, **13** (1996), 371–388.

[Lu 96g] Ü. Lumiste, Differential geometry in Estonia: history and recent developments, *Arkhimedes*, **4** (1996), 31–34 (in Finnish).

[Lu 97a] Ü. Lumiste, Martin Bartels as researcher: His contribution to analytical methods in geometry, *Historia Math.*, **24** (1997), 46–65.

[Lu 97b] Ü. Lumiste, Semi-parallel time-like surfaces in Lorentzian spacetime forms, *Differential Geom. Appl.*, **7** (1997), 59–74.

[Lu 99a] Ü. Lumiste, Isometric semiparallel immersions of two-dimensional Riemannian manifolds into pseudo-Euclidean spaces, in J. Szenthe, ed., *New Developments in Differential Geometry* (*Budapest* 1996), Kluwer, Dordrecht, the Netherlands, 1999, 243–264.

[Lu 99b] Ü. Lumiste, University of Tartu and geometry of the 19th century, in V. Abramov, M. Rahula, K. Riives, eds., *Ülo Lumiste: Mathematician: Development of Differential Geometry in Estonia*, Estonian Mathematical Society, Tartu, Estonia, 1999, 68–102.

[Lu 2000a] Ü. Lumiste, Submanifolds with parallel fundamental form, in F. Dillen and L. Verstraelen, eds., *Handbook of Differential Geometry*, Vol. I, Elsevier Science, Amsterdam, 2000, 779–864.

[Lu 2000b] Ü. Lumiste, 2-semiparallel surfaces in space forms 2: The general case, *Proc. Estonian Acad. Sci. Phys. Math.*, **49** (2000), 203–214.

[Lu 2001] Ü. Lumiste, Semiparallel submanifolds with plane generators of codimension two in a Euclidean space, *Proc. Estonian Acad. Sci. Phys. Math.*, **50** (2001), 115–123.

[Lu 2002a] Ü. Lumiste, Normally flat semiparallel submanifolds in space forms as immersed semisymmetric Riemannian manifolds, *Comm. Math. Univ. Carolinae*, **43**-2 (2002), 243–260.

[Lu 2002b] Ü. Lumiste, Semiparallelity, semisymmetricity, and Ric-semisymmetricity for normally flat submanifolds in Euclidean space, *Eesti TA Toim. Füüs. Mat. Proc. Estonian Acad. Sci. Phys. Math.*, **51** (2002), 67–85.

[Lu 2003] Ü. Lumiste, Semiparallel isometric immersions of 3-dimensional semisymmetric Riemannian manifold, *Czech. Math. J.*, **53**-128 (2003), 707–734.

[Lu 2004] Ü. Lumiste, Riemannian manifolds of conullity two admitting semiparallel isometric immersions, *Eesti TA Toim. Füüs. Mat. Proc. Estonian Acad. Sci. Phys. Math.*, **53** (2004), 203–217.

[LCh 81] Ü. Lumiste and A. Chakmazyan, Normal connection and submanifolds with parallel normal fields in spaces of constant curvature, in *Problems in Geometry*, Vol. 12, Vsesoyuzn. Inst. Nauchn. Tekhn. Inform., Akad. Nauk SSSR, Moscow, 1981, 3–30 (in Russian); *J. Soviet Math.*, **21** (1981), 107–127 (in English).

[LM 84] Ü. Lumiste and V. Mirzoyan, Submanifolds with parallel third fundamental form, *Tartu Ülik. Toim. Acta Comm. Univ. Tartuensis*, **665** (1984), 42–54 (in Russian; summary in English).

[LR 90] Ü. Lumiste and K. Riives, Three-dimensional semi-symmetric submanifolds with axial, planar or spatial points in Euclidean spaces, *Tartu Ülik. Toim. Acta Comm. Univ. Tartuensis*, **899** (1990), 13–28.

[LR 92] Ü. Lumiste and K. Riives, Semi-symmetric envelopes of some four-dimensional reducible symmetric submanifolds, *Tallinna Tehnikaülik. Toim. Trans. Tallinn Techn. Univ.*, **733** (1992), 49–58.

[Maa 74] I. Maasikas, Zur Riemannschen Geometrie der Grassmannschen Mannigfaltigkeiten von nichtisotropen Unterraume im pseudoeuklidischen Raum, *Tartu Ulik. Toim. Acta Comm. Univ. Tartuensis*, **342** (1974), 76–82 (in Russian; summary in German).

[Ma 81] S. Maeda, Imbedding of a complex projective space similar to Segre imbedding, *Arch. Math.*, **37** (1981), 556–560.

[Ma 83a] S. Maeda, Isotropic immersions with parallel second fundamental form, *Canad. Math. Bull.*, **26** (1983), 291–296.

[Ma 83b] S. Maeda, Isotropic immersions with parallel second fundamental form II, *Yokohama Math. J.*, **31** (1983), 131–138.

[Mag 84] M. A. Magid, Isometric immersions of Lorentz space with parallel second fundamental forms, *Tsukuba J. Math.*, **8** (1984), 31–54.

[Mag 85] M. A. Magid, Lorentz isoparametric hypersurfaces, *Pacific J. Math.*, **118** (1985), 165–197.

[Mat 83] Y. Matsuyama, Complete hypersurfaces with $R \cdot S = 0$ in E^{n+1}, *Proc. Amer. Math. Soc.*, **88** (1983), 119–123.

[Mat 85] Y. Matsuyama, On a hypersurface with recurrent or birecurrent second fundamental tensors, *Tensor*, **42** (1985), 168–172.

[Mat 90] Y. Matsuyama, On a hypersurface with birecurrent second fundamental tensor, *Tensor (N. S.)*, **49** (1990), 280–282.

[Me 91] F. Mercuri, Parallel and semi-parallel immersions into space forms, *Riv. Mat. Univ. Parma* (4), **17** (1991), 91–108.

[Mey 70] K. Meyberg, Jordan-Tripelsysteme und die Koecher-Konstruktion von Lie-Algebren, *Math. Z.*, **115** (1970), 58–78.

[Mi 78a] V. Mirzoyan, On submanifolds with parallel second fundamental form in spaces of constant curvature, *Tartu Ülik. Toim. Acta Comm. Univ. Tartuensis*, **464** (1978), 59–74 (in Russian; summary in English).

[Mi 78b] V. Mirzoyan, On submanifolds with parallel fundamental form of higher order, *Dokl. Akad. Nauk Armenian SSR*, **66** (1978), 71–75 (in Russian).

[Mi 83a] V. Mirzoyan, On canonical imbeddings of R-spaces, *Mat. Zametki*, **33** (1983), 255–260 (in Russian).

[Mi 83b] V. Mirzoyan, Submanifolds with commuting normal vector field, in *Problems in Geometry*, Vol. 14, Vsesoyuzn. Inst. Nauchn. Tekhn. Inform., Akad. Nauk SSSR, Moscow, 1983, 73–100 (in Russian).

[Mi 91a] V. Mirzoyan, *Ric*-semisymmetric submanifolds, in *Problems in Geometry*, Vol. 23, Vsesoyuzn. Inst. Nauchn. Tekhn. Inform., Akad. Nauk SSSR, Moscow, 1991, 29–66 (in Russian); *J. Math. Sci. New York*, **70**-2 (1994), 1624–1646 (in English).

[Mi 91b] V. Mirzoyan, Decomposition into a product of submanifolds with parallel fundamental form α_s $(s \geq 3)$, *Izv. Vyssh. Uchebn. Zaved. Mat.*, **35**-8 (1991), 44–53 (in Russian); *Soviet Math. (Iz. VUZ)*, **35**-8 (1991), 42–51 (in English).

[Mi 91c] V. Mirzoyan, Semi-symmetric submanifolds and their decomposition into a product, *Izv. Vyssh. Uchebn. Zaved. Mat.*, **35**-9 (1991), 29–38 (in Russian); *Soviet Math. (Iz. VUZ)*, **35**-9 (1991), 28–36 (in English).

[Mi 91d] V. Mirzoyan, On submanifolds with parallel fundamental form $\alpha_s (s \geq 3)$, *Tartu Ülik. Toim. Acta Comm. Univ. Tartuensis*, **930** (1991), 97–112 (in Russian; summary in English).

[Mi 91e] V. Mirzoyan, Submanifolds with semi-parallel Ricci tensor, *Tartu Ülik. Toim. Acta Comm. Univ. Tartuensis*, **930** (1991), 113–128 (in Russian; summary in English).

[Mi 92] V. Mirzoyan, Structural theorems for Riemannian *Ric*-semi-symmetric spaces, *Izv. Vyssh. Uchebn. Zaved. Mat.*, **36**-6 (1992), 80–89 (in Russian); *Russ. Math. (Iz. VUZ)*, **36**-6 (1992), 75–83 (in English).

[Mi 93] V. Mirzoyan, Submanifolds with parallel Ricci tensor in Euclidean spaces, *Izv. Vyssh. Uchebn. Zaved. Mat.*, **37**-9 (1993), 22–27 (in Russian).

[Mi 95] V. Mirzoyan, Structural theorems for Kaehler *Ric*-semisymmetric spaces, *Dokl. Akad. Nauk Armenii*, **95** (1995), 3–5 (in Russian).

[Mi 96] V. Mirzoyan, s-semiparallel submanifolds in spaces of constant curvature as envelopes of s-parallel submanifolds, *J. Contemp. Math. Anal. Armenian Acad. Sci.*, **31** (1996), 37–48.

[Mi 97] V. Mirzoyan, Submanifolds with symmetric fundamental forms of higher orders as envelopes, *Izv. Vyssh. Uchebn. Zaved. Mat.*, **41**-9 (1997), 35–40 (in Russian); *Russ. Math. (Iz. VUZ)*, **41**-9 (1997), 33–37 (in English).

[Mi 98a] V. Mirzoyan, On a class of submanifolds with a parallel fundamental form of higher order, *Izv. Vyssh. Uchebn. Zaved. Mat.*, **42**-6 (1998), 46–53 (in Russian); *Russ. Math. (Iz. VUZ)*, **42**-6 (1998), 42–48 (in English).

[Mi 98b] V. Mirzoyan, On generalizations of Ü. Lumiste theorem on semiparallel submanifolds, *J. Contemp. Math. Anal. Armenian Acad. Sci.*, **33** (1998), 48–58.

[Mi 99] V. A. Mirzoyan, Submanifolds with parallel and semi-parallel structures, *J. Contemp. Math. Anal. Armenian Acad. Sci.*, **34** (1999), 69–73.

[Mi 2000] V. A. Mirzoyan, Classification of Ric-semiparallel hypersurfaces in Euclidean spaces, *Mat. Sb.*, **191** (2000), 65–80 (in Russian); *Sb. Math.*, **191** (2000), 1323–1338 (in English).

[Mi 2002] V. A. Mirzoyan, Submanifolds with semiparallel tensor fields as envelopes, *Mat. Sb.*, **193** (2002), 99–112 (in Russian); *Sb. Math.*, **193** (2002), 1493–1505 (in English).

[Mi 2003] V. A. Mirzoyan, Warped products, cones over Einstein spaces, and classification of Ric-semiparallel submanifolds of a certain class, *Izv. Russ. Acad. Sci. Ser. Math.*, **67** (2003), 955–973.

[Mo 71] J. D. Moore, Isometric immersions of Riemannian products, *J. Differential Geom.*, **5** (1971), 159–168.

[Mu 61] R. Mullari, On submanifolds with fields of absolute principal directions, *Tartu Ülik. Toim. Acta Comm. Univ. Tartuensis*, **102** (1961), 275–288 (Russian; summary in English).

[Mu 62a] R. Mullari, On principal directions of m-dimensional submanifold, *Dokl. Akad. Nauk SSSR*, **144** (1962), 989–992 (in Russian); *Soviet Math. Dokl.*, **3** (1962) (in English).

[Mu 62b] R. Mullari, Über die maximal symmetrischen Flächen im n-dimensionalen Euklidischen Raum, *Tartu Ülik. Toim. Acta Comm. Univ. Tartuensis*, **129** (1962), 62–73 (Russian; summary in German).

[Mum 76] D. Mumford, *Algebraic Geometry*, Vol. I, Springer-Verlag, Berlin, Heidelberg, New York, 1976.

[Na 80] H. Naitoh, Isotropic submanifolds with parallel second fundamental forms in symmetric spaces, *Osaka J. Math.*, **17** (1980), 95–100.

[Na 81] H. Naitoh, Totally real parallel submanifolds in $P^n(c)$, *Tokyo J. Math.*, **4** (1981), 279–306.

[Na 84] H. Naitoh, Pseudo-Riemannian symmetric R-spaces, *Osaka J. Math.*, **21** (1984), 733–764.

[Na 86] H. Naitoh, Symmetric submanifolds of compact symmetric spaces, *Tsukuba J. Math.*, **10** (1986), 215–242.

[Na 90] H. Naitoh, Symmetric submanifolds and generalized Gauss maps, *Tsukuba J. Math.*, **14** (1990), 113–132.

[NT 89] H. Naitoh and M. Takeuchi, Symmetric submanifolds of symmetric spaces, *Sugaku*, **2** (1989), 157–188.

[NT 76] H. Nakagawa and R. Takagi, On locally symmetric Kaehler submanifolds in a complex projective space, *J. Math. Soc. Japan*, **28** (1976), 638–667.

[Ne 86] E. Neher, *Jordan Triple Systems by the Grid Approach*, Lecture Notes in Mathematics, Vol. 1280, Springer-Verlag, Berlin, New York, 1987, 1–193.

[Nö 96] S. Nölker, Isometric immersions of warped products, *Differential Geom. Appl.*, **6** (1996), 1–30.

[No 68] K. Nomizu, On hypersurfaces satisfying a certain condition on the curvature tensor, *Tohoku Math. J.*, **20** (1968), 46–59.

[NM 89] K. Nomizu and M. A. Magid, On affine surfaces whose cubic forms are parallel relative to the affine metric, *Proc. Japanese Acad.* A, **65** (1989), 215–222.

[NO 62] K. Nomizu and H. Ozeki, A theorem on curvature tensor fields, *Proc. Nat. Acad. Sci. U.S.A.*, **48** (1962) 206–207.

[NP 89] K. Nomizu and U. Pinkall, Cayley surfaces in affine differential geometry, *Tohoku Math. J.*, **41** (1989) 589–596.

[Nu 61] J. Nut (Nuut), *Lobachevskian Geometry in Analytical Treatment*, Izd. Akad. Nauk SSSR, Moscow, 1961 (in Russian).

[Ob 90] E. D. Oboznaya, On semi-symmetric spaces of the first affine class, *Ukr. Geom. Sb.*, **28** (1990), 95–102 (in Russian); *J. Soviet Math.*, **48** (1990), 77–82.

[Oh 83] Y. Ohnita, The degrees of the standard imbeddings of R-spaces, *Tohoku Math. J.*, **35** (1983), 499–502.

[O'N 65] B. O'Neill, Isotropic and Kaehler immersions, *Canad. J. Math.*, **17** (1965), 907–915.

[O'N 83] B. O'Neill, *Semi-Riemannian Geometry with Applications to Relativity*, Academic Press, New York, London, 1983.

[Os 2002] D. Osipova, Symmetric submanifolds in symmetric spaces, *Differential Geom. Appl.*, **16** (2002), 199–211.

[ÖA 2002a] C. Özgür, K. Arslan, and C. Murathan, On a class of surfaces in the Euclidean space, *Comm. Fac. Sci. Univ. Ankara Ser.* A1 *Math. Stat.*, **51** (2002), 47–54.

[ÖA 2002b] C. Özgür, K. Arslan, and C. Murathan, Surfaces satisfying certain curvature conditions in the Euclidean spaces, *Differential Geom. Dynam. Systems*, **4** (2002), 26–32.

[PK 86] J. S. Pak and J. J. Kim, Isotropic immersions with totally geodesic Gauss image, *Tensor*, **43** (1986), 167–174.

[Pa 2000a] A. Parring, Parallel and semiparallel symplectic submanifolds in the symplectic space, *Proc. Estonian Acad. Sci. Phys. Math.*, **49** (2000), 149–169.

[Pa 2000b] A. Parring, Semi-parallel and parallel symplectic surfaces in the four-dimensional symplectic space, *Acta Comm. Univ. Tartu Math.*, **4** (2000), 23–37.

[PS 86] A. Parring and A. Saarne, The two-dimensional symplectic surfaces of the symplectic space Sp_4, *Tartu Ulik. Toim. Acta Comm. Univ. Tartuensis*, **734** (1986), 80–101 (in Russian; summary in English).

[PR 86] R. Penrose and W. Rindler, *Spinors and Space-Time*, Vol. 2, Cambridge University Press, Cambridge, UK, 1986.

[Per 35] D. I. Perepelkin, Curvature and normal spaces of submanifold V_m in R_n, *Mat. Sb.*, **42** (1935), 81–120 (in Russian).

[Pe 98] P. Petersen, *Riemannian Geometry*, Springer-Verlag, New York, 1998.

[Pe 66] A. Z. Petrov, *New Methods in General Theory of Relativity*, Nauka, Moscow, 1966 (in Russian).

[Pe 69] A. Z. Petrov, *Einstein Spaces*, Gosudarstv. Izdat. Fiz.-Mat. Lit., Moscow, 1961 (in Russian); revised, corrected, and modified ed., Pergamon Press, Oxford, UK, 1969.

[Ph 79] E. Phillips, Karl M. Peterson: The earliest derivation of the Mainardi–Codazzi equations and the fundamental theorem of surface theory, *Historia Math.*, **6** (1979), 137–163.

[Pi 89] G. Pitiş, On parallel submanifolds of a Sasakian space form, *Rend. Mat. Appl.*, **7**-9 (1989), 103–111.

[Rä 94] A. Rääbis, *Semi-Symmetric Pseudo-Riemannian Surfaces in n-Dimensional de Sitter Spaces*, Master's thesis, University of Tartu, Tartu, Estonia, 1994 (in Estonian).

[Ra 53] P. K. Raschevski, *Riemannsche Geometrie und Tensoranalanalysis*, Gos. Izd. Tekhn.-Teor. Lit., Moscow, 1953 (in Russian); 2nd ed., Nauka, Moscow, 1964; translation of the 1st ed., VEB Deutscher Verlag, Berlin, 1959 (in German).

[Re 76] H. Reckziegel, Krümmungsflächen von isometrischen Immersionen in Räume konstanter Krümmung, *Math. Ann.*, **223** (1976), 169–181.

[Re 79] H. Reckziegel, Completeness of curvature surfaces of an isometric immersion, *J. Differential Geom.*, **14** (1979), 7–20.

[Re 81] H. Reckziegel, On the problem whether the image of a given differentiable map into a Riemannian manifold is contained in a submanifold with parallel second fundamental form, *J. Reine Angew. Math.*, **325** (1981), 87–104.

[Re 83] H. Reckziegel, A class of distinguished isometric immersions with parallel second fundamental form, *Results Math.*, **6** (1983), 56–63.

[Rei 73] K. Reich, Die Geschichte der Differentialgeometrie von Gauss bis Riemann (1828–1868), *Arch. History Exact Sci.*, **11** (1973), 273–382.

[Ri 86] K. Riives, Submanifolds V_3 with parallel third fundamental form in Euclidean space E_5, *Tartu Ülik. Toim. Acta Comm. Univ. Tartuensis*, **734** (1986), 102–110 (Russian; summary in English).

[Ri 88] K. Riives, About two classes of semi-symmetric submanifolds, *Tartu Ülik. Toim. Acta Comm. Univ. Tartuensis*, **803** (1988), 95–102 (Russian; summary in English).

[Ri 91] K. Riives, Second order envelope of congruent Veronese surfaces in E^6, *Tartu Ülik. Toim. Acta Comm. Univ. Tartuensis*, **930** (1991), 47–52.

[Ri 97] K. Riives, On a class of four-dimensional semiparallel submanifolds in Euclidean spaces, in *Proceedings of the* 4th *International Congress of Geometry (Thessaloniki, May* 26*–June* 1, 1996), Aristotle University of Thessaloniki, Thessaloniki, Greece, 1997, 351–357.

[Ri 99] K. Riives, On the special class of curves on some four-dimensional semiparallel submanifolds, in *Satellite Conference of ICM* (*Berlin, August* 10–14, 1998), Masaryk University, Brno, Czech Republic, 1999, 215–222.

[Ri 2000] K. Riives, On a function of reducibility of a class of four-dimensional semiparallel submanifolds, *Eesti TA Toim. Füüs. Mat. Proc. Acad. Sci. Estonia Phys. Math.*, **49** (2000), 3–11.

[Ros 84] A. Ros, On spectral geometry of Kaehler submanifolds, *J. Math. Soc. Japan*, **36** (1984), 433–448.

[Ros 85] A. Ros, A characterization of seven compact Kaehler submanifolds by holomorphic pinching, *Ann. Math.*, **121** (1985), 377–382.

[Ros 86] A. Ros, Kaehler submanifolds in the complex projective space, in *Differential Geometry* (*Pensicola* 1985), Lecture Notes in Mathematics, Vol. 1209 Springer-Verlag, Berlin, New York, 1986, 259–274.

[Ro 49a] B. A. Rosenfeld, Projective differential geometry of families of pairs $P^m + P^{n-m-1}$ in P^n, *Mat. Sb.*, **24** (1949), 405–428 (in Russian).

[Ro 49b] B. A. Rosenfeld, Symmetric spaces and their geometric applications, in E. Cartan, *Geometry of Lie Groups and Symmetric Spaces*, Inostr. Lit., Moscow, 1949, appendix, 331–368 (in Russian).

[RV 70] E. A. Ruh and J. Vilms, The tension field of a Gauss map, *Trans. Amer. Math. Soc.*, **149** (1970), 569–573.

[Ry 69] P. J. Ryan, Homogeneity and some curvature conditions for hypersurfaces, *Tohoku Math J.*, **21** (1969), 363–388.

[Ry 71] P. J. Ryan, Hupersurfaces with parallel Ricci tensor, *Osaka J. Math.*, **8** (1971), 251–259.

[Ry 72] P. J. Ryan, A class of complex hypersurfaces, *Colloq. Math.*, **26** (1972), 175–182.

[Saf 2001] E. Safiulina, Parallel and semiparallel space-like surfaces in pseudo-Euclidean spaces, *Eesti TA Toim. Füüs. Mat. Proc. Estonian Acad. Sci. Phys. Math.*, **50** (2001), 16–33.

[Sak 73] K. Sakamoto, Submanifolds satisfying the condition $K(X, Y) \cdot K = 0$, *Kodai Math. Sem. Rep.*, **25** (1973), 143–152.

[San 85] C. U. Sanchez, k-symmetric submanifolds of R^N, *Math. Ann.*, **270** (1985), 297–316.

[San 92] C. U. Sanchez, A characterization of extrinsic k-symmetric submanifolds in R^N, *Rev. Union Mat. Argent.*, **38** (1992), 1–15.

[Sch 24] J. A. Schouten, *Der Ricci-Kalkül*, Springer-Verlag, Berlin, 1924; 2nd ed., 1954.

[SchStr 35] J. A. Schouten and D. J. Struik, *Einführung in die neueren Methoden der Differentialgeometrie*, Vols. I and II, Noordhoff, Groningen, the Netherlands, 1935, 1938.

[Sek 72] K. Sekigawa, On some hypersurfaces satisfying $R(X, Y) \cdot R = 0$, *Tensor*, **25** (1972), 133–136.

[Sek 75] K. Sekigawa, On some 3-dimensional complete Riemannian manifolds satisfying $R(X, Y) \cdot R = 0$, *Tohoku Math. J.*, **27** (1975), 561–568.

[Sek 77] K. Sekigawa, On some 4-dimensional Riemannian manifolds satisfying $R(X, Y) \cdot R = 0$, *Hokkaido Math. J.*, **6** (1977), 216–229.

[ST 70] K. Sekigawa and S. Tanno, Sufficient conditions for a Riemannian manifold to be locally symmetric, *Pacific J. Math.*, **34** (1970), 157–162.

[Sha 88] I. R. Shafarevich, *Basic Algebraic Geometry*, Vol. I, Nauka, Moscow, 1988 (in Russian); Springer-Verlag, Berlin, 1994 (in English).

[She 2002] Y.-B. Shen, On complete submanifolds with parallel mean curvature in R^{n+p}, in *Topology and Geometry: Commemorating SISTAG*, Contemporary Mathematics, Vol. 314, American Mathematical Society, Providence, RI, 2002, 225–234.

[Shi 25] P. A. Shirokov, Constant vector fields and tensor fields of 2nd order in Riemannian spaces, *Izv. Kazan Fiz.-Mat. Obshchestva Ser.* 2, **25** (1925), 86–114 (in Russian). (See [Shi 66], 256–280.)

[Shi 61] P. A. Shirokov, *Tensor Calculus*, Kazan University Press, Kazan, Russia, 1961 (in Russian).

[Shi 66] P. A. Shirokov, *Selected Works on Geometry*, Kazan University Press, Kazan, Russia, 1966 (in Russian).

[SW 69] U. Simon and A. Weinstein, Anwendungen der De Rhamschen Zerlegung auf Probleme der lokalen Flächentheorie, *Manuscripta Math.*, **1** (1969), 139–146.

[Si 56] N. S. Sinyukov, On geodesic mapping of Riemannian spaces, in *Trudy* III *Vsesoyuzn. Mat. S'ezda*, Vol. 1, Izd. Akad. Nauk SSSR, Moscow, 1956, 167–168 (in Russian).

[Si 62] N. S. Sinyukov, Semisymmetric Riemannian spaces, in *Pervaya Vsesoyuznaya Geom. Konfer.* (*May* 1962, *Kiev*), Tezisy Dokl., Kiev, 1962, 84 (in Russian).

[Si 79] N. S. Sinyukov, *Geodesic Mppings of Riemannian Spaces*, Nauka, Moscow, 1979 (in Russian).

[So 61] A. S. Solodovnikov, Models of elliptic spaces, in *Trudy Semin. po Vektor. i Tensor. Analizu*, Vol. XI, University of Moscow, Moscow, 1961, 293–308 (in Russian).

[Ste 64] S. Sternberg, *Lectures on Differential Geometry*, Prentice–Hall, Englewood Cliffs, NJ, 1964; 2nd ed., Chelsea, New York, 1983.

[Str 79] W. Strübing, Symmetric submanifolds of Riemannian manifolds, *Math. Ann.*, **245** (1979), 37–44.

[Stru 33] D. J. Struik, Outline of a history of differential geometry, *Isis*, **19** (1933), 19–120, **20** (1934), 161–191.

[Sza 82] Z. I. Szabó, Structure theorems on Riemannian spaces satisfying $R(X, Y) \cdot R = 0$ I: The local version, *J. Differential Geom.*, **17** (1982), 531–582.

[Sza 84] Z. I. Szabó, Classification and construction of complete hypersurfaces satisfying $R(X, Y) \cdot R = 0$, *Acta Sci. Math.*, **47** (1984), 321–348.

[Sza 85] Z. I. Szabó, Structure theorems on Riemannian spaces satisfying $R(X, Y) \cdot R = 0$ II: Global version, *Geom. Dedic.*, **19** (1985), 65–108.

[Tai 68] S. S. Tai, Minimal imbeddings of R-spaces, *J. Differential Geom.*, **2** (1968), 203–215.

[Ta 72] H. Takagi, An example of Riemannian manifolds satisfying $R(X, Y) \cdot R = 0$ but not $\nabla R = 0$, *Tohoku Math. J.*, **24** (1972), 105–108.

[Tak 81] M. Takeuchi, Parallel submanifolds of space forms, in *Manifolds and Lie Groups: Papers in Honor of Y. Matsushima*, Birkhäuser, Basel, 1981, 429–447.

[TK 68] M. Takeuchi and S. Kobayashi, Minimal imbeddings of R-spaces, *J. Differential Geom.*, **2** (1968), 203–215.

[Tan 69] S. Tanno, Hypersurfaces satisfying a certain condition on the Ricci tensor, *Tohoku Math. J.*, **21** (1969), 297–303.

[Tan 71] S. Tanno, A class of Riemannian manifolds satisfying $R(X, Y) \cdot R = 0$, *Nagoya Math. J.*, **42** (1971), 67–77.

[Ts 85a] K. Tsukada, Parallel Kaehler submanifolds of Hermitian symmetric spaces, *Math. Z.*, **190** (1985), 129–150.

[Ts 85b] K. Tsukada, Parallel submanifolds in a quaternion projective space, *Osaka J. Math.*, **22** (1985), 187–241.

[Ts 85c] K. Tsukada, Parallel submanifolds of Cayley plane, *Sci. Rep. Niigata Univ.*, **A21** (1985), 19–32.

[Ts 96] K. Tsukada, Totally geodesic submanifolds and curvature-invariant subspaces, *Kodai Math. J.*, **19** (1996), 395–437.

[Ud 86] S. Udagawa, Spectral geometry of Kaehler submanifolds of a complex projective space, *J. Math. Soc. Japan*, **38** (1986), 453–472.

[Va 91] J. Varik, Extremal surfaces in Minkowski space ${}^1E_3$, *Tartu Ülik. Toim. Acta Comm. Univ. Tartuensis*, **930** (1991), 53–61.

[VdW 90] I. Van de Woestijne, Minimal surfaces of the 3-dimensional Minkowski space, in *Geometric Topology of Submanifolds* II, World Scientific, Singapore, 1990, 344–369.

[Ve 91] M. H. Vernon, Semi-symmetric hypersurfaces of anti-de Sitter spacetime that are S^1-invariant, *Tensor*, **50** (1991), 99–105.

[Ver 94] L. Verstraelen, Comments on pseudo-symmetry in the sense of Ryszard Deszcz, *Geometric Topology of Submanifolds* VI, World Scientific, River Edge, NJ, 1994, 199–209.

[Vi 70] J. Vilms, Totally geodesic maps, *J. Differential Geom.*, **4** (1970), 73–79.

[Vi 72] J. Vilms, Submanifolds of Euclidean space with parallel second fundamental form, *Proc. Amer. Math. Soc.*, **32** (1972), 263–267.

[Vr 88] L. Vrancken, Affine higher order parallel hypersurfaces, *Ann. Fac. Sci. Toulouse*, **9** (1988), 341–353.

[Vr 91] L. Vrancken, Affine surfaces with higher order parallel cubic form, *Tôhoku Math. J.*, **43** (1991), 127–139.

[vdWa 27] L. van der Waerden, Differentialkovarianten von n-dimensionalen Mannigfaltigkeiten in Riemannschen m-dimensionalen Räumen, *Abh. Math. Sem. Hamburg*, **5** (1927), 153–160.

[Wa 73] R. Walden, Untermannigfaltigkeiten mit paralleler zweiter Fundamentalform in euklidischen Räumen und Sphären, *Manuscripta Math.*, **10** (1973), 91–102.

[Wal 46] A. G. Walker, Symmetric harmonic spaces, *J. London Math. Soc.*, **21** (1946).

[Wal 50] A. G. Walker, On Ruse's spaces of recurrent curvature, *Proc. London Math. Soc.*, **51** (1950), 36–64.

[Wo 72] J. A. Wolf, *Spaces of Constant Curvature* (*University of California Berkley*, 1972), 4th ed., Publish or Perish, Berkeley, 1977.

[Won 52] Y.-C. Wong, A new curvature theory for surfaces in an Euclidean 4-space, *Comment. Math. Helv.*, **26** (1952), 152–170.

[YI 71] K. Yano and S. Ishihara, Submanifolds with parallel mean curvature vector, *J. Differential Geom.*, **6** (1971), 95–118.

[YI 72] K. Yano and S. Ishihara, Submanifolds of codimension 2 or 3 with parallel second fundamental tensor, *J. Korean Math. Soc.*, **9** (1972), 1–11.

[Yau 74] S. T. Yau, Submanifolds with constant mean curvature I, *Amer. J. Math.*, **96** (1974), 346–366.

Index